Deposition of Organic Facies

Edited by
A. Y. Huc

AAPG Studies in Geology #30

Published by
The American Association of Petroleum Geologists
Tulsa, Oklahoma, U.S.A.

Cover photographs by A.Y. Huc—

Kimmeridgian shales outcropping on the coastline of Dorset, United Kingdom.

ISBN: 0-89181-038-2
ISSN: 0149-1377

Association Editor: Susan Longacre
Science Director: Gary D. Howell
Publications Manager: Cathleen P. Williams
Special Projects Editor: Anne H. Thomas
Copy Editor: Donald W. Hutcheson

PREFACE

A great variety of organic-bearing sediments are acting as source rocks for the economic accumulations of hydrocarbons trapped in the subsurface. The occurrence of organic-rich intervals; their place in the stratigraphic column, their lateral extent, and the variation in their quality both stratigraphically and laterally are key parameters to be considered if one wants to evaluate the petroleum potential of a sedimentary basin.

Besides direct assessment of organic-rich horizons by means of the analysis of available samples, and the subsequent construction of regional databases, there is a need to incorporate improved models into a conceptual framework that enhances the ability to predict the occurrence and variation of source rocks as a function of a depositional environment in time and space. During the last decade, important progress has been made in our understanding of the processes responsible for deposition of organic-rich sediments (for example, organic productivity, anoxia, and sedimentation rate) and in recognition of the restricted distribution of the majority of these source rocks in the stratigraphic record (such as in the Upper Jurassic and the mid-Cretaceous).

The purpose of this volume is to present recent advances in organic sedimentology. The papers herein discuss a wide range of aspects of this field of research. The diverse nature of these papers includes modern environments, considered as present-day analogs of source rock formation; numerical modeling of paleoproductivity; and studies related to specific time periods during which organic matter accumulation has been particularly impressive (the Kimmeridgian, Cenomanian–Turonian, and others).

The papers of this volume initially were presented orally at an SEPM symposium on "Deposition of Organic Facies" held during the 1988 AAPG Mediterranean Basins Conference in Nice (France).

I am most grateful to the many individuals who have helped to make this volume possible. In particular, I thank the contributing authors, the colleagues who assisted as article reviewers, and Anne H. Thomas, AAPG special projects editor, who provided us with needed advice and editorial assistance.

A. Y. Huc

PUBLISHER'S NOTE

The AAPG Special Projects Department expresses its appreciation of the superior editorial efforts made by Dr. A. Y. Huc in directing the scientific aspects of this volume, in compiling and organizing the contributions, and in aiding the book's production phase.

ABOUT THE EDITOR

Alain Yves Huc is currently Head of the Organic Geochemistry group at the Institut Francais du Pétrole; also known as IFP (French Petroleum Institute).

He was educated at the University of Nancy (France), and he received his Ph.D. in Organic Geochemistry from the University of Strasbourg (France) in 1978. He spent one-and-a-half years as a post-doctoral fellow at Woods Hole Oceanographic Institution (USA) and two years as a CNRS research at the applied Geology Department of the University of Orléans (France). Following this, he joined IFP.

The major theme of his research during the past seven years has been directed at the study of the sedimentology of organic matter and its application to oil exploration.

Table of Contents

Understanding Organic Facies: A Key to Improved Quantitative Petroleum Evaluation of Sedimentary Basins

A. Y. Huc
Institut Francais du Pétrole
Rueil-Malmaison, France

As complexity of exploration problems increases, petroleum geologists tend to consider more carefully oil formation and migration in terms of quantity, quality, and timing, in order to assess more accurately the history of hydrocarbons in a given basin. The present knowledge concerning the behavior of source rocks (quantity of hydrocarbons generated, composition of hydrocarbons, kinetics, expulsion efficiency) is that differences can be attributed to a large extent to organic properties of sediments, such as organic-matter content, organic-matter quality, and the pattern of the microdistribution of organic matter in the mineral fabric (layered, lenslike, intergranular). Moreover, field observations show that organic facies change at various scales in space and time, in response to variation in depositional environments and sedimentary processes.

The effects of these factors on the source rocks emphasize, beyond the efforts to achieve comprehensive descriptions of source beds (Rock-Eval pyrolysis, microscopy, typing of organic matter, measurement of kinetic parameters of carefully selected kerogens, use of wireline logs, seismic information), the needs to increase our conceptual understanding of deposition of organic facies in order to improve our ability to predict the petroleum potential of sedimentary basins.

INTRODUCTION

Explorationists now accept that source-rock analysis is an important risk-reduction tool essential to more effective search for oil and gas prospects. Besides the fact that the occurrence of source rocks is a basic prerequisite for the presence of any hydrocarbon accumulation, we are aware that modern exploration requires a more quantitative approach. Such an approach needs to consider carefully the properties of source rocks, which are the materials where hydrocarbons are generated, through which they have to migrate, and from which they are eventually expelled. This requirement for a more quantitative approach is clearly illustrated by the growing success of basin modeling as a guide to exploration thinking. Current scientific understanding of the physico-chemical processes involved in the generative and expulsive behavior focuses attention on the organic content, quality, and distribution within the source bed. This situation demands a thorough characterization of source rocks in basins under exploration so that realistic quantitative reserve forecasting and basin modeling studies can be undertaken. Realization that source-rock properties, including the organic content, type, and distribution within the rock mineral matrix, are inherited from the time of deposition emphasizes why the study and understanding of organic facies are so important. Currently, considerable research effort is being devoted to this topic.

ORGANIC FACIES AND HYDROCARBON GENERATION

Petroleum generation results from the transformation of sedimentary organic matter in the subsurface under the influence of both temperature and geologic time. It is widely accepted that this transformation can be ascribed to the thermal cracking of the kerogen (a solid organic assemblage that is insoluble in organic solvents) which releases micropetroleum (a mixture of liquid and gaseous hydrocarbon-like products that are soluble in organic solvents) into the pore system of the source rock (Tissot and Welte, 1984).

It is self evident that the quantity of micropetroleum generated in a unit volume of source rock is directly controlled by the kerogen content of the sediment and by the type of organic matter. The diversity of kerogens, in terms of hydrocarbon proneness, is now a well-established concept. As an illustration, Figure 1 represents the atomic elemental ratios (H/C, O/C, S/C) of a set of thermally immature kerogens from various well-documented source rocks.These source rocks include the Green River Shale (Eocene, USA), a carbonate source rock from the Middle East (Cenomanian-Turonian), the Kimmeridge Formation (Kimmeridgian, Dorset, UK), the Monterey Formation (Miocene, USA), the Toarcian Shale of the Paris basin (lower Toarcian, France), and coals and organic shales of the Mahakam delta (Miocene, Indonesia). The considerable scatter of the data, summarized in this diagram, is a clear indication of the diversity in composition exhibited by these kerogens. Numerous studies have shown that this diversity in kerogen chemistry is a reflection of: (1) the biological composition of the living communities whose remains are eventually entrapped in the sediments (i.e., planktonic organisms, terrestrial higher plants, bacterial biomass) and (2) the type and extent of microbiological alteration affecting the organic matter during its transit from the precursors living habitat toward its final site of deposition, and during early diagenesis in the sediments.

The resulting difference in chemical structures is reflected by the ability of the various kerogens to yield hydrocarbon compounds upon thermal stress, either in the subsurface or by laboratory simulation. As determined by pyrolysis experiments, 15% by weight of an immature kerogen from the Mahakam delta series is convertible into hydrocarbon-like products. This value increases up to 50% for an immature kerogen from the Monterey Formation and reaches a value as high as 70% for an immature kerogen from the Green River Shale (Figure 1). The petroleum potential of a given source rock, which is a basic basin-evaluation criterion, is dependent on a combination of the amount of organic matter and the type of organic matter, both of which are inherited from the time of deposition (Demaison and Moore, 1980; Huc, 1988a). In addition to this overall quantitative aspect, the diversity in kerogen composition has important implications in two domains: the nature and composition of hydrocarbons formed as a result of the thermally induced kerogen transformation, and the kinetics of this thermogenic generative transformation.

Field observations and laboratory simulation support each other in demonstrating that the composition of hydrocarbons released by thermal cracking depends on the nature of the source kerogen. This is a consequence of the variability in the constituent chemical units contributing to the kerogen assemblage. For instance, it is widely believed that land-plant-derived oils are commonly characterized by a high wax content, as indicated by an abundance of n-alkanes in the C22 to C36 range. This feature is assumed to be a consequence of the contribution to the kerogen assemblage of chemical moieties derived from cuticles. These form the protective outer covering of leaves, rootlets, stems, and fruits of certain land plants (Brooks and Smith, 1969; Thomas, 1982).

Another example of the influence of the kerogen composition on the chemical properties of oils is provided by crudes exhibiting a high sulfur content. In most cases these oils can be ascribed to source rocks containing organic-sulfur-rich kerogens (exemplified in Figure 1 by the Middle East carbonate and the Monterey Formation). High organic-sulfur content in kerogens is most likely the result of early diagenetic reactions between hydrogen sulfide species (i.e., H_2S, polysulfurs) and sedimentary organic matter (Gransh and Posthuma, 1974; Orr, 1974; Dinur et al., 1980; Jones, 1984; Berner, 1984; Orr, 1986; Kenig and Huc, in press). The hydrogen sulfide is derived from the dissimilatory reduction of dissolved sulfate by microorganisms in oxygen-poor environments. If the amount of hydrogen sulfide produced in interstitial waters exceeds the amount of iron available in the sediment, the excess hydrogen sulfide is available to react with the organic constituents of the protokerogen. Such a mechanism implies anoxic conditions in order to initiate sulfate reduction, and a low iron content. Otherwise iron would have acted as a favorable sink for sulfur. This situation is most likely to be met in carbonate depositional environments, and explains why oils associated with carbonate sequences are commonly rich in sulfur.

Distribution of generated hydrocarbons in terms of molecular weight is also controlled to a large extent by the specific chemical properties of source kerogens. Compositional studies of pyrolysis pro-

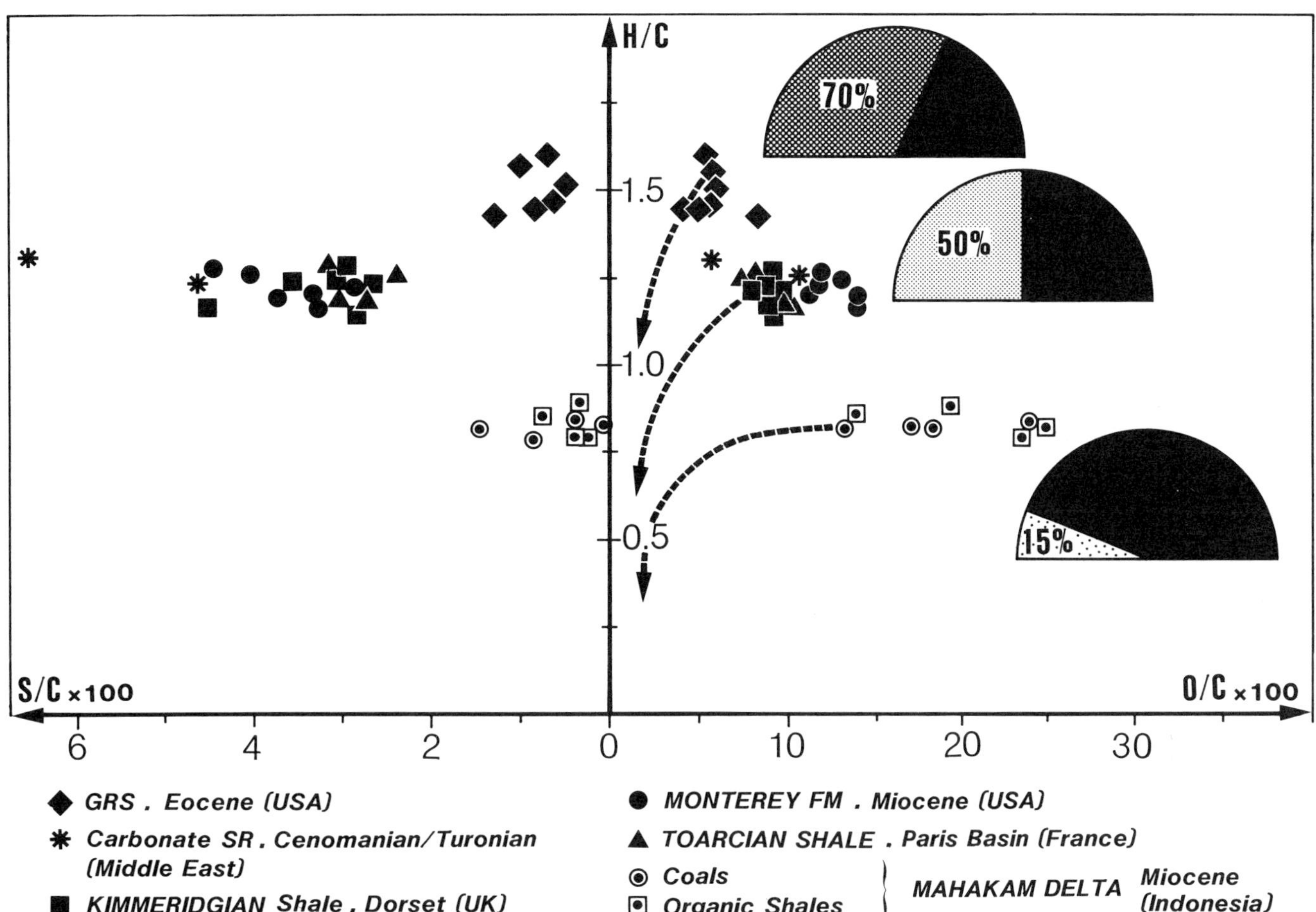

Figure 1. Elemental composition and hydrocarbon yield upon pyrolysis of various kerogens (percent of hydrocarbon-like compounds released by weight of kerogen). GRS: Green River Shale; SR: source rock.

ducts, and accumulated exploration experience, emphasize that the molecular-weight distribution of hydrocarbons generated upon primary cracking varies widely from one kind of kerogen to another. As an example, recent experimental results on the primary cracking of North Sea Jurassic kerogens, as reported by Espitalié et al. (1987), show that relative abundances of the four classes of hydrocarbons (denoted as C_1, C_2–C_5, C_6–C_{15}, and C_{15+}) are significantly different for the Brent and Viking groups. The Brent group contains dominantly terrestrial-derived organic matter (type III), whereas the organic matter in the Viking group is of mainly marine origin (type II). The proportion of gaseous hydrocarbons (C_1–C_5), normalized to the total hydrocarbons released, is higher for the Brent kerogen (24% versus 11%). These differences, expressed in terms of liquid hydrocarbons, condensate, and gaseous products, have significant impact on oil exploration.

It is commonly accepted that generation of hydrocarbons from kerogen is properly described by first-order kinetics. Numerical models based on this concept have been popularized for use as an aid to oil exploration. The application of these models relies, in part, on the reconstruction of the thermal history of the source rock. Additionally, it has been recently shown that calibration of kinetic parameters, taking into account the characteristics of individual kerogens, allows a much more accurate simulation of the natural process (Ungerer and Pelet, 1987; Tissot et al., 1987). The pre-exponential factor (A) and the activation energies (E) that are involved in the Arrhenius equation describing first-order kinetics: $dX_i/dt = -A_iX_i\exp(-E_i/RT)$ (X_i being the residual petroleum potential involved in reaction i) are clearly related to the chemical structure and chemical bonding characteristic of individual kerogens. Because of the complexity of the kerogen structure and the overall cracking reaction, it is currently impossible to formulate individually all the involved reactions into a comprehensive kinetic model. However, bulk methods based on laboratory-simulation experiments have been designed in order to calibrate the numerical parameters for use in hydrocarbon-generation models. For instance, the OPTKIN procedure (Ungerer and Pelet, 1987) can provide sets of activation energies derived from carefully controlled pyrolysis of immature kerogens. Figure 2 shows the distribution of activation energies, obtained through the OPTKIN method, for the

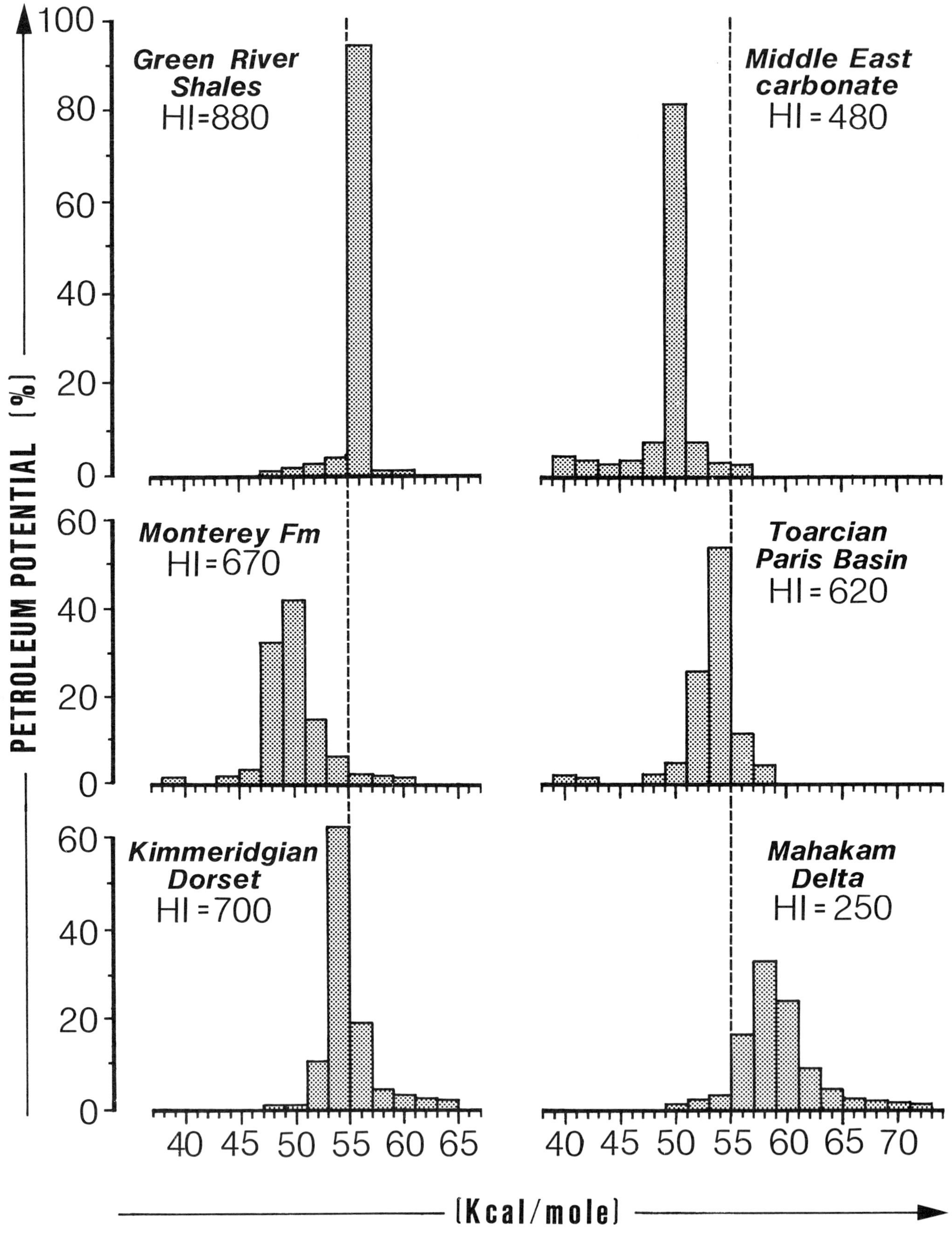

Figure 2. Examples of distribution of activation energies calibrated on various kerogens according to the OPTKIN method. HI: Rock-Eval Hydrogen Index.

reference set of kerogens in Figure 1. The distributions are highly variable in terms of mean value and range. The distribution is very narrow for the kerogens from the Middle East carbonate and from the Green River Shale; such a feature is probably a consequence of the relatively low variety of bonds involved in the chemical structure of these kerogens. This observation is supported by available analytical data on the chemical properties of the Green River Shale kerogen, which is described as a macromolecule featuring cross-linked aliphatic chains in which C-C bonds dominate (Yen, 1976).

On the other hand, the distribution of the activation energies is broader for kerogens from the Monterey Formation, the Toarcian shale of the Paris basin, the Kimmeridgian shale of Dorset, and the Mahakam delta series. In these examples such a range is probably a reflection of the wider diversity of constituent chemical bonds, as documented for several of them (Behar and Vandenbroucke, 1987). Another conspicuous difference between these kerogens is the magnitude of the mean value: low for the Monterey and the Middle East kerogens; intermediate for the Green River, the Paris basin, and the Dorset kerogens, and high for the Mahakam kerogen. Again, this feature is likely to be related to the types, and relative proportions, of chemical bonds present in the individual kerogen assemblages. It is remarkable that the two kerogens exhibiting the lower activation energies (Monterey and Middle East) are also those with the highest sulfur content. It has been suggested that this sulfur is incorporated as carbon-sulfur bonds which, upon thermal stress, may cleave more readily than most of the other bonds. This would produce hydrocarbons at an earlier stage of thermal evolution (Lewan, 1985; Orr, 1986).

A direct implication of this variation in chemical-bond strength is that kinetic control of hydrocarbon formation will be dependent on the organic facies involved. This is exemplified by Figure 3 which shows a comparison of the transformation ratio plotted against time for five different kerogen assemblages, each subjected to the same simulated thermal history with imposed first-order kinetics. From this simulation it is observed that the nature of the kerogen assemblage can give rise to a 40–50 m.y. range in the timing of hydrocarbon generation. This can be crucial when considered in the context of the timing of hydrocarbon generation versus the formation of prospective traps.

ORGANIC FACIES AND OIL EXPULSION

An understanding of, and ability to model, the generation of hydrocarbons from kerogen is insufficient to independently predict timing and quantity of hydrocarbon charging in a reservoir system. This realization has prompted investigations into the mechanism of primary migration of hydrocarbons, leading to the development of source-rock-expulsion-efficiency concepts (Meissner, 1978; Leythaeuser et al., 1984; Talukdar et al., 1986; Durand, 1988).

Although the mechanisms of migration are not yet fully understood, the role of pressure buildup in the pore system of the source rock has been clearly recognized as the main driving force for fluid expulsion. These investigations further suggest that primary migration of oil proceeds as a multiphase flow. As a consequence, and according to current concepts, there is a competition in the porous medium of the source rock between hydrocarbons and water in terms of relative permeability. Hydrocarbons are likely to be expelled preferentially to the water phase only when the hydrocarbon saturation is in excess of a minimum threshold. Hydrocarbon saturation is indeed controlled by the progressive expulsion of water during compaction and by the quantity of hydrocarbons delivered into the pore system as a result of the thermal alteration of kerogen. Because the amount of generated hydrocarbons is directly regulated by the organic-matter richness and type, makeup of the organic facies provides crucial control on the expulsion behavior of a source bed. Within these concepts, differences in migration efficiency should be expected for source rocks of differing organic facies makeup, at the level of internal heterogeneity (Comer and Hinch, 1987). In this respect, a rock rich in oil-prone organic matter is likely to reach effective oil saturation more rapidly, and consequently to promote a relatively earlier expulsion, as compared to an organic-leaner source rock.

This situation is simulated in Figures 4 and 5 which represent a very simplified source-reservoir-trap system for which the bidimensional numerical modeling procedure THEMIS (Doligez et al., 1987; Ungerer et al., 1990) has been applied. The schematic cross section (Figure 4) displays a shaly source rock as represented by one layer of mesh elements and a sandy reservoir rock represented by three layers. They are embedded in a barren shaly lithology which acts as a cap rock. The applied model is designed to simulate the evolution of a given basin, considering sediment compaction, expulsion of fluids, thermal transfer, and generation and migration of hydrocarbons. The formation of hydrocarbons is described by a system of parallel reactions controlled by first-order kinetics according to the Arrhenius equation. As far as migration is concerned, fluid flow is deduced from a version of Darcy's law extended to polyphasic flow by application of the concept of relative permeabilities. The relative permeabilities to water and to hydrocarbons are assumed to be dependent on the degree of saturation. Attributes of the source rock illustrated in Figure 4 are: 5.2% Total Organic Carbon content (TOC), Hydrogen Index value (HI) of 630, and a set of kinetic parameters corresponding to the previously discussed Kimmeridge clay kerogen. Based on these data, Figure 4 shows a time-related

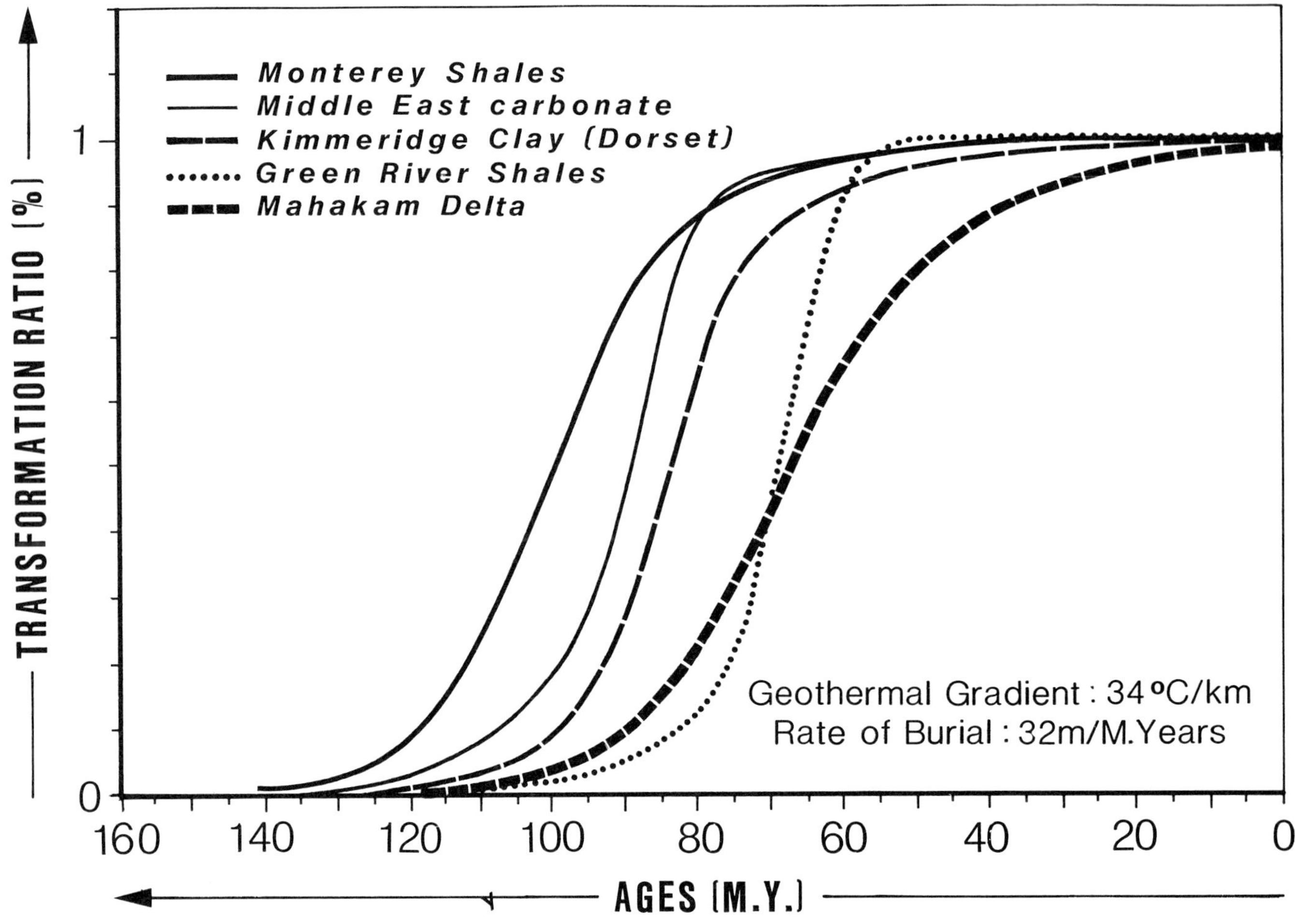

Figure 3. Comparison of calculated transformation ratio versus time for five different kerogens. These curves are obtained by simulating the same thermal history.

evolution of the computed hydrocarbon saturation in the schematic cross section at different steps of the basin history. In this situation (source rock A), hydrocarbon generation starts about 100 million years ago (Ma) in the part of the source unit which is located in the deepest part of the basin. At 70 Ma an increase in the hydrocarbon-saturation values, as well as the beginning of expulsion and loading of the reservoir rock, is noted. Subsequent accumulation in the recipient trap appears to become significant at 60 Ma. In order to comparatively test this scenario involving source rock A against an alternative scenario involving a source rock B containing the same type of organic matter (same HI and same set of kinetic parameters) but exhibiting a lower petroleum potential as a result of a lower organic content (TOC=2.3%), specific cell elements were selected in the grid. They include one element of the source rock (SR), one element of the reservoir rock (RR), and one element of the trap (T) (Figure 4). The monitoring of hydrocarbon saturation versus time in these elements for the two source-rock situations shows that expulsion and accumulation of hydrocarbons is delayed for the organic-leaner source rock B (Figure 5). The timing difference, which spans tens of millions of years (m.y.), has a considerable significance in terms of exploration implications. This example, although very schematic and not pretending to depict any real situation, should be viewed as a numerical experiment indicating the importance of organic facies to migrational concepts, and hence quantitative basin modeling.

Since expulsion from source rocks implies the movement of hydrocarbons within the sediment fabric, substantial progress in the understanding and quantification of the primary migration can be anticipated from current research related to the microhabitat of kerogen. In this respect the current repertoire of microscopic techniques, including examination using transmitted and reflected light, fluorescence under various wavelengths, scanning electron microscopy (secondary electrons), carbon mapping using EDS, infrared microscopy, and image analysis coupled with SEM (back-scatter electron mode), provide powerful tools. Progress concerning the recognition and analysis of organic matter and the spatial relationship between organic matter and mineral groundmass can be anticipated (Bertrand et

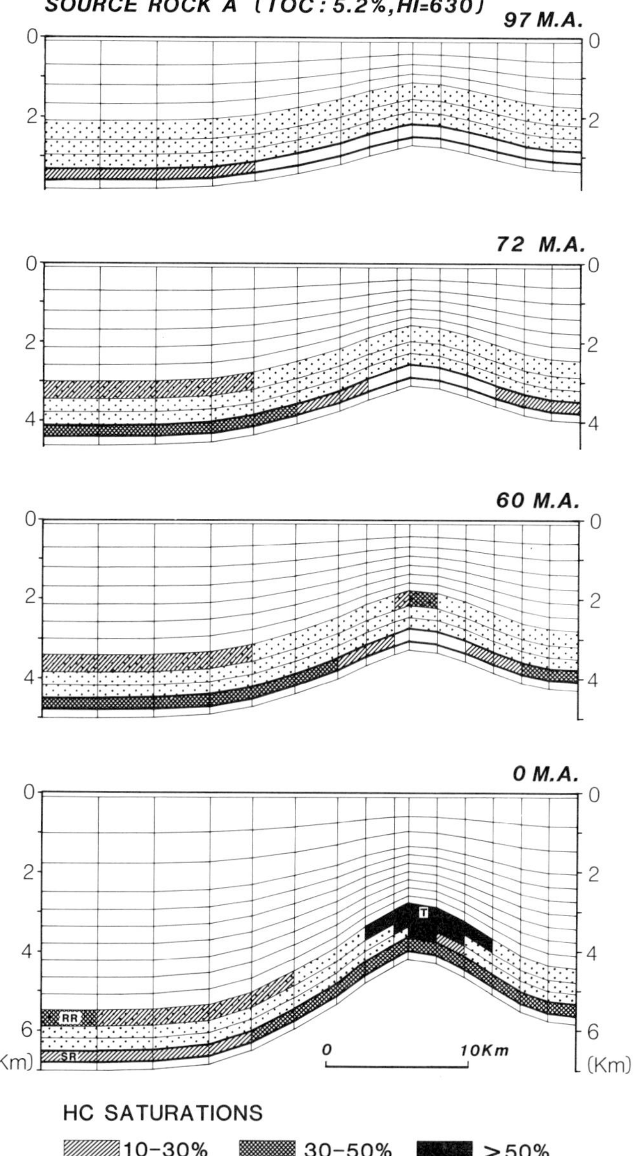

Figure 4. Evolution of the distribution of hydrocarbon saturation in a schematic cross section according to the prediction of the THEMIS model. SR: source rock; RR: reservoir rock; T: trap.

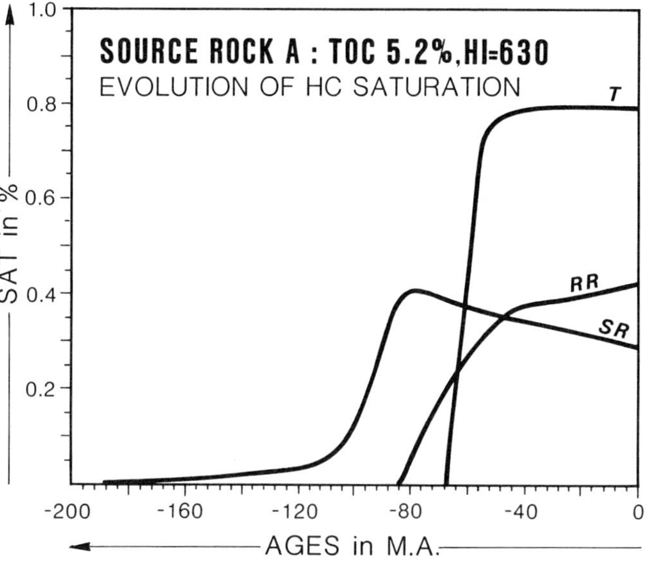

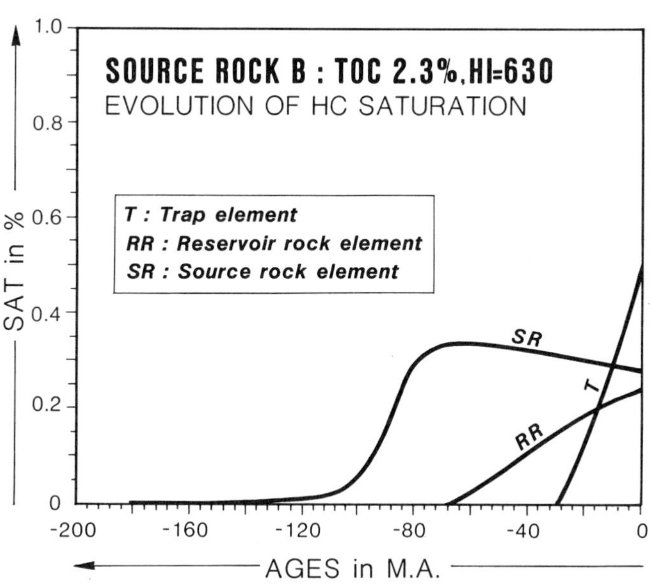

Figure 5. Comparison of the evolution of hydrocarbon saturation in selected mesh elements of Figure 4 between source rock A (TOC = 5.2%) and source rock B (TOC = 2.3%). SR: source rock; RR: reservoir rocks; T: trap.

al., in press; Kearsley et al., in press). Although the constraints of microsedimentological features on primary migration are presently poorly understood, most researchers realize that rock properties such as texture, pore-size distribution, pattern of micro-distribution of organic matter, surface to volume ratio, and nature of the grain surface are all likely to significantly influence expulsion efficiency (Comer and Hinch, 1987). It can also be inferred that these parameters have an effect on the distribution and pattern of the oil-saturated domains within the pore system. Similarly, surface interactions between fluids and solids are probably important parameters in these fine-textured sediments.

ORGANIC FACIES AND SOURCE ROCK EVALUATION STRATEGY

As a consequence of the previous remarks concerning the effective control of source rocks, and of the multiscale distribution of associated organic facies, on the petroleum potential of sedimentary basins, it is clear that organic-bearing intervals deserve to be considered with special care in order to evaluate the potential profitability of prospects. The methods (concepts and tools) currently available for source-rock evaluation have various applicability according to the state of regional knowledge and the degree of maturity of an exploration campaign.

In frontier basins where very few data are available, we have to rely on the best educated guesses. To help the explorationist in the difficult task of predicting the presence of subsurface source beds and their stratigraphic distribution, various semiexperimental prognosis guides are currently under development. They are based on several approaches: global stratigraphic regularities, numerical modeling, and tectono-sedimentary settings.

Organic-rich strata are not randomly distributed in the stratigraphic record but are commonly concentrated in temporally restricted periods. The Cenomanian-Turonian boundary, Aptian/Albian, Kimmeridgian, and Toarcian are well-documented examples of such intervals which have been favorable for the deposition of organic-matter-rich sediments (Schlanger and Jenkyns, 1976; Jenkyns, 1980; Dean and Gardner, 1982; DeGraciansky et al., 1984; Arthur et al., 1987; Herbin et al., 1986; Kuhnt et al., this volume; Lipson-Benitah et al., this volume; Miller, this volume; Baudin et al., this volume). A recent survey by Klemme and Ulmishek (1989) points out that source rocks for 90% of the discovered oil and gas reserves are restricted to six stratigraphic intervals (Silurian, Upper Devonian-Tournaisian, Pennsylvanian-lower Permian, Upper Jurassic, middle Cretaceous, and Oligocene-Miocene) which, in turn, account for only 30% of the Phanerozoic time span. These global stratigraphic regularities have to be kept in mind when one tries to anticipate the presence and the stratigraphic occurrence of source beds. This is especially true in undrilled basins or in geological situations where source rocks have not been penetrated by the drill.

Since the pioneering work performed by Parrish (1982), several authors have devoted their research efforts to the prediction of the paleodistribution of effective upwelling areas through the modeling of paleoclimates and paleocirculations in combination with palinspastic reconstructions (Barron, 1985; Scotese and Summerhayes, 1986). Since the seafloor beneath highly productive upwelled surface waters is recognized as a favorable setting for the deposition of large accumulations of organic matter (Summerhayes, 1981), these attempts should be seen as potential clues for providing indications as to the distribution of specific source rocks. In this respect, it is noteworthy that the origin of prolific source rocks, such as the Miocene Monterey Formation and the Permian Phosphoria Formation, has been linked to upwelling regimes (Demaison and Moore, 1980). Constant improvement of these models may provide a tool of routine application in the exploration process. However, the current status of source-rock data bases does not provide yet a confident test of these modeled simulations (Kruijs and Barron, this volume).

As a part of the sedimentary record, source rocks have to be considered within the scope of basin type and evolution. Such an approach can be exemplified by a tentative conceptual model which has been proposed to rationalize the formation of source rocks during the early phase of intracontinental rifting under tropical conditions (Huc et al., in press). During the initial spreading phase, the intracratonic rifting area is affected by a dense network of fault systems exhibiting small vertical displacements and resulting in widespread depressional settings. Under tropical conditions, these depressions are likely to be dominated by marshy wetlands. In the sediment record this environment can be preserved as a fluvio-deltaic facies. This facies is commonly recognized in early rift systems, in which organic matter can eventually accumulate as type III kerogen, including coal deposits. In the subsequent phases, only a limited number of normal faults, specifically orientated, will be active, exhibiting mainly large vertical displacements associated with block tilting. These tectonic processes will progressively build a topographic architecture similar to the one presently displayed by the Tanganyika graben. At this stage, the great depth and the elongated shape of the basin are factors favorable for the onset and persistence of water stratification, which, when associated with a sufficient algal productivity, favors the development of an oxygen-depleted water mass and results in the accumulation and preservation of organic matter. Moreover, the occurrence of uplifted shoulders limits the size of the drainage catchment, thereby, restricting the input of land-plant-derived organic matter and preventing dilution of the autochthonous lacustrine organic matter by large detrital sedimentary discharge. In such a situation the sediments accumulating on the lake floor will be characterized by a high content of organic matter, mainly derived from algal and bacterial biomass, which can be related to a type I and/or type II kerogen assemblage.

Accordingly, we perceive that an understanding of the potentially favorable periods of source-rock deposition, within the framework of the structural history of a given sedimentary basin, would be an invaluable exploration tool to use as a working hypothesis for predictive purposes. However, a clear overview of the relationship between the various types of structural situations and the potential occurrence and placement of source rocks in the associated sedimentary pile is currently lacking.

Further in the process of exploration, the availability of samples provides the means to establish a source-rock data set which can be used for diagnosis. It should be stressed that during the initial stages of exploration, field geologists should be encouraged to sample source rocks at outcrops. Early geochemical information identifying organic-rich rocks is of great value in formulating a pre-exploration prognosis. Direct assessment of organic-rich horizons, including their stratigraphic distribution and regional lateral variations (Huc, 1988b), is facilitated by rapid pyrolysis techniques (i.e., the ROCK-EVAL method) which are suited to high-frequency data acquisition (Espitalié et al., 1977, 1985a, 1985b). This data acquisition permits construction of geochemical logs and maps that are invaluable in the quantification of the regional distribution of source rocks and for assembly of the data sets essential to numerical basin modeling (Espitalié et al., 1984, 1986; Demaison, 1984). However, pyrolysis methods rely on samples that are most commonly provided as ditch cuttings. Special care in collection is essential to ensure quality and representativeness. Otherwise, well-data sets will be incomplete and/or biased. Moreover, although ROCK-EVAL-type pyrolysis is now widely used as the basic tool of source evaluation, analyses are often performed on a limited number of wells and too few returns. In order to supplement pyrolysis evaluations, methods of source-rock assessment based on wireline-log data have been developed in recent years (Meyer and Nederlof, 1984; Mendelson and Toksoz, 1985; Carpentier et al., 1989; Herron et al., this volume). Most of them rely on the usual logging tools for information. The main benefit of such wireline-log-based methods is the substantial increase, at low cost (using only material already collected), of source-rock data, providing the means for vertical profiling of organic-matter content. The resulting continuous profiles can be used for monitoring and correlating source rocks in cross-section context (Figure 6). In order to integrate this information on a basin-wide scale, seismic data can provide a convenient framework for constructing maps of the areal distribution of source rocks. Under favorable conditions, source-rocks sequences can even be traced by seismic lines, thereby helping in the regional assessment of petroleum potential.

An important conclusion to any comprehensive basinal source-rock survey is to provide guidance for subsequent detailing work. This includes: the selection of representative samples for crucial geochemical investigations, including kerogen typing by optical, spectroscopic, and chemical means; determination of kinetic parameters; specific laboratory experiments (preparative pyrolysis, hydrous pyrolysis); and bitumen extraction for molecular-marker information used in oil/source-rock correlation exercises.

CONCLUSIONS

The increasing cost of exploration stimulates increasing scrutiny by management of the potential value of petroleum prospects. In this respect, emphasis is placed on increased understanding of the source-rock system, which controls to a large extent the petroleum potential of sedimentary basins. As far as source rocks are concerned, conceptual guidelines and practical tools are currently available for prospect evaluation. However, source-rock prediction and quantitative assessment are still areas where more development is needed. Future advances in understanding the occurrence of specific organic-rich intervals at the global scale in time and space (so-called "anoxic events") and the distribution and volume of source rocks according to their tectonic setting and climatic environment of deposition, should significantly help in improving a predictive capability.

A better assessment of the relative influence of the basic factors controlling organic-matter accumulation (carbon flux to the water-sediment interface, total sediment flux rate, and degree of preservation) is expected to provide a possible quantitative approach to predicting organic-carbon content. To some extent, the rates of sediment accumulation can be determined by seismic/stratigraphic interpretation, but both the productivity and the preservation factors have to be calibrated for different depositional environments as a function of geologic age. This can most probably be achieved through studying a variety of well-characterized recent and ancient sedimentary systems.

Progress allowing better prediction in the lateral distribution of organic facies at the basin scale, and in the vertical distribution of organic facies at the sedimentary-sequence scale, will aid in upgrading the quality and the statistical representativeness of the geochemical inputs to numerical basin modeling. Advances in the interpretation of seismic-reflection data in terms of source-rock recognition and the sequential stratigraphic/seismic stratigraphic approach need to be integrated in source-rock-appraisal studies for fullest exploitation. Such an approach can be exemplified by the models which have been proposed for the deposition of the Upper Cretaceous coals in the Rocky Mountains and Great Plains regions (Ryer, 1984; Cross, 1988). Based on the concepts of sequential stratigraphy, these models show that the occurrence of major coal beds is related to vertically stacked progradational events which characterize trangressive and regressive maxima. Extension of this approach to other types of organic-rich beds is indeed very promising.

One of the main challenges of geochemical prospect appraisal remains in reaching a reliable quantification of the migration efficiency over drainage areas. To achieve this, a better understanding of the distribution of organic matter at the sedimentary-sequence scale and at the sediment-fabric scale (microhabitat of the kerogen), including the intimate influence on primary migration, is certainly an important clue. More work is definitely needed to further quantify these considerations.

ACKNOWLEDGMENTS

I thank Drs. E. Brosse, B. Doligez, B. Durand, R. Eschard, and P. Forbes (IFP) for valuable discussions. The manuscript was improved considerably by the critical comments of the reviewers: Drs. R. Burwood

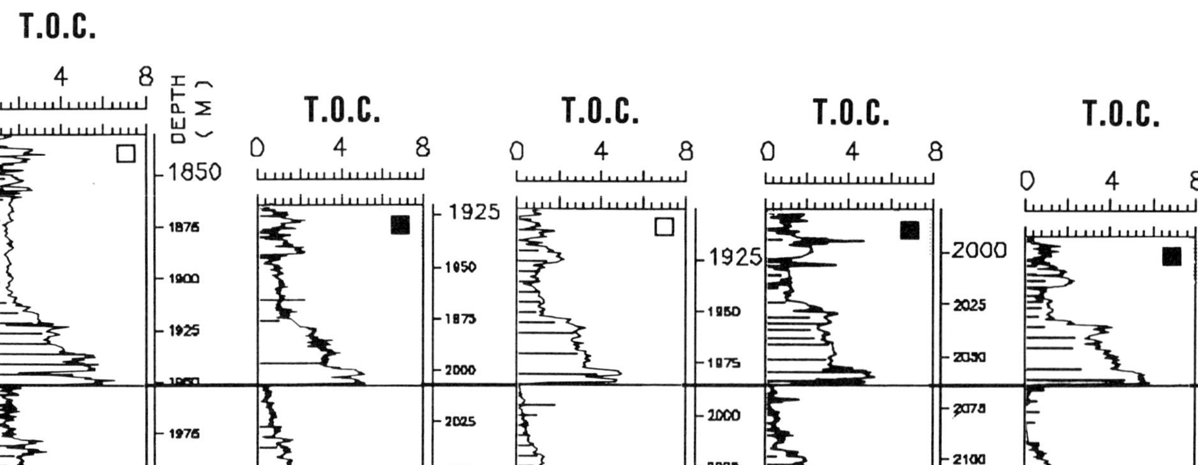

Figure 6. Application of the CARBOLOG method (Carpentier et al., 1989) for providing TOC estimation from sonic/resistivity logs.

(Petrofina), G. Claypool (Mobil), and J. P. Houzay (Elf-Aquitaine).

REFERENCES CITED

Arthur, M. A., S. O. Schlanger, H. C. Jenkyns, 1987, The Cenomanian-Turonian oceanic anoxic event, II. Paleoceanographic controls on organic matter production and preservation, *in* J. Brooks, A. Fleet, eds., Marine petroleum source rocks: Geological Society of London Special Publication 26, p. 401-420.

Barron, E. J., 1985, Numerical climate modeling, a frontier in petroleum source rock prediction: results based on Cretaceous simulations: AAPG Bulletin, v. 69, p. 448-459.

Behar, F., and M. Vandenbroucke, 1987, Chemical modeling of kerogens: Organic Geochemistry, v. 11, p. 15-24.

Berner, R. A., 1984, Sedimentary pyrite formation: an update: Geochimica et Cosmochimica Acta, v. 48, p. 605-615.

Bertrand, P., E. Lallier-Verges, L. Martinez, B. Pradier, P. Tremblay, A. Y. Huc, R. Jouhannel, and J. P. Tricard, in press, Spatial relationships between organic matter and mineral groundmass in the microstructure of the Dorset formation organic-rich rocks (Great Britain), *in* Advances in organic geochemistry: 14th International Meeting on Organic Geochemistry, Paris, September 18-22, 1989.

Brooks, J. D., and J. W. Smith, 1969, The diagenesis of plant lipids during the formation of coal, petroleum and natural gas. II. Coalification and the formation of oil and gas in the Gippsland Basin: Geochimica et Cosmochimica Acta, v. 33, p. 1183-1194.

Carpentier, B., G. Bessereau, and A. Y. Huc, 1989, Diagraphies et roches-meres, estimation des teneurs en carbone organique par la methode CARBOLOG: Revue de l'Institut Francais du Petrole, v. 44, p. 6.

Comer, J. B., and H. H. Hinch, 1987, Recognizing and quantifying expulsion of oil from the Woodford Formation and age-equivalent rocks in Oklahoma and Arkansas: AAPG Bulletin, v. 71, p. 844-858.

Cross, T. A., 1988, Controls on coal distribution in trangressive-regressive cycles, Upper Cretaceous, Western Interior, USA, *in* C. K. Wilgus et al., eds., Sea level changes. An integrated approach: SEPM Special Publicaton 42, p. 371-380.

Dean, W. E., and J. V. Gardner, 1982, Origin and geochemistry of redox of Jurassic to Eocene age, Cape Verde basin (DSDP site 367), Continental Margin of North West Africa, *in* S. O. Schlanger and M. B. Cita, eds., Nature and origin of Cretaceous carbon rich facies: London, Academic Press, p. 57-78.

DeGraciansky, P. C., G. Deroo, J. P. Herbin, L. Montadert, C. Muller, A. Schaaf, and J. Sigal, 1984, Ocean wide stagnation episode in the Late Cretaceous: Nature, v. 308, p. 346-349.

Demaison, G. J., and G. T. Moore, 1980, Anoxic environments and oil source bed genesis: AAPG Bulletin, v. 64, p. 1179-1209.

Demaison, G. J., 1984. The generative basin concept, *in* G. Demaison and R. J. Murris, eds., Petroleum geochemistry and

basin evaluation: AAPG Memoir 35, p. 1-14.
Demaison, G. J., and G. T. Moore, 1980, Anoxic environments and oil source bed genesis: AAPG Bulletin, v. 64, p. 1179-1209.
Dinur, D., B. Spiro, and Z. Aizenshtat, 1980, The distribution and isotopic composition of sulphur in organic-rich sedimentary rocks: Chemical Geology, v. 31, p. 37-51.
Doligez, B., P. Ungerer, P. Y. Chenet, F. Bessis, and G. Bessereau, 1987, Numerical modeling of sedimentation, heat transfer, hydrocarbon formation and fluid migration in the Viking Graben, North Sea, *in* J. Brooks and K. Glennie, eds., Petroleum Geology of North West Europe: Graham and Trotman. p. 1039-1048.
Durand, B., 1988, Understanding of hydrocarbon migration in sedimentary basins (present state of knowledge): Organic Geochemistry, v. 13, p. 445-459.
Espitalié, J., J. L. Laporte, M. Madec, F. Marquis, P. Leplat, and J. Paulet, 1977, Methode rapide de caracterisation des roches meres, de leur potentiel petrolier et de leur degre d'evolution: Revue de l'Institut Francais du Petrole, v. 28, p. 37-66.
Espitalié, J., F. Marquis, and I. Barsony, 1984, Geochemical logging, *in* K. J. Voorhees, ed., Analytical Pyrolysis: Butterworth and Co., p. 276-304.
Espitalié, J., G. Deroo, and F. Marquis, 1985a, La pyrolyse ROCK-EVAL et ses applications. Partie I: Revue de l'Institut Francais du Petrole, v. 40, p. 563-579.
Espitalié, J., G. Deroo, and F. Marquis, 1985b, La pyrolyse ROCK-EVAL et ses applications. Partie II: Revue de l'Institut Francais du Petrole, v. 40, p. 755-784.
Espitalié, J., G. Deroo, and F. Marquis, 1986, La pyrolyse ROCK-EVAL et ses applications. Partie III: Revue de l'Institut Francais du Petrole. v. 41, p. 73-89.
Espitalié, J., P. Ungerer, I. Irwin, F. Marquis, 1987, Primary cracking of kerogens. Experimenting and modeling C_1, C_2-C_5, C_6-C_{15} and C_{15+} classes of hydrocarbons formed: Organic Geochemistry, v. 13, p. 893-899.
Gransh, J. A., and J. Posthuma, 1974, On the origin of sulphur in crudes, *in* B. Tissot and F. Bienner, eds., Advances in organic geochemistry, 1973: Paris, Editions Technip, p. 727-739.
Herbin, J. P., F. Magniez-Jannin, and C. Muller, 1986, Mesozoic organic-rich sediments in the south Atlantic: distribution in time and space, *in* E. T. Degens, P. A. Meyers, and S. C. Brassel, eds., Biogeochemistry of black shales: Mitteilungen aus dem Geologisch-Palaeontologischen, Institut der Universitaet Hamburg, p. 71-97.
Huc, A. Y., 1988a, Sedimentology of organic matter, *in* F. H. Frimmel and R. F. Christman, eds., Humic substances and their role in the environment: Dahlem Konferenzen, John Wiley, p. 215-243.
Huc, A. Y., 1988b, Aspects of depositional processes of organic matter in sedimentary basins: Organic Geochemistry, v. 13, p. 263-272.
Huc, A. Y., J. Le Fournier, M. Vandenbroucke, and G. Bessereau, in press, Northern Lake Tanganyika: an example of organic sedimentation in an anoxic rift lake, *in* B. J. Katz and B. Rosendahl, eds., Lacustrine basin research conference proceedings, AAPG.
Jenkyns, H. C., 1980, Cretaceous anoxic events: From continents to oceans: Journal of the Geological Society of London, v. 137, p. 171-188.
Jones, R. W., 1984, Comparison of carbonate and shale source rocks, *in* J. G. Palacas, ed., Petroleum geochemistry and source rocks: AAPG Studies in Geology 18, p. 163-180.
Kearsley, A. T., A. N. Bishop, and R. L. Patience, in press, Fine scale heterogeneity of organic matter distribution and composition in potential hydrocarbon source rocks: an electron microscopy study, *in* Advances in organic geochemistry: 14th International Meeting on Organic Geochemistry, Paris, September 18-22, 1989.
Kenig, F., and A. Y. Huc, in press, Incorporation of sulfur into recent organic matter in a carbonate environment, Abu Dhabi, U.A.E., *in* W. L. Orr and C. M. White, eds., Geochemistry of sulfur in fossil fuels: ACS Symposium Series.
Klemme, H. D., and G. F. Ulmishek, 1989, Depositional controls, distribution and effectiveness of world's petroleum source rocks (abs.): AAPG Bulletin, v. 73, p. 372.
Lewan, M., 1985, Evaluation of petroleum generation by hydrous pyrolysis experimentation: Philosophical Transactions of the Royal Society of London, v. 315, p. 123-134.
Leythaeuser, D., A. S. Mackenzie, R. G. Schaefer, and M. Bjoroy, 1984, A novel approach for recognition and quantification of hydrocarbon migration effects in shale-sandstone sequences: AAPG Bulletin, v. 68, p. 196-219.
Meissner, F. F., 1978, Petroleum geology of the Bakken Formation: Williston Basin Symposium, Montana Geological Society, p. 207-227.
Mendelson, J. D., and M. N. Toksoz, 1985, Source rock characterization using multivariate analysis of log data: Society of Professional Well Log Analysts 26th logging symposium, June 17-20, 1985, p. UU1-UU21.
Meyer, B. L., and M. H. Nederlof, 1984, Identification of source rocks on wireline logs by density/resistivity and sonic transit time/resistivity crossplots: AAPG Bulletin, v. 68, p. 121-129.
Orr, W. L., 1974, Changes in sulfur content and isotopic ratios of sulfur during petroleum maturation: Study of Big Horn Paleozoic oils, AAPG Bulletin, v. 50, p. 2295-2318.
Orr, W. L., 1986, Kerogen/asphaltene/sulfur relationships in sulfur-rich Monterey oils: Organic Geochemistry, v. 10, p. 499-516.
Parrish, J. T., 1982, Upwelling and petroleum source beds with reference to the Paleozoic: AAPG Bulletin, v. 66, p. 750-774.
Ryer, T. A., 1984, Transgressive-regressive cycles and occurrence of coal in some Upper Cretaceous strata of Utah, USA, *in* R. A. Rahmani and R. M. Flores, eds., Sedimentology of coal and coal-bearing sequences: International Association of Sedimentologists Special Publication 7, Oxford, Blackwell Scientific Publications, p. 217-227.
Schlanger, S. O., and H. C. Jenkyns, 1976, Cretaceous oceanic events, causes and consequences: Geologie en Mijnbouw, v. 55, p. 179-184.
Scotese, C. R., and C. P. Summerhayes, 1986, Computer model of paleoclimate predicts coastal upwelling in the Mesozoic and the Cenozoic: Geobyte, summer, p. 28-42.
Summerhayes, C. P., 1981, Sedimentation of organic matter in upwelling regimes, *in* E. Suess and J. Thiede, eds., Coastal upwelling in sediment record: NATO Conference Series, Series IV: Marine Sciences, p. 29-72.
Talukdar, S., O. Gallando, and M. Chin-a-Lien, 1986, Generation and migration of hydrocarbons in the Maracaibo Basin, Venezuela: an integrated basin study: Organic Geochemistry, v. 10, p. 261-280.
Thomas, B. M., 1982, Land-plant source rocks for oil and their significance in Australian basins: Australian Petroleum Exploration Association Journal, v. 22, p. 164-178.
Tissot, B. P., and D. H. Welte, 1984, Petroleum formation and occurrence: Berlin, Springer-Verlag, 699 p.
Tissot, B., R. Pelet, and P. Ungerer, 1987, Thermal history of sedimentary basins, maturation indices, and kinetics of oil and gas generation: AAPG Bulletin, v. 71, p. 1445-1466.
Ungerer, P., and R. Pelet, 1987, Extrapolation of the kinetics of oil and gas formation from laboratory experiments to sedimentary basins: Nature, v. 327, p. 52-54.
Ungerer, P., J. Burrus, B. Doligez, P. Y. Chenet, and F. Bessis, 1990, Basin evaluation by integrated two dimensional modelling of heat transfer, fluid flow, hydrocarbon generation and migration: AAPG Bulletin, v. 74, p. 309-335.
Yen, T. F., 1976, Structural aspects of organic components in oil shales, *in* T. F. Yen and G. V. Chilingarian, eds., Oil shale: Amsterdam, Elsevier, v. 7, p. 129.

A Paleoceanographic Approach to the Kimmeridge Clay Formation

Richard G. Miller
BP Research Centre
Sunbury-on-Thames, United Kingdom

The Kimmeridge Clay Formation is a Late Jurassic/Early Cretaceous clastic source rock which extends from offshore Norway to offshore Canada and the English Channel. It is the principal source of the North Sea oils. The base is diachronous but the top is synchronous over the North Sea. In the North Sea it is typically 150 m thick with TOC = 4–5% and HI = 350 to 450; accumulation rate was about 15 m/Ma.

Deposition apparently occurred in a stratified sea. An upper layer flowed southward from the polar Boreal Ocean to Tethys at perhaps 1 km day^{-1}, over a warm saline bottom water (WSBW). Much evidence exists for a negative water balance, but none for unusual organic productivity. Deposition of the organic-rich facies probably ceased following a slight climatic change; this produced a basin overturn, and consequently a cool, oxygenated bottom current flowing southward.

INTRODUCTION

The Kimmeridge Clay Formation (KCF) in its broadest sense is northern Europe's premier source rock, having provided the source of hydrocarbons for the bulk of the North Sea oil fields. It was deposited in a series of basins in the "KCF sea" which extended from eastern Greenland and Canada to offshore Norway, and south to the English Channel, in Late Jurassic and Early Cretaceous times.

Understanding its genesis is important not only for further exploration within this extensive area, but also in the search for analogous source rocks elsewhere, and in the study of source rocks in general. The type section, onshore at Kimmeridge Bay in the Wessex Basin, has therefore been previously extensively studied and regarded as the key for constructing depositional models applicable across the whole KCF sea. Unfortunately, this type section is not representative of the whole formation. It is only one corner of a carpet with a complex pattern.

This paper attempts to integrate observations from many disciplines into a single internally consistent model of the KCF sea and the deposition of the KCF. The model involves the gradual formation of warm, saline, oxygen-deficient bottom water (WSBW) beneath a sluggish surface current in a simple two-layer stratified shelf sea. It implies normal rates of marine organic productivity, low rates of sedimentation, and very efficient organic preservation. The stratification was probably ultimately destroyed following a slight climatic shift.

KIMMERIDGE CLAY FORMATION

Three problems confront any study of the Kimmeridge Clay Formation. The first is that the type section in the Wessex Basin is essentially synchronous with the Kimmeridgian age, whereas throughout much of the North Sea, KCF deposition extended into the Early Cretaceous (Figures 1, 2). The second is that the KCF overlaps two broad faunal provinces with three quite different chronostratigraphic nomenclatures (Figure 2). The third is that a confusing variety of local names for the formation

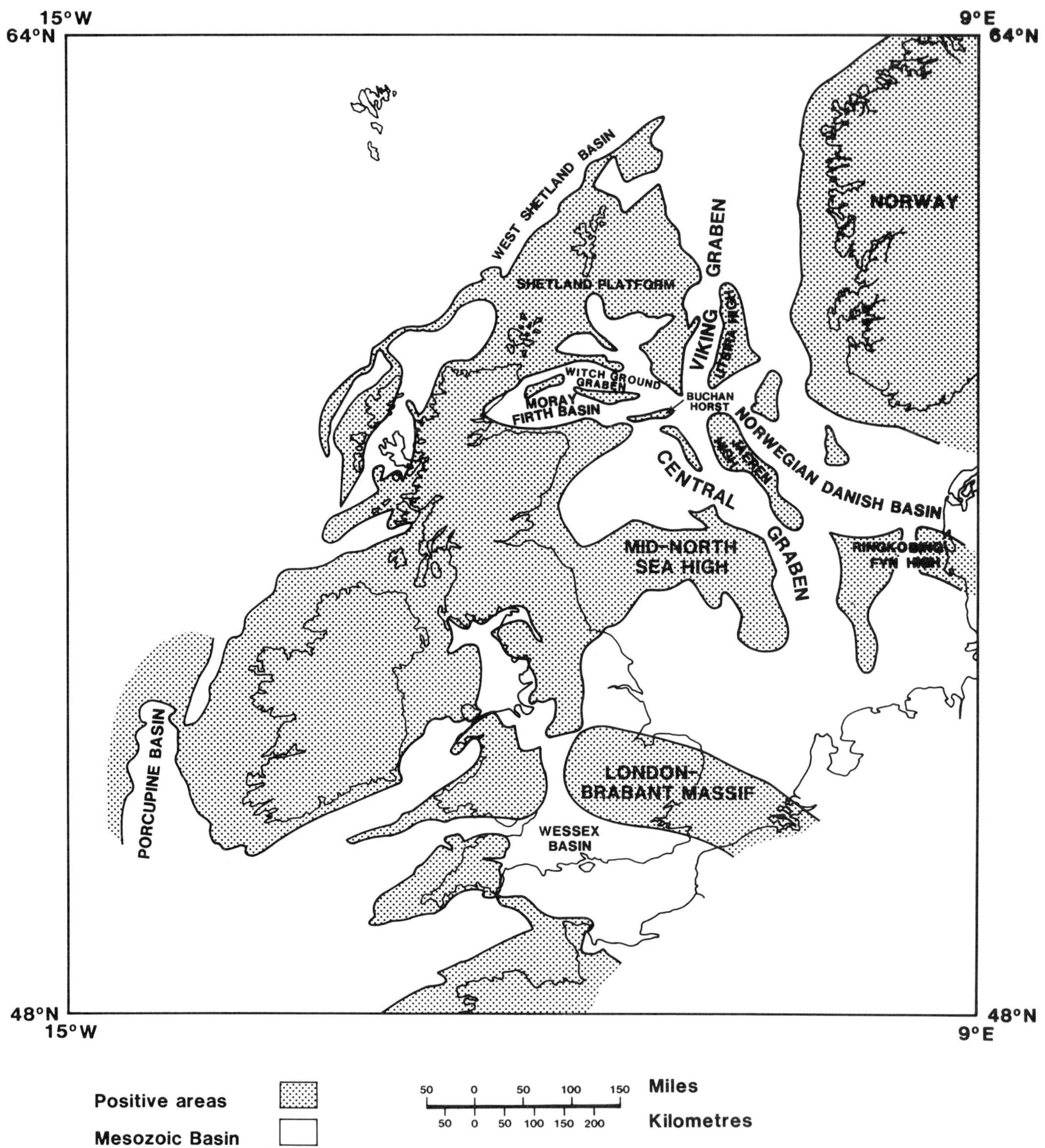

Figure 1. Tectonic elements of the North Sea and the U. K. continental shelf. The type section of the Kimmeridge Clay Formation lies in the Wessex Basin, some distance from the North Sea occurrences.

exists (e.g. Dore et al., 1985; Powell, 1985; Brown, 1984; Birkelund, 1975). This paper will therefore:

1. follow informal industry practices which define the KCF as an oil-prone, organic-rich mudstone, siltstone, or shale extending from the late Ryazanian down to a somewhat arbitrary base at the Oxfordian-Kimmeridgian boundary;
2. use the age and stage names in the literature, following Harland et al. (1982) for absolute ages;
3. use the name Kimmeridge Clay Formation (KCF) throughout in preference to local formation names.

The KCF type section in the Wessex Basin is 508 m thick and of Kimmeridgian age (Cox and Gallois, 1981). It overlies cycles of Oxfordian sandstones, limestones, and claystones, and is overlain by Portlandian sandstones, Portlandian limestones, and the Purbeck limestones and evaporites.

In this study the organic geochemistry and palynology of the KCF was examined in 29 offshore wells from northern Norway to the southern North Sea and the West Shetlands Basin. In the North Sea the KCF can exceed 1200 m in thickness, but 50–250 m is a more typical range and 150 m a probable mean value (Figure 3). The base is diachronous, reflecting subsidence or transgression, and at many places is gradational into leaner but superficially similar claystones or siltstones such as the Bathonian to Oxfordian Heather Formation. The base of the KCF is usually taken at the first major log breaks indicative of an upward sharp rise in the organic-matter content (Deegan and Scull, 1977; Brown, 1984). A gamma-ray intensity of 100 APIγ has been regarded as diagnostic of the KCF (Deegan and Scull, 1977) but this definition can fail, particularly in the central and southern North Sea sectors or in the lower KCF. The North Sea KCF extends into the upper Ryazanian where there is an abrupt and synchronous change to oxic calcareous mudstones of the Cromer Knoll Group (unpublished data; Rawson and Riley, 1982). The Cromer Knoll Group has a thin basal sandstone (Deegan and Scull, 1977), but there is no immediate change in the clay mineralogy relative to the KCF. The detailed geochemistry implies that the KCF/Cromer Knoll contact represents only a sudden change to oxidizing conditions (Johnson, 1975).

Published data on the KCF elsewhere are relatively scarce, but the formation is present in the Porcupine Basin to the west of Ireland, in the Avalon Basin east of Newfoundland, and in East Greenland. In the Porcupine Basin the mudstones typically exceed 200 m in thickness but are locally diluted by sands; the organic-rich intervals are early Kimmeridgian and early Tithonian, and the sequence is overlain in places by hundreds of meters of Berriasian halite and organic-poor mudstone (A.S. Pepper, personal communication). In the Avalon Basin the source interval comprises up to 200 m of calcareous Kimmeridgian shales overlain by carbonates, lean shales, and sandstones (Powell, 1985; Creaney and Allison, 1987). The onshore Greenland deposits are complicated by continual tectonism in small near-shore sub-basins, resulting in sediments including lagoonal coals and thick canyon sandstones; however, 200–500 m of silty black shale accumulated between the late Oxfordian and the mid Volgian (Birkelund, 1975). To the south, the formation is replaced by limestones along a facies boundary roughly equivalent to the modern north coast of continental Europe.

The basic lithology of the KCF is thus organic-rich siltstone or argillite, modified according to local conditions. For example the onshore British exposures of Wessex and eastern England have marls, laminated oil shales, bioturbated mudstones and shell pavements, and a few coccolithic limestones less than a meter thick but probably continuous throughout the onshore English KCF; the latter have exercised a disproportionate influence on previous depositional models (Oschmann, 1988; Tyson et al., 1979; Gallois and Medd, 1979; Gallois, 1976). In the North Sea and farther west, intercalated sandstones, generally ascribed to submarine fans associated with local tectonic activity, are present (Brown, 1984). Other sandstones, such as the Ryazanian-Valanginian Sandringham Sands of the southern North Sea, are present toward the paleo-shorelines and around intrabasinal highs. Limestones and marls become

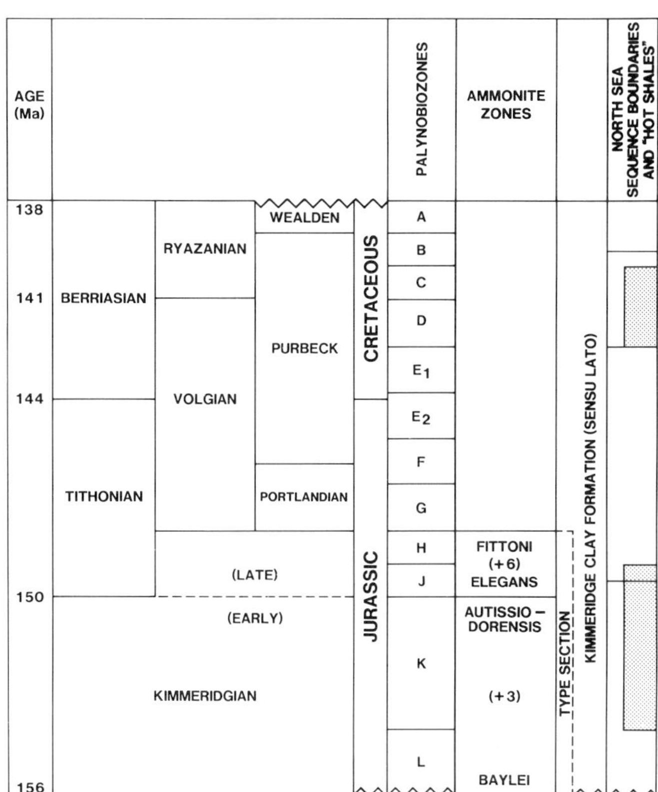

Figure 2. Chronostratigraphy of the Kimmeridge Clay Formation. Absolute ages are taken from Harland et al. (1982). The palynobiozones demonstrate the fine resolution in time that is now possible, although there are only three "tie-points," e.g. base Ryazanian, base Volgian, and base Tithonian. Note that the type section covers less than half the total time during which the KCF was deposited elsewhere. In some cases the KCF extends far down into the Oxfordian, overlying more mudstones such as the Oxford Clay or the Heather Formation. In the right-hand column the horizontal lines represent depositional sequence boundaries, while the shaded areas represent the deposition of "hot shales," or particularly radioactive units.

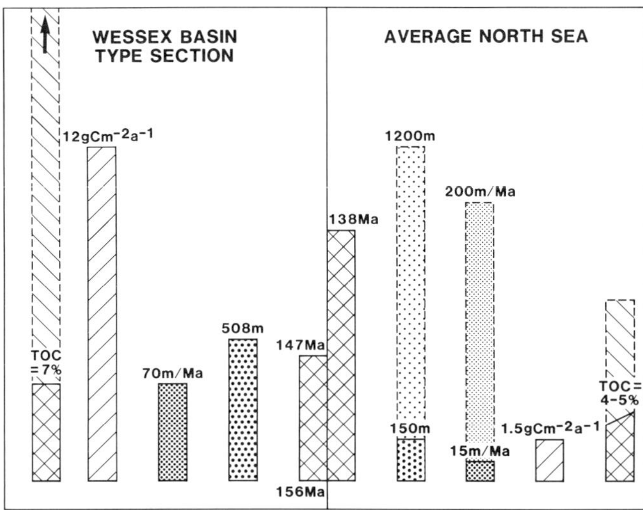

Figure 3. Properties of the Kimmeridge Clay Formation in the type section compared with those of the North Sea KCF. Note the differences in duration at each locality and the much higher rates of deposition and of organic carbon accumulation in the type section. Dashed lines indicate maximum values. Clearly the type section is not representative of the formation in the North Sea.

typical in the Dutch sector KCF. However, evidence for deltas is generally lacking except off East Greenland, and little clastic indication of significant fluvial input has been found except locally along the southern margin, in sharp contrast to the Middle Jurassic (e.g. Anderton et al., 1979). This has major implications for fluvial runoff, rainfall, and climate during the deposition of the KCF.

South of the KCF sea, the typical siltstones and argillites give way relatively abruptly to limestones except wherever nearshore sandstones predominate (Ager, 1975; Anderton et al., 1979). This facies boundary fluctuated with time, for example extending north into the Wessex Basin in post-Kimmeridgian time, so that the Portlandian and Purbeck limestones are chronologically equivalent to the upper KCF elsewhere.

No "average depth" of deposition can be given for the KCF. Good arguments can be made for depths on the order of 50–100 m in the Wessex Basin (Myers and Wignall, 1987; Oschmann, 1988). The depths certainly varied with the changing sea level, which reached a transgressive maximum in the early Volgian (Rawson and Riley, 1982; Brown, 1984) or late Volgian (Goff, 1983). The evaporites noted are related to the subsequent regression. However, this was also a period of continual faulting and graben subsidence. Oschmann (1988) suggested reasonable paleo-depths of 200–500 m in the grabens.

TECTONISM AND TRANSGRESSION

The North Sea contains a complex graben system with a basic triple junction pattern (Brown, 1984) (Figure 1). The stem of the triple junction is the Central Graben; the western arm is the Witch Ground Graben and its extension into the Moray Firth Basin; the northern arm is the Viking Graben. Rifting was initiated in the Permian (Glennie, 1984) and continued during the Jurassic (Brown, 1984). The seaway from the polar Boreal Ocean to the central Atlantic had opened in the Early Jurassic (Glennie, 1984) (Figure 4). Similar rifting was also producing the basins of the western KCF sea, west of the British Isles.

Pre-existing faults in the North Sea underwent continual readjustment during the Late Jurassic, culminating in the major "Late Cimmerian" uplifts in the Volgian and Ryazanian (Rawson and Riley, 1982). This tectonism resulted in numerous sub-basins, linked by grabens or (like Wessex) separated by shallows and by a shifting archipelago. Certain structural blocks, notably the Buchan Horst, may have acted as temporary circulation barriers, restricting water movement during their uplift.

The minor volcanism of this period was summarized by Woodhall and Knox (1979). Minor tuffs of Kimmeridgian to Berriasian age are patchily distributed from Holland to north Norway. Although volcanogenic clays may in fact be quite extensive in the northern KCF (A. Aplin, personal communication) they are not obviously relevant to the depositional model proposed here. Volcanism may, however, have produced effects related to mass mortality or nutrient enrichment.

Four major stratigraphic sequences are present in the North Sea KCF within the early Kimmeridgian to late Ryazanian interval. This interval is divisible into about 12 palynobiozones (in-house data, C. Stronach, W. Paley et al., personal communication), identified by letters on Figure 2 and in subsequent figures. Sequences may be missing or not apparent due to the interplay of tectonic uplift and subsidence and marine transgression/regression. The sequence boundaries do not necessarily coincide with obvious lithological or geochemical changes—for example, the KCF/Heather Formation contact is not a sequence boundary. The palynozonation provides a high-resolution time frame which identifies numerous, and seemingly random, depositional hiatuses within the KCF. It also permits estimates of process rates. Consensus concerning the absolute or relative ages shown on Figure 2 has not yet been reached (e.g. Haq et al., 1987). Process rates estimated over the whole 18 Ma of deposition of the KCF are probably accurate to ± 10%, but the duration of a single palynobiozone or ammonite zone is not known to within better than a factor of two.

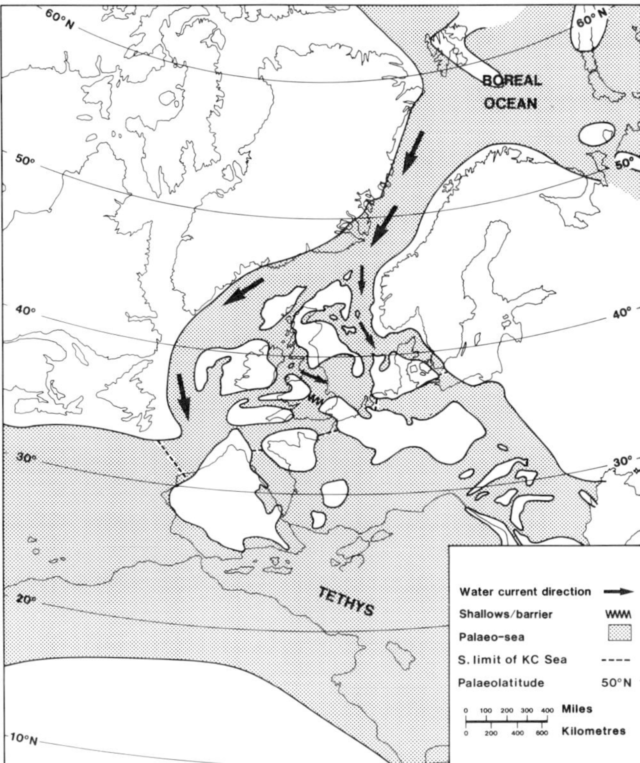

Figure 4. Late Kimmeridgian paleogeography, based on in-house data by R. V. Tyson, with additional information from Ronnevik (1984), Ziegler (1981), Hallam and Sellwood (1976), Arkell (1956), and the author. The southern limit of the KCF sea is the facies change between argillites to the north and carbonates to the south, which at this time followed the north coast of continental Europe. Note that the Wessex Basin containing the type section is effectively isolated from the North Sea area; the argillite/carbonate boundary had moved up to this line of shallows by the Portlandian. Water currents are based on a southward flow from the Boreal Ocean, modified by the Coriolis effect which concentrates this flow down the west side of the basin.

ORGANIC GEOCHEMISTRY

The description of the KCF as "geochemically fairly homogeneous" (Cornford et al., 1983) is true only insofar as it is equally variable everywhere. Typical previous estimates of the mean TOC are in the range 5–10%. The TOC range in the Avalon Basin is typically 2–5% (Creaney and Allison, 1987); in the Porcupine Basin the TOC is in the range 1.1–2.4% (A. Pepper, personal communication); in the Wessex Basin the TOC is about 7% overall but can reach 35% (Oschmann, 1988), 50% (Myers and Wignall, 1987), or 70% (Goff, 1983). Unpublished data for the entire North Sea and West Shetlands area give an overall mean TOC of 4–5% and a maximum of 12% over 6 meters, with values tending to rise upward (Figure 3).

Our data show that in the North Sea TOC values may vary rapidly and unpredictably, both in time and space. For example the Ryazanian section in two Central Graben wells 30 km apart has mean TOC values of 2.5% and 7.6%, respectively, while similar differences actually reverse with time in two Viking Graben wells. Within individual wells there is a slight tendency for TOC to increase as the sedimentation rate decreases, suggestive of a clastic dilution effect. This effect, however, cannot be extrapolated between wells; for example, in two wells in apparently similar settings within the South Viking Graben, the Ryazanian sections are respectively 203 m thick, TOC = 6.5%, and 9 m thick, TOC = 1.4%. Any patterns in the TOC distribution seem to be on a very small scale, and may stem from very local variations in the depositional environment such as tidal current streams.

The organic matter type as reflected by the calculated pre-maturation hydrogen index, or HI (= $100 \times S_2/TOC$)[1], shows unsurprising variations in the North Sea and Wessex Basin. The HI falls with increasing proximity to the paleo-shoreline, and rises with inferred depositional depth, reaching values of around 700 in some graben bottoms. However, Thomas et al. (1985) and unpublished data record occasional anomalously low HI values in some basinal deeps, notably the South Viking Graben, which may be due to turbiditic influxes of terrestrial organic matter with simultaneous replenishment of the pore-water sulfate. A small but distinct rise occurs in the pre-maturation HI with time, with mean values in the North Sea KCF of 365 in the Kimmeridgian, 422 in the Volgian, and 452 in the Ryazanian.

There are numerous published and commercial palynofacies maps of the North Sea KCF (e.g. Barnard and Cooper, 1981; Demaison et al., 1983; Fisher and Miles, 1983; Baird, 1986). The North Sea KCF contains little terrigenous organic material, even in the relatively enclosed Moray Firth (Pearson and Watkins, 1983). The shoreward fall in HI is probably due as much to poorer preservation in shallower water as to increased terrigenous input. The onshore British KCF is similarly dominated by amorphous algal or bacterial kerogen (Williams, 1986), although Ioannides et al. (1976) suggested a swamp source for much of the Wessex Basin organic matter. Molecular geochemistry also strongly suggests that the kerogen in the KCF is predominantly marine (A. Mann, personal communication; Farrimond et al., 1984).

It is instructive at this point to consider the rates of sedimentation and of the accumulation of organic

[1]S_2 is the whole-rock pyrolysis yield of petroleum in kg t^{-1}; TOC is the total organic carbon content in %; HI is therefore in units of g petroleum per kg organic carbon (Tissot and Welte, 1978).

carbon. The detailed palynozonation applied to the 29 study wells reveals unexpected variation in the rate of sedimentation, from zero to more than 200 m/Ma after compaction. Although this partly reflects normal condensed sequences on highs, I suspect that some of this variation is due to both tidal-current turbulence causing sediment redistribution or resuspension, and to slumping into the grabens. Thick sections would therefore represent material missing in depositional hiatuses elsewhere. The mean depositional rates in m/Ma during the whole period of KCF deposition including the hiatuses are 2.0-2.4 off midwest Norway, 4-26 in the northern North Sea, 5-26 in the central North Sea, 12-41 in the semienclosed Moray Firth, 19-32 in the southern North Sea, and 4-13 in the West Shetland region. A typical mean rate is therefore some 15 m/Ma (Figure 3), deposited for up to 18Ma (but with a mean thickness of only 155 m, reflecting the diachronous or transgressive base of the North Sea KCF). Prior to compaction, the rates may have been 3-10 times greater. The rates are distinctly lower in the late Volgian and Ryazanian of the central North Sea, but otherwise no clear time dependence is apparent.

This sediment accumulation rate is rather low compared to typical subsidence rates of passive continental margins (Blatt et al., 1980). Tyson (1987) noted that rather low accumulation rates are in fact typical of many marine source rocks. The typical calculated rate of deposition in the Wessex Basin is, in contrast, 70 m/Ma after a slow start (using the data of Cox and Gallois, 1981), demonstrating that the type section does not resemble the North Sea KCF in this important respect. Short-term accumulation rates may have been even higher in east Greenland, although probably augmented by a significant terrigenous input (Birkelund, 1975).

From these sedimentation rates, the average TOC, and an assumed density of 2.5 g cm^{-3}, one may now calculate the rate of organic carbon accumulation. This is about 1.5 g C m^{-2} a^{-1} in the North Sea and about 12 g C m^{-2} a^{-1} in the Wessex Basin, again emphasizing the nonrepresentative nature of the type section (Figure 3).

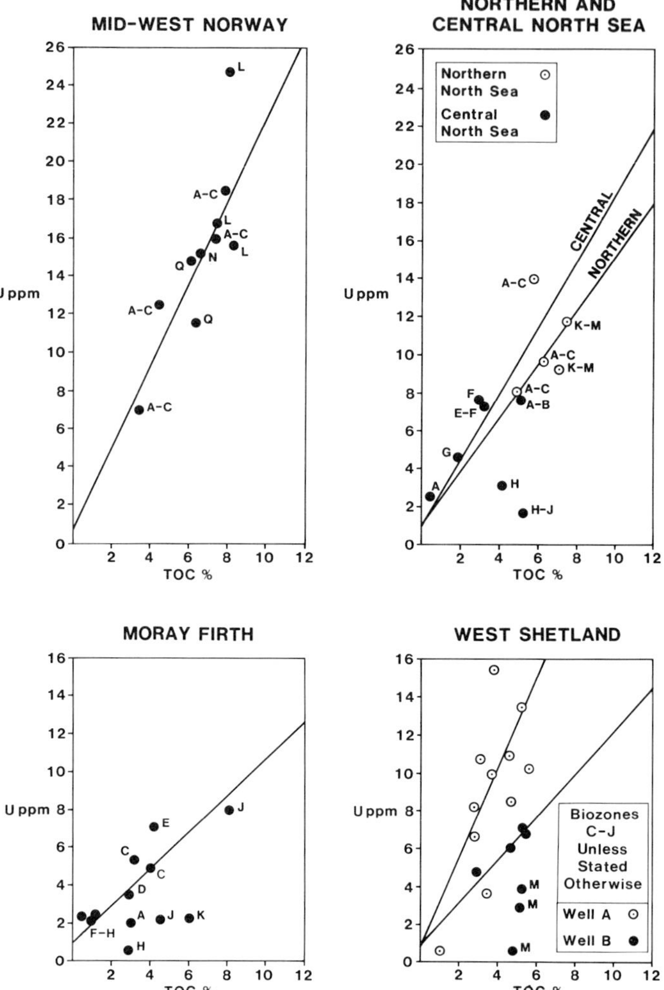

Figure 5. Correlations between TOC and uranium in well cuttings of the Kimmeridge Clay Formation. Letters designate the palynobiozones (see Figure 2). The regression lines are estimates forced through an assumed 1 ppm of uranium due to the clastic component, and considering only biozones C to G, where the clearest correlations occur. The steepest slopes indicate the highest ratio of authigenic uranium to organic carbon.

RADIOACTIVITY AND URANIUM CONTENT

Although the KCF is typically more radioactive than the formations above and below, its radioactivity is very variable, and in many instances less than the >100 APIγ units previously suggested (Deegan and Scull, 1977). Significant thicknesses of otherwise unremarkable North Sea KCF are in fact notably "cold," being less radioactive and uraniferous than average middle North American shale (100 APIγ and 6.55 ppm U) (Bjorlykke et al., 1975; Hearst and Nelson, 1985; unpublished data). Excess radioactivity, where present in the KCF, is due to uranium correlated with the organic matter (Figure 5).

Two notably radioactive intervals occur in the North Sea KCF (Figure 2). The younger, typically referred to as "the hot shale," is of late Volgian/early Ryazanian age (palynobiozones C and D). It is restricted to areas east of the UK/Shetland landmass, extending southward to at least the Mid North Sea High and probably beyond to the London-Brabant landmass (Figure 1). This unit overall has the highest radioactivity, TOC, and HI, and according to Goff (1983) represents the maximum extent of transgression.

The older interval is less radioactive, and of middle to upper Kimmeridgian age (palynobiozones J and K). It occurs in the Moray Firth, Viking Graben, and Norwegian-Danish Basin, i.e. north of the Ringkobing-Fyn/Jaeren Highs and the Buchan Horst. The Buchan Horst was possibly a positive feature at this time which prevented open communication between the Viking and Central Grabens. The full northward extent of this unit is unknown because this section is missing in many areas.

In addition, numerous local minor peaks of radioactivity are present. The West Shetland region has a different pattern with a single radioactivity maximum in the middle-Volgian. It is suggested here that each hot shale unit marks the extent of a particular and distinct body of seawater, perhaps a bottom pool of oxygen-depleted warm weak brine or possibly a surface current stream.

The uncorrected γ-ray intensity of the younger "hot shale" decreases from north to south, being about 300 APIγ off midwest Norway, 200 APIγ in the northern North Sea, 190 APIγ in the central North Sea, 150 APIγ in the southern North Sea, and dropping to 130 APIγ in the semi-enclosed Moray Firth Basin. This trend is confirmed by the U/TOC ratios of selected North Sea wells. Figure 5 shows crossplots of TOC against the uranium content for six wells in the North Sea KCF, the uranium being measured by neutron activation analysis at the London Universities Reactor Centre. All six show a degree of correlation, although a seventh in the northern North Sea (not illustrated) does not. The correlation may break down below biozone G and above biozone C (early Volgian and mid Ryazanian respectively). About 1 ppm uranium resides in the clastic component; the remainder is authigenic and is associated with kerogen. For the correlated intervals, the ratio of authigenic uranium to TOC falls from about 215×10^{-6} off midwest Norway, to 142-175 $\times 10^{-6}$ in the northern and central North Sea, and to 100×10^{-6} in the semi-enclosed Moray Firth. Further data from the southern North Sea are required to confirm a progressive southward fall.

The two West Shetland wells (Figure 5) also suggest U-TOC correlations, with U/TOC ratios of 234×10^{-6} and 112×10^{-6}, or a $\times 2$ variation within essentially a single sub-basin. However, if the data are combined, these two wells could equally imply that the uranium content is independent of the organic matter. In fact a comparison of the γ-ray logs with the TOC in several West Shetland wells demonstrates that both conclusions are valid: there is a progressive 4-fold rise in radioactivity from the middle Kimmeridgian to biozone F (early or middle Volgian) with no corresponding increase in TOC, after which time γ-ray intensity and TOC are correlated and falling (Figure 6). Evidently a high U content and γ-ray signature indicate a high TOC, but the reverse is not necessarily true in this example. The trends noted here are independent of the kerogen type, although some wells show a slight negative correlation between HI and U/TOC (Figure 7).

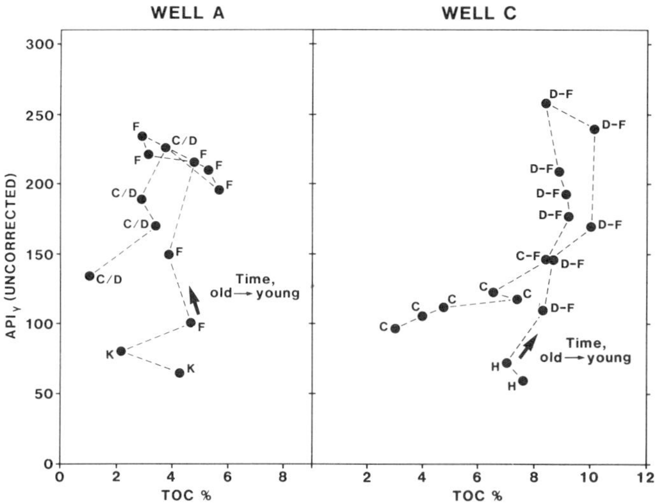

Figure 6. Crossplots of total radioactivity (from well logs) against TOC (from cuttings) in wells from the West Shetland Basin. The overall patterns are remarkably similar. Both show an initial 4-fold rise in radioactivity (and hence in authigenic uranium) which is completely independent of TOC, during biozones K to F. This implies increased levels of U adsorption, possibly resulting from a rise in the concentration of dissolved uranium. Both wells then show a correlated fall in TOC and uranium during biozones C and D. Evidently the authigenic U content of the KCF is not necessarily controlled by the amount of organic carbon present.

The Wessex Basin KCF rarely exceeds 100 APIγ units of radioactivity despite TOC values of 50% or more, and Myers and Wignall (1987) recorded authigenic U/TOC values of only 10-50 $\times 10^{-6}$ in the most organic-rich facies. These authors showed that fluctuating benthic oxygen levels are the likely local control on the authigenic uranium content. It is also possible that the high sedimentation rate of this section reduced its capacity to concentrate dissolved uranium from the sea. Nevertheless this supports a trend of falling U contents and U/TOC ratios from north to south, which might reflect a regional depletion in the dissolved marine uranium content.

PALEOGEOGRAPHY

The KCF sea was triangular, its apex extending into the polar Boreal Ocean to the north and its base entering Tethys. This sea was littered with an archipelago of shifting islands whose pattern changed with changing sea level, sedimentary fill, and fault reactivation. Figure 4 is effectively a "snapshot" at 150Ma, showing the distribution of land and sea. The

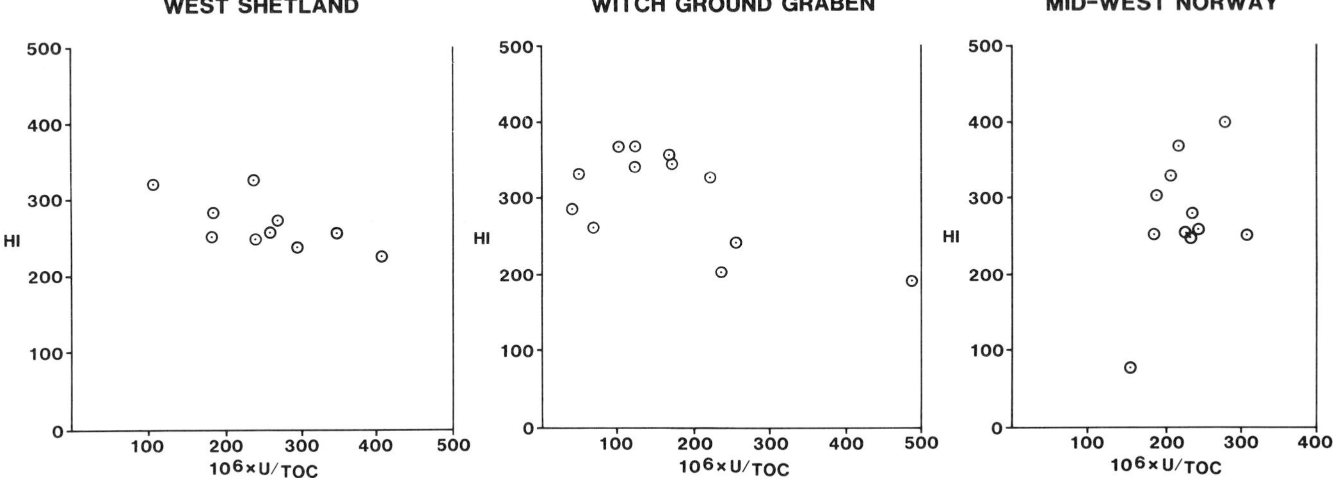

Figure 7. Crossplots of U/TOC against HI for the Kimmeridge Clay Formation in three wells. Two show a slight trend toward negative correlation, effectively indicating that radioactivity may increase where source quality decreases. This may reflect the higher affinity of uranium for terrigenous kerogen compared to algal kerogen (e. g. Leventhal 1981). However, the trend is not strong, and is not apparent in the third well.

extent to which these islands changed attests to relatively shallow coastal waters and low relief. East Greenland and west Norway, however, had coastal mountain belts.

Climatic interpretations are difficult and contradictory. Britain lay in paleolatitudes 35–45°N in a warm and unglaciated world (Ager, 1975). Parrish et al. (1982) suggested that the KCF sea lay within the second driest of four climatic zones. Unpublished data suggest that seasonal high and low atmospheric pressure systems could have been centered over the KCF sea, which would not produce the persistent southerly winds suggested by Oschmann (1988). However, the small size of the KCF sea, and the flanking mountains in the northern part, place the region at the limits of resolution of paleoatmospheric models.

Evidence for the rainfall comes from both palynology and sedimentology. In Wessex, Partington (1983; personal communication 1988) identified a humid subtropical flora in the early Kimmeridgian, which by late *Hudlestoni* time (late Kimmeridgian) had changed to an arid/semiarid assemblage. Aridity prevailed until the late Ryazanian when a wetter climate returned and KCF deposition ceased. As noted previously there are no sedimentological indications of major fluvial inputs except in East Greenland, where a local deltaic sequence begins in the mid Volgian (Birkelund, 1975), and along the southern margins of the sea where small rivers are known. In general the flanking mountains of East Greenland and west Norway would be expected to divert much of the runoff inland. This apparent lack of rivers makes it difficult to identify the source and transport mechanism of the clastic bulk of the KCF. Flash-flood erosion of deeply weathered hinterlands may be indicated.

The presence of terrigenous organic matter in the KCF has been cited as evidence for high runoff rates from a well-vegetated hinterland (e.g., Barnard and Cooper, 1981). However, as noted earlier, there is little terrigenous organic matter, and the geochemical evidence is strongly in favor of a marine algal and bacterial source for the bulk of the kerogen.

The presence of Tasmanitid algae noted onshore by Williams (1986) and offshore by Barnard and Cooper (1981) is not necessarily indicative of a brackish sea, as has sometimes been suggested. They are only indicative of environmental stress (W. Paley, personal communication). The Porcupine halites and Purbeck evaporites are convincing indicators of hypersaline conditions and a negative water balance where evaporation exceeded runoff and rainfall, in some southern sub-basins of the KCF sea, while KCF deposition was still continuing elsewhere.

Consequently, it is suggested here that, at least from late Kimmeridgian to late Ryazanian times, the KCF sea experienced a negative water balance in a dry climate. Modern analogs for the vegetation are probably savannahs with coastal salt-water marshes and mangrove swamps. Wetter conditions prevailed in the early and middle Kimmeridgian but, in a hot climate with no evidence for fluvial runoff or inputs of terrigenous organic material into the KCF sea, the water balance was still possibly negative.

A final climatic note concerns paleoupwelling. Oschmann (1988) suggested seasonal upwelling in the Boreal Ocean; Parrish and Curtis (1982) identified possible summer upwelling over Denmark, and Scotese and Summerhayes (1986) identified possible summer upwelling along northeast Greenland. In-house modelling (Miller, 1989) indicates the possibility of upwelling in the northern KCF sea, of coastal type in summer and of open ocean divergence type in winter, but as noted earlier such modelling is approaching the limits of resolution. No lithological indicators of upwelling have been reported. The mean accumulation rate of 1.5 g C_{org} m^{-2} a^{-1} in the North Sea KCF is not high; typical modern primary production rates of organic carbon are 50 g m^{-2} a^{-1} in the open ocean, 25–150 g m^{-2} a^{-1} along coasts, and 150–500 g m^{-2} a^{-1} along upwelling zones (Duxbury and Duxbury, 1984). Published data for modern sediments being deposited beneath the Namibian upwelling system, for example, result in calculated C_{org} accumulation rates of 1 g m^{-2} a^{-1} (Summerhayes, 1983), 12.5 g m^{-2} a^{-1} (Baturin, 1983), and 44 g m^{-2} a^{-1} (Bremner, 1983). Upwelling in the KCF sea is therefore not ruled out, but neither does it appear to be necessary. Tyson (1987) pointed out that the importance of high primary productivity in the formation of source rocks has probably been overemphasized, and I suggest that the mean productivity of the KCF sea was not unusually high. This does not preclude the seasonal formation of algal blooms (Gallois, 1976; Tyson et al., 1979).

OCEAN CURRENTS

Ager (1975) proposed a detailed Late Jurassic world ocean current system, constructed by analogy with the present day. He suggested that one current flowed southward from the Boreal Ocean down East Greenland, to turn and flow back up Norway, and that a second flowed northward from Tethys up eastern Canada, to turn by the Shetlands and flow back southward down the North Sea into Europe. A third current flowed westward to the south of Newfoundland, from Tethys to the Pacific. In this reconstruction the KCF sea had a clockwise circulation.

Oschmann (1988) suggested that a wind-driven surface current flowed from Tethys to the Boreal Ocean, although this proposed persistent south wind should have generated an eastward-flowing current according to the principles of Ekman transport (e.g., Harvey, 1976). He suggested that a bottom current in the reverse direction upwelled over the English Channel to complete the circulation. In this model the central Boreal Ocean experienced divergence upwelling in winter and downwelling in summer (with coastal upwelling), and the net water flow was from Tethys northward.

Both of the above current systems apparently fail to account for the basically Boreal nature of the KCF fauna (Ager, 1975; Anderton et al., 1979; Cox and Gallois, 1981). Consequently, I suggest that the overall flow was from north to south, forming a "Boreal current," with no countercurrent (Figure 4). The Coriolis force would concentrate this current down the west side of the KCF sea, off Greenland and Canada. This western bias and the constricted "bottleneck" of the seaway from the Boreal Ocean imply extremely sluggish rates of flow over the British Isles and North Sea, probably at least an order of magnitude slower than the main western stream. The East Greenland Volgian marine delta shows consistent southeast current directions (Birkelund, 1975), but otherwise no direct sedimentary current indicators have apparently been reported. Under such a regime, tidal currents probably predominated east of the main Boreal current stream.

A decrease in the authigenic uranium content from north to south has already been noted. If this stems from progressive uranium depletion in a sluggish sea with little fluvial input, it would also imply an overall southward flow. One may even propose mean flow rates from this hypothesis, as follows. Bulk KCF contains approximately 2%K, 10 ppm Th, and 1-2 ppm U in the clastic component (Bjorlykke et al., 1975; Myers and Wignall, 1987; this study). In terms of γ-activity, 1 ppm U is equivalent to approximately 0.95% K or 2.13 ppm Th, and has an activity of about 6.9 APIγ units (Hearst and Nelson, 1985). The typical background radioactivity of the KCF due to the clastic component is therefore about 54 APIγ, and a "hot shale" which registers 200 APIγ contains about 20 ppm of authigenic uranium. This agrees with the maximum uranium contents actually measured (Figure 5). If (1) the Jurassic undepleted sea contained 3 ppb U as it does today (Riley and Skirrow, 1965), (2) the KCF accumulated at 15 m/Ma, and (3) its density is 2.5 g cm^{-3}, then it is easily calculated that the necessary uranium could have been supplied by a 25-cm column of seawater annually. Therefore, in order for uranium depletion effects to be observable, the rate of water replacement over the North Sea area could not have been equivalent to more than a layer a meter or so thick annually. This is admittedly speculative, and even more speculative are calculations based on the rate of phosphorus input required to sustain the organic productivity. Nevertheless, assuming no fluvial phosphorus input and a reasonable degree of phosphorus recycling after death, values of less than 10 meters of water replacement annually appear to be adequate to maintain the rate of organic carbon accumulation. Translated into paleocurrents of plausible dimensions, velocities of the order of 1 km per day are implied over the North Sea region. For comparison, typical open ocean currents today flow at rates of 10–40 km per day (Duxbury and Duxbury, 1984).

DEPOSITIONAL MODEL

The components assembled to form this model are not new, many having been suggested previously for this or other source rocks. There appears to be no plausible alternative to a stratified water system (e.g. Gallois, 1976; Tyson et al., 1979; Myers and Wignall, 1987). The lowest layer is necessarily slow moving or stagnant and oxygen depleted. This prevents aerobic degradation of deposited organic matter by benthos or bacteria. All the key components for a simple two-layer model seem to have been present in the KCF sea.

In a two-layer model the Boreal current flowing southward to Tethys must have formed the surface layer, as it would have been oxygenated. Oschmann (1988), however, proposed a southward-flowing *bottom* current, deoxygenated in summer " . . . during the mid-Arctic downwelling, bottom flow and especially by the planktonic productivity in the North Atlantic surface waters." This seems unlikely, by analogy with modern oceanic bottom water, which is polar in origin and is not deoxygenated (e.g., Harvey, 1976).

The archipelago in the KCF sea defined extensive shallow bays. With sluggish surface currents, low latitudes in a warm world, an arid climate in at least the late Kimmeridgian to late Ryazanian, and evidently little runoff, evaporation in these bays would have led to enhanced salinities, and in some cases ultimately to strong brines and evaporites. While the more humid conditions of the early and middle Kimmeridgian would be less favorable, the apparent lack of major fluvial input still argues for a negative water balance in the sea. Furthermore, weak brine formation need only have proceeded in one or two localities. From here the brine would sink and spread, following the deep grabens. The phrase "warm saline bottom water" or WSBW (Brass et al., 1982) is perhaps preferable to "brine" as only slight increases in the salinity are envisaged. The gradual onset of WSBW formation and the stabilization of different WSBW pools in different sub-basins at different times are perhaps reflected in the diachronous base of the North Sea KCF. The halocline could ultimately reach a height dictated by the deepest channel off the shelf and into the depths of Tethys or the Boreal Ocean.

This model is similar to that proposed on a global scale for the Late Cretaceous by Brass et al. (1982). Extensive deoxygenation is envisaged in the WSBW pools themselves, but the newly created warm brines flowing in from their various source bays could have contained sufficient oxygen to maintain a benthos. WSBW generation may have been continual, seasonal, or restricted to particularly arid years. Without implying precise values for the exact temperatures and salinities involved, hypothetical and plausible values can be used to illustrate the dynamics and sensitivity of the stratification. The Boreal Ocean can be assumed to have had a surface temperature of 6°C (Oschmann, 1988, suggested values in the range 0-10°C), and a salinity of 34‰ in keeping with modern high-latitude seas (Figure 8). It therefore had a density of 1.02678 g cm^{-3} (*see* Riley and Skirrow, 1965, for density tables). The isotopic data of Irwin et al. (1977) for surface-dwelling coccoliths suggest a mean surface temperature of 26.4°C in the Wessex Basin KCF sea, and I assume a temperature of 30°C for the WSBW. At a salinity of 42‰, this water would have had a density of 1.02700 g cm^{-3} (cf. Red Sea surface water with T = 21.7°C and salinity = 40.6‰ (Brewer et al., 1969). The warm brine would have been slightly denser than any surface water in the system, and may therefore have sunk, resulting in stable stratification.

Little is known of the final destabilization except that all of the North Sea area which was still in the KCF facies (sand deposition was then predominant in the southern North Sea area) experienced a single irreversible overturn in the late Ryazanian. Logically the destruction of a single WSBW pool is represented by this event. After overturn, a flowing, oxygenated bottom water current would be required over the North Sea area to maintain oxygenation, presumably the Boreal current. Simple cooling of the model Boreal Ocean surface water by 2°C (to 4°C) would have raised its density to 1.02701 g cm^{-3}, sufficient to displace the slightly lighter WSBW below. Other equally plausible climatic mechanisms for the overturn can be suggested, but their common feature is the relatively small shift in climate or oceanography required to destabilize the stratification, causing the Boreal current to become a bottom current at the end of the Ryazanian.

The Wessex Basin had changed long before the Ryazanian, falling under a Tethyan regime, while the Porcupine Basin may have become a hypersaline lagoon by this time. The precise timing of the end of KCF deposition in eastern Canada is unclear, but it may be analogous to the Wessex Basin in this respect. In east Greenland the end of the KCF appears to be due to the appearance of a local delta system in the middle Volgian which overran the sub-basin.

A consequence of the warm saline bottom water model is that the sea-floor temperature should be higher than the average sea-surface temperature. However, the oxygen isotope temperatures calculated by Irwin et al. (1977) from calcite nodules and by Oschmann (1988) from benthic molluscs in the Wessex Basin KCF are about 15°C, or 11° cooler than the surface. The solution lies in the isotopic fractionation which accompanies evaporation, and which raises the $\delta^{18}O$ of seawater by between 0.2‰ and 0.35‰ for every 1‰ rise in salinity (Epstein and Mayeda, 1953). False apparent temperatures result if this change in the seawater is not taken into account, as Brass et al. (1982) predicted. In this instance, following Irwin et al. (1977) but increasing the $\delta^{18}O$ of the water by 1.6‰–2.8‰, corresponding to an 8‰ increase in salinity, the calculated temperatures rise by 7–13°C, sufficient to make them higher than the surface water temperature.

Other values of temperature and salinity can be chosen which are equally plausible and internally consistent. It is re-emphasized that the model values used here, while apparently reasonable, are not regarded as being anything more than illustrative.

The model can be used to interpret various observations concerning the KCF. The facies boundary between the KCF and the limestones to the south, previously interpreted as a climatic effect (Ager, 1975), more probably marks the contact between limpid flowing Tethyan water and muddier sluggish water

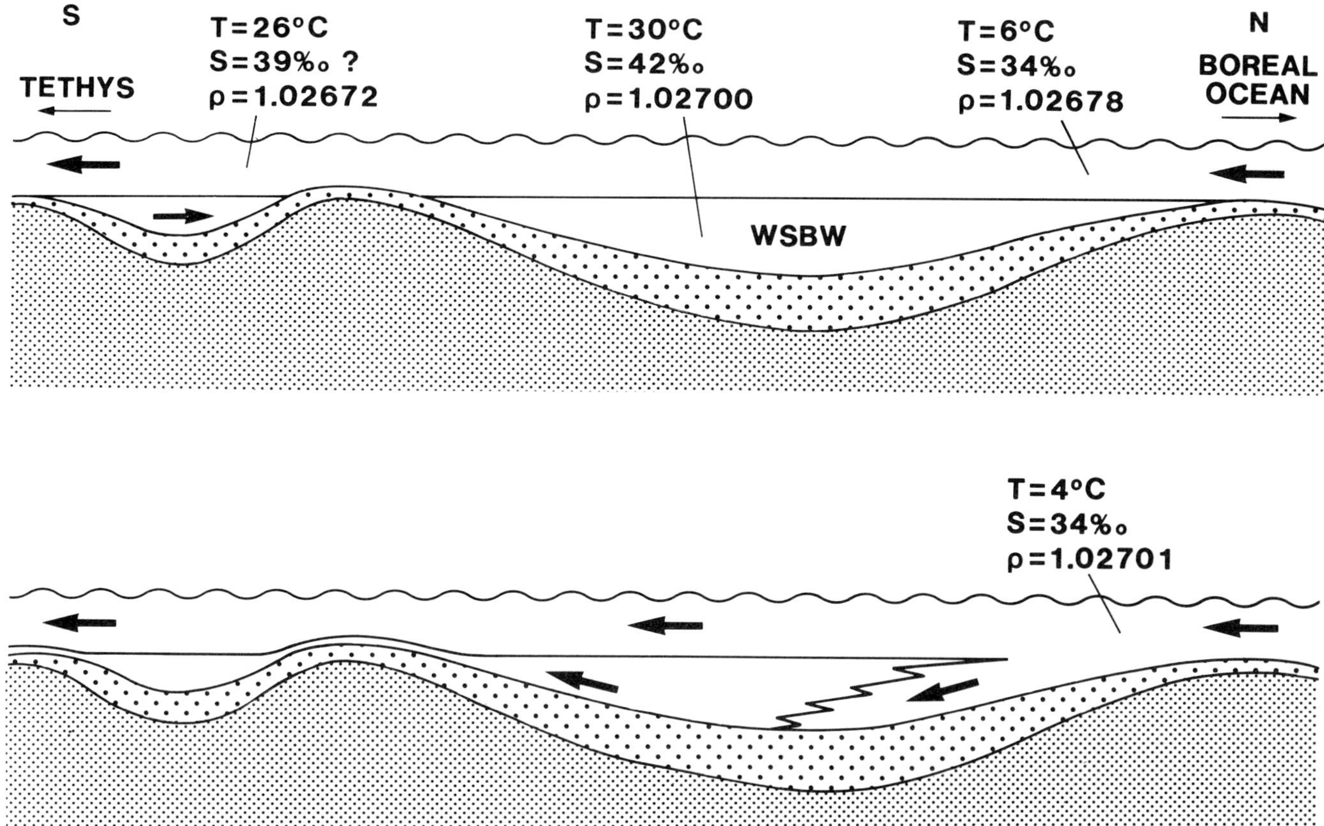

Figure 8. Diagrammatic representation of the proposed salinity-stratification model for the Kimmeridge Clay sea. (Top) During stratification a southerly current from the Boreal Ocean forms the surface layer. It originates as cool water of normal salinity. Solar heating of the water as it drifts southward raises its temperature to 26°C (on isotopic evidence), and its mean salinity could therefore rise to as much as 39‰. Beneath this layer, in a series of interconnecting sea-floor basins, is warm saline bottom water (WSBW), modelled here as being at 30°C and 42‰ salinity. (Bottom) Stratification breaks down, induced by a 2°C drop in the initial surface temperature. The surface current now extends down to sea floor, flushing out the weak bottom brines, and finally becoming established as an oxygenated bottom current.

in the KCF sea. Mud entering Tethys was presumably diluted to insignificance. The shifting position of the facies boundary would reflect the hydrodynamic balance between currents in the two water masses. Features suggesting that the halocline in the Wessex Basin was periodically disrupted by storms suggest a halocline at perhaps 50 m depth (Gallois, 1976; Myers and Wignall, 1987; Oschmann, 1988). Intervals of pale, organically lean siltstones and mudstones within the KCF, for example in the Moray Firth Basin, suggest a temporary emergence of the sea floor through the halocline. Isolated WSBW pools separated by topographic highs could have had different trace-element chemistries, leading to the geographically restricted deposition of the various "hot-shale" units. The model requires no changes in the clastic input at the time of the Ryazanian overturn. The local basal sandstone in the Cromer Knoll Group is compatible with the onset of a sea-floor current and a period of winnowing.

The simplest alternative to stratification resulting from evaporation is stratification by influxes of fresh water, the so-called "fresh-water cap" model. This envisages a bottom layer of deoxygenated, normal-salinity seawater, a top layer of less dense, less saline water, with little mixing or oxygen transport across the interface. The model requires a humid climate or major river inputs, and certainly the lower KCF was deposited in an area having a humid climate. The perceived problems with this model can be summarized as follows.

1. Fresh-water capping was not possible during deposition of the upper KCF, in an arid climate where evaporites were forming. No clear-cut change in the organic geochemistry is evident which would suggest any change in the water-stratification mechanism from an earlier period of fresh-water capping. Nevertheless, there is a slight gradual upward increase in HI, which is compatible with a steady reduction in terrigenous kerogen inputs.
2. Notwithstanding the above, there is a marked lack of the terrigenous kerogen expected from extensive runoff. The underlying Heather Formation mud-

stones (Bathonian to early Oxfordian) are geochemically more likely to be terrigenous (Field, 1985; Thomas et al., 1985).

3. The faunal evidence indicates a continual southerly flow of oxygenated seawater, capable for example of carrying and sustaining ammonites. This would have to form a current beneath the hypothetical fresh-water cap, and become the lower layer in a simple two-layer model. It is difficult to equate this with a stagnant and deoxygenated bottom layer.

Consequences of the Model

This model has several significant implications for petroleum exploration.

1. In general the entire area of the KCF sea, as far south as the facies boundary with carbonates, is prospective for petroleum as regards the availability of source rocks. The KCF is unlikely to recur south of this facies boundary.
2. The least prospective portions of the KCF sea are around the basin margins and on highs which projected above the halocline, and perhaps the offshore Canada and Greenland sections where the surface current velocities were highest (and thus conducive to water mixing and prevention of brine generation). This does not preclude the existence of secluded, and perhaps relatively short-lived, sub-basins off Canada and Greenland where the KCF could accumulate.
3. There are no known modern analogs to the paleogeographic, paleoclimatic, and paleoceanographic setting of the KCF, so that comparisons with modern settings are probably overused. However, there may well be ancient analogs.

DISCUSSION AND CONCLUSIONS

The model presented here is intended to explain the regional characteristics of the KCF without conflicting with other models which explain more local details. For example the coccolithic horizons which extend from Wessex to Yorkshire may well be algal blooms (Gallois and Medd, 1979), although more likely as a consequence of storm disturbance of the halocline than of inputs of terrigenous nutrients. Stratification and subsequent fluctuation of an O_2/H_2S interface in the water column have been invoked by Tyson et al. (1979) and Oschmann (1988) to explain other features of the Wessex Basin. Wignall and Myers (1988) have presented evidence for fluctuations in the Wessex Basin benthic oxygen levels which suggest the periodic disruption of a stratified system. The cyclicity observed in Wessex by Oschmann (1988) and Myers and Wignall (1987) at scales of 1 to 14,000 years is a feature superimposed by seasonal and long-term climatic fluctuation on the overall model.

Perhaps the only models which are quite specifically excluded are those involving surface brackish water stratification and implying a positive water balance (e. g., Gallois, 1976; Barnard and Cooper, 1981). Models relying on thermal stratification have yet to be presented in detail, and are seemingly incompatible with a southward surface flow of oxygenated cool water from the Boreal Ocean (e. g., Tyson et al., 1979; Hallam, 1987). The detailed model suggested by Oschmann (1988) requires improbable wind and current directions and does not apparently result in an oxygen-depleted bottom water layer.

In conclusion, the type section of the Kimmeridge Clay Formation in the Wessex Basin is not representative of this formation elsewhere, especially much of the North Sea where deposition of this facies continued for much longer. Consequently, it cannot form the sole basis of any viable depositional model. It does, however, contain features and provide constraints which must be accommodated by acceptable models. There is no evidence that the organic productivity of the sea was unusually high, and one must turn to settings which allow exceptionally efficient preservation of the organic matter produced. Stratification of the water system, with an oxygen-depleted stagnant bottom layer, is appropriate, particularly in the light of low average sedimentation rates in the order of 15 m/Ma. Most climatic evidence, the presence of evaporites, and the lack of sedimentary evidence for fluvial input suggest that weak brines could have formed by evaporation in shallow restricted bays. These could have sunk to form a warm saline bottom water, the lower layer of a simple two-layer stratified sea. The surface layer was probably an extremely sluggish current which flowed from the polar Boreal Ocean in the north to Tethys in the south. A surface flow rate on the order of 1 km per day seems appropriate. The evaporation which produced the WSBW also increased the $\delta^{18}O$ of this water. Consequently, the shells of benthic fauna in the shallow pans where the WSBW formed may well be unusually isotopically heavy, resulting in apparent isotopic temperatures which are too low. Deposition of the KCF in the North Sea ceased following a possibly catastrophic overturn of the basin. This could have been caused by climatic changes in the temperature and/or salinity of either the WSBW or, more probably, of the surface current. The climatic change required would be slight. After KCF deposition, the stagnant warm bottom brine was replaced by a cold southward-flowing bottom current. This introduced sufficient oxygen to prevent the preservation of any further organic material.

ACKNOWLEDGMENTS

This paper has matured during 2 years of sometimes heated discussions with Chris Clayton, Andy Fleet, and Keith Myers. Useful and sometimes critical information has also come from Andy Aplin, Alastair Mann, Bill Paley, Mike Partington, Andy Pepper, and Clare Stronach, and the BP Research Centre paleoreconstructors David Smith and Helen Turnell. I also thank BP Research International for the opportunity to write and present this paper.

REFERENCES

Ager, D. V., 1975, The Jurassic world ocean *in* K. C. Finstad and R . C. Selley, eds., Jurassic northern North Sea symposium: Norwegian Petroleum Society, p. 1-1-1-43.

Anderton, R., P. H. Bridges, M. R. Leeder, and B. W. Sellwood, 1979, A dynamic stratigraphy of the British Isles: London, George, Allen and Unwin, 301 p.

Arkell, W. J., 1956, Jurassic geology of the world: London, Oliver and Boyd Ltd., 681 p.

Baird, R. A., 1986, Maturation and source rock evaluation of Kimmeridge Clay, Norwegian North Sea: AAPG Bulletin, v. 70, p. 1-11.

Barnard, P. C., and B. S. Cooper, 1981, Oils and source rocks of the North Sea area, *in* L. V. Illing and G. D. Hobson, eds., Petroleum geology of the continental shelf of northwest Europe: London, Institute of Petroleum, p. 169-175.

Baturin, G. N., 1983, Some unique sedimentological and geochemical features of deposits in coastal upwelling regions, *in* J. Thiede and E. Suess, eds., Coastal upwelling, its sediment record, part B: New York, Plenum Press, p. 11-27.

Blatt, H., G. Middleton, and R. Murray, 1980, Origin of sedimentary rocks: New Jersey, Prentice-Hall, 782 p.

Brass, G. W., J. R. Southam, and W. H. Peterson, 1982, Warm saline bottom water in the ancient ocean: Nature, v. 296, p. 620-623.

Bremner, J. M., 1983, Biogenic sediments on the south west African (Namibian) continental margin, *in* J. Thiede and E. Suess, eds., Coastal upwelling, its sediment record, part B: New York, Plenum Press, p. 73-103.

Brewer, P. G., C. D. Densmore, R. Munns, and R. J. Stanley, 1969, Hydrography of the Red Sea brines, *in* E. T. Degens and D. A. Ross, Hot brines and recent heavy metal deposits in the Red Sea: Berlin, Springer-Verlag, p. 138-147.

Brown, S., 1984, Jurassic, *in* K. W. Glennie, ed., Introduction to the Petroleum Geology of the North Sea: Oxford, England, Blackwell Scientific Publications, p. 103-131.

Cornford, C., J. A. Morrow, A. Turrington, J. A. Miles, and J. Brooks, 1983, Some geological controls on oil composition in the UK North Sea, *in* J. Brooks, ed., Petroleum geochemistry and exploration of Europe: Geological Society of London Special Publication 12, p. 175-194.

Cox, B. M., and R. W. Gallois, 1981, The stratigraphy of the Kimmeridge Clay of the Dorset type area and its correlation with some other Kimmeridgian sequences: Institute of Geological Sciences report 80/4, London, Her Majesty's Stationery Office, 44 p.

Creaney, S., and B. H. Allison, 1987, An organic geochemical model of oil generation in the Avalon/Flemish Pass sub-basins, east coast Canada: Bulletin of Canadian Petroleum Geology, v. 35, p. 12-23.

Deegan, C. E., and B. J. Scull, 1977, A standard lithostratigraphic nomenclature for the central and northern North Sea: Institute of Geological Sciences report 77/25, London, Her Majesty's Stationery Office, 36 p. + enclosures.

Demaison, G., A. J. J. Holck, R. W. Jones, and G. T. Moore, 1983, Predictive source bed stratigraphy: A guide to regional petroleum occurrence, North Sea basin and eastern North American continental margin, *in* Eleventh World Petroleum Congress, section PD1(2): New York, John Wiley, p. 1-13.

Dore, A. G., J. Vollset, and G. P. Hamar, 1985, Correlation of the offshore sequences referred to the Kimmeridge Clay Formation—relevance to the Norwegian sector, *in* B. M. Thomas, ed., Petroleum geochemistry in exploration of the Norwegian shelf: London, Graham and Trotman Ltd. (Norwegian Petroleum Society), p. 27-37.

Duxbury, A. C. and A. Duxbury, 1984, An introduction to the world's oceans: Reading, Massachusetts, Addison Wesley, 549 p.

Epstein, S., and T. Mayeda, 1953, Variation of O^{18} content of waters from natural sources: Geochimica et Cosmochimica Acta, v. 4, p. 213-224.

Farrimond, P., P. Comet, G. Eglinton, R. P. Evershed, M. A. Hall, D. W. Park, and A. M. K. Wardroper, 1984, Organic geochemical study of the upper Kimmeridge Clay of the Dorset type area: Marine and Petroleum Geology, v. 1, p. 340-354.

Field, J. D., 1985, Organic geochemistry in exploration of the northern North Sea, *in* B. M. Thomas, ed., Petroleum geochemistry in exploration of the Norwegian shelf: London, Graham and Trotman Ltd. (Norwegian Petroleum Society), p. 39-57.

Fisher, M. J., and J. A. Miles, 1983, Kerogen types, organic maturation and hydrocarbon occurrences in the Moray Firth and South Viking Graben, North Sea basin *in* J. Brooks, ed., Petroleum geochemistry and exploration of Europe: Geological Society of London Special Publication 12, p. 195-201.

Gallois, R. W., 1976, Coccolith blooms in the Kimmeridge Clay and origin of North Sea oil: Nature, v. 259, p. 473-475.

Gallois, R. W., and A. W. Medd, 1979, Coccolith-rich marker bands in the English Kimmeridge Clay: Geological Magazine, v. 116, p. 247-260.

Glennie, K. W., 1984, The structural framework and the pre-Permian history of the North Sea area, *in* K. W. Glennie, ed., Introduction to the petroleum geology of the North Sea: Oxford, England, Blackwell Scientific Publications, p. 17-39.

Goff, J. C., 1983, Hydrocarbon generation and migration from Jurassic source rocks in the E. Shetland basin and Viking Graben of the northern North Sea: Journal of the Geological Society of London, v. 140, p. 445-474.

Hallam, A., 1987, Mesozoic marine organic-rich shales, *in* J. Brooks and A. J. Fleet, eds., Marine petroleum source rocks: Geological Society of London Special Publication 26, p. 251-261.

Hallam, A., and B. W. Sellwood, 1976, Middle Mesozoic sedimentation in relation to tectonics in the British area: Journal of Geology, v. 84, p. 301-321.

Haq, B. U., J. Hardenbol, and P. R. Vail, 1987, Chronology of fluctuating sea levels since the Triassic: Science, v. 235, p. 1156-1167.

Harland, W. B., A. V. Cox, P. G. Llewellyn, C. A. G. Pickton, A. G. Smith, and R. Walters, 1982, A geologic time scale: Cambridge, England, Cambridge University Press, 131 p.

Harvey, J. G., 1976, Atmosphere and ocean: Sussex, England, Artemis Press Ltd., 143 p.

Hearst, J. R., and P. H. Nelson, 1985, Well logging for physical properties: New York, McGraw-Hill, 571 p.

Ioannides, N. S., G. N. Stavrinos, and C. Downie, 1976, Kimmeridgian microplankton from Clavell's Hard, Dorset, England: Micropaleontology, v. 22, p. 443-478.

Irwin, H., C. Curtis, and M. J. Coleman, 1977, Isotopic evidence for source of diagenetic carbonates formed during burial of organic-rich sediments: Nature, v. 269, p. 209-213.

Johnson, R. J., 1975, The base of the Cretaceous—a discussion, *in* A. W. Woodland, ed., Petroleum and the continental shelf of northwest Europe: Barking, England, Applied Science Publishers, p. 389-399.

Leventhal, J. S., 1981, Pyrolysis gas chromatography-mass spectrometry to characterize organic matter and its relationship to uranium content of Appalachian Devonian black shales: Geochimica et Cosmochimica Acta, v. 45, p. 883-889.

Miller, R. G., 1989, Prediction of ancient coastal upwelling and related source rocks from palaeo-atmospheric pressure maps: Marine and Petroleum Geology, v. 6, p. 277-283.

Myers, K. J., and P. B. Wignall, 1987, Understanding Jurassic organic-rich mudrocks—new concepts using gamma-ray spectrometry and palaeo-ecology: Examples from the Kimmeridge Clay of Dorset and the Jet Rock of Yorkshire, *in* J. K. Leggett and G. G. Zuffa, eds., Marine clastic sedimentology concepts and case studies: London, Graham and Trotman, p. 172-189.

Oschmann, W., 1988, Kimmeridge Clay sedimentation—a new cyclic model: Palaeogeography, Palaeoclimatology, Palaeoecology, v. 65, p. 217-251.

Parrish, J. T., and R. L. Curtis, 1982, Atmospheric circulation, upwelling, and organic-rich rocks in the Mesozoic and

Cenozoic eras: Palaeogeography, Palaeoclimatology, Palaeoecology, v. 40, p. 31-66
Parrish, J. T., A. M. Ziegler, and C. R. Scotese, 1982, Rainfall patterns and the distribution of coals and evaporites in the Mesozoic and Cenozoic: Palaeogeography, Palaeoclimatology, Palaeoecology, v. 40, p. 67-101.
Partington, M. A., 1983, The stratigraphy and distribution of circumpollen in southern England: PhD thesis, University of Aberdeen, Aberdeen, Scotland.
Pearson, M. J., and D. Watkins, 1983, Organofacies and early maturation effects in upper Jurassic sediments from the Inner Moray Firth Basin, North Sea, *in* J. Brooks, ed., Petroleum geochemistry and exploration of Europe: Geological Society of London Special Publication 12, p. 147-160.
Powell, T. G., 1985, Palaeogeographic implications for the distribution of Upper Jurassic source beds: Offshore eastern Canada: Bulletin of Canadian Petroleum Geology, v. 33, p. 116-119.
Rawson, P. F., and L. A. Riley, 1982, Latest Jurassic-Early Cretaceous events and the "Late Cimmerian Unconformity" in North Sea area: AAPG Bulletin, v. 66, p. 2628-2648.
Riley, J. P., and G. Skirrow, eds., 1965, Chemical oceanography, volume 1: London, Academic Press, 712 p.
Ronnevik, H. C., 1984, Geology of the Barents Sea *in* L. V. Illing and G. D. Hobson, eds., Petroleum geology of the continental shelf of Northwest Europe: Institute of Petroleum, London, p. 395-406.
Scotese, C. R., and C. P. Summerhayes, 1986, Computer model of palaeoclimate predicts coastal upwelling in the Mesozoic and Cenozoic: Geobyte, Summer 1986, p. 28-42.
Summerhayes, C. P., 1983, Sedimentation of organic matter in upwelling regimes, *in* J. Thiede and E. Suess, eds., Coastal upwelling, its sediment record, part B: New York, Plenum Press, p. 29-72.
Thomas, B. M., P. Møller-Pedersen, M. F. Whitaker, and N. D. Shaw, 1985, Organic facies and hydrocarbon distributions in the Norwegian North Sea, *in* B. M. Thomas, ed., Petroleum geochemistry in exploration of the Norwegian shelf: London, Graham and Trotman Ltd. (Norwegian Petroleum Society), p. 3-26.
Tissot, B. P., and D. H. Welte, 1978, Petroleum formation and occurrence: Berlin, Springer-Verlag, 538 p.
Tyson, R. V., 1987, The genesis and palynofacies characteristics of marine petroleum source rocks, *in* J. Brooks and A. J. Fleet, eds., Marine petroleum source rocks: Geological Society of London Special Publication 26, p. 47-67.
Tyson, R. V., R. C. L. Wilson, and C. Downie, 1979, A stratified water column environmental model for the type Kimmeridge Clay: Nature, v. 277, p. 377-380.
Wignall, P. B., and K. J. Myers, 1988, Interpreting benthic oxygen levels in mudrocks: A new approach: Geology, v. 16, p. 452-455.
Williams, P. F. V., 1986, Petroleum geochemistry of the Kimmeridge Clay of onshore southern and eastern England: Marine and Petroleum Geology, v. 3, p. 258-281.
Woodhall, D., and R. W. O'B. Knox, 1979, Mesozoic volcanism in the northern North Sea and adjacent areas: Bulletin of the Geological Survey of Great Britain, v. 70, p. 34-56.
Ziegler, P. A., 1981, Evolution of sedimentary basins in Northwest Europe, *in* L. V. Illing and G. D. Hobson, eds., Petroleum geology of the continental shelf of Northwest Europe, Institute of Petroleum, Heyden, London, p. 3-39.

Dysoxic Sedimentation in the Cenomanian-Turonian Daliyya Formation, Israel

S. Lipson-Benitah
The Israel Institute of Petroleum and Energy
Tel Aviv, Israel

A. Flexer
Tel Aviv University
Tel Aviv, Israel

A. Rosenfeld
Geological Survey of Israel
Jerusalem, Israel

A. Honigstein
Tel Aviv University
Tel Aviv, Israel

B. Conway
Geological Survey of Israel
Jerusalem, Israel

H. Eris
Tel Aviv University
Tel Aviv, Israel

The Daliyya Formation accumulated in four basins near the Cenomanian-Turonian shelf edge of Israel. The formation includes bituminous marls, interbedded black laminated shales enriched in marine organic matter (up to 2.4% TOC), and pyrite, implying oxygen-poor paleoenvironments. The age of the formation, based on foraminifers and ostracodes, is late Early Cenomanian-Late Turonian. The maximum range is found in the southern Gaza basin. Palynological analysis reveals blooms of dinoflagellate cysts. The assemblages suggest subtropical-tropical marine waters. The overall period of dysoxic sedimentation may be divided into three phases (Cenomanian, Cenomanian/Turonian transition, and Turonian) which correspond to, but extend over a longer period than, the global middle Cretaceous Oceanic Anoxic Event. The northward migration of the dysoxic facies with time is related to gradual advance of the transgression. Mechanisms involved in the formation and preservation of the organic-rich rocks are discussed for each phase. We suggest that, during the Cenomanian-Turonian transgression, a combination of global oceanic changes—increased upwelling, expansion and intensification of the oxygen-minimum layer, and local oceanographic, climatic, and tectonic conditions—was responsible for the occurrence and distribution of dysoxic sediments. Oxygen depletion affected the microfossil assemblages. The organic-rich facies might be a potential source rock for petroleum.

INTRODUCTION

One of the principal topics of renewed interest in the Cretaceous is the interaction between geodynamic events, sea level changes and the deposition of black shales (Reyment and Bengtson, 1986). The Cretaceous Period bears witness to one of the largest transgressive phases in earth history (Gignoux, 1955; Kummel, 1970), worldwide orogenic events (Schwan, 1980), seafloor spreading, and accretion of oceanic ridges (Vail et al., 1977).

During sea level highstands, organic-rich black shales were deposited under anoxic or nearly anoxic (dysoxic) conditions in deep basinal areas and marginal sea basins. The three worldwide oceanic anoxic events (OAE) recognized are OAE-1 during the Late Barremian-Albian, OAE-2 during the Late Cenomanian-Early Turonian, and a third, somewhat restricted, event during the Coniacian-Santonian (Schlanger et al., 1987). In the Cretaceous succession of Israel, four events of organic-rich sedimentation under dysoxic conditions have been recognized (Flexer et al., 1986) (Figure 1).

The present study deals with the organic-rich marls of the Daliyya Formation (Kashai, 1966) deposited during a time span including OAE-2. The age of the Daliyya Formation has been redefined here as late Early Cenomanian to Late Turonian. Geological information about this formation has been compiled from 100 oil and water wells in the coastal plain and from outcrops in the Mount Carmel area (Figure 2). The onshore part of the Daliyya Formation seems to be restricted to the Gaza, Ashqelon, Carmel-Caesarea, and Western Galilee basins. The maximum thickness of the formation measured to date is 202 m (Figure 3). A dark organic facies, with a maximum thickness of about 30 m, occurs in all of the basins.

Detailed biostratigraphic determinations, based on planktonic foraminifera and ostracodes, palynological examinations, and total organic carbon content, are provided for sections from each basin.

Mechanisms involved in the formation and preservation of organic-rich facies are reviewed, and models that integrate global and local factors are presented.

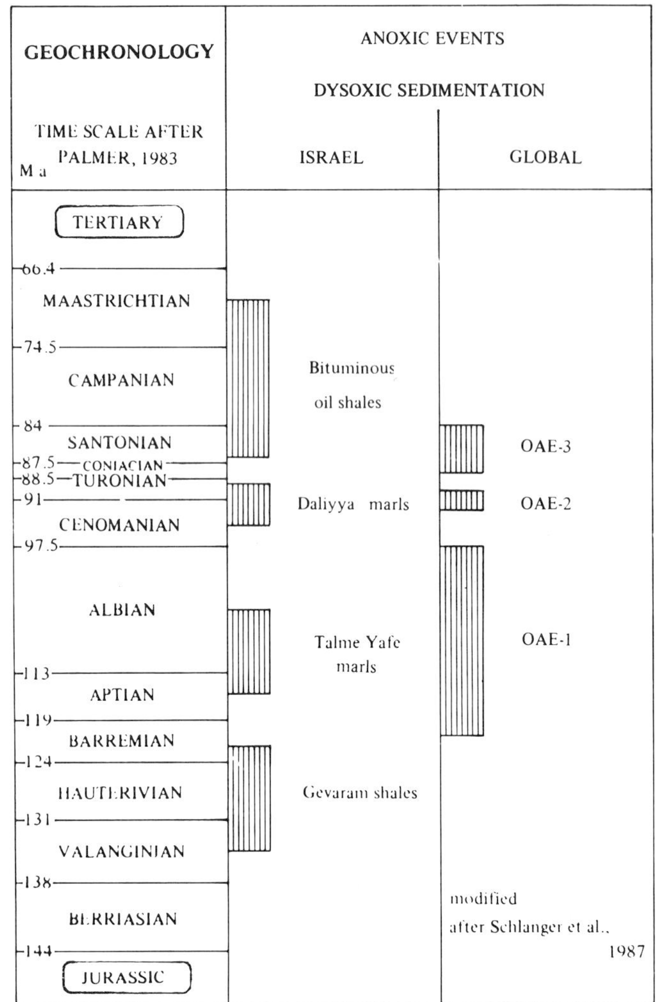

Figure 1. Comparison of Cretaceous dysoxic sedimentation in Israel with global oceanic anoxic events (modified after Flexer et al., 1986).

LITHOSTRATIGRAPHIC FRAMEWORK OF THE DALIYYA FORMATION

During the Cenomanian-Turonian stage, Israel and its offshore area can be divided into two main provinces of deposition. Shallow shelf carbonates were deposited on a shelf platform that extended from the Arabo-Nubian Massif in the southeast to the Tethys Sea in the northwest. The edge of the platform was located close to the present coast of Israel. Chalks and marls accumulated in shelf basins and on the continental slope (Sass and Bein, 1982).

The Daliyya Formation, which accumulated in a shelf basin, was initially defined at the Dalyat el Karmil outcrop in the Carmel (Kashai, 1966). There it occurs above the limestones of the Muhraqa Formation. The formation was also recognized in the subsurface and described by Gvirtzman and Reiss (1965). In the studied wells, this subsurface lithostratigraphic unit consists of soft limestones, commonly chalky, and argillaceous limestones and marls overlying and underlying limestone and dolomite units (Figures 4, 5). Rhythmic intercalations of dark shales and bituminous marls are commonly present in the lower part of the formation as well as scattered flint. Its upper part is generally richer in carbonates.

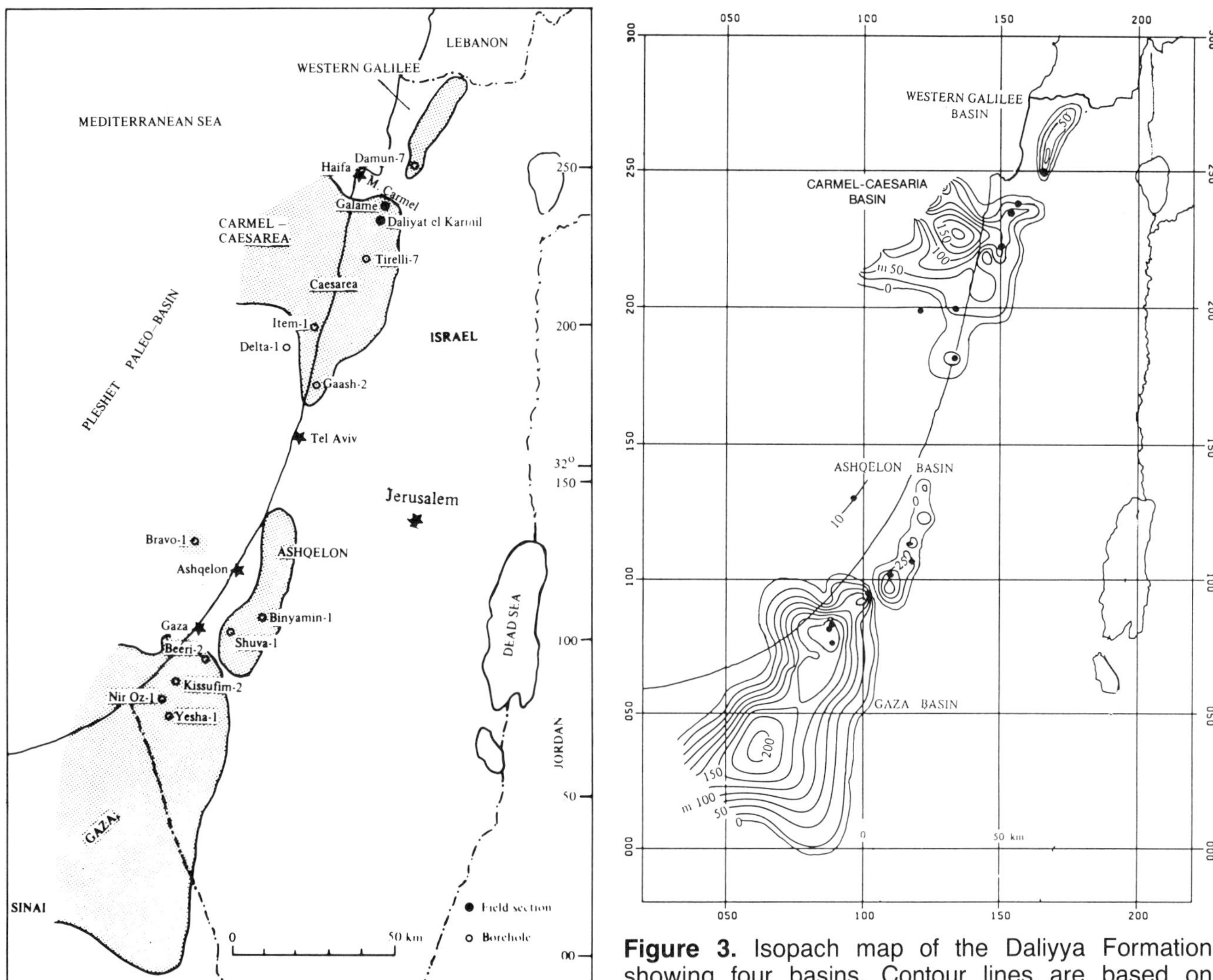

Figure 2. Location map.

Figure 3. Isopach map of the Daliyya Formation showing four basins. Contour lines are based on information in the computerized data bank (over 100 control points) of Oil Exploration (Investments) Ltd. Sections examined in this study are indicated with dots.

BIOSTRATIGRAPHIC FRAMEWORK OF THE DALIYYA FORMATION

The stratigraphic framework adopted here is based mainly on planktonic foraminifera and ostracodes (Figures 4, 5). Two foraminiferal zones cover most of the Cenomanian: the *Rotalipora globotruncanoides/ Thalmanninella brotzeni* zone for the lower part and the *Rotalipora cushmani* zone for the upper part (Lipson-Benitah, 1980). In the absence of planktonic markers, the benthonic foraminifera *Gavelinella aumalensis* is used to cover the late Early to early Late Cenomanian (Hamaoui, 1979). The Cenomanian-Turonian boundary falls within the *Whiteinella archaeocretacea* zone. The lower *Helvetoglobotruncana helvetica* zone (below the entry of *Marginotruncana sigali*) covers most of the Early Turonian and its upper zone covers the middle Turonian (Robaszynski and Caron, 1979; Robaszynski et al., 1982). The lower *Marginotruncana angusticarinata* zone (Lipson-Benitah, 1987) is assigned to the Late Turonian.

Three assemblage zones based on ostracodes are discernible: the Cenomanian UC-2 zone named *Amphicytherura distincta*, the Early-middle Turonian UC-6 zone named *Cythereis rawashensis kenaanensis* (Rosenfeld and Raab, 1974; Lipson-Benitah et al., 1988), and the Late Turonian UC-7 zone (lower part) named *Oertliella dextrospinata*. The UC-7 zone was hitherto only recognized in sediments of the Zihor Formation in southern Israel (Honigstein and Rosenfeld, 1985).

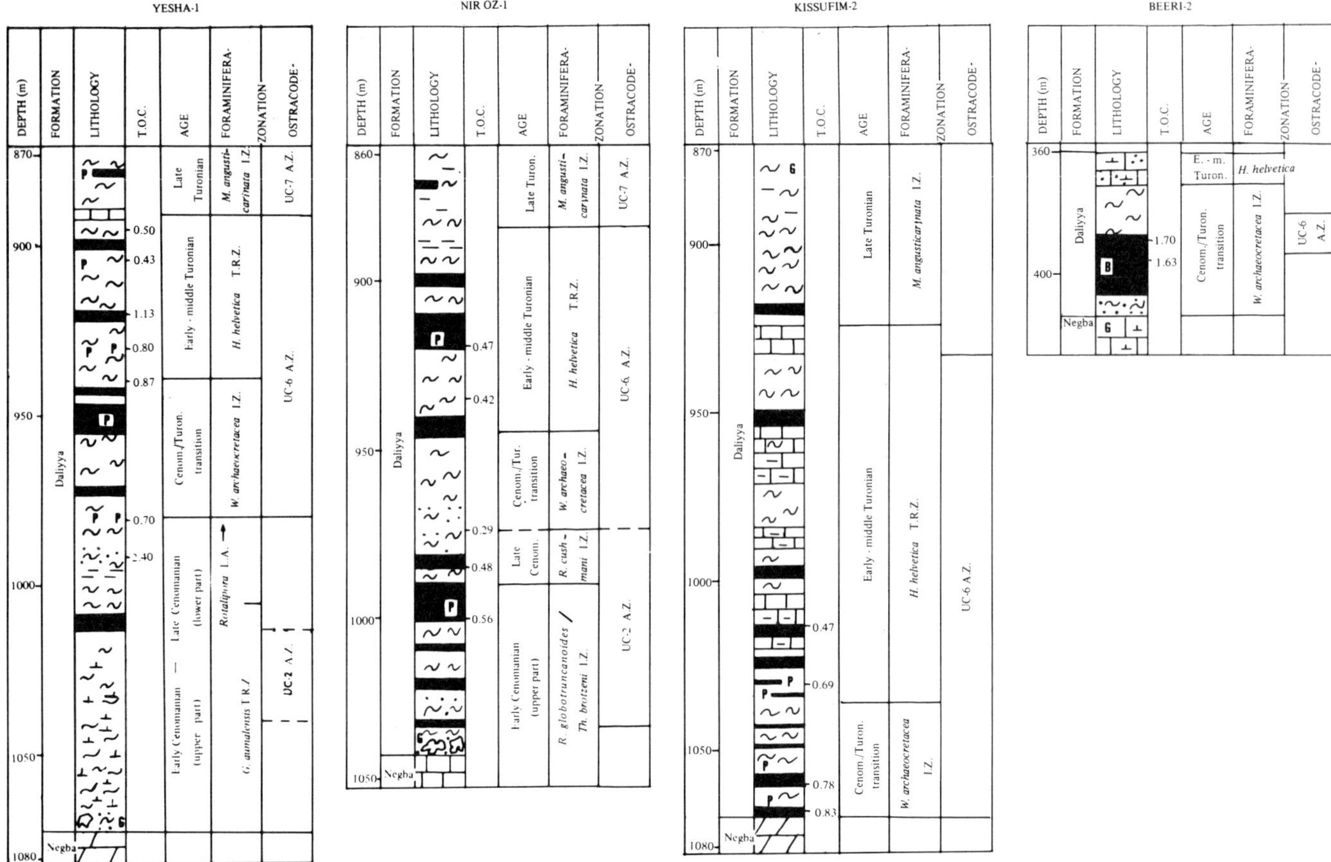

Figure 4. Lithology and biostratigraphy of selected drill holes from the Gaza basin (for legend see Figure 5).

The relative abundance of important palynomorphs from selected wells is shown in Figure 6. Palynomorphs show color indices of TAI 1 (immature) implying that the sediments did not reach temperatures above 60°C. Most of the kerogen is of marine origin. Photomicrographs of some representative specimens are shown in Figure 7.

EXTENT OF THE DALIYYA FORMATION

The Gaza Basin

The Gaza basin is 100 by 50 km and has been penetrated by 28 wells. The maximum thickness of the Daliyya Formation drilled to date is 202 m. Four representative wells from this basin were studied in detail (Figures 2,4).

Yesha-1 Drill Hole

The lower 67 m of the Daliyya Formation consist of soft marls and chalks and the upper 135 m consist of alternating gray pyritic shales and marls which are sandy at the bottom. Three foraminiferal zones from the late Early Cenomanian to the Late Turonian are identified. The lower segment, which contains few indicative planktonic forms (few *Rotalipora* specimens were recovered), is attributed to the *Gavelinella aumalensis* zone. The upper section belongs mostly to the *W. archaeocretacea* and *H. helvetica* zones. Both are characterized by many small *Hedbergella* and *Heterohelix* associated with *Pithonella. Pithonella* refers to planktonic organisms of uncertain origin (detached part of algae?), possessing a resistant calcitic test (Bein and Reiss, 1976). The uppermost 20 m are assigned to the *M. angusticarinata* zone. Three ostracode zones are discernible: the UC-2 zone within the lower section, and the UC-6 and UC-7 zones in the upper section. Pyritized ostracodes account for 30 to 50% of the population. Palynological results here, as in all the other wells analyzed, show that the kerogen is predominantly of marine origin. Dinoflagellate cysts, recovered from the 899–974-m interval, are both abundant and diversified. The three types (chorate, cavate, and proximate) are present, although not equally represented: chorate forms dominate, cavate are well represented, and proximate forms are rare. TOC (Total Organic Carbon) values range from 0.43 to 2.4%.

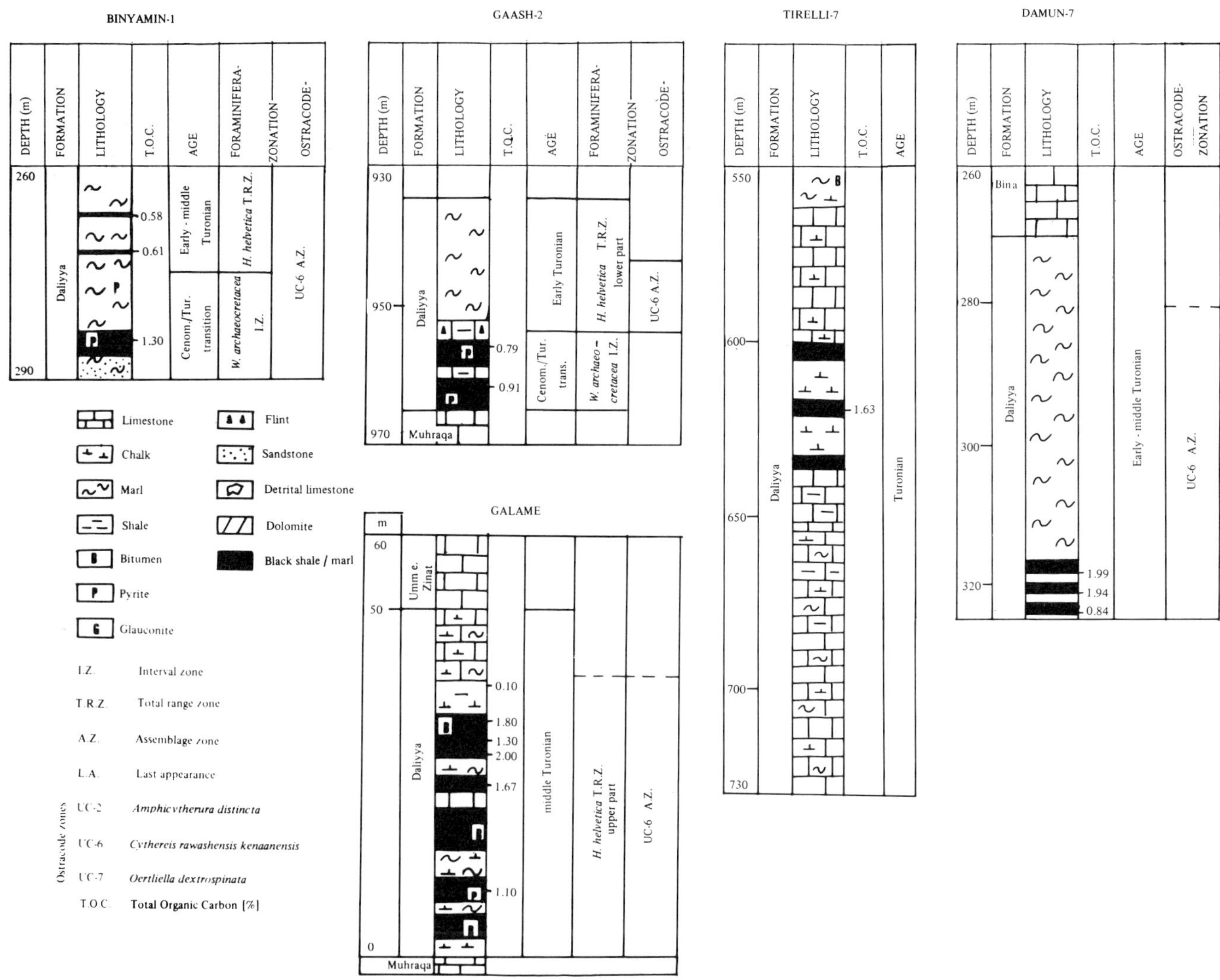

Figure 5. Lithology and biostratigraphy of selected drill holes and an outcrop from the Ashqelon (Binyamin-1), Carmel-Caesarea (Gaash-2, Tirelli-7, Galame), and Western Galilee (Damun-7) basins.

Nir Oz-1 Drill Hole

The 180-m-thick Daliyya Formation is made up of soft marls, silty and sandy in places, alternating with predominantly thin streaks of dark pyritic shales. The late Early to Late Cenomanian is represented by the *R. globotruncanoides/T. brotzeni* zone and part of the *R. cushmani* zone. The incomplete Late Cenomanian section is consistent with earlier observations (Flexer et al., 1986). The assemblages are rich in small *Hedbergella* and *Heterohelix*, and the *Rotalipora* specimens recovered tend to be small. Three zones, from the Cenomanian/Turonian transition to Late Turonian, are also present: the *W. archaeocretacea*, *H. helvetica*, and *M. angusticarinata* zones. Some Radiolaria are present. The *Hedbergella-Heterohelix* facies which also contains *Clavihedbergella* is present in the lower *W. archaeocretacea* zone. Benthos, mainly represented by lenticulinid, gavelinid, and agglutinated forms, is sparse. Ostracode assemblages belong to the UC-2, UC-6, and UC-7 zones. Partly pyritized specimens are present mostly in the lower section. TOC values range from 0.29 to 0.56%. These low values probably represent an average of the mixed dark and light shales in the cuttings rather than the proper value of each lithology.

Kissufim-2 Drill Hole

The Daliyya Formation consists of 35 m of alternating soft, gray pyritic shales and light-cream, argillaceous marls, 113 m of predominantly shaly limestones and marls, and 47 m of marls. The *W. archaeocretacea*, *H. helvetica*, and *M. angusticarinata* zones from the Cenomanian/Turonian transition to the Late Turonian have been identified. The small *Hedbergella-Heterohelix* facies is also restricted here to the *W. archaeocretacea* zone. Ostracodes recovered

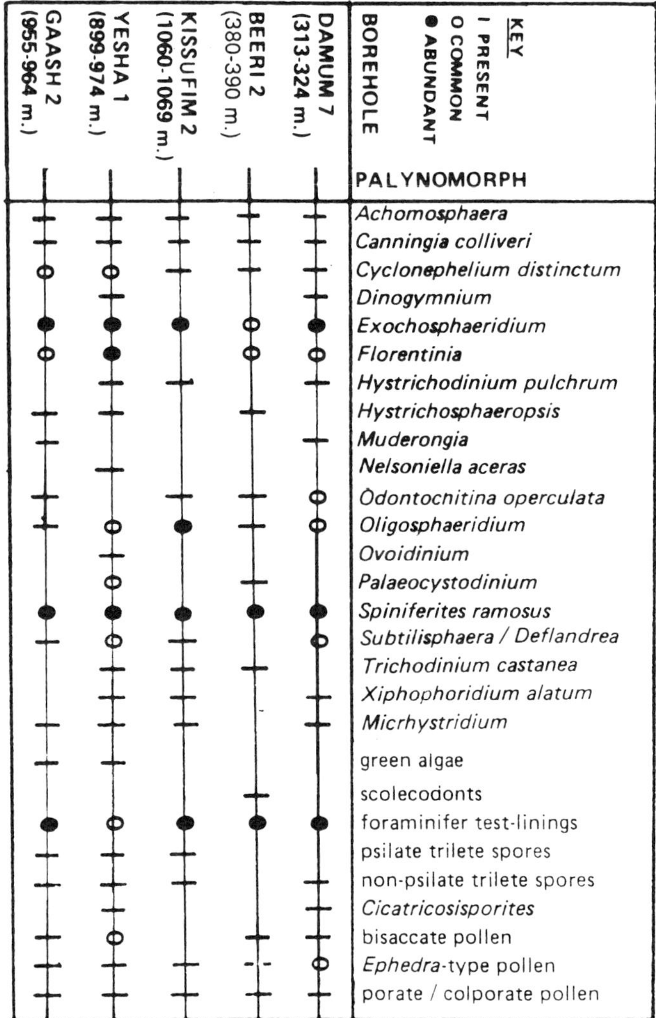

Figure 6. Relative abundance of palynomorphs recovered from the Daliyya Formation in selected drill holes.

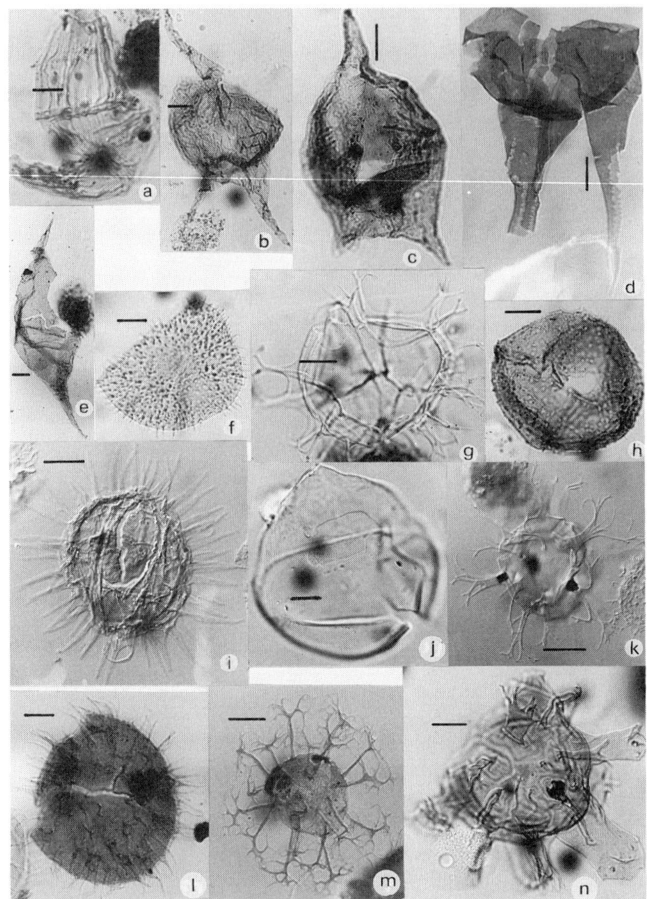

Figure 7. Representative specimens of Late Cretaceous dinoflagellates from the Daliyya Formation in the Yesha-I drill hole. Scale bar = 10 microns. (a) *Dinogymnium* sp., 904 m. (b) *Deflandrea diebeli* Alberti, 904 m. (c) *Deflandrea speciosa* Alberti, 913 m. (d) *Odontochitina operculata* (O. Wetzel), 947 m. (e) *Palaeocystodinium* sp., 904 m. (f) *Cyclonephelium distinctum* Deflandre and Cookson, 947 m. (g) *Spiniferites ramosus* (Ehrenberg), 904 m. (h) *Trichodinium castanea* (Deflandre), 947 m. (i) *Hystrichodinium pulchrum* Deflandre, 928 m. (j) *Nelsoniella aceras* Cookson and Eisenack, 928 m. (k) *Oligosphaeridium* sp., 938 m. (l) *Exochosphaeridium phragmites* Davey et al., 928 m. (m) *Oligosphaeridium* sp., 947 m. (n) *Florentinia* sp., 947 m.

belong to the UC-6 zone. The diversity of species is greater in the shaly limestone section. In the lower part of the section, 50% of the ostracodes are pyritized. Dinoflagellate cysts from the 1060–1069-m interval are abundantly represented but are not diversified. They are dominated by chorate types, while cavate types are rare and proximate types are absent. The Cenomanian/Turonian transition beds contain the highest TOC values (0.78–0.83%), which decrease upward (0.69–0.47%).

Beeri-2 Drill Hole

This is the northernmost well in the Gaza basin. The Daliyya Formation is thinner here and includes 40 m of dark bituminous marls, sandy at the bottom, covered by 10 m of weathered beige sandy limestones. The marls are attributed to the *W. archaeocretacea* zone, and the upper limestones belong to the Early-middle Turonian *H. helvetica* zone. Numerous dinoflagellate cysts from the 380–390-m interval are represented by a relatively small number of species of chorate types along with a few cavate types. Rare acanthomorphid acritarchs are present. Relatively high TOC values have been discerned at the Cenomanian/Turonian transition zone (1.63–1.70%).

History of Sedimentation in the Gaza Basin

Shallow shelf sedimentation, represented by hard dolomites and limestones of the Negba Formation, prevailed during the Early Cenomanian. In the late Early Cenomanian, soft marls and chalks were laid down in the Yesha area, implying a lower neritic to upper bathyal environment. Palynological remains are diverse and numerous, and indicate open marine conditions. In the Nir Oz area pyritic shales alternating with marls accumulated. The presence of quartz grains and fragments of glauconitic sandstone found in the top of the Negba Formation and base of Daliyya Formation in Yesha-1 and Nir Oz-1 could indicate tectonic activity. The resulting sea-floor topography contributed to the facies differences between the Yesha and Nir Oz areas.

In the Kissufim area, the Cenomanian part of the Daliyya Formation is absent because of either nondeposition or erosion. On the other hand, the section drilled (Cenomanian/Turonian transition and Turonian) is comparatively thicker than the sections in Nir Oz-1 and Yesha-1, suggesting that the extent of the formation was controlled by structural processes.

The Beeri-2 well is located on a ridge between the Gaza and Ashqelon basins (Figures 2, 3). This would explain the reduced thickness of the section.

Oxygen levels started to decline in the late Early Cenomanian (Nir Oz-1). The pyritization of the ostracodes probably took place shortly after death when the individuals were buried a few millimeters or centimeters below the surface, where reducing conditions prevented the complete decomposition of the organic matter (Oertli, 1971). Dysoxic conditions on the floor of the basin are also inferred from poor benthonic fauna and some chondrite-like burrows. The alternation of dark shales and lighter marls indicates fluctuating environmental conditions. The maximum extent of dysoxic sedimentation occurs during the Cenomanian/Turonian transition (*W. archaeocretacea* zone).

The Ashqelon Basin

The Ashqelon basin, a relatively small depression which can also be considered an extension of the large Gaza basin, trends NNE and is 50 km long and 6–10 km wide. The maximum thickness of Daliyya Formation drilled is 92 m at Shuva-1. The Daliyya Formation was studied at Binyamin-1 (Figure 5).

Binyamin-1 Drill Hole

The Daliyya Formation consists of soft gray pyritic marls, and is sandy at the bottom. Two foraminiferal zones are discernible: the Cenomanian/Turonian transition *W. archaeocretacea* zone in the lower 15 m and the Early to middle Turonian *H. helvetica* zone in the upper 15 m. An association of small *Hedbergella-Heterohelix* and *Pithonella* characterizes the *W. archaeocretacea* zone. The entire section belongs to the UC-6 ostracode zone. Pyritized ostracodes constitute up to 25% of the population in the lower 15 m. TOC values are higher in the *W. archaeocretacea* zone (1.30%) than in the *H. helvetica zone* (0.58–0.61%).

The Carmel-Caesarea Basin

The Carmel-Caesarea basin is most probably open to the west and attains, with the Gaash extension, a length of 70 km (Figures 2, 3). It includes several sub-basins. The Daliyya Formation is exposed in Mount Carmel, an ancient structural high (Ginzburg et al., 1975), which was uplifted in the Cenozoic Era. The outcrops at Daliyat el Karmil and Galame have been studied by Lipson-Benitah et al. (1988) and Honigstein et al. (1989). Three sections from two wells (Tirelli-7, Gaash-2) and one outcrop (Galame) are described (Figure 5).

Tirelli-7 Drill Hole

The Daliyya Formation in this well overlies Cenomanian dolomite and is 180 m thick. The lower half is composed of limestones and chalky limestones. The upper half is composed of white chalks alternating with dark shales, limestones, and dark-gray bituminous marls. The shales are finely laminated (Figure 8a-c), indicating that biological reworking was at a minimum, or absent. The bivalves were probably epifaunal. Planktonic foraminiferal remains consist of rare tests of *Marginotruncana* sp. No ostracodes were recovered. The TOC reaches 1.63%.

Gaash-2 Drill Hole

The Daliyya Formation comprises 13 m of alternating dark marls and beige argillaceous limestones with some flint, and 17 m of marls. The dark marls owe their color to disseminated pyrite. The section, which unconformably overlies Cenomanian limestones (Lipson, 1988), is attributed to the Cenomanian/Turonian transition *W. archaeocretacea* zone and the Early Turonian lower *H. helvetica* zone, as well as to the ostracode UC-6 zone. Twenty percent of the ostracode specimens are pyritized. Abundant, but not diversified, cavate type, dinoflagellate cysts were recovered, as well as rare acanthomorphid acritarchs (955–964-m interval). TOC values of the dark marls range from 0.79 to 0.91%.

The Galame Outcrop

The Galame outcrop is located on the eastern flank of Mount Carmel, and 50 m of the Daliyya Formation are exposed (Honigstein et al., 1989). The section can be subdivided into two units: a lower 35 m of soft bituminous, nonbioturbated gray to black marls, shales, and chalks (Figure 8d) and an upper 15 m

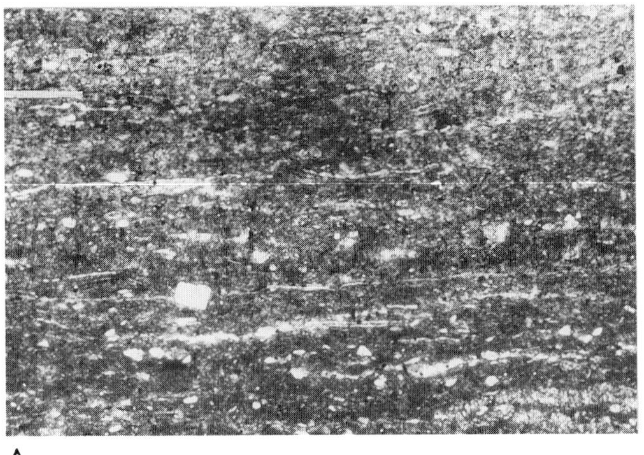

A

B

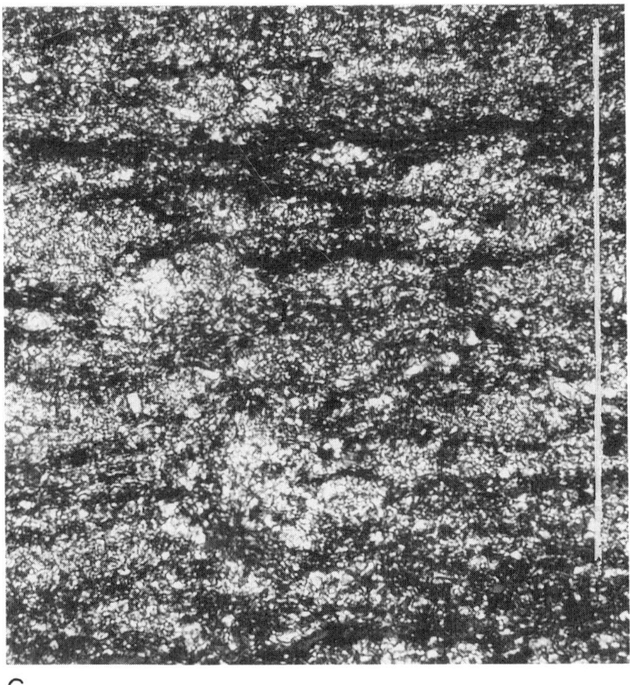

C

D

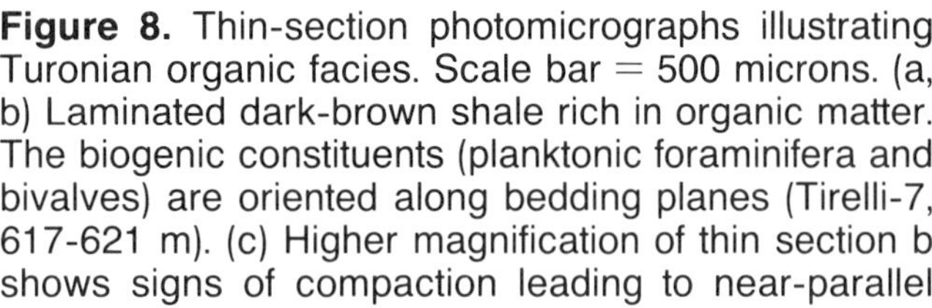

Figure 8. Thin-section photomicrographs illustrating Turonian organic facies. Scale bar = 500 microns. (a, b) Laminated dark-brown shale rich in organic matter. The biogenic constituents (planktonic foraminifera and bivalves) are oriented along bedding planes (Tirelli-7, 617-621 m). (c) Higher magnification of thin section b shows signs of compaction leading to near-parallel layering (Tirelli-7, 617-621 m). (d) Dark calcilutite rich in organic matter. The matrix is made up of weakly orientated particles derived from the platform. Minute quartz grains are present. Plankton include some *Pithonella* and foraminiferal tests (mainly *Hedbergella* sp. and *Heterohelix* sp.(Galame).

of yellow, mostly harder, chalky limestones. The Daliyya Formation overlies the Muhraqa Formation and is overlain by limestones of the Umm e Zinat Formation (Kashai, 1966). The section belongs to the middle Turonian upper *H. helvetica* foraminiferal zone and the UC-6 ostracode zone. TOC of the lower part of the section varies between 1.10 and 2.0%.

History of Sedimentation in the Carmel Basin

The Carmel basin was studied at Daliyat el Karmil and Galame (Figure 9). The soft marls and limestones of the Daliyya Formation in these outcrops are dated

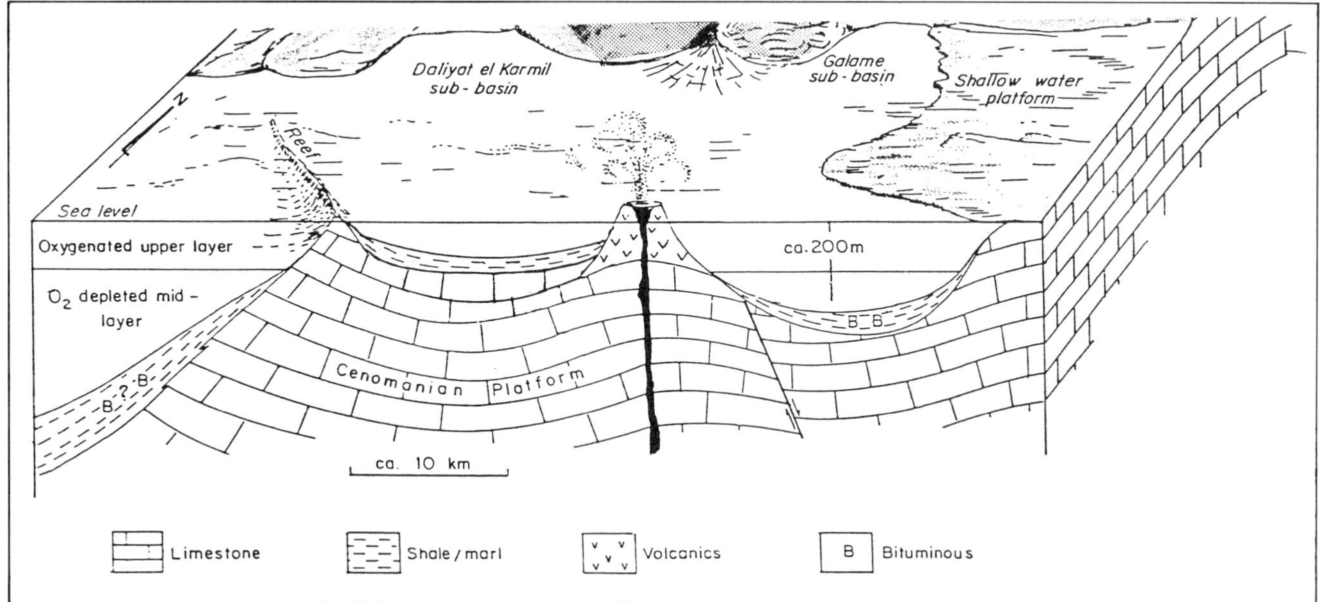

Figure 9. Paleogeographic reconstruction of the middle Turonian in the Carmel basin (modified after Honigstein et al., 1989).

as middle Turonian. Bituminous marls occur only in the deeper Galame sub-basin.

The Albian to Turonian sediments of the Carmel basin were deposited in three cycles, with each cycle followed by partial exposure (Bein, 1976). Erosional processes (Bein, 1977) and/or a pre-Turonian tectonic phase (Freund, 1961) were probably responsible for the formation and location of basins in which the marls of the Daliyya Formation accumulated during the Turonian transgressive cycle. In the Daliyat el Karmil sub-basin, the marls accumulated under well-oxygenated conditions while in the deeper Galame sub-basin they accumulated under oxygen-poor conditions.

History of Sedimentation of the Gaash Extension

The Gaash-2 well is located in the vicinity of the paleoshelf edge (Bein, 1977, Figure 1). The reduced thickness of oxygen-poor sediments can be related to the location of this well in an area of active wave mixing.

The Western Galilee Basin

The Western Galilee basin is approximately 35 km long. It might be a northern extension of the Carmel-Caesarea basin. Exposures of marls in this basin similar to those of the Daliyya Formation have been named the Yirka Formation (Freund, 1959; Kafri, 1972). They have been studied by Freund (1961) who investigated the ammonite-rich Turonian beds and equated the upper Yirka Formation with the Daliyya marls. Micropaleontological examination and TOC measurements have been performed in this study at the Damun-7 water well.

Damun-7 Drill Hole

The Damun-7 water well is located in the southern part of the Western Galilee basin. The Daliyya Formation, composed of a monotonous marly sequence that is darker at the bottom, attains a thickness of 55 m, but its base was not reached in this well. Planktonic foraminifera consist of scarce representatives of the *Whiteinella-Praeglobotruncana* group. Ostracodes are more abundant and indicate that most of the formation belongs to the UC-6 zone. TOC values range from 0.84 to 1.99%. Palynological analysis of the 313–324-m interval shows numerous, but not diverse, dinoflagellate cysts represented by chorate and cavate types.

AGE OF THE DALIYYA FORMATION

The Daliyya Formation was first attributed to the Late Cenomanian-Early Turonian on the basis of ammonites and foraminifera (Vroman, 1958; Reiss, 1958), and later to the Early Turonian based on ammonites and ostracodes (Freund, 1961; Rosenfeld and Raab, 1974). The age of the Daliyya Formation in its type locality at the Daliyat el Karmil has been redefined as middle Turonian (Lipson-Benitah et al., 1988), while in a nearby, correlative but somewhat thicker, section the age also includes the early Late Turonian (Kashai, 1966).

In this work the age of the Daliyya Formation has been extended to range from the late Early Cenomanian to the Late Turonian. The complete range is found in the Gaza basin.

DURATION AND MIGRATION OF THE DYSOXIC FACIES

Detailed study of eight wells and one section has shown that pulses of dysoxic sedimentation occurred from the late Early Cenomanian to the middle Turonian, and might extend locally to the Late Turonian. This time interval corresponds to the maximum duration of the middle Cretaceous Anoxic Event (OAE-2, Figure 1) as compiled by Schlanger et al. (1987). The global event, typified in the pelagic deposits of the Bonarelli bed of Gubbio (Italy), has been recorded from shallow seas to deep oceanic settings. In epicontinental seas, the duration of the event (which can be more accurately evaluated) seems longer than in deeper environments, where it is restricted to the Cenomanian-Turonian boundary (Kuhnt et al., 1986).

Correlations of OAE with transgressions have been suggested by Schlanger and Jenkyns (1976). In Israel, during Cenomanian-Turonian times, a transgressive phase linked to an eustatic rise of sea level covered the low-relief platform. This was interrupted by a short lowering of sea level in the Late Cenomanian (*R. cushmani* zone) (Flexer et al., 1986). Figure 10 depicts the spatial and temporal distribution of the Cenomanian-Turonian dysoxic event and the northward migration of the facies with time. Three main pulses are recognized in the Cenomanian, Cenomanian/Turonian transition, and Turonian.

The Gaza basin, presumably located on a subsiding shelf or as suggested by Bein (1977) beyond the shelf edge, records the first evidence of dysoxic sedimentation in late Early Cenomanian. Sedimentation continued until the middle and Late Turonian. With the renewal of transgression (Cenomanian/Turonian transition), the oxygen-minimum column of water encroached on the shelf edge, and a black shale facies belonging to the *W. archaeocretacea* zone was deposited in the Gaash area. By the middle Turonian, oxygen-poor conditions also developed on the northern shelf, and the basins cut in the Carmel uplifted area were flooded by the progressing transgression (Galame).

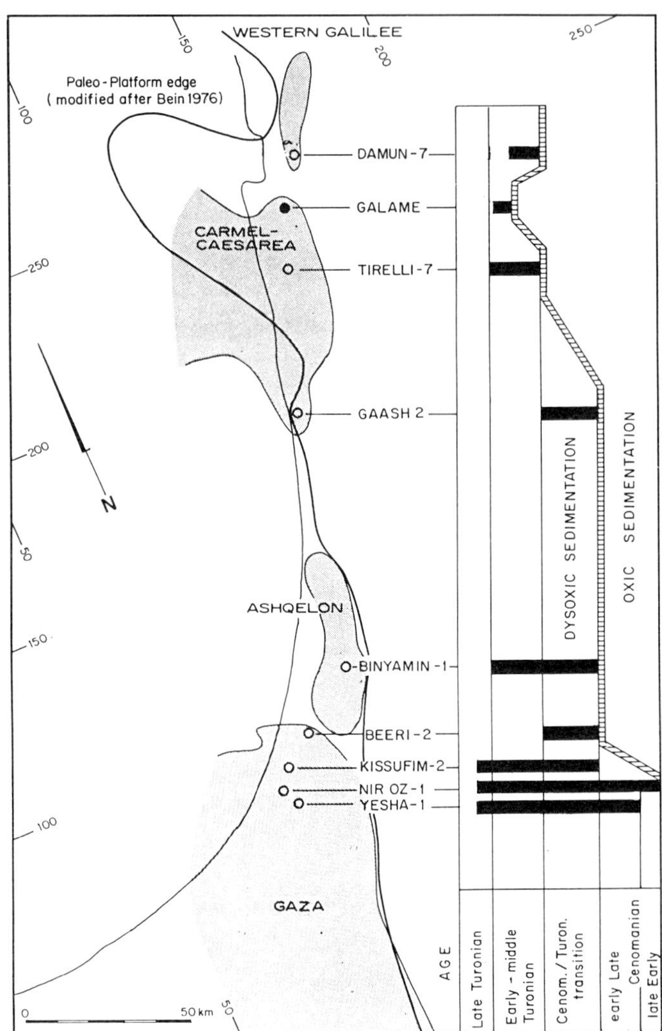

Figure 10. Migration of dysoxic sedimentation along the shelf edge during the Cenomanian and Turonian.

ORIGIN OF HIGHLY ORGANIC FACIES AND APPLICATION OF MODELS

Sporadic indications of oxygen-deficient conditions necessary to induce and preserve organic-rich sediments have been found throughout much of Cretaceous marine record. These recurring and widely recognized events are termed Oceanic Anoxic Events (Jenkyns, 1980) and refer to oxygen depletion in large parts of the ocean, and to the deposition of organic-rich facies. Criticism expressed about the criterion of globality (Einsele and Wiedmann, 1982; Waples, 1983) does not invalidate the concept, and regional modifications of the general scenario have been suggested (Bralower, 1988). However, the subject of their origin remains controversial. Explanations that account for temporal and spatial distribution of organic-rich facies were first proposed in 1976 (Schlanger and Jenkyns, 1976). There are two models: a preservational model and a productivity model (Zimmerman et al., 1987), and each addresses itself to one of the aspects of the problem.

The first model suggests that salinity-stratification and stagnation enabled organic matter to be incorporated in the sediments (Ryan and Cita, 1977). The second model proposes that increased surface productivity (Arthur and Silva, 1982) and an expanded and intensified oxygen-minimum zone contributed to the deposition of anoxic or near-anoxic (dysoxic) sediments (Schlanger and Jenkyns, 1976).

Stein (1986) calculated Cenomanian-Turonian paleoproductivity from Atlantic deep sea data, and relatively low values were obtained. However, he did not take into account deeper circulation rates which can affect upwelling and surface fertility. Cretaceous oceans, especially the low latitude Tethys, were more convective than originally assumed (Brass et al., 1982; Saltzman and Barron, 1982). Expansion and intensification of the oxygen-minimum zone resulted from lack of balance between oxygen demand at the surface and poor oxygen renewal at the bottom. However, the accumulation of organic-rich sediments also depends on other factors such as sedimentation and burial rates (Calvert, 1987). A model involving feedback between volcanism, tectonism, and sea-level, climatic, and oceanographic changes has been proposed by Arthur et al. (1987) for the Cenomanian-Turonian anoxic phase.

Detailed study of organic-rich sequences in Israel has revealed different facies patterns, diachronism, and a longer period of dysoxic sedimentation than is generally described for similar sequences. This implies that no simple or single model can account for their occurrence and that interactions between described global changes and local conditions probably contributed to the complexity of the process. Furthermore, not all the mechanisms involved are fully known.

The Cenomanian occurrences of oxygen-poor sedimentation in the Nir Oz area, Gaza Basin, could be related to local conditions, such as the position or shape of the local depression, or they might be precursors of the main phase of dysoxia. The alternating dark shales and light marls, found in Nir Oz-1, are similar to the short-term cycles that often characterize the strata deposited during the Cenomanian-Turonian anoxic event (Schlanger et al., 1987). The origin of these cycles is not clear and may be related to climatic pulses.

Dysoxic sedimentation during the Cenomanian/Turonian transition phase is more widespread. It is found in Gaza, Ashqelon, and the southern part of the Carmel-Caesarea basins. It corresponds to the maximum development of the oceanic anoxic event and could be attributed to the advance of oxygen-depleted oceanic water as a result of the transgression. Blooms of small, simple species of *Hedbergella*, known to have a shallow habitat and high reproductive potential, and the presence of *Pithonella* commonly characterize these sediments. The increased populations of these plankton, termed opportunistic, occur in areas of mixed nutrient-rich surface water (Caron and Homewood, 1982). Pyritized ostrocodes (up to 50%) and the decrease of benthonic foraminifera (mainly lenticulinids and gavelinids) indicate oxygen-poor water. The association of chorate and cavate types of dinoflagellate cysts and an almost complete absence of proximate types suggest a subtropical-tropical water mass and, accordingly, a warm-water upwelling system rather than the cold-water system (Kuhnt et al., 1986).

The Turonian dysoxic phase is a continuation of the preceding phase in the south; in the north, it is a new one. In the Galame area, the only Turonian section accurately dated, this phase is limited to the middle Turonian (part of the upper *H. helvetica* zone). Here, a transgressive pulse filled the newly formed depression with waters belonging to the upper oxygen-depleted layer, which was still expanded after its global peak at the Cenomanian/Turonian transition. A return to normally oxygenated waters started at the end of the middle Turonian.

Late Turonian-early Senonian folding resulted in a more pronounced relief of the shelf and affected circulation. Consequently, the depositional pattern of oxygen-poor facies during the Senonian was more dependent on regional factors (Sass and Bein, 1982).

RESPONSE OF THE PLANKTONIC FORAMINIFERA TO THE CENOMANIAN-TURONIAN DYSOXIC CONDITION

The faunal changes associated with the Cenomanian-Turonian OAE-2 have been widely described (e.g., Wonders, 1980; Arthur et al., 1987). These changes, observed in the planktonic foraminifera of the *Rotalipora cushmani*, *Whiteinella archaeocretacea*, and *Helevetoglobotruncana helvetica* zones, illustrate some of their adaptive strategies.

R. cushmani Zone

The *Rotalipora* specimens in the Gaza basin are reduced in size and outnumbered by *Hedbergella*. A reduction of test size and low ratios of keeled/unkeeled species have also been observed in other dysoxic paleoenvironments (Kuhnt et al., 1986).

W. archaeocretacea Zone

The assemblages of the *W. archaeocretacea* zone are characterized worldwide by the presence of nonkeeled and weakly keeled forms and the absence of strongly keeled forms, whether the environment is dysoxic or not. However, the predominance of thin-shelled, low-trochospire *Hedbergella* and *Whiteinella* species over rugose, incipiently keeled or high conical species, observed in our material, could be related to oxygen depletion as suggested by Jarvis et al. (1988). The presence of clavate forms of *Hedbergella* and *Clavihedbergella* in this epipelagic assemblage is in agreement with their depth range as reinterpreted by Leckie (1987).

H. helvetica Zone

Morphological changes and the decreases in the population density and diversity of the planktonic foraminifers belonging to the middle Turonian *H. helvetica* zone in the Galame section have been discussed (Honigstein et al., 1989).The intermediate and "deeper" species of *Marginotruncana* and *Helvetoglobotruncana* did not disappear completely under oxygen-poor conditions. They are present as smaller forms with a diminutive last chamber, a low average number of chambers in the last coil, and poorly calcified tests (or without calcareous shells). In Tirelli-7 the few *Marginotruncana* recovered also showed poorly calcified tests.

IMPLICATIONS FOR OIL EXPLORATION

This study has shown that organic-rich marls and shales are commonly present within the Cenomanian-Turonian Daliyya Formation. TOC values range from 0.6 to 2% and reach a maximum of 2.4%. The lowest values obtained (0.3–0.5%) were from mixed white and dark cuttings, while hand-picked dark pieces gave a TOC value of 1.6%.

It seems that the Daliyya Formation (or at least its organic-rich layers) could be a potential source rock if buried to an adequate depth. The existence of a deep offshore basin (13.5 km of sediments), as implied from magnetic data, has been suggested by Cohen et al. (1988). During the Cenomanian-Turonian transgression, black shales could have accumulated in this basin, and if conditions for maturation and entrapment existed, the generally fine-grained, dark, organic-rich units of the Daliyya Formation could have generated oil. Additional data are required to test this hypothesis.

ACKNOWLEDGMENTS

The authors thank Y. Natan and E. Kashai for reviewing the manuscript and A. Almogi-Labin for her helpful comments. Oil Exploration (Investments) Ltd. provided Figure 3. L. Diav of the Geological Survey of Israel performed TOC measurements. R. Duani typed the draft. The Open University of Israel typed the manuscript. The manuscript was also reviewed by R.G. Miller of BP Research Centre and M. Stefani of University of Ferrare who contributed valuable and constructive criticism.

REFERENCES

Arthur, M. A., and I. Premoli Silva, 1982, Development of widespread organic carbon-rich strata on the Mediterranean Tethys, *in* S. O. Schlanger and M. B. Cita, eds., Nature and origin of Cretaceous carbon-rich facies: London, Academic Press, p. 7-54.

Arthur, M. A., S. O. Schlanger, and H. C. Jenkyns, 1987, The Cenomanian-Turonian Oceanic Anoxic Event, II. Palaeoceanographic controls on organic-matter production and preservation, *in* J. Brooks and A. J. Fleet, eds., Marine petroleum source rocks: Geological Society of London Special Publication 26, p. 401-420.

Bein, A., 1976, Rudistid fringing reefs of the Cretaceous shallow carbonate platform of Israel: AAPG Bulletin, v. 60, p. 258-272.

Bein, A., 1977, Shelf basin sedimentation: Mixing and diagenesis of pelagic and clastic Turonian carbonates, Israel: Journal of Sedimentary Petrology, v. 47, p. 382-391.

Bein, A., and Z. Reiss, 1976, Cretaceous *Pithonella* from Israel: Micropaleontology, v. 22, p. 83-91.

Bralower, T. J., 1988, Calcareous nannofossil biostratigraphy and assemblages of the Cenomanian-Turonian boundary interval: Implications for the origin and timing of oceanic anoxia: Palaeoceanography, v. 3, p. 275-316.

Brass, G. W., J. R. Southam, and W. H. Peterson, 1982, Warm saline bottom water in the ancient ocean: Nature, v. 296, p. 620-623.

Calvert, S. E., 1987, Oceanographic controls on the accumulation of organic matter in marine sediments, *in* J. Brooks and A. J. Fleet, eds., Marine petroleum source rocks: Geological Society of London Special Publication 26, p. 137-151.

Caron, M., and P. Homewood, 1983, Evolution of early planktic foraminifers: Marine Micropaleontology, v. 7, p. 453-462.

Cohen, Z., A. Flexer, and V. Kaptsan, 1988, The Pleshet basin: A newly discovered link in the peripheral chain of basins of the Arabian Craton: Journal of Petroleum Geology, v. 11, p. 403-414.

Einsele, G., and J. Wiedmann, 1982, Turonian black shales in the Moroccan coastal basins: First upwelling in the Atlantic Ocean?, *in* U. Von Rad, K. Hinz, M. Sarnthein, and E. Seibold, eds., Geology of the northwest African margin: New York, Springer-Verlag, p. 396-414.

Flexer, A., A. Rosenfeld, S. Lipson-Benitah, and A. Honigstein, 1986, Relative sea level changes during the Cretaceous in Israel: AAPG Bulletin, v. 70, p. 1685-1699.

Freund, R., 1959, On the stratigraphy and tectonics of the Upper Cretaceous in Western Galilee: Bulletin of the Research Council of Israel, v. 8G, p. 43-50.

Freund, R., 1961, Distribution of Lower Turonian ammonites in Israel and neighbouring countries: Bulletin of the Research Council of Israel, v. 10G, p. 79-100.

Gignoux, M., 1955, Stratigraphic geology: San Francisco, W.H. Freeman, 682 p.

Ginzburg, A., S. Cohen, H. Hay-Roe, and A. Rosenzweig, 1975, Geology of the Mediterranean Shelf of Israel: AAPG Bulletin, v. 59, p. 2142-2160.

Gvirtzman, G., and Z. Reiss, 1965, Stratigraphic nomenclature in the Coastal Plain and Hashefela regions: Geological Survey of Israel Report OD 1/65, 13 p.

Hamaoui, M., 1979, Contribution a l'etude du Cenomanien-Turonien d'Israel. Comparisons micropaleontologiques avec quelques regions mesogeennes: Ph.D. thesis, Universite P. et M. Curie, Paris, 121 p.

Honigstein, A., and A. Rosenfeld, 1985, Late Turonian-early Coniacian ostracodes from the Zihor Formation, southern Israel: Revista Espanola de Micropaleontologia, v. 17, p. 447-466.

Honigstein, A., S. Lipson-Benitah, B. Conway, A. Flexer, and A. Rosenfeld, 1989, Mid-Cretaceous anoxic event in Israel—a multidisciplinary approach: Palaeogeography, Palaeoclimatology, Palaeoecology, v. 69, p. 103-112.

Jarvis, I., G. A. Carson, M. K. E. Cooper, M. B. Hart, P. N. Leary, B. A. Tocher, D. Horne, and A. Rosenfeld, 1988, Microfossil assemblages and the Cenomanian-Turonian (Late Cretaceous) Oceanic Anoxic Event: Cretaceous Research, v. 9, p. 3-103.

Jenkyns, H.C., 1980, Cretaceous anoxic events—from continents to oceans: Journal of the Geological Society of London, v. 137, p. 171-188.

Kafri, U., 1972, Lithostratigraphy and environment of deposition, Judean Group, western and central Galilee, Israel: Israel Geological Survey Bulletin 54, p. 1-56.

Kashai, E., 1966, The geology of the eastern and southwestern Carmel: Ph.D. thesis, Hebrew University, Jerusalem, 115 p. [in Hebrew, with English abstract].

Kuhnt, W., J. Thurow, J. Wiedmann, and J. P. Herbin, 1986, Oceanic anoxic conditions around the Cenomanian/Turonian boundary and the response of the biota: Mitteilungen des Geologisches Palaeontologischen Instituts der Universitaet Hamburg, p. 205-246.

Kummel, B., 1970, History of the Earth: San Francisco, Freeman, 707 p.

Leckie, R. M., 1987, Paleoecology of mid-Cretaceous planktonic foraminifera: A comparison of open ocean and epicontinental sea assemblages: Micropaleontology, v. 38, p. 164-176.

Lipson, S., 1988, The Cenomanian/Turonian boundary over the carbonate platform: Israel Institute of Petroleum and Energy, Report 1/88, 5 p.

Lipson-Benitah, S., 1980, Albian to Coniacian zonation of the western coastal plain of Israel: Cretaceous Research, v. 1, p. 3-12.

Lipson-Benitah, S. 1987, The planktonic foraminiferal biostratigraphy of the Coniacian in Israel—A Correction: Cretaceous Research, v. 8, p. 187-188.

Lipson-Benitah, S., A. Rosenfeld, A. Honigstein, A. Flexer, and E. Kashai, 1988, The middle Turonian Daliyya Formation in Israel, biostratigraphy, paleoenvironment and sea level changes: Cretaceous Research, v. 9, p. 321-336.

Oertli, H. J., 1971, The aspect of ostracode faunas—A possible new tool in petroleum sedimentology: Bulletin Centre Recherche Pau SNPA, v. 5 suppl., p. 137-151.

Reiss, Z., 1958, Remarks on the age of the so-called "Dalia-Marls" in northern Israel: Bulletin of the Geological Society of Israel, v. 17, p. 3-4.

Reyment, R. A., and P. Bengtson, eds., 1986, Events of the mid-Cretaceous: Physics and Chemistry of the World-16, Oxford, Pergamon, 231 p.

Robaszynski, F., and M. Caron, 1979, Atlas des foraminiferes planctoniques du Cretace Moyen (2 Volumes): Cahiers de Micropaleontologie I, p. 1-185; II, p. 1-181.

Robaszynski, F., G. Alcayde, F. Amedro, G. Badillet, R. Damotte, J. C. Foucher, S. Jardine, O. Legoux, H. Manivit, C. Monciardini, and J. Sornay, 1982, L'etage Turonien dans sa region-type "de Saumur a Montrichard." Biostratigraphie, Sedimentologie, Coupures: Memoires du Museum National d'Histoire Naturelle, Serie C, v. 49, p. 15-28.

Rosenfeld, A., and M. Raab, 1974, Cenomanian-Turonian ostracodes from the Judea Group in Israel: Israel Geological Survey Bulletin 62, p. 1-64.

Ryan, W. F. B., and M. B. Cita, 1977, Ignorance concerning episodes of oceanwide stagnation: Marine Geology, v. 23, p. 197-215.

Saltzman, E. S., and E. J. Barron, 1982, Deep circulation in the Late Cretaceous: Oxygen isotope paleotemperatures from *Inoceramus* remains in D.S.D.P. cores: Palaeogeography, Palaeoclimatology, Palaeoecology, v. 40, p. 167-181.

Sass, E., and A. Bein, 1982, The Cretaceous carbonate platform in Israel: Cretaceous Research, v. 3, p. 135-144.

Schlanger, S. O., and H. C. Jenkyns, 1976, Cretaceous oceanic anoxic events—Causes and consequences: Geologie en Mijnbouw, v. 55, p. 179-184.

Schlanger, S. O., M. A. Arthur, H. C. Jenkyns, and P. A. Scholle, 1987, The Cenomanian-Turonian Oceanic Anoxic Event, I. Stratigraphy and distribution of organic carbon-rich beds and the marine 13C excursion, *in* J. Brooks and A. J. Fleet, eds., Marine petroleum source rocks: Geology Society of London Special Publication 26, p. 371-399.

Schwan, W., 1980, Geodynamic peaks in Alpinotype orogenies and changes in ocean-floor spreading during Late Jurassic-Late Tertiary time: AAPG Bulletin, v. 64, p. 359-373.

Stein, R., 1986, Surface-water paleo-productivity as inferred from sediments deposited in oxic and anoxic deep-water environments of the Mesozoic Atlantic Ocean: Mitteilungen des Geologisches Palaentologischen Instituts der Universitaet Hamburg, p. 55-70.

Vail, P. R., R. M. Mitchum Jr., and S. Thompson III, 1977, Seismic stratigraphy and global changes of sea level: part 4, Global cycle of relative changes of sea level, *in* C. E. Payton, ed., Seismic stratigraphy—applications to hydrocarbon exploration: AAPG Memoir, 26, p. 83-97.

Vroman, A. J., 1958, The Cenomanian-Turonian boundary in northern Israel (preliminary note): Bulletin of the Geological Survey of Israel, v. 17, p. 1-2.

Waples, D. W., 1983, A reappraisal of anoxia and organic richness with emphasis on the Cretaceous North Atlantic: AAPG Bulletin, v. 67, p. 963-978.

Wonders, A. A. H., 1980, Middle and Late Cretaceous planktonic foraminifera of the western Mediterranean area: Utrecht Micropaleontological Bulletin, v. 24, p. 1-157.

Zimmerman, H. B., A. Boersma, and F. W. McCoy, 1987, Carbonaceous sediments and palaeoenvironment of the Cretaceous South Atlantic, *in* J. Brooks and A. J. Fleet, eds., Marine petroleum source rocks: Geological Society of London Special Publication 26, p. 271-286.

Organic-Carbon-Rich Sediments and Paleoenvironment: Results from Baffin Bay (ODP-Leg 105) and the Upwelling Area off Northwest Africa (ODP-Leg 108)

Ruediger Stein
Institut für Geowissenschaften und Lithosphärenforschung (IGL)
Universität Giessen
Giessen, Federal Republic of Germany

Ralf Littke
Institut für Erdöl und Organische Geochemie
Jülich, Federal Republic of Germany

Detailed organic geochemical investigations were performed on organic-carbon-rich sediments from Baffin Bay (ODP-Site 645) and the upwelling area off Northwest Africa (ODP-Site 658). Main objectives of this study were (1) to reconstruct the depositional environment of these sediments and (2) to point out factors controlling the formation of organic-carbon-rich sediments.

Sediments with high total organic carbon contents of up to more than 3% were deposited at both Site 645 and Site 658. Mass accumulation rates of total organic carbon are also similar, ranging from 0.05 to 0.4 gC cm^{-2} ky^{-1}.

The types of organic matter, however, differ between Site 645 and Site 658, reflecting different depositional environments. In the narrow Baffin Bay, climate-controlled changes in terrigenous carbon supply have been the prime influence on organic-matter accumulation at Site 645 throughout the last 20 Ma, with maximum rates of input in the middle Miocene and the late Pliocene/Pleistocene. During the former period, additional quantities of marine organic carbon accumulated. These are thought to reflect a phase of higher productivity caused by the first inflow of cold Arctic water masses.

In the upwelling area off Northwest Africa, variations in marine organic carbon input, caused by changes in surface-water productivity, were the primary control on organic-carbon accumulation at Site 658. Nevertheless, terrigenous organic carbon may still represent a significant proportion of the total organic carbon. Estimated paleoproductivity values are in the same range as those measured in the area today. Between 3.6 and 3.1 Ma, the supply of terrigenous organic carbon reached its maximum, probably due to a phase of dominantly humid climatic conditions in Northwest Africa. The higher oceanic productivity also recorded for this interval may have been caused by increased fluvial nutrient supply.

INTRODUCTION

In the modern oceanic environment, sediments are typically impoverished, rather than enriched, in organic carbon. Typical recent deep-sea sediments have organic-carbon contents of less than 0.3%. Sediments with organic-carbon contents of more than 1% are largely restricted to minor parts of the modern ocean, such as the upper continental slope. However, ancient intervals are known in which carbon-enriched sediments were much more common and widespread, for example the Aptian/Albian and Cenomanian/Turonian black shales (Schlanger and Jenkyns, 1976; Arthur et al., 1984; Stein et al., 1986). The deposition of these carbon-rich sediments requires special environmental conditions such as (1) high surface-water productivity, (2) increased preservation of organic matter, (3) enhanced supply of terrigenous organic matter, and (4) rapid burial of organic matter (Müller and Suess, 1979; Demaison and Moore, 1980).

Two of these mechanisms, the influence of high oceanic productivity and increased supply of terrigenous organic matter, were investigated in detail regarding sediments from Baffin Bay (Figure 1A) and the upwelling area off Northwest Africa (Figure 1B). The results of these investigations not only yield information concerning the evolution of paleoenvironments in Baffin Bay and off Northwest Africa, but may also give more insight about the general formation of other organic-carbon-rich sediments, such as those of the Cretaceous.

MODERN ENVIRONMENTS AND GEOLOGICAL SETTING IN BAFFIN BAY AND THE UPWELLING AREA OFF NORTHWEST AFRICA

Three sites were drilled in Baffin Bay and the Labrador Sea during ODP-Leg 105 (Figure 1A, Sites 645, 646, and 647) in order to study the tectonic, paleoclimatic, and paleoceanographic evolution of the North Atlantic Ocean and its Arctic connection (Srivastava et al., 1987). At Site 645 (70° 27.49′N, 64° 39.42′W; water depth of 2018 m), a total depth of 1147.1 m was penetrated, and sediments of Miocene to Holocene age were recovered. Lithologies are predominantly terrigenous clay, silt, and sand; biogenic carbonate and siliceous components are of only minor importance. Furthermore, the sediments are characterized by relatively high organic-carbon contents (Figure 2) (Srivastava et al., 1987; Stein et al., 1989a).

Today, surface-water circulation in Baffin Bay is dominated by currents entering the bay from the Arctic Ocean through the narrow Nares Strait (Baffin Island Current, BIC) and from the North Atlantic Ocean (West Greenland Current, WGC) (Figure 1A) (Tchernia, 1982; Williams, 1986). Between December and April, Baffin Bay is covered by sea ice, whereas during the rest of the year drifting icebergs are of major importance (Marko et al., 1982; Williams, 1986).

During ODP-Leg 108, Sites 657 through 661 were drilled on the northwest African continental margin between 21° and 10° N at water depths between 2270 and 4330 m (Figure 1B). The principal objective of Leg 108 was to study the evolution of Cenozoic paleoclimate and paleoceanography in the eastern subtropical/tropical Atlantic Ocean (Ruddiman et al., 1988). Site 658 (20° 44.95′N, 18° 34.85′W; water depth of 2263 m) was drilled directly underneath a major coastal upwelling cell (Figure 1B) (Shaffer, 1976; Tomczak and Hughes, 1980). The sediments recovered are of early Pliocene to Holocene age and are characterized by cyclic changes between light nannofossil ooze and olive to gray calcareous, siliceous, and siliciclastic sediments. Organic-carbon contents may be as great as 4% (Figure 3) (Ruddiman et al., 1988; Stein et al., 1989b). Sediments deposited at nearby Sites 657 and 659 (Figure 1B) were not influenced by coastal upwelling, and detailed study of these latter lithologies allows comparison of the "high-oceanic productivity facies" of Site 658 with the normal open-ocean environment (Ruddiman et al., 1988; Stein et al., 1989b). Furthermore, at Site 657, several turbiditic sequences are intercalated (Faugères et al., 1989), giving the opportunity to study organic-carbon preservation in turbidites (Stein et al., 1989b).

METHODS

Methods used for studying organic geochemistry are summarized in Figure 4 (for more details, see Stein et al., 1989a, b). In this paper, we concentrate on data from elemental analyses, Rock-Eval pyrolysis, and kerogen microscopy. Gas chromatography (GC) and gas chromatography/mass spectrometry (GC/MS) data are also available for samples from Legs 105 and 108. These are described and discussed by ten Haven and Rullkötter (1989) and ten Haven et al. (1989).

Organic and inorganic carbon and total nitrogen were determined using a HERAEUS CHN-Analyser. In addition, carbon analyses were performed with a LECO IR-112 Carbon Analyser (for details and comparison of data derived from both instruments, see Stein et al., 1989a). Rock-Eval pyrolysis was performed according to the method described by Espitalié et al. (1977).

Kerogen-microscopy analyses were carried out after concentration of kerogen by treating the samples with HCl and HF to remove carbonates and silicates, respectively (Durand, 1980). Macerals were

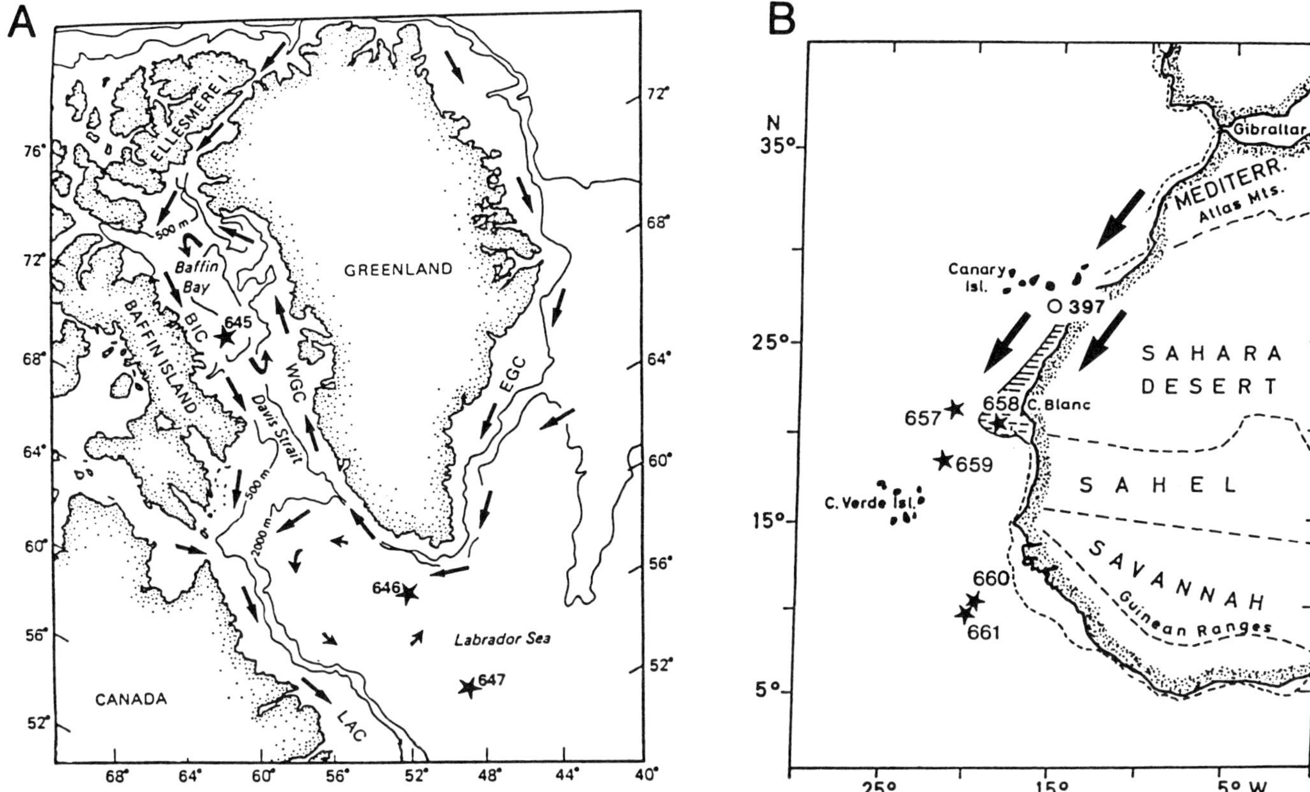

Figure 1. A. Location map of Sites 645, 646, and 647. Arrows indicate surface-water circulation (according to Tchernia, 1982). BIC = Baffin Island Current; LAC = Labrador current; EGC = East Greenland Current; WGC = West Greenland Current. B. Location map of ODP-Sites 657 to 661. Hatched pattern marks the upwelling area off Northwest Africa. Arrows indicate northeasterly trade winds.

classified according to the nomenclature described by Stach et al. (1982). For maturity assessments, vitrinite reflectance was measured on particles larger than 10 μm in oil immersion (see Stach et al., 1982).

Total-organic-carbon data are difficult to interpret since changes in carbon content may result from variations in both organic matter input and dilution by carbonates, biogenic opal, and/or siliciclastics. Consequently, percent values were converted to accumulation rates (cf., van Andel et al., 1975), using the physical-property and sedimentation-rate data from Srivastava et al. (1987), Sarnthein and Tiedemann (1989), and Tiedemann et al. (1989):

$$ARBS = LSR \cdot (WBD - 1.026\ (PO/100))$$

$$AROC = (C_{org}/100) \cdot ARBS$$

where

ARBS - accumulation rates of bulk sediment ($g \cdot cm^{-2} \cdot ky^{-1}$)

AROC - accumulation rates of organic carbon ($g \cdot cm^{-2} \cdot ky^{-1}$)

C_{org} - organic carbon content (weight percent)

LSR - linear sedimentation rate ($cm \cdot ky^{-1}$)

WBD - wet bulk density ($g \cdot cm^{-3}$)

PO - porosity (percent)

Use of these accumulation rates makes it possible to exclude dilution effects from non-organic carbon components, and the data can be interpreted in terms of carbon supply/preservation. Compaction effects also are considered in the calculation procedure allowing a comparison of organic-carbon data from different sites and different time intervals.

Surface-water paleoproductivity was estimated using the equation

$$PP = 5.31 \cdot (C \cdot (WBD - 1.026\ PO/100))^{0.71} \cdot LSR^{0.07} \cdot DEP^{0.45}$$

where

PP paleoproductivity ($gC \cdot m^{-2} \cdot yr^{-1}$)

C (marine) organic carbon (weight percent)

DEP paleowater depth of sea floor (m)

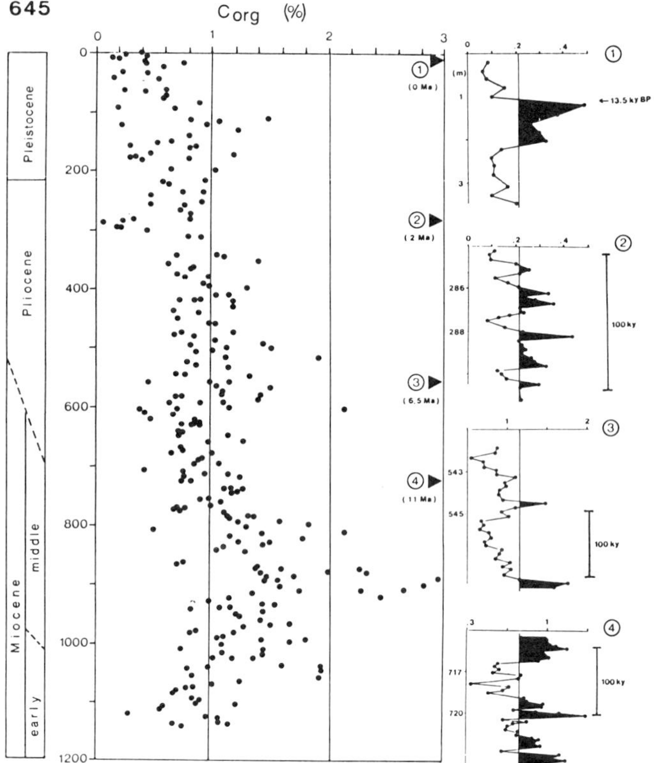

Figure 2. Total organic-carbon content of sediment samples from ODP-Site 645. On the right-hand side, short-term fluctuations in organic-carbon contents are shown for four densely sampled time intervals.

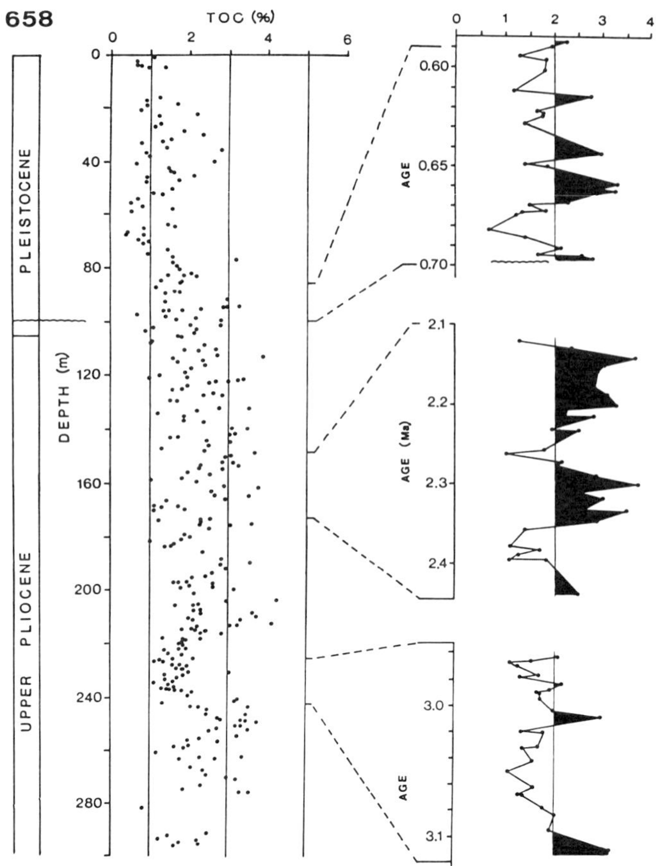

Figure 3. Total organic-carbon content of sediment samples from ODP-Site 658. On the right-hand side, three intervals are enlarged to better illustrate short-term fluctuations.

This formula is based on empirical relationships between organic-carbon flux to the sea floor, water depths, organic-carbon accumulation rates in surface sediments, sedimentation rates, and measured Holocene surface-water productivity. It combines the original Müller and Suess (1979) formula with data on carbon flux rates to the sea floor (Betzer et al., 1984) (for deduction and limits of use of the equation, see Stein, 1986a). A similar equation for estimating paleoproductivity, which is also based on the original work of Müller and Suess (1979)and includes new sedimentation-rate and organic-carbon data, is presented by Sarnthein et al. (1987) (for detailed discussion of methods for reconstruction of paleoproductivity, see Berger et al., 1989).

RESULTS

Quantity of Organic Carbon

The sediment sequences of both Site 645 and Site 658 are characterized by organic-carbon contents distinctly higher than those observed in normal open-marine environments. At Site 645, most of the organic-carbon values are greater than 0.5%, ranging from 0.5 to almost 3% (Figure 2). Minimum values in the upper 300 m (i.e., in the late Pliocene and Pleistocene) are somewhat lower. At Site 658, similar high organic-carbon contents of 0.5 to 4% are typical, with the lower values concentrated in the upper 100 m (i.e., in the Pleistocene) (Figure 3). At both sites, short-term variations in organic-carbon content with periods of 20,000 to 100,000 years occur (Figures 2, 3). These short-term variations are also reflected in the carbonate records (Stein et al., 1989a, b).

In comparison, the pelagic sediments of the two non-upwelling Sites 657 and 659 from the northwestern Africa continental margin (Figure 1B) are characterized by low organic-carbon contents of less than 0.4%. At Site 657, however, several turbidites and slumps with higher organic-carbon contents of 1 to 3% are intercalated (Stein et al., 1989b).

Based on bulk accumulation rates (Figures 5b, 6b), accumulation rates of organic carbon (AROC) have been calculated for Sites 645 and 658 (Figures 5d, 6d) (see methods). The record is much more detailed for Site 658 (Figure 6) than for Site 645 (Figure 5), since at the former locality, detailed oxygen isotope stratigraphy and sedimentation-rate records are available (Sarnthein and Tiedemann, 1989; Tiede-

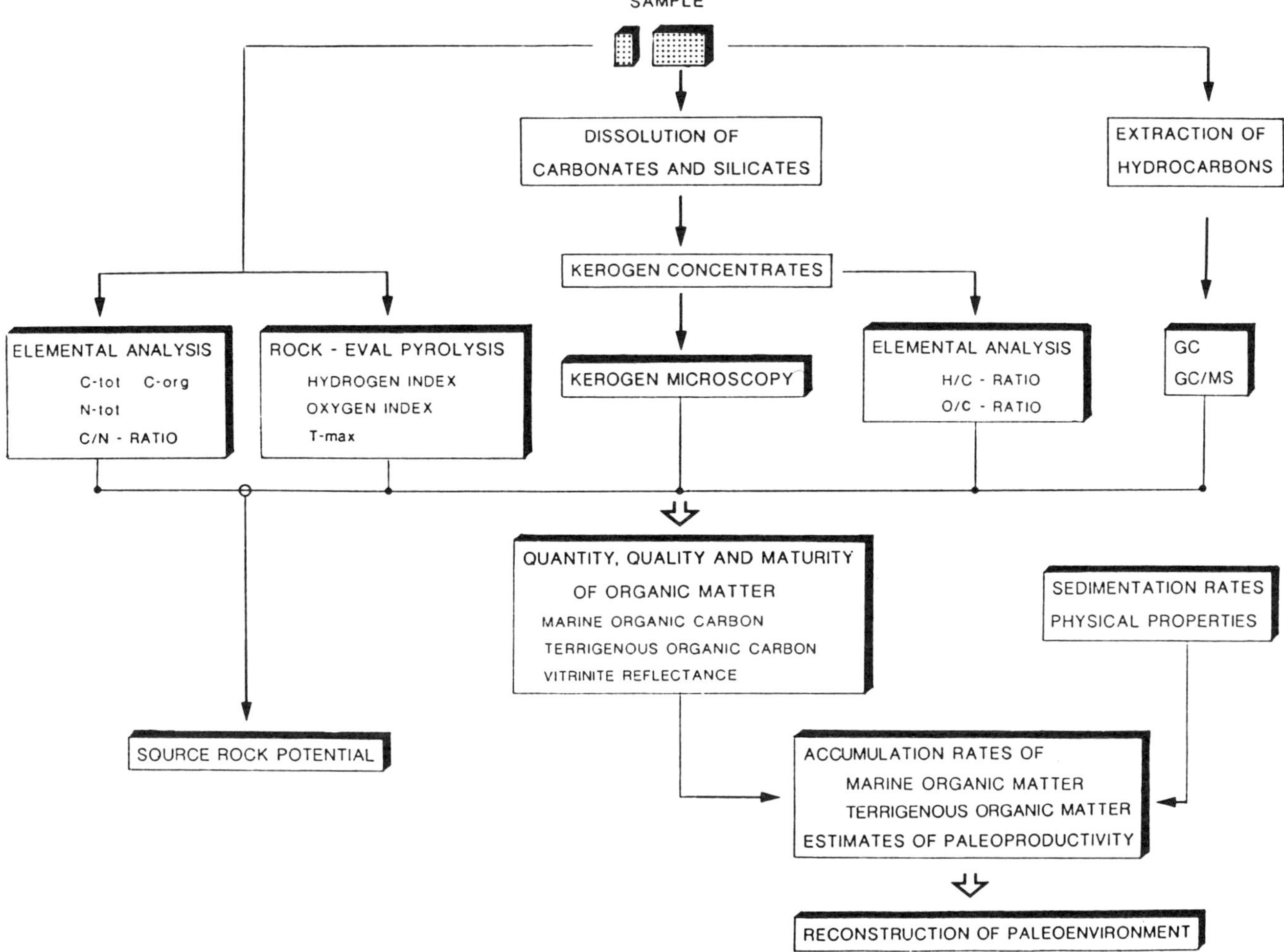

Figure 4. Summary of methods used for studying organic geochemistry in the investigation of sediments from Legs 105 and 108.

mann et al., 1989). The upper part of the succession at Site 658 (i.e., above the hiatus) especially demonstrates a high-resolution record representing glacial/interglacial fluctuations in organic-carbon accumulation (Figure 6). On the other hand, the record of Site 645 is based only on relatively weak bio- and magnetostratigraphies (Srivastava et al., 1987).

At both sites, AROC are about one order of magnitude higher than those recorded in the open-ocean Sites 646, 647, and 659. At Sites 645 and 658, AROC vary between 0.04 and 0.50 gC · cm^{-2} · ky^{-1} (Figures 5, 6), whereas at Sites 646, 647, and 659 AROC vary between 0.002 and 0.06 gC · cm^{-2} · ky^{-1} (Stein et al., 1989a, b).

According to AROC data, the record of Site 645 can be divided into three intervals. The middle Miocene was apparently characterized by relatively high AROC of 0.10 to 0.15 gC · cm^{-2} · ky^{-1} (Figure 5d). During the following late Miocene and early Pliocene, AROC decreased to values around 0.04 gC · cm^{-2} · ky^{-1}. Finally, during the last 3.4 Ma, maximum AROC of up to 0.20 gC · cm^{-2} · ky^{-1} were reached (Figure 5d).

At upwelling Site 658, AROC are normally about twice as high as those recorded at Site 645, ranging from 0.10 to 0.50 gC · cm^{-2} · ky^{-1} (Figure 6d). This record can also be divided into three intervals. Between 3.6 and 3.1 Ma, high AROC of 0.30 to 0.40 gC · cm^{-2} · ky^{-1} are typical. During the late Pliocene (3.1 to 1.6 Ma), AROC varied between 0.20 and 0.35 gC · cm^{-2} · ky^{-1}, whereas the interval above the major hiatus (the last 0.7 Ma) was characterized by high-amplitude variations between 0.10 and 0.50 gC · cm^{-2} · ky^{-1} (Figure 6d). The maximum values of 0.50 and 0.40 gC · cm^{-2} · ky^{-1} (Figure 6d) were recorded at about 0.22 and 0.53 Ma.

Source of the Organic Matter

If organic-carbon data are to be used in reconstructing depositional environments, it is necessary to

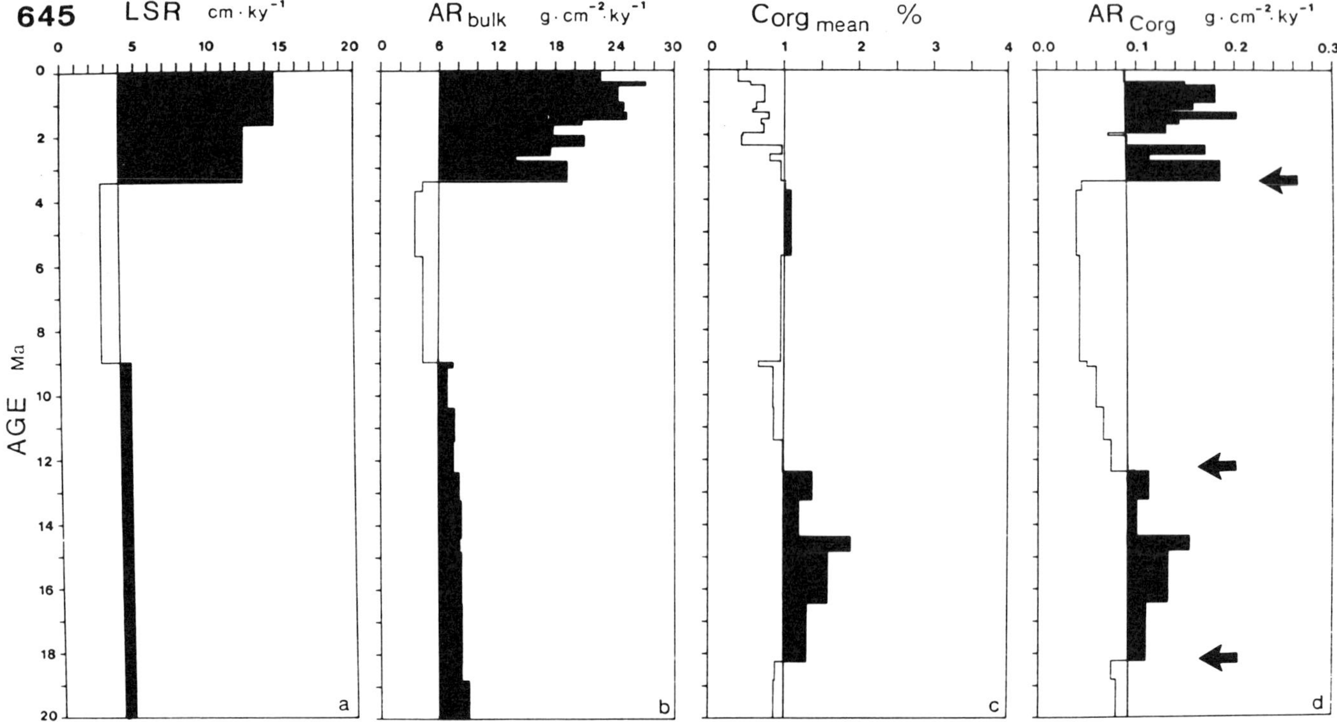

Figure 5. Analyses of sediments from Site 645. Linear sedimentation rates (a) and accumulation rates of bulk sediment (b) are based on data from Srivastava et al. (1987). Mean organic-carbon contents (c) are calculated for intervals where accumulation rates of bulk sediments have been predetermined. Accumulation rates of organic carbon (d) are calculated from (b) and (c).

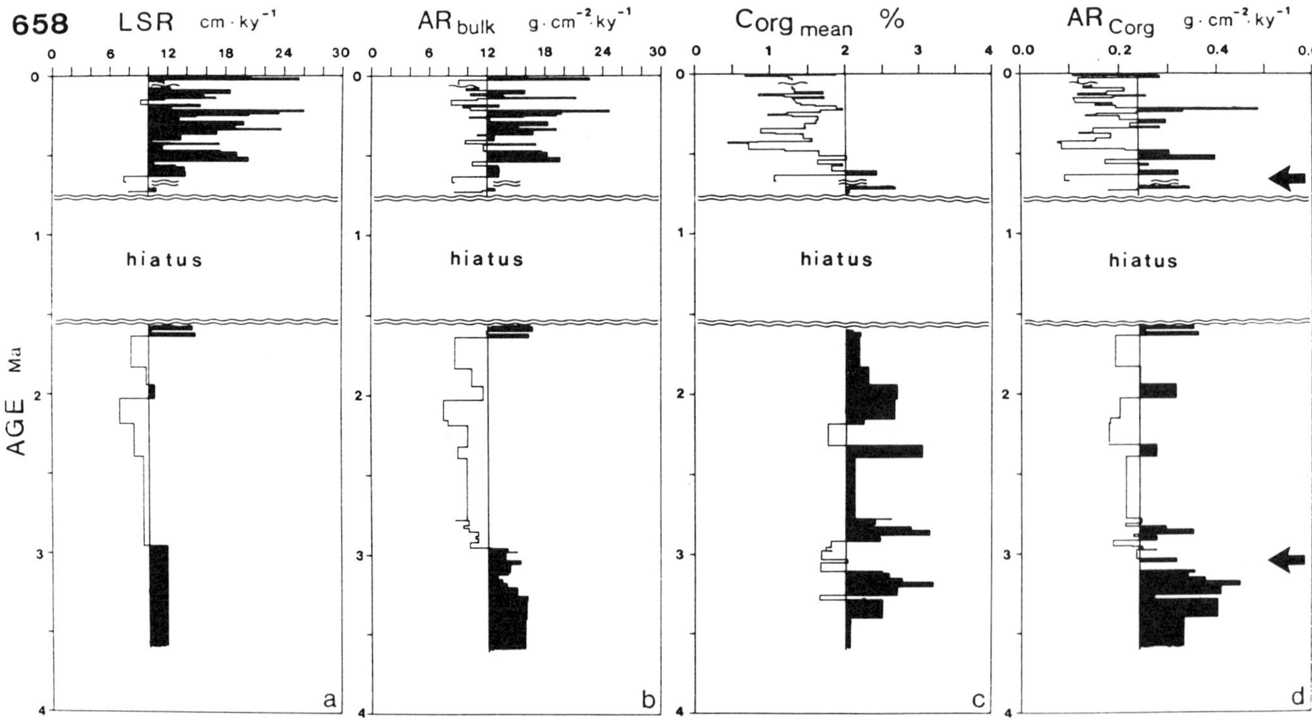

Figure 6. Analyses of sediments from Site 658. Linear sedimentation rates (a) and accumulation rates of bulk sediments (b) are based on data from Sarnthein and Tiedemann (1989) and Tiedemann et al. (1989). Mean organic-carbon contents (c) are calculated for intervals where accumulation rates of bulk sediments have been predetermined. Accumulation rates of organic carbon (d) are calculated from (b) and (c).

know not only the quantity but also the type of organic matter. Consequently, it is important to know how much of the organic carbon is marine and how much is of terrigenous origin. There are several techniques available to obtain such information (cf. Figure 4). The use of several different independent methods is desirable because factors other than the primary source may also have influenced organic-carbon composition. Furthermore, some quality parameters may change with increasing maturity. Thus, information concerning the thermal history of the organic matter is most helpful in interpretation of the data. According to maximum pyrolysis yield temperatures (T_{max} values of less than 435 °C) and vitrinite reflectance values ranging from 0.20 and 0.31%, the organic carbon at both Sites 645 and 658 is very immature (Figure 7).

Although the quantity of organic carbon appears similar at Sites 645 and 658 (Figures 5, 6), there are distinct differences in the types of organic matter.

In samples from Site 645 low hydrogen index values of less than 100 mgHC/gC_{org} dominate the upper 500 m (late Pliocene/Pleistocene), whereas below 500 m (middle Miocene to early Pliocene) a few values of more than 200 occur (Figure 8). This is thought to indicate that, during the late Pliocene/Pleistocene, terrigenous organic matter (kerogen type III) predominated. Consequently, in the middle Miocene to early Pliocene, marine organic carbon may have contributed a more significant proportion of the total organic carbon (Figure 8).

These Rock-Eval pyrolysis results are supported by data from kerogen microscopy. In general, terrigenous macerals (vitrinites, sporinites, and inertinites) are dominant at Site 645 (cf., Stein et al., 1989a). Only in the middle Miocene to early Pliocene are higher quantities of brightly fluorescing alginite observed (Figure 8).

Ratios of organic carbon to total nitrogen (C/N) that are mostly higher than 10 also corroborate the dominance of terrigenous organic carbon in sediments from Site 645 (Stein et al., 1989a), as well as GC/MS results of saturated hydrocarbons (ten Haven and Rullkötter, 1989). However, some lower C/N values may suggest admixtures of marine compounds in some intervals (Stein et al., 1989a). The occurrence of large amounts of marine organic debris in the middle Miocene to lower Pliocene sediments is, however, not supported by GC/MS data (ten Haven and Rullkötter, 1989). Further analysis is necessary to resolve this discrepancy.

Hydrogen index values at Site 658 are relatively high throughout. These vary between 200 and 400 mgHC/gC_{org} (Figure 9), suggesting that the organic matter is mainly composed of kerogen types II and III, i.e., a mixture of marine phytoplankton, zooplankton and/or bacteria, and terrigenous material.

Kerogen-microscopy analyses were performed on a selected set of samples from Site 658. The main macerals in the organic matter of these sediments are alginite, liptodetrinite, vitrinite, and inertinite

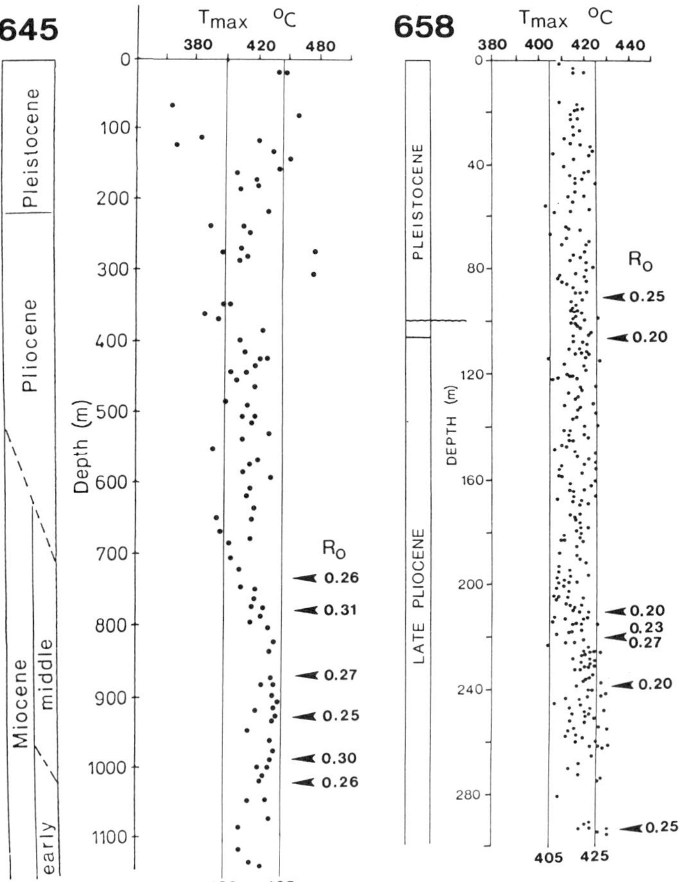

Figure 7. Maximum pyrolysis yield temperatures (T_{max} values) and vitrinite reflectance values for sediments from Sites 645 and 658 (Stein et al., 1989a, b; supplemented).

(Figure 9) (cf., Stein et al., 1989b). The amounts of alginite and liptodetrinite vary between 12 and 60% and 11 and 46%, respectively (Figure 9). Since (marine) alginite is the only liptinite identified in large quantities (i.e., terrigenous liptinites such as sporinite only occur in very minor amounts), the origin of the liptodetrinites is also interpreted as marine. Thus, the total marine components (alginite plus liptodetrinite) vary between 30 and 80%, with the minimum values (30 to 40%) occurring in sediments from the older part of the record between 3.6 and 3.1 Ma (Figure 9), i.e., during that time interval the terrigenous macerals vitrinite and inertinite may represent 60 to 70% of the total macerals.

Both hydrogen indices and maceral composition suggest a dominance of marine organic carbon in the Site 658 sediments. This is also supported by C/N ratios around 10, atomic H/C ratios (determined on kerogen concentrates) ranging from 1.1 to 1.5, and the dominance of marine-derived, long-chain, unsaturated ketones and steroids in the total lipid fraction (ten Haven et al., 1989; Stein et al., 1989b).

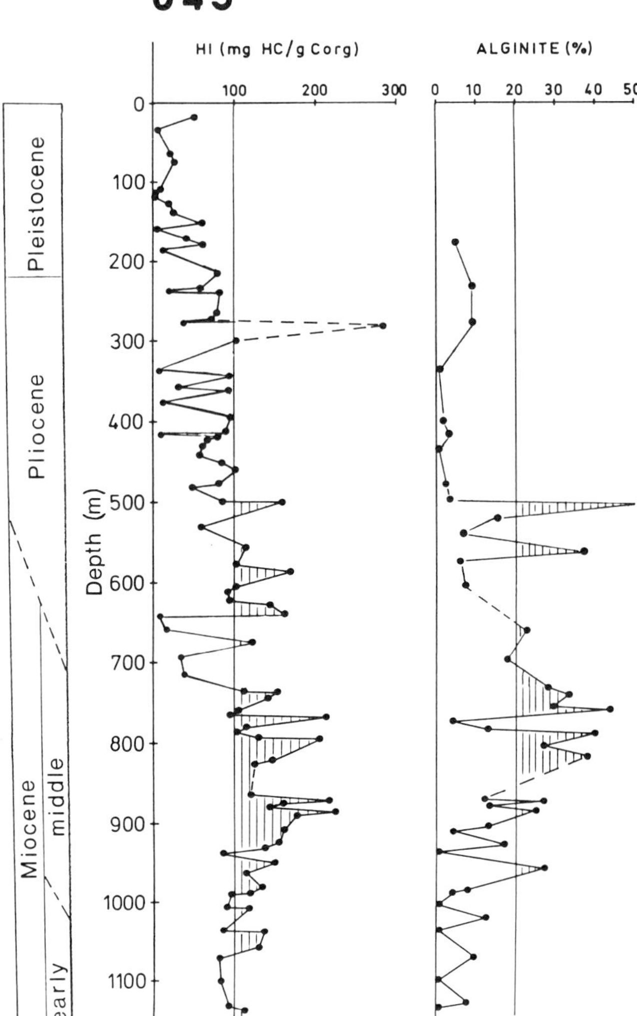

Figure 8. Hydrogen index values (mgHC/gC_{org}) and content of alginites in sediments from Site 645.

However, the results of the different methods show that terrigenous organic components may become important in sediments from Site 658.

DISCUSSION

Quantity of Organic Carbon and Paleoenvironment

Sediments with high organic-carbon contents (up to 4%) can accumulate in very different environments, such as small intracontinental basins dominated by terrigenous sediment supply (Baffin Bay; Figure 2) and high-productivity upwelling areas (e.g., off Northwest Africa; Figure 3). However, much higher values of up to 20% have been recorded from other upwelling areas (e.g., off Peru or Namibia; Calvert and Price, 1983; Reimers and Suess, 1983) and in the anoxic Holocene Black Sea (Degens and Ross, 1974). From these data it is obvious that depositional environments cannot be reconstructed simply on the basis of total organic-carbon values. Changes in the quantity of organic carbon may result from very different factors, i.e., variations in (1) supply of marine and/or terrigenous organic matter, (2) preservation of organic matter, or (3) dilution by nonorganic-carbon components.

Calculation of organic-carbon accumulation rates is one approach that may help to solve the problem since dilution effects can be excluded. Consequently, the record gives information about changes in supply/preservation of organic carbon. These accumulation rates for different environments may, however, be very similar although the depositional environments are very different. In Baffin Bay (Figure 5d) and the upwelling area off Northwest Africa (Figure 6d) accumulation rates of about 0.1 to 0.2 and 0.1 to 0.4 $gC \cdot cm^{-2} \cdot ky^{-1}$, respectively, are recorded. Thus, a comprehensive interpretation requires additional information concerning the composition of the organic matter. Based on quality parameters (see above), accumulation rates of organic carbon can be divided into marine and terrigenous portions. In Figures 10 and 11, this distinction is based on kerogen-microscopy data (cf., Littke et al., 1988; Stein et al., 1989a, b). These records can now be interpreted in terms of depositional environment. It should be pointed out, however, that the record of Site 658 especially is still preliminary, due to the incomplete data on organic-carbon composition (Figure 11). Thus, for detailed interpretation of the short-term fluctuations in organic-carbon accumulation, further data on quality of the organic matter are needed.

Terrigenous Organic-Carbon Input and Its Paleoenvironmental Significance

Since terrigenous organic matter that has survived long-distance transport to the ocean is relatively resistant to oxic decomposition (Tissot et al., 1979; Waples, 1983), changes in accumulation rates of such terrigenous material are primarily controlled by changes in rates of supply rather than variations in preservation. Consequently, the terrigenous organic-carbon accumulation rates may reveal information concerning contemporary paleoclimates. Furthermore, the size of organic particles derived from higher land plants is mainly a function of transport distance and transport energy, and can, therefore, also be used as a paleoenvironmental indicator (e.g. Stein et al., 1988).

At Site 645, accumulation rates of terrigenous organic matter varied between 0.03 and 0.20 $gC \cdot cm^{-2} \cdot ky^{-1}$. However, distinct long-term, as well as short-term, changes can be identified for the last 20 Ma (Figure 10). According to the long-term record,

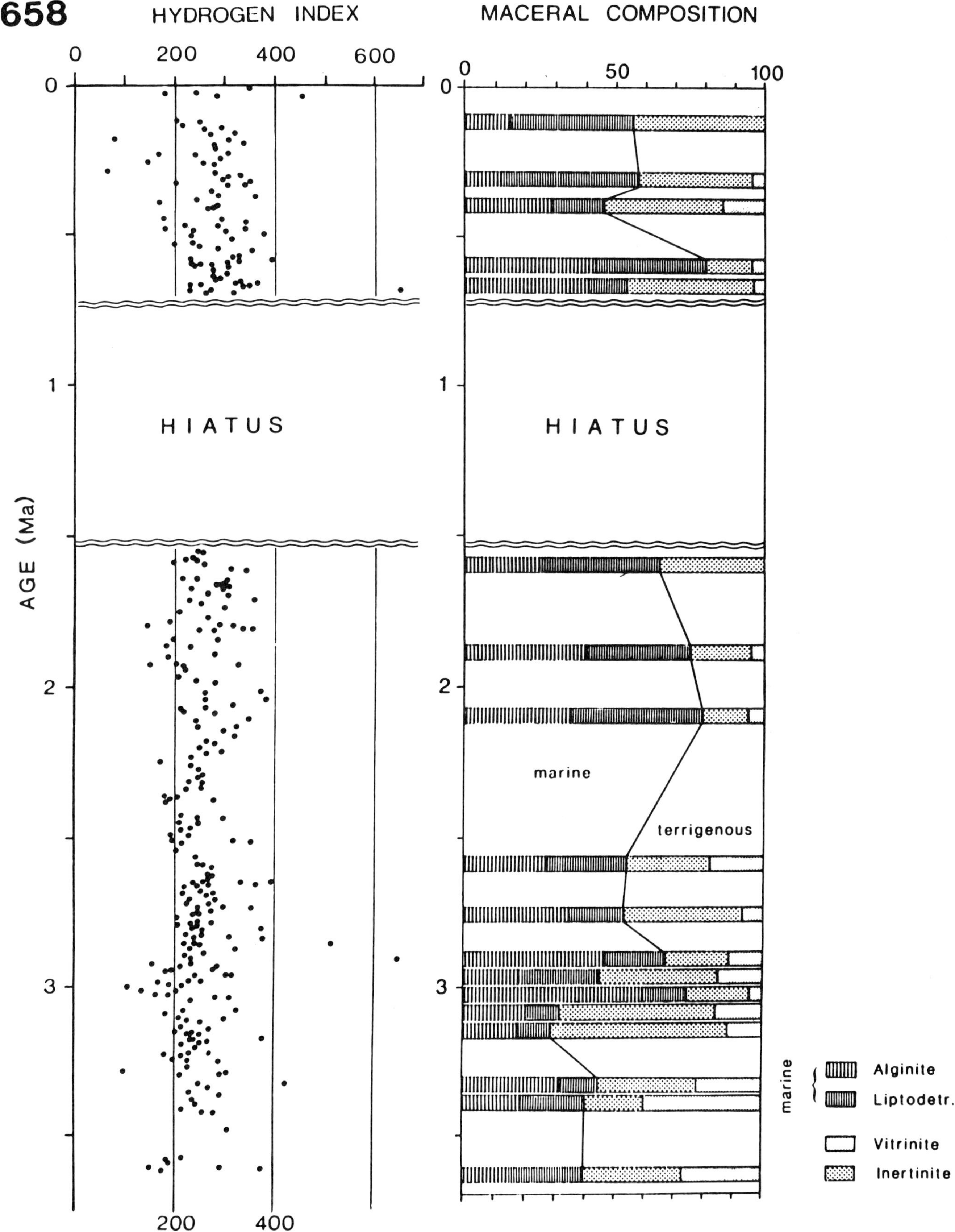

Figure 9. Hydrogen index values ($mgHC/gC_{org}$) and maceral composition in sediments from Site 658.

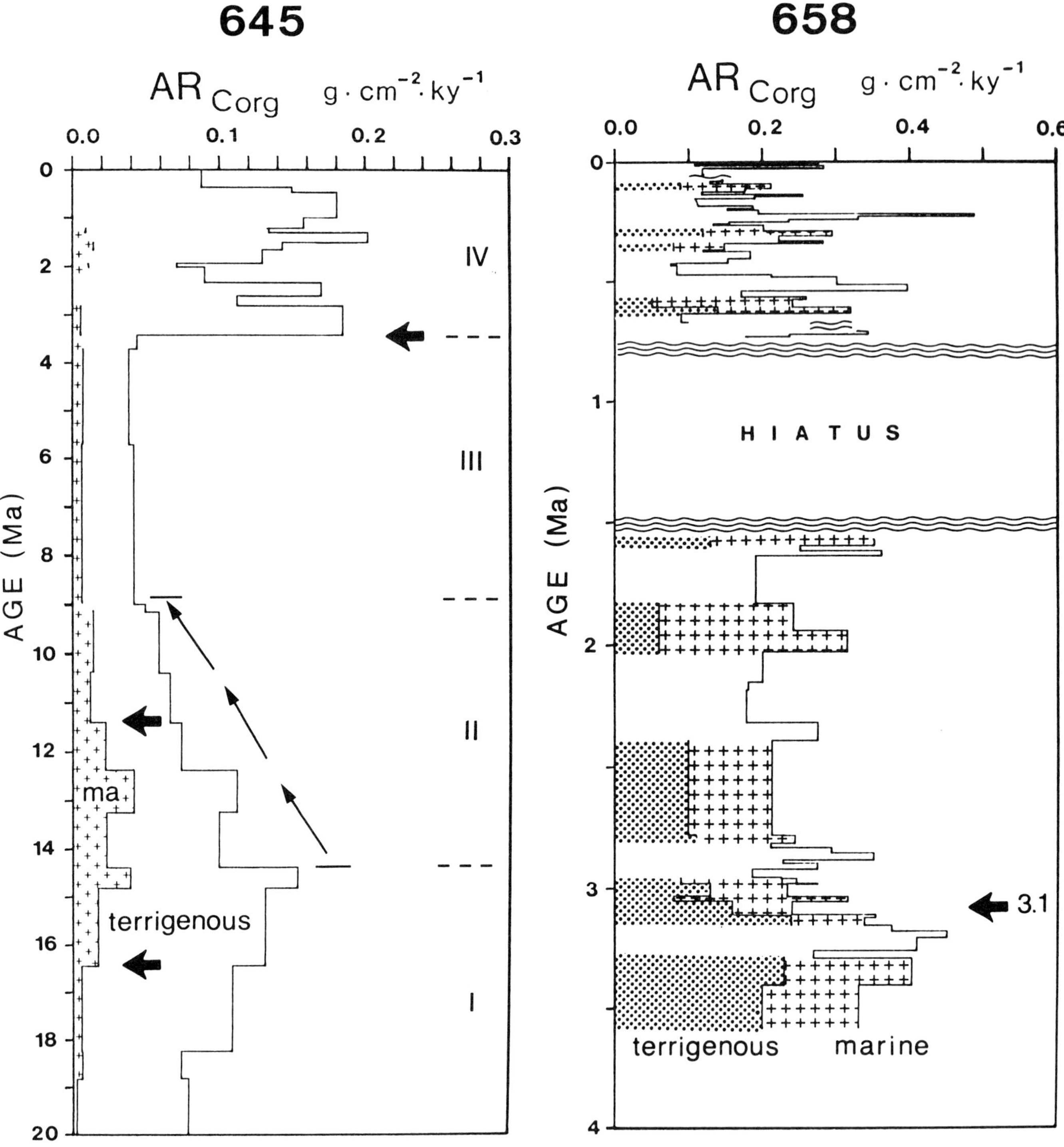

Figure 10. Accumulation rates of organic carbon at Site 645. Distinctions between marine and terrigenous organic-carbon fractions are based on kerogen-microscopy data (cf., Stein et al., 1989a). Crosses indicate quantity of marine organic carbon at Site 645.

Figure 11. Accumulation rates of organic carbon at Site 658. Distinctions between marine and terrigenous organic-carbon fractions are based on kerogen microscopy (cf., Stein et al., 1989b). Dots indicate terrigenous organic carbon, and crosses indicate marine organic carbon.

four major climatic intervals affected the Baffin Bay area.

Interval I (early to middle Miocene, 20 to about 14.5 Ma) was characterized by relatively high accumulation rates of terrigenous organic matter—around 0.1 gC · cm^{-2} · ky^{-1} (Figure 10). This is explained by dense vegetation cover and fluvial sediment supply from Baffin Island and/or Greenland areas. An increased fluvial supply is also supported by the large quantities of terrigenous clay, at least in the upper part of this interval (Srivastava et al., 1987; Stein et al., 1989a).

During Interval II (middle Miocene, 14.5 to 9 Ma), accumulation rates of terrigenous organic carbon fell significantly—to less than 0.04 gC · cm^{-2} · ky^{-1} (Figure 10). This decrease coincided with the global middle Miocene climatic cooling that ultimately produced the first glacial deposits (tillites) on Iceland and in South Alaska around 10 Ma (Denton and Amstrong, 1969; Mudie and Helgason, 1983) and the formation of major Antarctic ice caps (Kennett et al., 1977). Such a climatic deterioration also resulted in the widespread expansion of high-latitude steppe and tundra vegetation (Mudie and Helgason, 1983; Traverse, 1988) which may have caused the decrease in supply of terrigenous organic matter recorded at Site 645.

Interval III (late Miocene to early Pliocene, 9 to 3.4 Ma) was characterized by very low input of terrigenous organic matter (about 0.04 gC · cm^{-2} · ky^{-1}) (Figure 10), indicating continued dominance of cold climatic conditions and reduced vegetation cover in the Baffin Island/Greenland areas.

Near 3.4 Ma, the supply of terrigenous organic carbon increased by a factor of about 5 (Figure 10). Furthermore, this climatic Interval IV (3.4 Ma to Recent) was characterized by high-amplitude, short-term variations in climatically modulated terrigenous organic-carbon input between 0.07 and 0.20 gC · cm^{-2} · ky^{-1}. The important increase in supply of terrigenous organic carbon near 3.4 Ma is paralleled by an even more obvious increase in accumulation rates of terrigenous inorganic (i.e., siliciclastic and carbonate) material and the onset of major ice-rafted debris at Site 645 (Srivastava et al., 1987; Stein et al., 1989a). This may indicate that glacial erosion and fluvial (melt-water) outwash of the adjacent islands caused the synchronous changes in supplies of organic and inorganic matter. This eroded material was partly derived from pre-Quaternary sedimentary formations cropping out around Baffin Bay, as indicated by the abundance of reworked palynomorphs (Cretaceous and Tertiary) recorded in the upper 22 m of sediment at Site 645 (Hillaire-Marcel et al., 1989). Several high T_{max} values of 435 to 480°C, as determined by Rock-Eval pyrolysis (Figure 7), also support the local occurrence of reworked, more mature organic carbon in late Pliocene/Pleistocene sediments from Site 645.

According to the preliminary record of terrigenous organic-carbon input for Site 658 (Figure 11), the evolution of paleoclimate in Northwest Africa during the last 3.6 Ma can be separated into two intervals.

Maximum accumulation rates of terrigenous organic carbon of 0.25 gC · cm^{-2} · ky^{-1} were reached between 3.6 and 3.1 Ma (Figure 11). Maximum input of fine-grained siliciclastic materials (mainly clay minerals) occurred coevally (Tiedemann et al., 1989). Both maxima are interpreted as signals caused by humid climatic conditions that dominated in Northwest Africa during that time interval and caused an increased fluvial input of organic and inorganic terrigenous material (cf., Sarnthein et al., 1982; Stein, 1985). The climatic situation was probably similar to that suggested for Quaternary interglacials (Figure 12A).

Close to 3.1 Ma, input of terrigenous organic carbon—0.05 to 0.13 gC · cm^{-2} · ky^{-1} (Figure 11), as well as accumulation of terrigenous clays (Tiedemann et al., 1989)—distinctly decreased, suggesting expansion of more arid climatic conditions in Northwest Africa. This is also supported by a coeval increase in eolian dust supply recorded at the northern Site 397 (Figure 1B) (Stein, 1985). During the late Pliocene/Pleistocene, arid climatic conditions with reduced supplies of terrigenous organic matter (Figures 11, 12B) were dominant, although short humid periods occurred during interglacials throughout this time interval (Stein, 1985; Tiedemann et al., 1989). Further data on terrigenous organic-carbon input are necessary for reconstruction of these short-term glacial/interglacial climatic cycles.

Organic-Carbon Content and Surface-Water Productivity

Accumulation rates of marine organic carbon are mainly controlled by the rate of supply (i.e., the surface-water productivity) and/or the preservation rate (Müller and Suess, 1979; Tissot and Welte, 1984). Anoxic deep-water conditions (<0.1 ml O_2/1 H_2O), such as those found in the modern Black Sea (Degens and Ross, 1974), result in a strongly enhanced preservation rate of organic carbon. However, such a phenomenon was unlikely in Baffin Bay and off Northwest Africa, since bioturbation is common throughout the successions at Sites 645 and 658 (Srivastava et al., 1987; Ruddiman et al., 1988). Consequently, changes in accumulation rates at Sites 645 and 658 were probably caused by paleoproductivity variations.

Low accumulation rates of marine organic carbon suggest that Baffin Bay was an area of relatively poor productivity for much of the last 20 Ma (Figure 10). Only very small amounts of dinoflagellate cysts were generally found in Pleistocene sediments at Site 645, supporting low productivity for this interval (Hillaire-Marcel et al., 1989). A few samples, however, contained abundant dinocysts, suggesting short but distinct episodes of high phytoplankton productivity in Baffin Bay during the late Quaternary (Hillaire-Marcel et al., 1989). Furthermore, according to Williams (1986), it is likely that the western side of Baffin Bay has a relatively high surface-water productivity, at least seasonally, as today the Baffin Island Current sweeps along the fast ice edge and causes upwelling. Such "high-productivity events" are not documented in the long-term record of Site 645 (Figures 10, 13).

Based on the studies of Müller and Suess (1979) and Betzer et al. (1984), paleoproductivity in Baffin

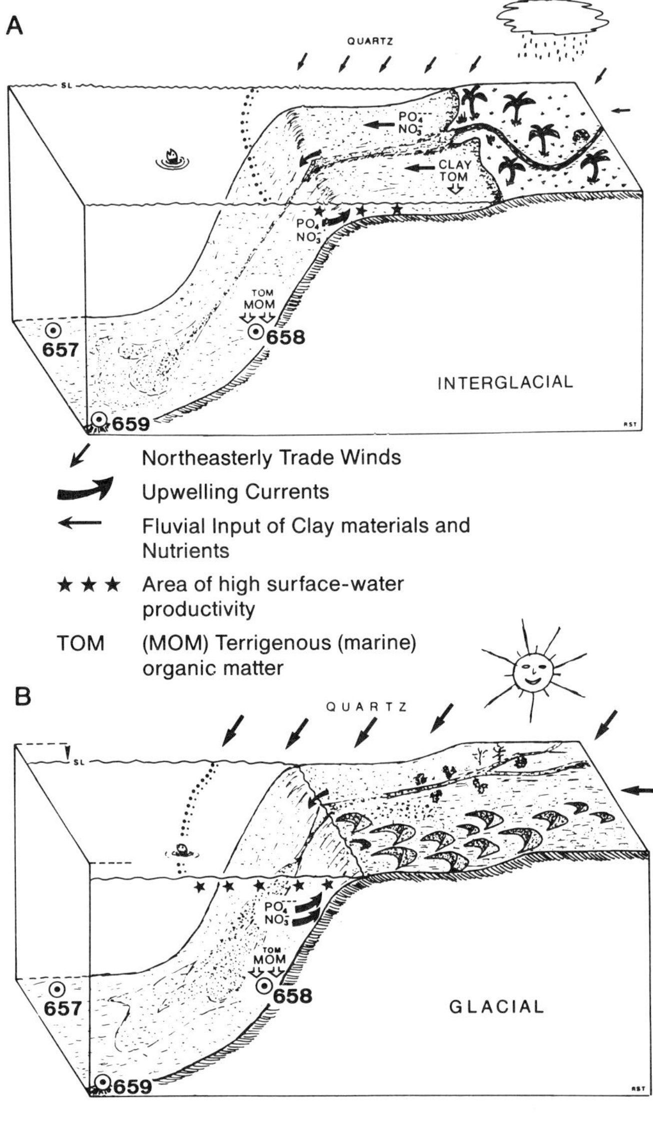

Figure 12. Depositional model for Pliocene-Pleistocene sediments at the northwestern continental margin of Africa. During interglacial stages (A), the fluvial input of clay materials and nutrients is increased because of the humid climate. Supply of terrigenous organic matter (TOM) is increased. Trade-wind intensity and coastal upwelling are also reduced. Glacial stages (B) result in increased areas of high surface-water productivity. The amount of terrigenous organic matter is decreased, and the fluvial input of clay materials and nutrients ceases because of the arid climate. Trade-wind intensity and the upwelling of coastal currents are increased. (For relationships between climate in Northwest Africa, trade-wind intensity, and upwelling intensity see Sarnthein et al., 1981, 1982; Stein, 1985; Tiedemann et al., 1989).

Bay was estimated from the marine organic-carbon content of sediments at Site 645 (for procedure, see the section on "Methods"; cf. Stein, 1986a). Use of this model demonstrates that periods with paleoproductivity rates of about 150 gC · m^{-2} · y^{-1} occurred during middle Miocene to early Pliocene times (Figure 13). Such an increase in surface-water productivity close to 16.5 Ma may have been caused by early middle Miocene inflow of cold (nutrient-rich ?) Arctic Ocean-derived water through the Nares Strait into Baffin Bay (Srivastava et al., 1981). Since this productivity increase is accompanied by an increased clay content in apparently coeval sediments (Srivastava et al., 1987; Stein et al., 1989a) an alternate source of enhanced productivity might have been intensified fluvial nutrient supply.

Productivity was probably relatively low (less than 100 gC · m^{-2} · y^{-1}) during late Pliocene/Pleistocene times, i.e., similar to that measured in western Baffin Bay today (Figure 13) (Romankevich, 1984). Since that time, it is thought that Baffin Bay has been seasonally covered by sea ice which has caused this decrease in average surface-water productivity (cf., Srivastava et al., 1987; Macko, 1989).

In areas where the organic carbon is dominated by marine components (such as in the upwelling area off Northwest Africa), an approximation of paleoproductivity can be determined through comparison of marine organic carbon content and bulk sedimentation rate (cf., Müller and Suess, 1979; Stein, 1986b, 1990). In the "organic-carbon/sedimentation-rate diagram" (Figure 14), the data from upwelling Site 658 fall into the "high-productivity field" A′, whereas the data points from northeast Atlantic non-upwelling sites lie in the "open-ocean field" A (Figure 14). The influence of high productivity at Site 658 is also supported by abundant biogenic opal recorded in sediments from this site (Mienert et al., 1988; Tiedemann et al., 1989).

For upwelling Site 658, a first preliminary record of productivity, based on the paleoproductivity equation discussed previously, is shown in Figure 15. The values (100–350 gC · m^{-2} · y^{-1}) are similar to those measured in the upwelling area off Northwest Africa today (150 to >325 gC · m^{-2} · y^{-1}) (Schemainda et al., 1975).

In the upper part of the record in Figure 15 (above the hiatus), it appears that the lower values are from interglacial intervals and the higher values from glacial intervals (according to the correlation with oxygen isotope stages from Sarnthein and Tiedemann, 1989). This observation agrees with the studies of Müller and Suess (1979) and Sarnthein et al. (1981, 1987) who suggest an increased productivity off Northwest Africa during glacial times due to enhanced upwelling activity caused by increased trade winds. Nevertheless, further paleoproductivity values from upwelling Site 658 are necessary to clarify whether this is a general phenomenon for glacial/interglacial cycles in the coastal upwelling area off Northwest Africa or whether there are variations between different

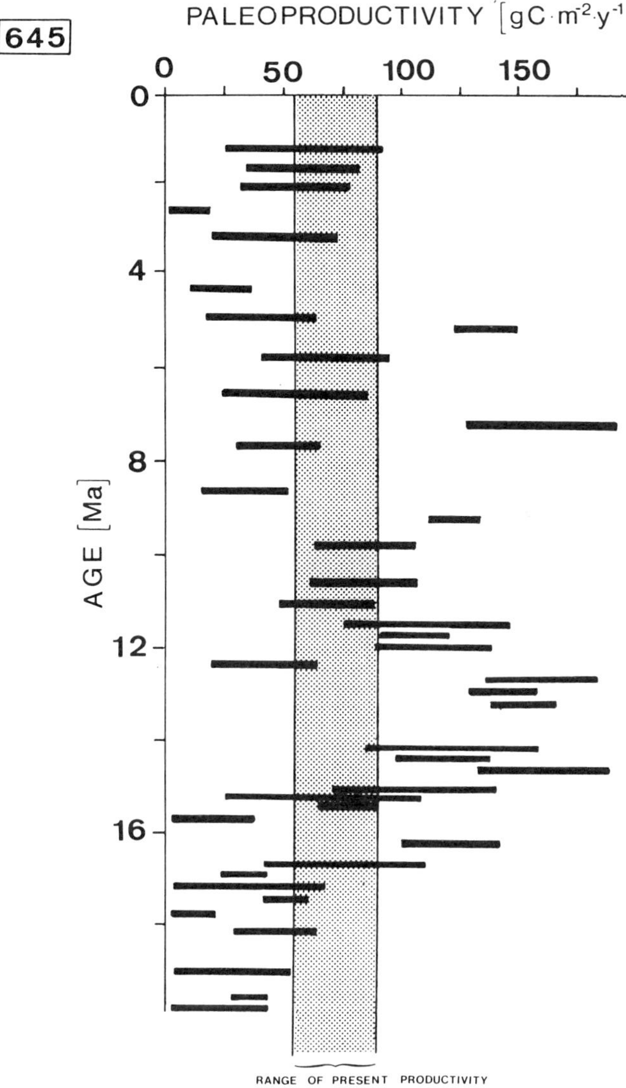

Figure 13. Estimates of paleoproductivity from marine organic-carbon content at Site 645 (for calculation procedure see section on "Methods"). Minimum and maximum productivity values (e.g., the width of the black bar) are based on minimum and maximum estimates of marine organic-carbon contents from kerogen microscopy (cf., Stein et al., 1989a). Range of present productivity in western Baffin Bay is taken from Romankevich (1984).

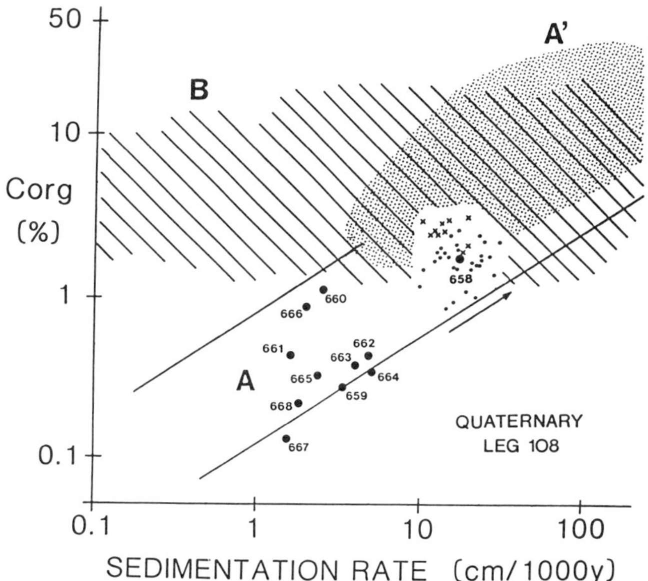

Figure 14. Correlation between organic-carbon contents and sedimentation rates for sediments from ODP-Leg 108. Site 658 is located in the coastal upwelling area off Northwest Africa; all other sites (659 to 668) are not influenced by coastal upwelling. The distinction between fields A, A′, and B is based on data derived from Holocene to Miocene sediments deposited in oxic open-ocean environments (field A), coastal upwelling high-productivity areas (field A′), and anoxic deep-water environments (field B) (according to Stein, 1986b). Large black dots represent mean values for Quaternary sediments at the numbered sites. In the detailed record for Site 658, using sedimentation rates based on oxygen isotope stratigraphy of Sarnthein and Tiedemann (1989), small black dots represent Quaternary intervals and crosses represent late Pliocene intervals.

glacials and interglacials. This is important because paleoproductivity values for the late Pliocene interval may have reached levels similar to those estimated for late Pleistocene glacials (Figure 15). During Pliocene times, when upwelling was less intense and humid climatic conditions dominated in Northwest Africa (Stein, 1985; Tiedemann et al., 1989), fluvial nutrient supply may have been the main factor causing such increased surface-water productivity (cf., Figure 12A, interglacial situation). According to the results above, nutrient supply affected by upwelling and fluvial input may have yielded similarly high productivity in coastal areas. This is also in agreement with the modern situation off Northwest Africa where surface-water productivity in the central upwelling area off Cap Blanc is very similar to the fluvially dominated region south of 12°N (Schemainda et al., 1975; cf., Diester-Haass, 1983).

CONCLUSIONS

The following conclusions are based on the investigation of sediments from Baffin Bay Site 645 and the coastal-upwelling Site 658 off Northwest Africa.

At both Site 645 and Site 658, sediments with high organic-carbon contents of more than 3% were deposited. Mass accumulation rates of total organic carbon are similar, ranging from 0.05 to 0.2 and from 0.08 to 0.4 gC $\cdot$ cm^{-2} $\cdot$ ky^{-1}.

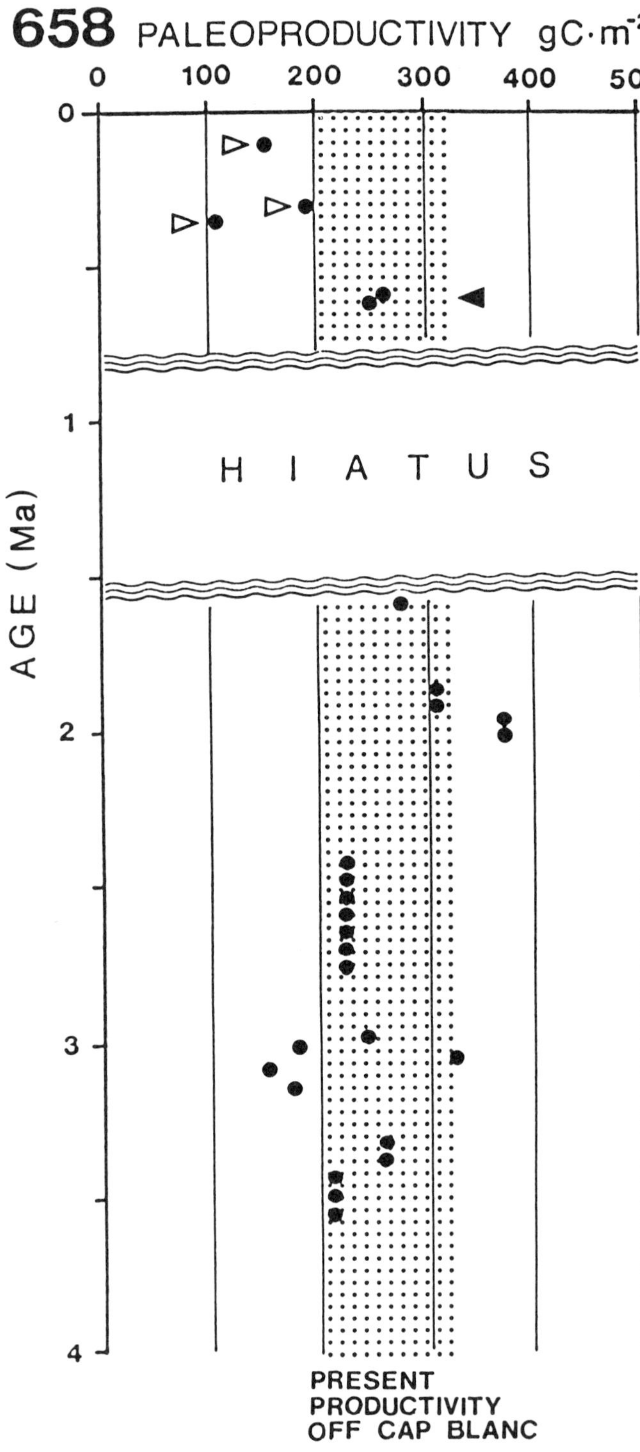

Figure 15. Estimates of paleoproductivity at Site 658, based on marine organic-carbon contents determined by kerogen microscopy (for calculation procedures, see section on "Methods"). Solid triangles indicate data from glacial intervals; open triangles indicate data from interglacial intervals (according to oxygen-isotope stratigraphy of Sarnthein and Tiedemann, 1989). The dotted bar represents the range of present productivity off Cap Blanc (Schemainda et al., 1975).

Distinct differences in the type of organic matter are obvious, reflecting the very different depositional environments at the two sites. At Site 645, located in a narrow intracontinental basin, supply of terrigenous organic carbon was dominant throughout the last 20 Ma. Two distinct maxima can be identified: (1) a middle Miocene maximum which may reflect dense vegetation cover and fluvial sediment supply from adjacent islands, decreasing during the late Miocene and early Pliocene through global climatic deterioration, and (2) a late Pliocene/Pleistocene maximum which was probably caused by glacial erosion and meltwater outwash. Significant amounts of marine organic carbon were deposited in Baffin Bay during middle Miocene times, suggesting higher productivity triggered by inflow of cold (nutrient-rich) Arctic water masses.

At upwelling Site 658, marine organic carbon was dominant during the last 3.6 Ma, probably as a consequence of high oceanic productivity. Estimated paleoproductivity values are similar to those measured off Northwest Africa today. Based on a limited data set, glacial productivity appears to have been higher than that estimated for interglacial intervals. Significant amounts of terrigenous organic matter were additionally deposited at Site 658, with maximum input between 3.6 and 3.1 Ma, suggesting dominantly humid climatic conditions in Northwest Africa prior to 3.1 Ma. During these humid intervals, fluvial nutrient supply may have also caused increased surface-water productivity off Northwest Africa.

The foregoing results demonstrate that paleoenvironmental interpretation of organic-carbon data suites requires information concerning both the quantity *and* type of organic matter.

ACKNOWLEDGMENTS

For technical assistance, we thank D. Brosinsky, M. Dersch, U. Mann, and R. Stax. We gratefully thank Drs. Burchell and Curial for their numerous constructive suggestions for improvement of the manuscript. Financial support by the Deutsche Forschungsgemeinschaft (Grant No. STE 412/1) is gratefully acknowledged.

REFERENCES

Arthur, M. A., W. E. Dean, and D. A. V. Stow, 1984, Models for the deposition of Mesozoic-Cenozoic fine-grained organic-carbon-rich sediment in the deep sea, *in* D. A. V. Stow and D. J. W. Piper, eds., International workshop on deep-sea, fine-grained sediments: Geological Society of London Special Publication 15, p. 527–560.

Berger, W. H., V. Smetacek, and G. Wefer, 1989, Ocean productivity and paleoproductivity—An overview, *in* W. H. Berger et al., eds., Productivity of the ocean: Present and past: New York, John Wiley, p. 1–34.

Betzer, P. R., W. J. Showers, E. A. Laws, C. D. Winn, G. R. Ditullio, and P. M. Kroopnick, 1984, Primary productivity and particle fluxes on a transect of the equator at 153°W in the Pacific Ocean: Deep-Sea Research, v. 31, p. 1-11.

Calvert, S. E., and N. B. Price, 1983, Geochemistry of Namibian Shelf sediments, *in* E. Suess, and J. Thiede, eds., Coastal Upwelling, Its Sediment Record, Part A: New York, Plenum Press, p. 337-375.

Degens, E. T., and D. A. Ross, 1974, The Black Sea—geology, chemistry and biology. AAPG Memoir 20, 633 p.

Demaison, G. T., and G. T. Moore, 1980, Anoxic environments and oil source bed genesis: AAPG Bulletin, v. 64, p. 1179-1209.

Denton, G. H., and R. L. Amstrong, 1969, Miocene-Pliocene glaciations in southern Alaska: American Journal of Science, v. 267, p. 1121-1142.

Diester-Haass, L., 1983, Differentiation of high oceanic fertility in marine sediments caused by coastal upwelling and/or river discharge off northwest Africa during the late Quaterary, *in* J. Thiede, and E. Suess, eds., Coastal upwelling, its sediment record, Part B: New York, Plenum Press, p. 399-419.

Durand, B., 1980, Kerogen: Paris, Editions technip, 513 p.

Espitalié, J., J. L. Laporte, M. Madec, F. Marquis, P. Leplat, J. Paulet, and A. Boutefeu, 1977, Methode rapide de characterisation des roches-mere, de leur potential petrolier et de leur degre d'evolution: Revue del' Institut Francais Petrole, v. 32, p. 23-42.

Faugères, J. C., P. Legian, N. Maillet, M. Sarnthein, and R. Stein, 1989, Turbidite characteristics and distribution in Neogene deposits at Site 657 (Leg 108, Cap Blanc Continental Rise, NW Africa): Evidence of sources and climatic variations, *in* W. Ruddiman, M. Sarnthein, et al., eds., Proceedings of the Ocean Drilling Program, Scientific Results, 108: College Station, Texas, Ocean Drilling Program.

ten Haven, H. L., and J. Rullkötter, 1989, Unusual terrigenous triterpenoid biological markers in Baffin Bay sediments, *in* S. Srivastava, et al., eds., Proceedings of the Ocean Drilling Program, Scientific Results, 105: College Station, Texas, Ocean Drilling Program.

ten Haven, H. L., J. Rullkötter, and R. Stein, 1989, Preliminary analysis of extractable lipids in sediments from the eastern North Atlantic (ODP-Leg 108): Comparison of a coastal upwelling area (Site 658) and a non-upwelling area (Site 659), *in* W. Ruddiman, M. Sarnthein, et al., eds., Proceedings of the Ocean Drilling Program, Scientific Results, 108: College Station, Texas, Ocean Drilling Program.

Hillaire-Marcel, C., A. de Vernal, A. Aksu, and S. Macko, 1989, High resolution isotopic and micropaleontological studies of upper Pleistocene sediments at ODP-Site 645, Baffin Bay, *in* S. Srivastava, et al., eds., Proceedings of the Ocean Drilling Program, Scientific Results, 105: College Station, Texas, Ocean Drilling Program.

Kennett, J., et al., 1977, Cenozoic paleoceanography in the southwest Pacific Ocean, Antarctic glaciation, and the development of the circum-Antarctic current, *in* J. Kennett, et al., Initial Reports of the Deep Sea Drilling Program DSDP, v. 29, Washington D.C., p. 1155-1169.

Littke, R., D. R. Baker, and D. Leythaeuser, 1988, Microscopic and sedimentologic evidence for the generation and migration of hydrocarbons in Toarcian source rocks of different maturities: Organic Geochemistry, v. 13, p. 549-559.

Macko, S., 1989, Stable isotope organic geochemistry on sediments from the Labrador Sea (Sites 646, 647) and Baffin Bay (Site 645)—ODP-Leg 105, *in* S. Srivastava, et al., eds., Proceedings of the Ocean Drilling Program, Scientific Results, 105: College Station, Texas, Ocean Drilling Program.

Marko, J. R., J. R. Birch, and M. A. Wilson, 1982, A study of long-term, satellite-tracked iceberg drifts in Baffin Bay and Davis Strait: Arctic, v. 35, p. 234-240.

Mienert, J., R. Stein, P. Schultheiss, and Shipboard Scientific Party., 1988, Relationship between grain density and biogenic opal in sediments from Sites 658 and 660, *in* W. Ruddiman, M. Sarnthein, et al., eds., Proceedings of the Ocean Drilling Program, Scientific Results, 108: College Station, Texas, Ocean Drilling Program, p. 1047-1053.

Mudie, P. J., and J. Helgason, 1983, Palynological evidence for Miocene climatic cooling in eastern Iceland about 9.8 Myr ago: Nature, v. 303, p. 689-692.

Müller, P. J., and E. Suess, 1979, Productivity, sedimentation rate and sedimentary organic matter in the oceans. I.—Organic matter preservation: Deep-Sea Research, v. 26, p. 1347-1362.

Reimers, C. E., and E. Suess, 1983, Late Quaternary fluctuations in the cycling of organic matter off central Peru: A proto-kerogen record, *in* E. Suess, and J. Thiede, eds., Coastal upwelling, its sediment record, Part A: New York, Plenum Press, p. 497-526.

Romankevich, E. A., 1984, Geochemistry of organic matter in the ocean: Heidelberg, Springer-Verlag, 334 p.

Ruddiman, W., et al., 1988, Proceedings of the Ocean Drilling Program, Initial Reports, 108: College Station, Texas, Ocean Drilling Program.

Sarnthein, M., B. Koopmann, K. Wolter, and U. Pflaumann, 1981, Glacial and interglacial wind regimes over the eastern subtropical Atlantic and Northwest Africa: Nature, v. 293, p. 193-196.

Sarnthein, M., J. Thiede, U. Pflaumann, H. Erlenkeuser, D. Fütterer, B. Koopmann, H. Lange, and E. Seibold, 1982, Atmospheric and oceanic circulation patterns off Northwest Africa during the past 25 million years, *in* U. von Rad, K. Hinz, M. Sarnthein, and E. Seibold, eds., Geology of the northwest African continental margin: Heidelberg, Springer, p. 545-604.

Sarnthein, M., and R. Tiedemann, 1989, Towards a high-resolution stable isotope stratigraphy of the last 3.4 Ma, ODP-Sites 658 and 659 off Northwest Africa, *in* W. Ruddiman, et al., eds., Proceedings of the Ocean Drilling Program, Scientific Results, 108: College Station, Texas, Ocean Drilling Program.

Sarnthein, M., K. Winn, and R. Zahn, 1987, Paleoproductivity of oceanic upwelling and the effect on atmospheric CO_2 and climatic change during deglaciation times, *in* W. H. Berger, and L. D. Labeyrie, eds., Abrupt climatic change: D. Reidel Publishing Company, p. 311-337.

Schemainda, R., D. Nehring,, and S. Schulz, 1975, Ozeanologische Untersuchungen zum Produktionspotential der nordwestafrikanischen Wasserauftriebsregion 1970-1973: Geod. Geophys. Veröff., Reihe IV, 16: 85 p.

Schlanger, S. O., and H. C. Jenkyns, 1976, Cretaceous oceanic anoxic event: Causes and consequences: Geol. en Mijnbouw, 55: p. 179-184.

Shaffer, G., 1976, A mesoscale study of coastal upwelling variability off NW-Africa. "Meteor" Forschungsergebn., Reihe A, 17: p. 33-70.

Srivastava, S., et al., 1987, Proceedings of the Ocean Drilling Program, Initial Reports, Pt. A, 105: College Station, Texas, Ocean Drilling Program.

Srivastava, S., R. K. H. Falconer, and B. MacLean, 1981, Labrador Sea, Davis Strait, Baffin Bay: Geology and geophysics—a review, *in* J. W. Kerr, et al., eds., Geology of the North Atlantic borderlands, Canadian Society of Petroleum Geologists Memoir, v. 8, p. 333-398.

Stach, E., M.-TH. Mackowsky, M. Teichmüller, G. H. Taylor, D. Chandra, and R. Teichmüller, 1982, Stach's textbook of coal petrology: Berlin, Borntraeger Verlag, 535 p.

Stein, R., 1985, Late Neogene changes of paleoclimate and paleoproductivity off northwest Africa (DSDP- Site 397): Palaeogeography, Palaeoclimatology, Palaeoecology, v. 49, p. 47-59.

Stein, R., 1986a, Surface-water paleo-productivity as inferred from sediments deposited in oxic and anoxic deep-water environments of the Mesozoic Atlantic Ocean, *in* E. T. Degens, et al., eds., Biochemistry of black shales: Mitt. Geol. Paläont. Inst. Univ. Hamburg, 60: p. 55-70.

Stein, R., 1986b,. Organic carbon and sedimentation rate—Further evidence for anoxic deep-water conditions in the Cenomanian/Turonian Atlantic Ocean: Marine Geology, v. 72, p. 199-209.

Stein, R., 1990, Organic carbon content/sedimentation rate relationship and its paleoenvironmental significance for marine sediments: Geo-Mar. Lett.

Stein, R., R. Littke, R. Stax, and D. H. Welte, 1989a, Quantity, provenance, and maturity of organic matter at ODP-Sites 645, 646, and 647: Implications for reconstruction of paleoenvironments in Baffin Bay and Labrador Sea during Cenozoic times, *in* S. Srivastava, et al., eds., Proceedings of the Ocean Drilling Program, Scientific Results, 105: College Station, Texas, Ocean Drilling Program.

Stein, R., H. L. ten Haven, R. Littke, J. Rullkötter, and D. H.

Welte, 1989b, Accumulation of marine and terrigenous organic carbon at upwelling-Site 658 and non-upwelling-Sites 657 and 659: Implications for the reconstruction of paleoenvironments in the eastern subtropical Atlantic through late Cenozoic times, *in* W. Ruddiman, et al., eds., Proceedings of the Ocean Drilling Program, Scientific Results, 108: College Station, Texas, Ocean Drilling Program.

Stein, R., J. Rullkötter, R. Littke, R. G.Schaefer, and D. H. Welte, 1988, Organofacies reconstruction and lipid geochemistry of sediments from the Galicia Margin, Northeast Atlantic (ODP Leg 103), *in* G. Boillot, et al., eds., Proceedings of the Ocean Drilling Program, Scientific Results, 103: College Station, Texas, Ocean Drilling Program, p. 567-585.

Stein, R., J. Rullkötter, and D. H. Welte, 1986, Accumulation of organic-carbon-rich sediments in the Late Jurassic and Cretaceous Atlantic Ocean—A synthesis: Chemical Geology, v. 56, p. 1-32.

Tchernia, P., 1982, Descriptive regional oceanography: Pergamon Marine Series, v. 3, 253 p.

Tiedemann, R., M. Sarnthein, and R. Stein, 1989, Climatic changes in the western Sahara: Aeolo-marine sediment record of the last 8 my (ODP-Sites 657-661), *in* W. Ruddiman, et al., eds., Proceedings of the Ocean Drilling Program, ODP, Scientific Results, 108: College Station, Texas, Ocean Drilling Program.

Tissot, B., G. Deroo, and J. P. Herbin, 1979, Organic matter in Cretaceous sediments of the North Atlantic: Contribution to sedimentology and paleoceanography, *in* M. Talwani, W. Hay, and W. B. R. Ryan, eds., Deep drilling results in the Atlantic Ocean: Continental margin and paleoenvironment: M. Ewing Series, 3, p. 362-374.

Tissot, B. and D. H. Welte, 1984, Petroleum formation and occurrence: Heidelberg, Springer Verlag, 699 p.

Tomczak, M., and P. Hughes, 1980, Three-dimensional variability of water masses and currents in the Canary current upwelling region: "Meteor" Forsch. Ergeb., A, 21, p. 1-24.

Traverse, A., 1988, Palynology: London, Unwin Hyman, 600 p.

van Andel, T. H., G. R. Heath, and T. C. Moore, 1975, Cenozoic history and paleoceanography of the central equatorial Pacific Ocean: Geological Society of America Memoir 143, 134 p.

Waples, D. W., 1983, Reappraisal of anoxia and organic richness, with emphasis on Cretaceous of North Atlantic: AAPG Bulletin, v. 61, p.963-978.

Williams, K. M., 1986, Recent Arctic marine diatom assemblages from bottom sediments in Baffin Bay and Davis Strait: Marine Micropaleontology, v. 10, p. 327-342.

Wireline Source-Rock Evaluation in the Paris Basin

Susan L. Herron
Schlumberger-Doll Research
Ridgefield, Connecticut, U.S.A.

Louis Le Tendre
Elf Aquitaine Offshore Asia b. v.
Bandar Seri Bagawan
Brunei Darussalam

Source-rock identification and quantification using wireline logs and interpretation techniques are examined for a well in the Paris Basin which penetrates the organic-rich lower Toarcian shale. An 18-m interval of the source rock was cored and logged with the Gamma Ray Spectrometry Tool (GST*) and Litho-Density* tool in order to determine the total organic carbon content (TOC). The GST carbon/oxygen ratio and calcium log are incorporated into a new interpretation which treats the formation as a mixture of solid, fluid, and kerogen. This analysis yields a carbon log with concentrations ranging from 4 to 11% and a total organic carbon log ranging from 2 to 6.5%, both of which are in excellent agreement with core data. Of four discriminant functions previously proposed for source-rock identification from conventional logs, only one correctly identifies the interval as a source rock; two others only identify intervals with total organic carbon (TOC) greater than 6%.

INTRODUCTION

A total organic carbon (TOC) log has been derived from geochemical wireline measurements for the Toarcian interval of a well in the Paris Basin. The ability to quantitatively evaluate TOC from geochemical logs has been demonstrated for shales which have generally low organic contents and which contain only small amounts of inorganic carbon in the form of carbonate minerals (Herron, 1986). The technique, which uses the carbon/oxygen ratios and calcium concentrations from the Gamma Ray Spectrometry Tool (GST), has now been improved to provide more accurate carbon estimations for formations where the organic carbon content exceeds about 3 wt %. The use of these geochemical logs represents a departure from the more traditional methods which attempt to identify source rock by correlating or calibrating TOC with some combination of gamma ray, density, neutron, sonic, or resistivity measurements. The source-rock interval investigated here is also fully cored. The primary goal of this study is to use the log data to compute a total organic carbon log and to verify the results with core analyses. A secondary goal is to compare the results with existing log-interpretation procedures, and for this the techniques of Meyer and Nederlof (1984) are chosen because although they are only semiquantitative, they represent the best published techniques that do not require calibration with core data.

GEOLOGICAL SETTING

The well which was selected for this study is located southeast of Paris in the Paris Basin, as shown in Figure 1. The Paris Basin is a large, semicircular intracratonic basin with a radius of approximately 250 km (Espitalié et al., 1987). It is bounded by Hercynian massifs, with the Armorican massif in the west and southwest, the Central massif in the south, the Vosges in the east, and the Ardennes

* Registered trademark of Schlumberger.

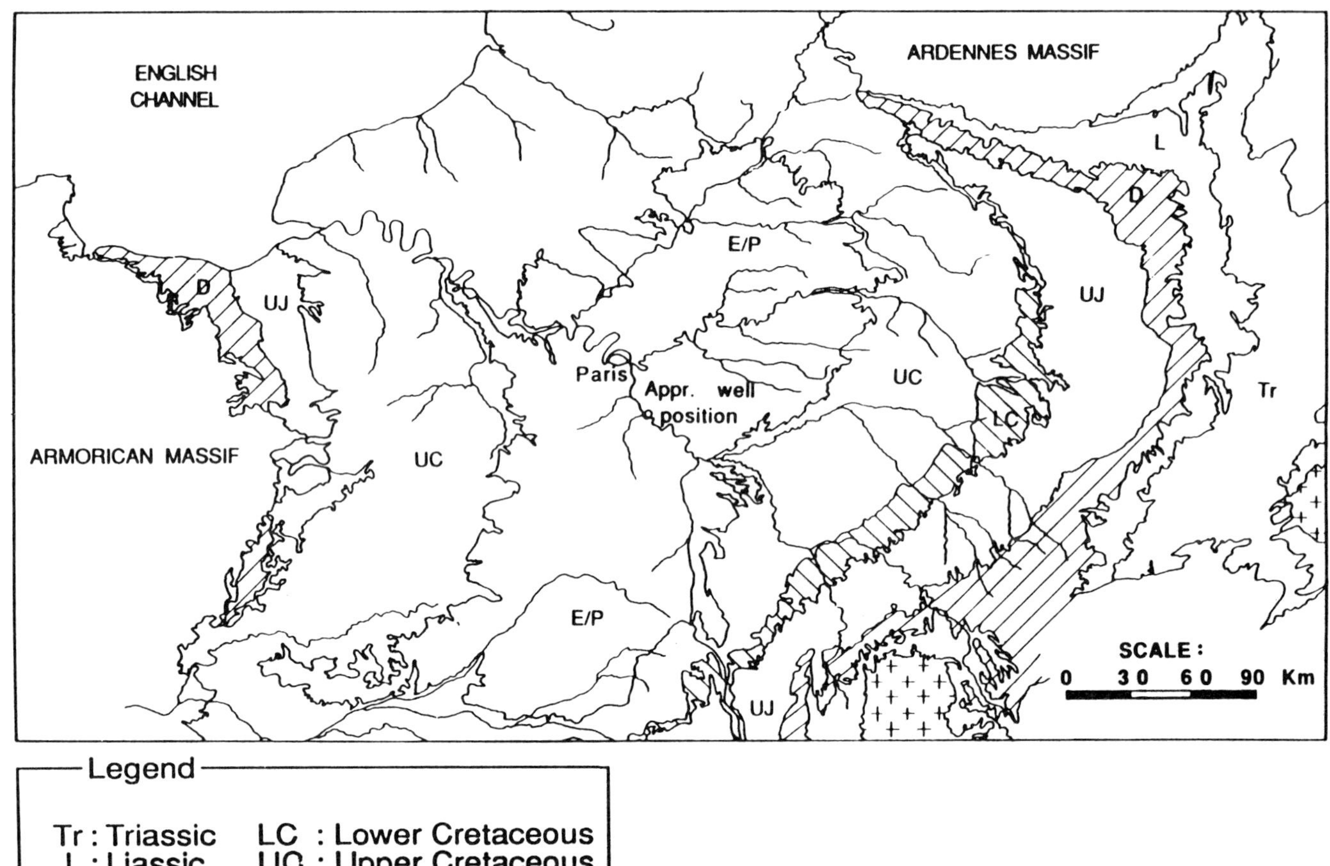

Figure 1. Location of study well in the Paris Basin southeast of Paris.

massif in the northeast. Geologic formations crop out in circular bands, decreasing in age from Triassic in the outer border to Tertiary sediments in the center (Figure 1). The maximum thickness of post-Paleozoic sediments nowhere exceeds 3500 m.

During the early stages of basin formation (Permian-Carboniferous phases of the Hercynian orogeny), important normal faulting occurred, accompanied by coarse detrital sedimentation. Basin extension continued during the Triassic, with transgressive deposition taking place from the east (Buntsandstein formations), followed by deposition of regressive evaporitic deposits of Muschelkalk and Keuper ages.

During Liassic time, transgressions occurred in a variably subsiding basin. In the early Liassic, strong subsidence took place in the area of the Ardennes block, and more than 150 m of mainly shaly sediments were deposited. On the northern edge of this block, sandy deposits are common, whereas in the southeast, shaly deposits are replaced by carbonate platform deposits.

In the middle Liassic, an important transgression occurred on the western edge of the basin resulting in the deposition of conglomeratic, oolitic, and sandy limestones. The Ardennes border is still an area of detrital sedimentation, while shaly deposits dominate in the center of the basin.

In the late Liassic (Toarcian), maximum subsidence occurred along a north-south axis in the center of the basin. Organic-rich argillaceous sediments were deposited in shallow water under anoxic conditions, with silty interbeds becoming more common in the middle and late Toarcian. Figure 2 presents an isopach map of the lower Toarcian interval in the relevant part of the Paris Basin.

Both the upper and lower Toarcian intervals are considered to have a very good source-rock potential. They are characterized by very high TOC contents, up to 10%; high H/C ratios, 1 to 1.4; and high hydrogen index values of greater than 500. The lower Toarcian "Shistes Carton" kerogen has been chosen as the reference for organic matter of marine origin, or type II, kerogen as described by Tissot et al. (1974).

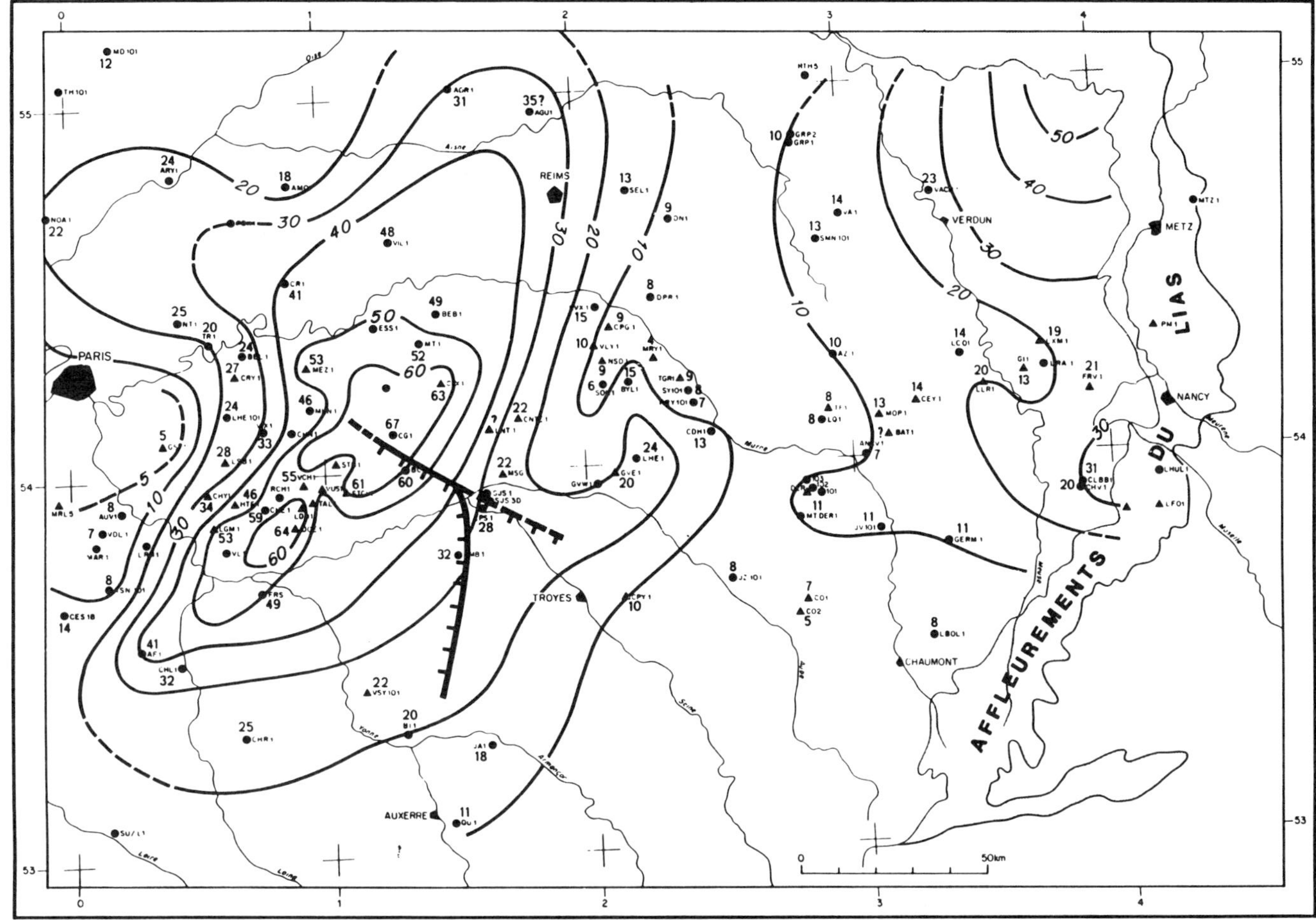

Figure 2. Isopach map of the lower Toarcian in part of the Paris Basin. Contour interval, 10 m. Adapted from Espitalié et al. (1987).

LOGGING AND CORE ANALYSIS

Eighteen meters of whole core were recovered from the study well, from 1722 to 1740 m, using a fiberglass-lined core barrel; 100% core recovery was achieved. The cored interval was logged with the GST tool in both stationary and continuous modes and also the Formation MicroScanner (FMS*) tool, in addition to running a full suite of conventional logs.

The core was dispatched to Elf Aquitaine laboratories where continuous gamma-gamma density and natural radioactivity measurements were made in order to precisely depth-match the core with log measurements (Figure 3). Although the core was immediately shipped to the laboratory, shale transformations occurred rapidly, causing severe disking of the material which produced the "noise" on the high-resolution density measurement made over the core.

This cored interval of early Toarcian and Domerian age consists mainly of slightly silty, calcareous shales (Figure 4) which generally contain 10 to 20% quartz and 10 to 20% calcite. Pyrite is commonly present, as are thin interbeds of bioclastic limestone and shell fragments. The base of the core consists of more massively bedded shaly limestone of Domerian age. The relatively abrupt change between the Domerian and Toarcian units is clearly visible in both the core and the FMS images, thus providing an extra depth-matching opportunity.

The core-sampling scheme was carefully designed to optimize the agreement between core analyses and log measurements. After the core was precisely depth-matched with log measurements, it was slabbed continuously and divided into 30 60-cm-long samples which were centered on depths where GST stationary measurements were taken. The samples were ground to 40 μm and split for organic and inorganic geochemical analyses. For the purpose of this experiment, this sampling technique proved to be far superior to simply using core plugs for the analyses

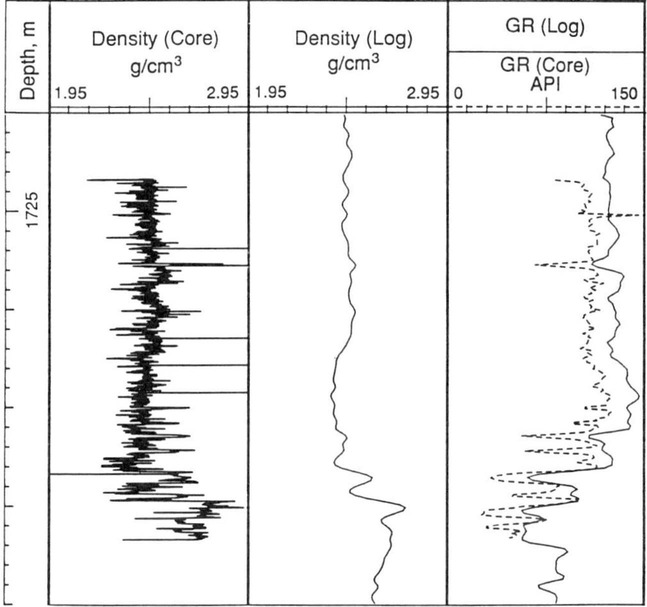

Figure 3. Comparison of open-hole logs and logs run on the core after depth-matching. "Noise" on the HR continuous gamma-gamma density measurement is due to the severe disking that took place between the time the core was recovered at the rig site and the time measurements were made. Open-hole gamma ray values have not been environmentally corrected, and the mud used contained KCl.

because it provides samples that are more representative of the formation in terms of what the GST tool measures. For a separate study, composite samples were prepared by taking a 1-cm slice every 4 cm and mixing (Figure 5). The results of TOC measurements on both sets of samples are shown in Figure 6 along with the density log. From these results it can be seen that the formation is actually quite heterogeneous on the 1-cm scale and that the larger sampling interval is much more representative of the larger vertical resolution of the logs.

Chemical analyses were performed on the core samples for both total carbon and total organic carbon. For total carbon, the samples were analyzed by XRAL (X Ray Assay Laboratories) in the following manner. Samples were combusted in oxygen to convert organic and inorganic forms of carbon into CO_2 at 950°C. The samples' combustion gases were swept through a barium chromate catalyst and scrubber to ensure complete oxidation of carbon into CO_2, and noncarbon combustion products were removed from the gas stream by a series of chemical scrubbers. The CO_2 was then measured with a CO_2 coulometer. Total organic carbon was measured at XRAL by combusting at 500°C and measuring with the CO_2 coulometer. TOC was also measured at Elf Aquitaine laboratories by treating the samples with dilute acid, combusting in oxygen, and measuring the CO_2.

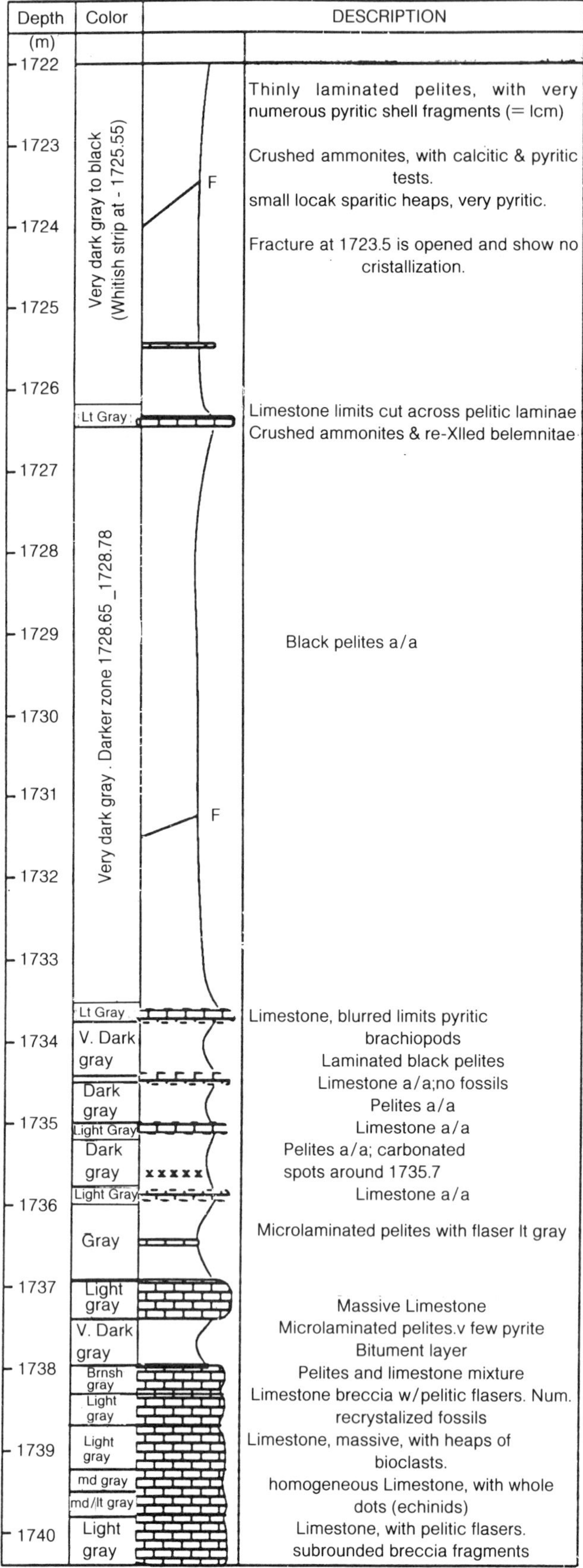

Figure 4. Description of core selected for study.

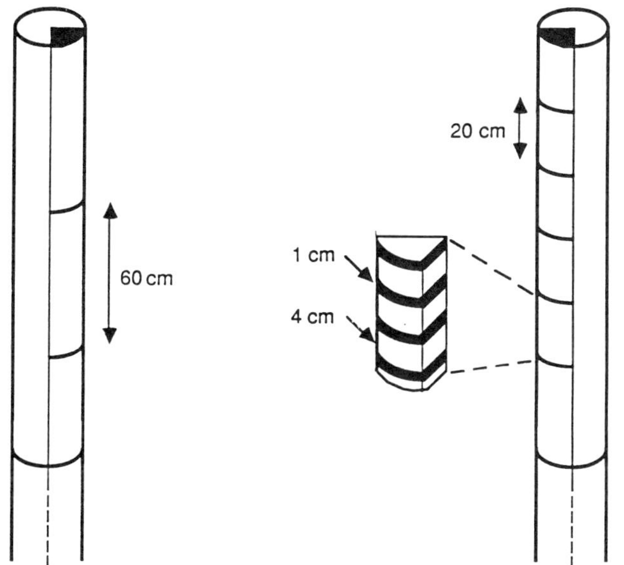

Figure 5. Core-sampling schemes. For this study the core was slabbed continuously and divided into 60-cm-long samples which were ground to 40 μm and divided for geochemical analyses. For the alternate scheme shown on the right, 1-cm slabs were taken approximately every 4 cm and combined in 20-cm intervals.

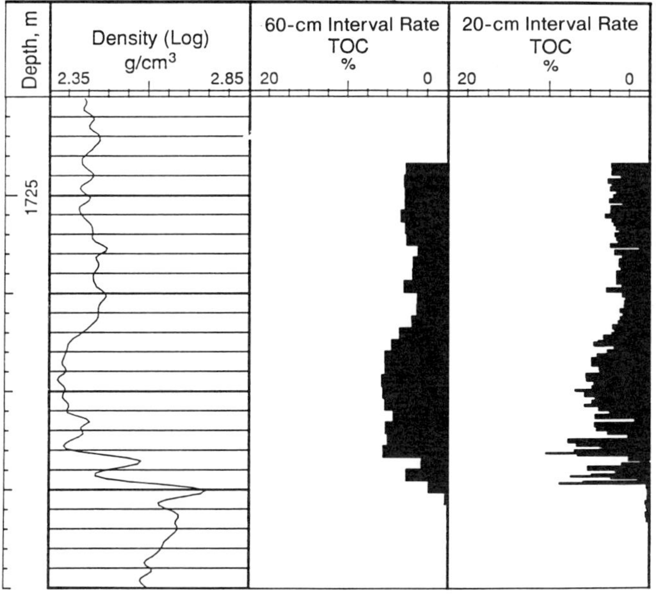

Figure 6. Comparison of results from different sampling intervals. TOC results for the 60-cm sampling (left) are far more homogeneous and representative of the formation volume typically measured by the wireline tools than the 20-cm interval (right).

GST INTERPRETATION OF TOTAL ORGANIC CARBON

Two-Component Model

A method has been described to determine the total organic carbon (TOC) of a formation by multiplying the measured carbon/oxygen ratio (C/O) by an estimated oxygen content to obtain the total amount of carbon, both organic and inorganic. The inorganic carbon is then estimated using either calcium and/or magnesium concentration logs or geochemically derived mineralogy, and is subtracted from the total carbon to obtain total organic carbon (Herron, 1986).

The general formula for determining total carbon (C) for a formation composed of a solid matrix and fluid-filled pore space is:

$$C = 0.75\,(C/O)\left[O_{ma} + O_{fl}\frac{\rho_{fl}\Phi}{\rho_{ma}(1-\Phi)}\right]$$

where carbon is in g C/g solid, ϕ is porosity, ρ is density, O represents the decimal fraction of oxygen, and ma and fl represent the matrix and fluid, respectively. The 0.75 constant represents the ratio of the molecular weights of carbon and oxygen and is required as part of the unit conversion from atomic concentrations to weight percent.

The key to solving this equation is the selection of values for the input parameters of density, porosity, and oxygen contents. A full description of parameter selection and the sensitivity of total carbon to each of the parameters is presented in Herron (1986), but the most salient points will be reviewed here. The first step in the analysis is the partitioning of the formation into its two components: matrix and pore space. This is accomplished with density porosity which is calculated using a matrix density (ρ_{ma}) which varies between a sandstone matrix with ρ_{ma} = 2.65 g/cm^3 and a limestone matrix with ρ_{ma} = 2.71 g/cm^3. This variable grain density is calculated on a level-by-level basis using the calcium log to determine the percent calcite or "limestone" in the formation according to the following formula:

$$\rho_{ma} = 2.65 + 0.06\left(\frac{Ca}{40}\right)$$

where Ca is the measured weight percent calcium and 40 is the weight percent calcium in calcite. The fluid density is assumed to be 1 g/cm^3. The mineral matrix is assumed to have an average oxygen content (O_{ma}) of about 50 wt %; this selection is governed by the fact that for most common sedimentary minerals there is only a small variation in oxygen content, as shown in Table 1 (Hurlbut, 1971; Westaway et al., 1980). The only difficult parameter to select is the oxygen content of the fluid which could vary between zero in a formation fully

Table 1. Oxygen content of common sedimentary minerals.

Mineral	Wt % O
Quartz	53
Orthoclase	46
Albite	49
Anorthite	46
Kaolinite	56
Illite	51
Montmorillonite	59
Calcite	48
Dolomite	52
Siderite	41
Anhydrite	47
Gypsum	56

saturated by hydrocarbons to 89 wt % in a formation fully saturated by water. For this application, it will be assumed that the pore space is water filled, and therefore the oxygen content is 89 wt %.

Figure 7 summarizes the results of a sensitivity analysis to determine the effect of parameter selection on carbon determination (Herron, 1986). In each example, the value of the independent variable increases as others are held constant at the default values described above; the calculations are made for a 15 p.u. formation and a measured C/O ratio of 0.01. Figures 7a and 7b reveal that the input values for ρ_{fl} and ρ_{ma} do not have a significant effect on calculated carbon. The carbon calculation is more sensitive to porosity (Figure 7c): a 10 p.u. error in porosity would result in a relative error on total carbon content of 8%. Errors in porosity of this magnitude are not common, and the use of the variable grain density in carbonate-bearing rocks can reduce a potential error by several porosity units. The greatest uncertainty in the calculation of carbon comes from the selection of values for oxygen contents of the mineral and fluid phases, as can be seen in Figures 7d and 7e. For anything other than a pure carbonate source rock, the mineral oxygen content is most likely to be within the 0.50–0.55 range. A 0.05 error in this parameter would again result in a relative error of 8% on calculated carbon content. The remaining parameter, oxygen content of the fluid, is the only totally unknown value; it can vary from zero to 0.89. By assuming complete water saturation in the formation, i.e., $O_{fl} = 0.89$, any error due to the fluid oxygen content will lead to an overestimation of carbon. The maximum error will be about 10 relative percent, and that will occur for a formation which is fully oil saturated (Figure 7e).

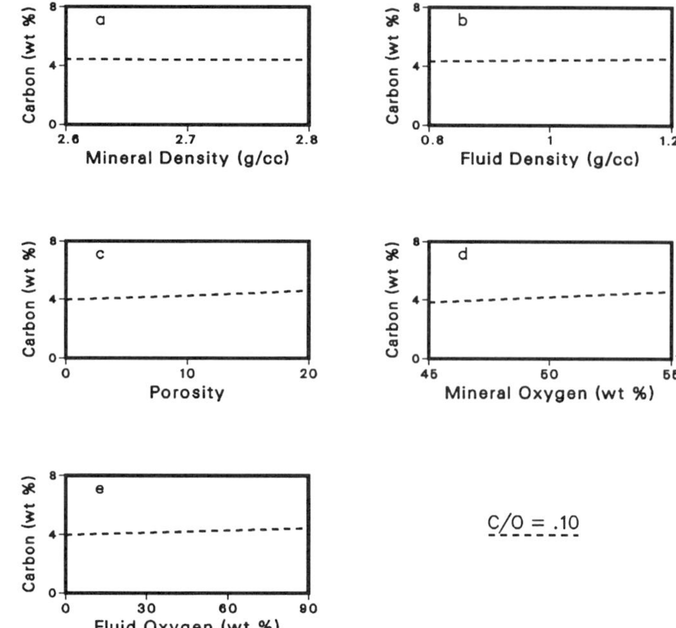

Figure 7. Comparison of the effects of changes in various input parameters on derived carbon content.

Application of Two-Component Model to the Paris Basin

The two-component model was applied to the stationary C/O measurements from the Paris Basin study well to obtain total carbon, and the results are presented in Figure 8. In the upper 9 m where the carbon content is around 6%, the agreement between core and log-derived carbon is quite good, but in the lower half of the interval where the carbon is about 10%, the log derivations tend to overestimate carbon by 1 wt % or more.

The source of this overestimation lies in the oversimplified representation of the formation. In reality, organic matter has a density close to 1 g/cm^3, the same as water. Consequently, where a large quantity of kerogen is present in the formation, the calculated density porosity is also large. In the two-component solution for carbon, the pores are assumed to be filled with water which has an oxygen content of 89 wt %, as opposed to kerogen which generally has an oxygen content of less than 20 wt %. As a result, the formation oxygen is overestimated, and the calculated carbon in the formation is too high. The magnitude of this effect depends on the level of organic carbon. This is illustrated in Figure 9 where total organic carbon is shown varying with oxygen in the pores. Here the organic-carbon content is derived for values of C/O ranging from 0.01 to 0.20 as the oxygen content of the materials occupying the pore space increases from 0 wt % ($S_w = 0$) to 89 wt % ($S_w = 100$). For this calculation, it is assumed that all carbon is organic, and therefore the porosity increases as the C/O increases. For low TOC contents (<3 wt %), the effect of assuming that water occupies the pore space is very small, but as TOC increases so does the bias introduced by the two-component

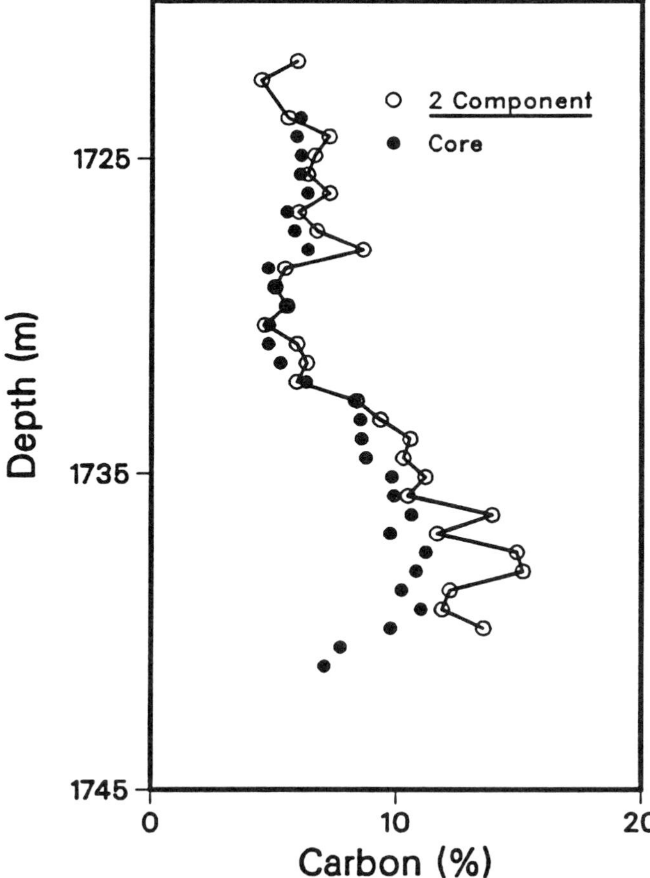

Figure 8. Total carbon estimated from stationary C/O measurements using a two-component formation model (open circles). Core data (solid dots) are provided for comparison.

model. For a formation with 8 wt % TOC, the two-component model would predict close to 10 wt %, as shown for the C/O = .20 curve in Figure 9.

Three-Component Model

In a more realistic model the formation is composed of three components: a solid mineral matrix, a water-filled pore space, and kerogen. The porosity (ϕ) calculated from the bulk density log now represents a volume containing both kerogen and fluid: $\phi = \phi_{ker} + \phi_{fl}$. If the kerogen has a density of ρ_{ker} and an oxygen content of O_{ker}, the total carbon (g C/g solid) in the formation can be expressed as:

$$C = \frac{0.75\,(C/O)}{(1 - \Phi)\rho_{ma}} \left[O_{ma}\,\rho_{ma}\,(1 - \Phi) + O_{ker}\rho_{ker}\Phi_{ker} + O_{fl}\rho_{fl}\Phi_{fl} \right]$$

Since the volume of kerogen is not known, the two-component model is used to approximate TOC, and the solution for carbon becomes an iterative one. The two-component TOC is converted to weight of kerogen and then to volume. The conversion factor for weight of organic carbon to weight of kerogen varies with type of kerogen and degree of maturation; for this purpose a medium value of 1.25 has been chosen (Tissot and Welte, 1978). The volume of kerogen is then:

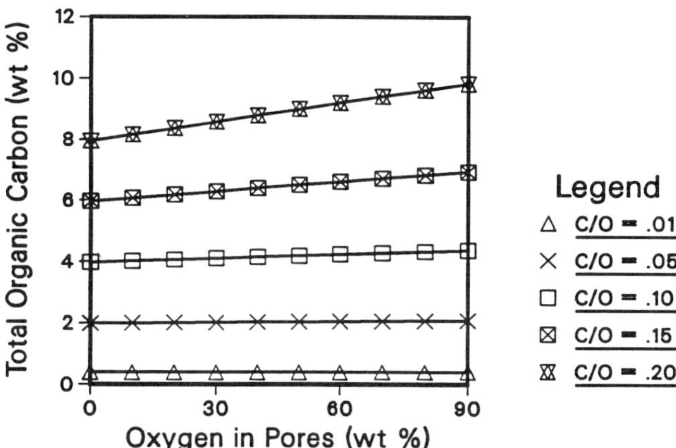

Figure 9. Relationship of total organic carbon to oxygen content of material occupying pore space. Water has an oxygen content of 89%, whereas kerogen averages about 10% oxygen by weight.

$$\Phi_{ker} = \frac{1.25\ TOC}{\rho_{ker}}\rho_b$$

It is also necessary to assign an oxygen content to the kerogen (O_{ker}). In nature, this value varies from about 2 to 20 wt % depending on kerogen type and maturity. In immature sediments, type I has the lowest oxygen and type III has the highest; for all types, as the kerogen matures, oxygen is lost (Tissot and Welte, 1978). For this application a middle value of 10 wt % was selected.

This three-component model was applied to the stationary C/O measurements from the Paris Basin well using the default parameters presented in Table 2. The results are presented in Figure 10. There is now very good agreement between log and core data through the entire zone where the carbon content ranges from about 2 to 6.5%. This is a significant improvement over the two-component solution shown in Figure 8.

Correction for Inorganic Carbon

The final step in obtaining TOC is the removal of inorganic carbon which resides primarily in carbonate minerals. If the mineralogy has been derived, the carbon contributions of individual carbonate species can be summed and subtracted from total carbon. If mineralogy is not available, the Ca or Ca and Mg logs (Hertzog et al., 1988) can be used to estimate inorganic carbon. If both Ca and

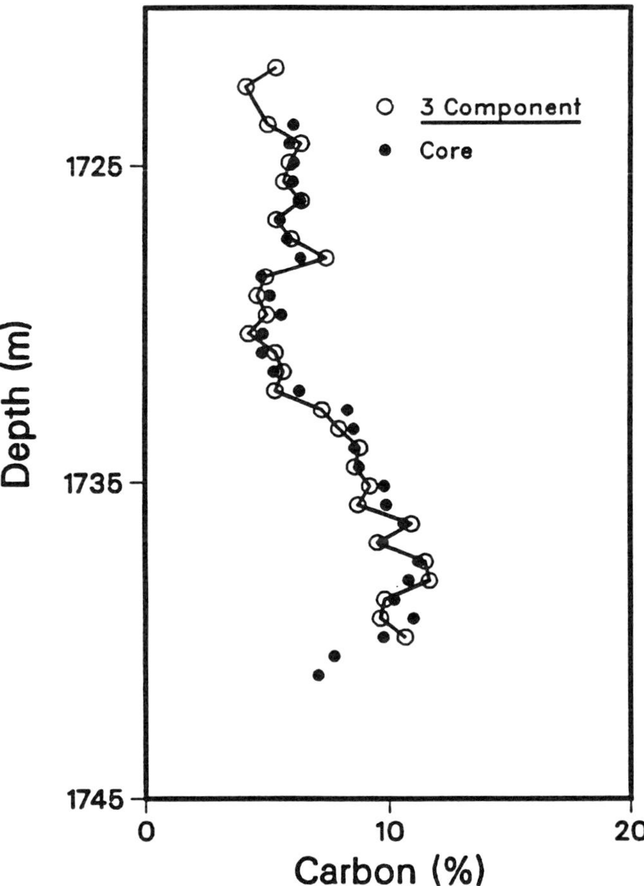

Figure 10. Total carbon estimated from stationary C/O measurements using a three-component formation model (open circles). Core data (solid dots) are provided for comparison.

Mg are present in the formation, the Mg can be attributed to dolomite, the appropriate amount of Ca can be apportioned, and the remaining Ca can be attributed to calcite. The amount of inorganic carbon is then:

$$C_{inorganic} = 12 \frac{Ca - \left(\frac{21.74\ Mg}{13.18}\right)}{40} + 13 \frac{Mg}{13.18}$$

Alternatively, if there is no magnesium, the inorganic carbon can be estimated by simply assuming that all the calcium belongs to calcite and that there is no other source of inorganic carbon:

$$C_{inorganic} = 12 \frac{Ca}{40}$$

In the Paris Basin well, the geochemical logs indicated the presence of little or no magnesium in most samples, and therefore the inorganic carbon was estimated using just calcium. The validity of deriving inorganic carbon from calcium is illustrated using core data in Figure 11. Here the measured value of inorganic carbon represents the difference between total organic carbon and total carbon measured on the core, and the calculated value is from calcium measured on the core (Figure 12). In general, the inorganic carbon estimated from the calcium is quite close to that from the carbon measurements.

The calcium data from the GST stations are presented in Figure 12 with core data for comparison. The values range from about 4.5 to 26.5 wt %, reflecting the large amount of calcite (11-66 wt %) in the formation. This corresponds to a substantial inorganic carbon content ranging from 1.4 to 8.0 wt %. The calcium log data were used to calculate the inorganic carbon, and this was then subtracted from the total carbon to produce total organic carbon.

The final result is the total organic carbon log which is presented along with core data as a function of depth in Figure 13. Over the depth interval covered by both core and log data, the organic carbon content ranges from about 2.5 to 6.5 wt %, and as can be seen in Figure 13, the agreement between core and log data is quite good. Where the core and log data do not agree, the disagreement can generally be related to a disagreement between the core and log estimates of calcium. This can be attributed to the fact that the carbonate minerals are not distributed homogeneously throughout the core, but rather they occur in discrete layers. Since these carbonate layers are not distributed symmetrically above and below the GST stations, the GST estimations of calcium are not in perfect agreement with the core samples. The core and log data for both total carbon and total organic carbon are compared in crossplot format in Figure 14. Perfect depth matching is assumed to exist between log measurements and core samples, and in fact the precautions taken to acquire representative core samples are partially responsible for the remarkably good agreement between the core data and the log interpretations presented in this figure. This new model was also applied to GST stationary data from a well in Oklahoma (Herron, 1986), and it was found to significantly improve the total organic carbon estimation in a thin, organic-rich layer (Herron and Herron, 1990).

It is important to emphasize that this TOC log has been obtained directly from a fundamentally sound formation model, and that it does not require calibration with core data, as do all other wireline techniques for quantitative estimation of organic carbon. A number of input parameters are required for this analysis, but it is demonstrated by the sensitivity analysis (Figure 7) that these input parameters do not drive the calculation of carbon, and therefore the default parameters listed in Table 2 serve well for general purposes. External knowledge, such as the type or maturity of organic matter, can be used to optimize the analysis, but it is not required.

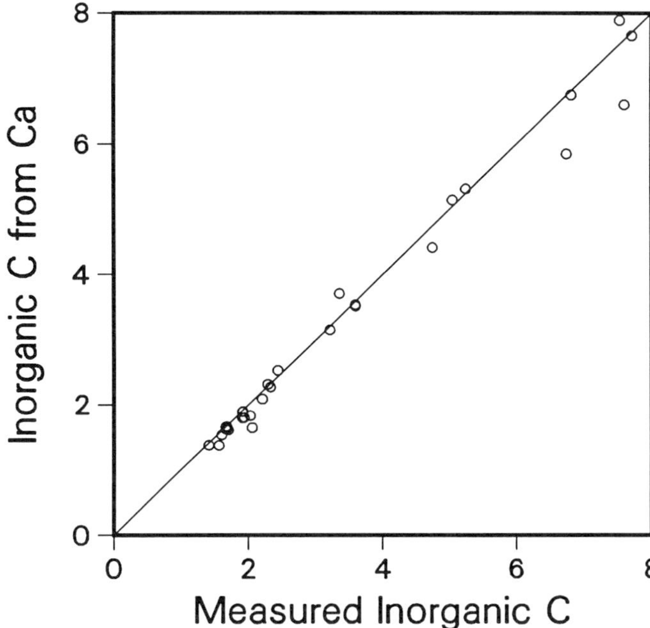

Figure 11. Comparison of inorganic carbon determined by measurements on core (total carbon minus organic carbon) with inorganic carbon estimated from measurements of calcium on the core assuming that all calcium resides in the mineral calcite and that there is no other source of inorganic carbon.

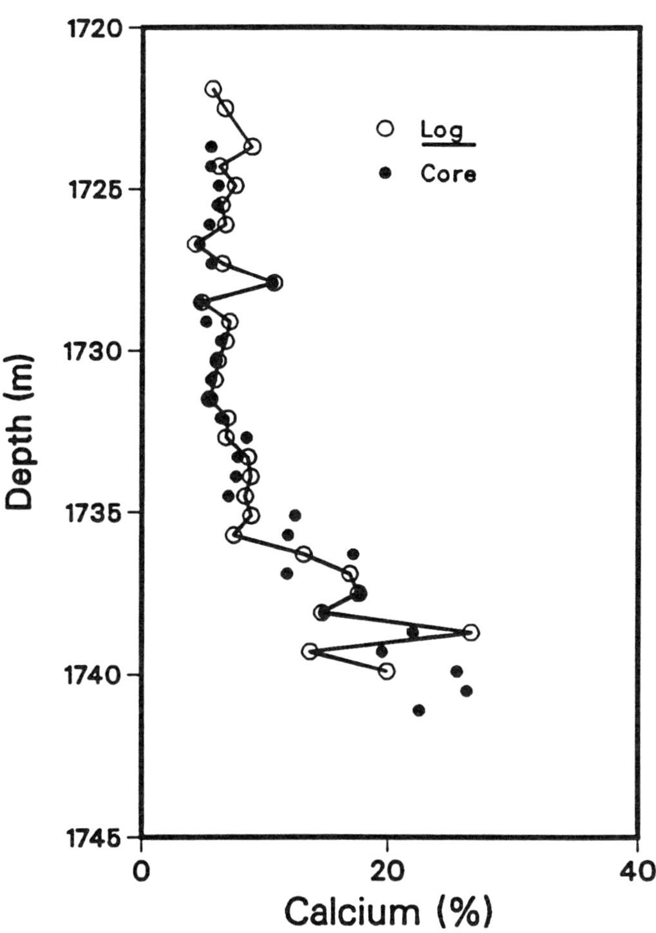

Figure 12. Calcium data from the GST stationary measurements (open circles). Core data (solid dots) are provided for comparison.

CONTINUOUS LOGGING FOR TOC

The GST data presented so far have been from stationary measurements which offer the obvious advantage of high precision in determining the carbon/oxygen ratio and the disadvantage of slow data acquisition. The measured error on these stationary carbon/oxygen measurements ranges from 0.006 to 0.008 which roughly translates to 0.3–0.4 wt % carbon; the length of data acquisition for each station was approximately 5 minutes. In addition to being used to acquire the stationary data, the GST tool was also run continuously at 600 ft/hr for three passes over the zone of interest. This represents only 3 seconds of data acquisition every 6 inches (0.152 m). The corresponding error on the carbon/oxygen ratio for the three passes averages 0.079, an order of magnitude higher than for the stationary measurements. The two possible techniques for improving the precision of these continuous measurements are laterally averaging the three passes and vertically averaging the data. The reduction in the error on the measurement is illustrated in Table 3 for several combinations of laterally and vertically averaged data from the Paris Basin well.

In order to determine the carbon content to within about 1 wt %, the error on the C/O ratio should be roughly 0.02. Therefore, for a single pass at 600 ft/hr in this borehole, it would be necessary to vertically average the data over 11 points, or 1.67 m. If three passes are laterally averaged, they could then be vertically averaged over only five levels, or 0.76 m. In practice, TOC is frequently measured on mud cuttings collected over intervals as large as 3 m, which would correspond to about a 20-point vertical average. If similar smoothing is applied to the log data, a single pass should produce carbon estimates within a percent of those obtained from the cuttings. Three combined passes would produce accuracies similar to those observed for the stationary data.

The three-component model has been applied to the continuous carbon/oxygen ratios to obtain continuous TOC logs. The results for a single pass with 11-point vertical smoothing are presented in Figure 15, and the results for three combined passes with 5-point vertical smoothing are presented in Figure 16. As expected, there is a combined loss of precision and vertical resolution in the continuous logs as compared to the stationary results, but they

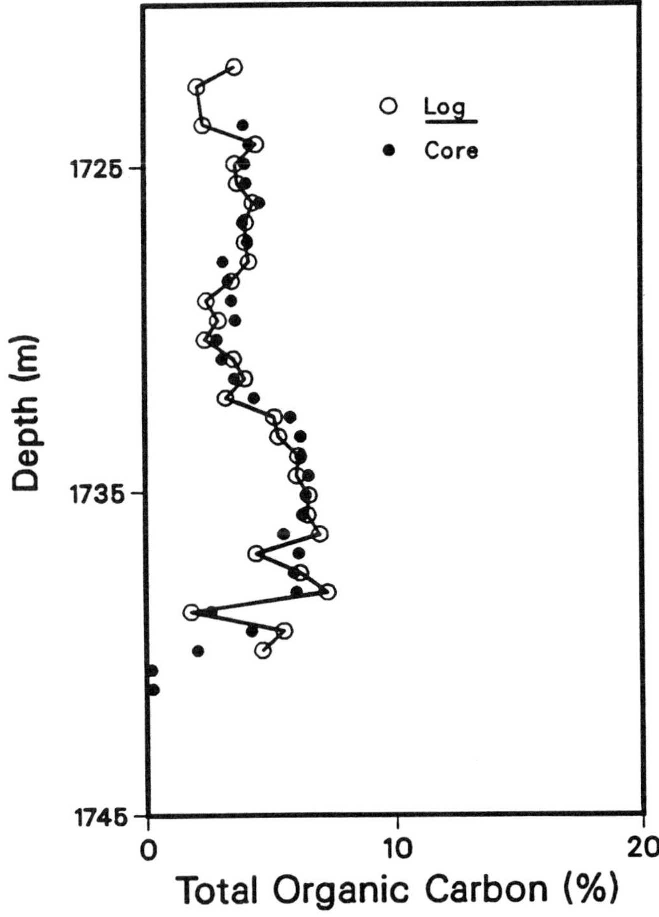

Figure 13. Total organic carbon estimated from stationary C/O measurements using the three-component formation model (open circles). Core data (solid dots) are provided for comparison.

Table 2. Input parameters for determining total carbon using the three-component model.

Variable	Default Value	Comments
C/O	Log Value	Environmentally corrected
ρ_{ma}	2.65–2.71 g/cm^3	Varies with Calcium content
ρ_{fl}	1.0 g/cm^3	Negligible impact on calculation
ρ_{ker}	1.0 g/cm^3	Negligible impact on calculation
O_{ma}	0.50 g O/g solid	A middle value
O_{fl}	0.89 g O/g H_2O	Oxygen content of water
O_{ker}	0.10 g O/g kerogen	A middle value

Table 3. Variations in error on C/O with vertical and lateral averaging.

Logging Speed 600 ft/hr	Single Pass Error on C/O	Three Passes Error on C/O
Single point (0.152 m)	0.075	0.044
5-point vertical average	0.032	0.019
11-point vertical average	0.022	0.013
20-point vertical average	0.017	0.010

still predict the level of organic carbon quite well. It is clear that for this application the logging program has to be designed to address individual requirements: if high vertical resolution is not a strict requirement, then vertically averaged data from one or more continuous passes would probably suffice. However, in situations where a higher vertical resolution is desired and particularly if low carbon contents are suspected, then stationary data should be obtained.

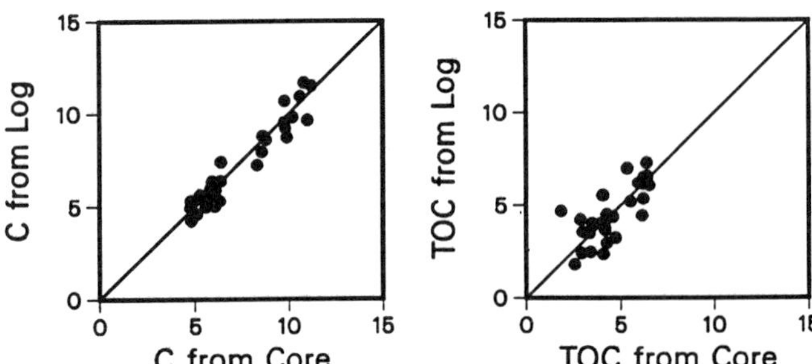

Figure 14. Comparison of values of total carbon (C) and total organic carbon (TOC) made from stationary GST measurements with measurements made on core samples.

COMPARISON WITH CONVENTIONAL TECHNIQUES

Conventional log measurements generally used for source-rock identification include gamma ray, density, sonic transit time, and resistivity. Source rocks commonly produce high gamma-ray values due to a uranium enrichment associated with the deposition of organic matter in marine environments; they generally have a low density and high sonic transit time due to the low density and slow velocity of kerogen. Many have a high resistivity compared to water-bearing sediments. In order to quantify the organic content of rocks, a number of studies have relied on correlations such as TOC and density (Tixier and Curtis, 1967; Schmoker, 1979; Schmoker and Hester, 1983) or TOC and gamma ray (Supernaw

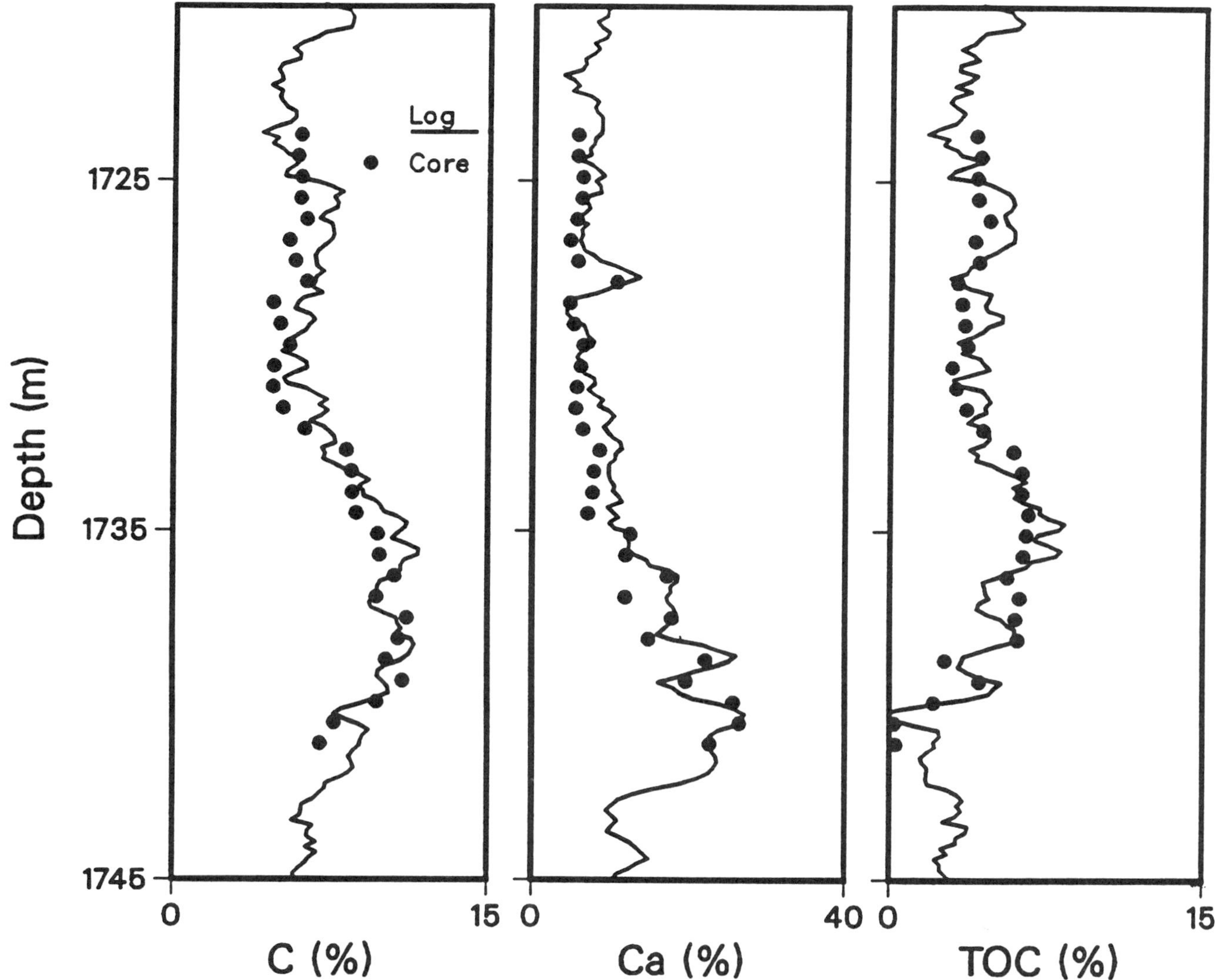

Figure 15. Total organic carbon from continuous GST measurements, based on a single pass with 11-point vertical averaging of the carbon/oxygen ratio. Core data are provided for comparison.

et al., 1978; Fertl and Rieke, 1980; Schmoker, 1981). Others have used sonic and resistivity logs to qualitatively determine organic content (Dellenbach et al., 1983; Autric and Dumesnil, 1985), and Mendelson and Toksoz (1985) have taken a multivariate statistical approach using sonic, neutron, density, and gamma-ray logs. These and other techniques were recently reviewed by Fertl and Chilingar (1986). As a general rule, the relationships which have been developed have required calibration with core data on a well-by-well or regional basis, and they have worked well only for zones with high organic-carbon contents.

A more generally applicable model, and one that does not require calibration with core data, was proposed by Meyer and Nederlof (1984). They provided a family of four functions, based on sonic, resistivity, density, and gamma-ray logs, designed to discriminate between organic-rich sediments with greater than 1.5% organic carbon and organic-poor sediments. Each of the four discriminant functions was derived twice: once for shales only and once for all lithologies. When applied to data, if the function produces a positive value, the data should represent source rocks with greater than 1.5% TOC, and when it produces a negative value, the formation should not have significant organic carbon. Although this technique is only semiquantitative in nature, it has been shown to work in many environments, and it offers the significant advantages over other published techniques of not requiring core data and not demanding complex, subjective interpretive input.

The Meyer and Nederlof discriminant functions were applied to the Paris Basin log data using the input data presented in Figure 17, and the results are presented in Figure 18. The solid lines represent the functions derived for shales, and the dashed lines represent those derived for all lithologies. In spite

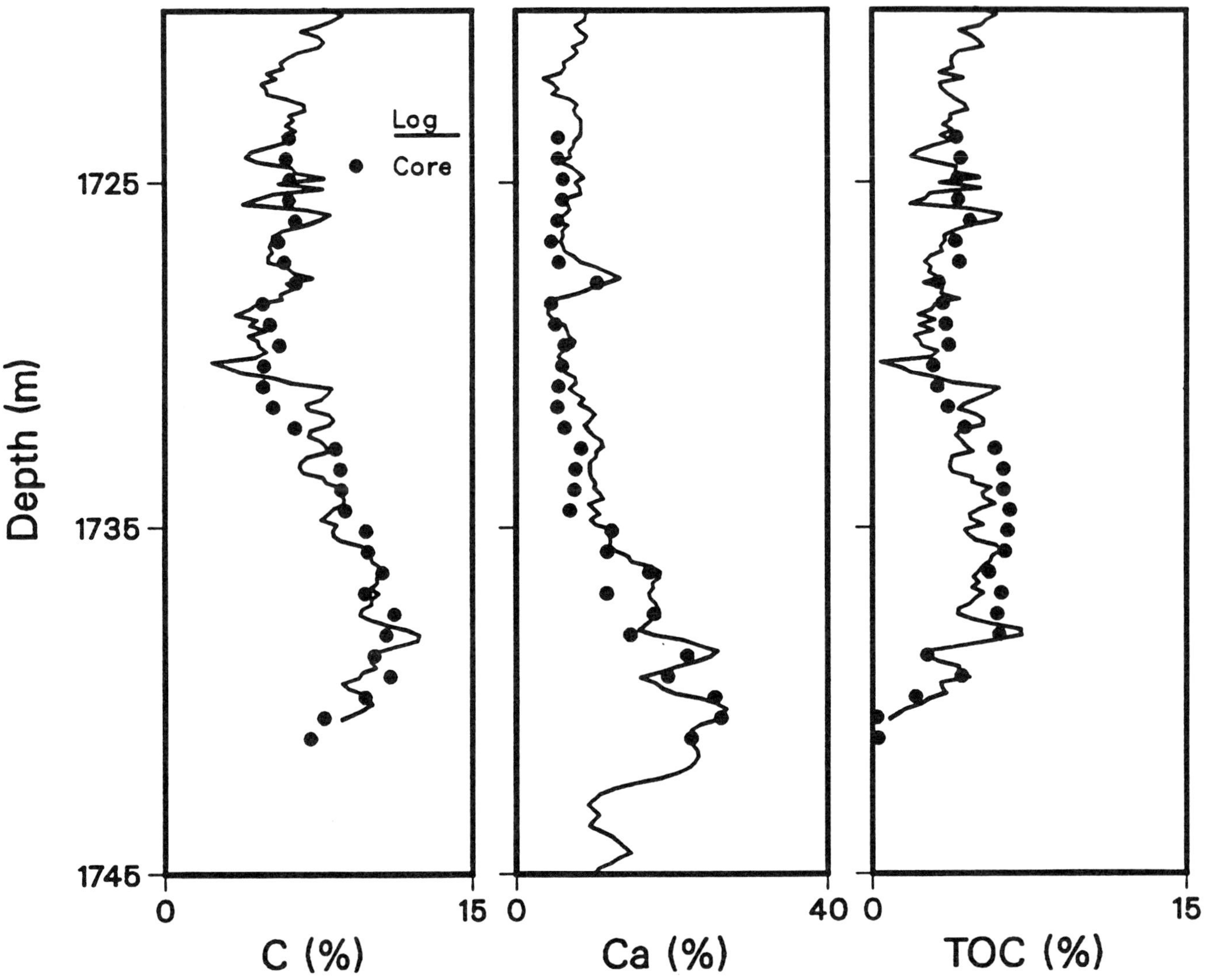

Figure 16. Total organic carbon from continuous GST measurements, based on three passes with 11-point vertical averaging of the carbon/oxygen ratio. Core data are provided for comparison.

of the fact that this is a very rich organic interval, only the gamma ray-sonic-resistivity combination correctly identifies the presence of organic matter. The sonic-resistivity and gamma ray-density-resistivity combinations identify only the sections where TOC is greater than 6 wt % as being potential source rocks, and the density-resistivity combination fails over the entire interval.

CONCLUSIONS

The GST data acquired in the Paris Basin well provide excellent estimates of the total organic carbon content of the formation. These estimates are made using a physically sound model of the formation, and they do not require any statistical manipulation or calibration with core data. Although there are input parameter values which can be altered, such as the oxygen content of the various components, the values used here represent good default values, and changing them within reasonable limits will not significantly alter the results.

The previously published two-component model for deriving carbon was found to be accurate only for low levels of organic carbon, and for higher organic contents it is necessary to expand to a three-component model which treats kerogen separately from the pore fluid and minerals. The two models are run sequentially; the two-component model is run to roughly estimate the amount of kerogen, and the three-component model is run for the final answer.

The GST carbon/oxygen ratios can be taken as stationary or continuous measurements for this application. Predictably, the stationary measurements provide the best precision, but when multiple

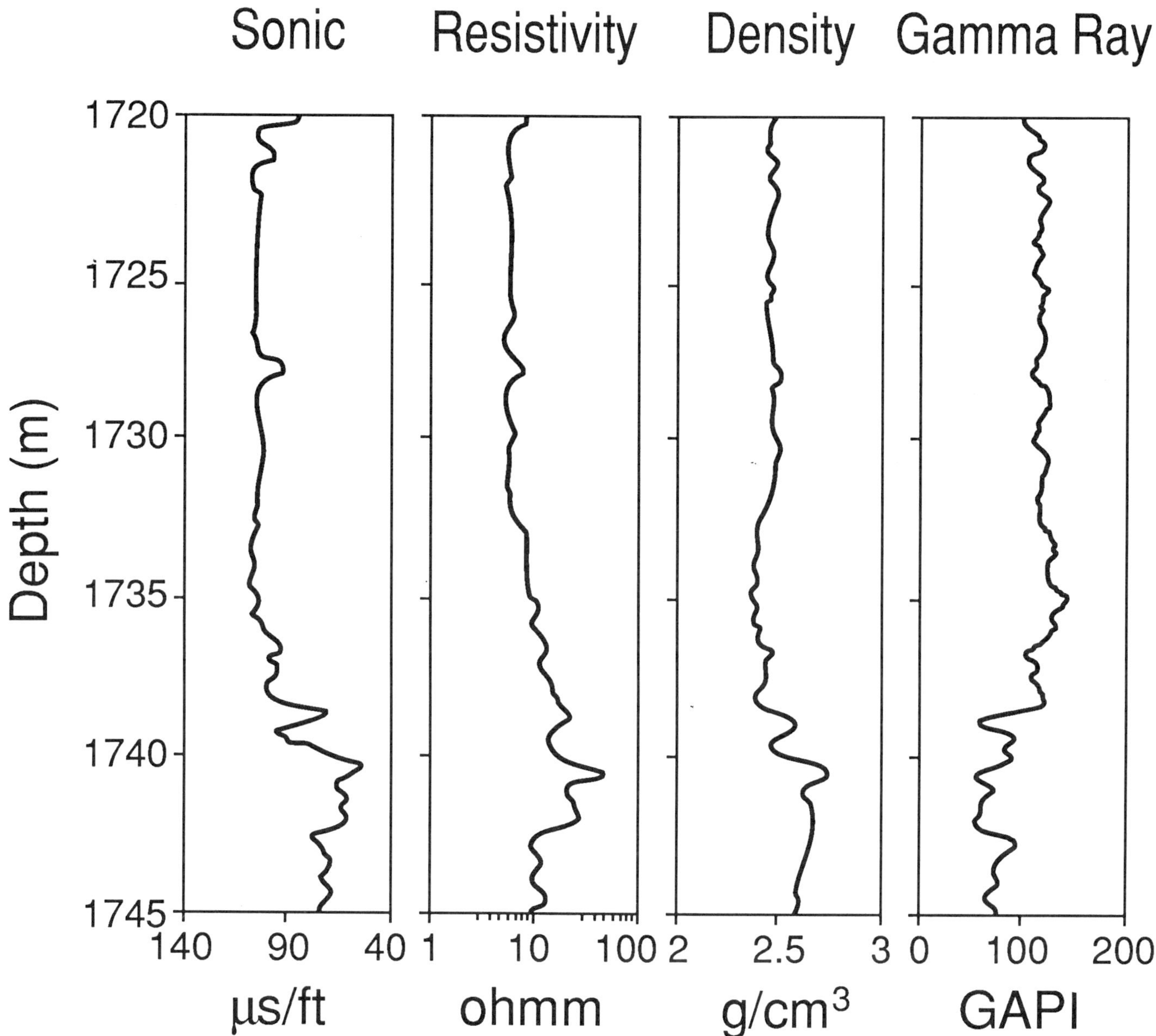

Figure 17. Input data for discriminant functions: sonic, resistivity, density, and gamma-ray logs for a portion of the Paris Basin well.

continuous passes are combined or continuous data are vertically averaged, the results can be quite good.

Finally, the GST data used with either the two-component model or the three-component model are far superior to the Meyer and Nederlof (1984) discriminant function analysis applied to the Paris Basin well. Only one of four functions identifies this very rich interval as a source rock, while two of the functions only recognize the intervals with greater than 6 % TOC as source rocks.

REFERENCES

Autric, A., and P. Dumesnil, 1985, Resistivity, radioactivity, and sonic transit time logs to evaluate the organic content of low permeability rocks: Log Analyst, May-June, p. 36-45.

Dellenbach, J., J. Espitalié, and F. Lebreton, 1983, Source rock logging: Society of Professional Well Log Analysts Eighth European Formation Evaluation Symposium Transactions, London, 1983.

Espitalié, J., F. Marquis, L. Sage, and I. Barsony, 1987, Géochimie organique du Bassin de Paris: Revue de l'Institut Francais du Pétrole, v. 42, no. 3, Mai-Juin, p. 271-302.

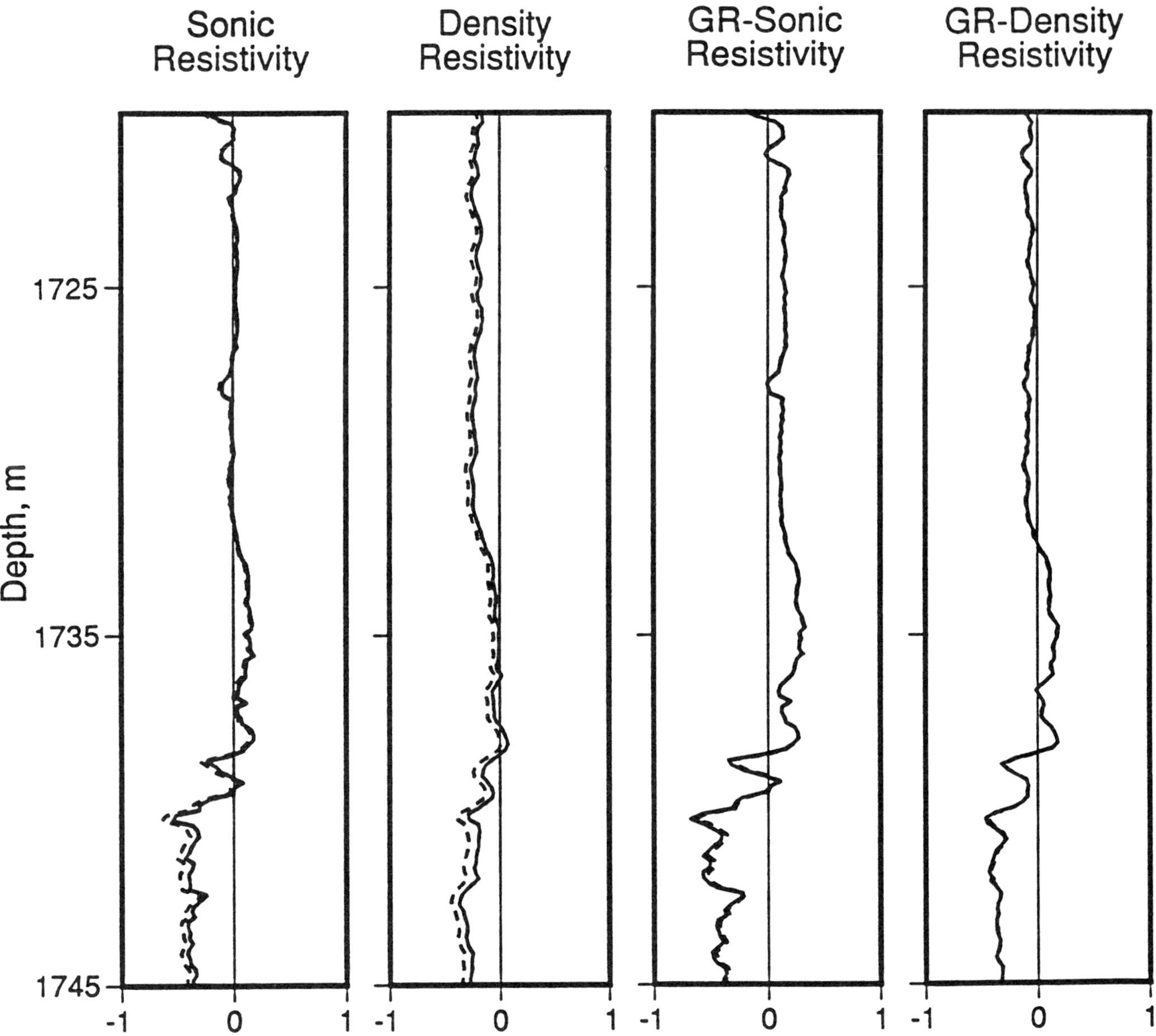

Figure 18. Discriminant scores derived from combinations of sonic, resistivity, density, and gamma-ray logs. Positive values indicate the probable existence of source rocks, whereas negative values indicate barren rock. In three of the four cases, the technique fails to identify rocks with total organic carbon contents of 4-6 wt %.

Fertl, W. H., and G. V. Chilingar, 1986, Total organic carbon content determined from well logs: Society of Petroleum Engineers, SPE-15612.

Fertl, W. H., and H. H. Rieke III, 1980, Gamma ray spectral evaluation techniques identify fractured shale reservoirs and source-rock characteristics: Journal of Petroleum Technology, November 1980, p. 2053-2062.

Herron, S. L., 1986, A total organic carbon log for source rock evaluation: Transactions of Twenty-Seventh Society of Professional Well Log Analysts Annual Meeting, Houston, paper HH, 1986; *revised in* 1987, Log Analyst, v. 28, no. 6, p. 520-527.

Herron, M. M., and S. L. Herron, 1990, Geological applications of geochemical well logging, *in* A. Hurst, M. A. Lovell, and A. C. Morton, eds., Geological applications of wireline logging: Geological Society Special Publication 48, p. 165-175.

Hertzog, R., L. Colson, B. Seeman, M. O'Brien, H. Scott, D. McKeon, P. Wraight, J. Grau, D. Ellis, J. Schweitzer, and M. Herron, 1987, Geochemical logging with spectrometry tools: Society of Petroleum Engineers, SPE-16792.

Hurlbut, C. S., 1971, Dana's manual of mineralogy: New York, John Wiley, 579 p.

Mendelson, J. D., and M. N. Toksoz, 1985, Source rock characterization using multivariate analysis of log data: Society of Professional Well Log Analysts Twenty-sixth Annual Logging Symposium, June 1985.

Meyer, B. L., and M. H. Nederlof, 1984, Identification of source rocks on wireline logs by density/resistivity and sonic transit time/resistivity crossplots: AAPG Bulletin, v. 68, p. 121-129.

Schmoker, J. W., 1979, Determination of organic content of Appalachian Devonian shales from formation-density logs: AAPG Bulletin, v. 63, p. 1504-1537.

Schmoker, J. W., 1981, Determination of organic-matter content of Appalachian Devonian shales from gamma-ray logs: AAPG

Bulletin, v. 65, p. 1285–1298.
Schmoker, J. W., and T. C. Hester, 1983, Organic carbon in Bakken Formation, United States portion of Williston Basin: AAPG Bulletin, v. 67, p. 2165–2174.
Supernaw, I. R., D. M. Arnold, and A. J. Link, 1978, Method for in situ evaluation of the source rock potential of earth formations: U.S. Patent 4,071,755.
Tissot, B., B. Durand, J. Espitalié, and A. Combaz, 1974, Influence of nature and diagenesis of organic matter in formation of petroleum: AAPG Bulletin, v. 58, p. 499–506.
Tissot, B. P., and D. H. Welte, 1978, Petroleum formation and occurrence: A new approach to oil and gas exploration: New York, Springer-Verlag, 538 p.
Tixier, M. P., and M. R. Curtis, 1967, Oil shale yield predicted from well logs: Seventh World Petroleum Congress, Mexico City, v. 3, p. 713–715.
Westaway, P., R. Hertzog, and R. W. Plasek, 1980, The gamma spectrometer tool inelastic and capture gamma-ray spectroscopy for reservoir analysis: Society of Petroleum Engineers, SPE-9461.

Distribution of Organic Matter during the Toarcian in the Mediterranean Tethys and Middle East

Francois Baudin
Université P. et M. Curie
Paris, France

Jean-Paul Herbin
Institut Francais du Pétrole
Rueil-Malmaison, France

Jean-Paul Bassoullet
Université de Poitiers
Poitiers, France

Jean Dercourt
Université P. et M. Curie
Paris, France

Georges Lachkar
Université P. et M. Curie
Paris, France

Hélène Manivit
Bureau des Recherches Géologiques et Minières et CNRS-URA 1315
Orléans, France

Maurice Renard
Université P. et M. Curie
Paris, France

Geochemical studies to characterize the amount, origin, and petroleum potential of organic matter in Toarcian rocks have been performed on samples from outcrops and cores within the Mediterranean area and the Middle East. Maps were prepared to show the concentration and nature of organic matter during Toarcian time. These maps show great heterogeneity in the quantity and type of organic matter: (1) marine planktonic material deposited in reducing environments in shallow-water basins in northwestern Europe, or in deep basins in the Tethys; (2) terrestrial organic matter from higher plants, that is moderately degraded along the northern Tethyan margin and in the northern North Sea; and (3) residual or oxidized organic matter that occurs in most of the Tethys basins. These data show the existence of several environments where organic matter was preserved during the Toarcian anoxic event.

INTRODUCTION

Lower Toarcian organic-rich mudstones are well known for their petroleum potential in northwestern Europe. They are variously known as Posidonienschiefer in Germany and Switzerland, Schistes Cartons in the Paris Basin, Posidonia Shales in the North Sea area, and Jet-Rock in Great Britain. These units range in thickness from 10 to 50 m and their organic-carbon content ranges from 2 to 20%, with an average of 5% (Espitalié et al., 1987; Fleet et al., 1987; Huc, 1976, 1977; Küspert, 1983; Mann et al., 1986; Moldovan et al., 1985; Myers and Wignall, 1987; Thomas, 1977). The hydrogen indices quoted by these authors for lower Toarcian shales typically lie between 400 and 700 mg HC/g TOC. These values and numerous other geochemical parameters (elemental analysis of kerogen, chromatography of rock extract, and mass spectrometry) (Tissot et al., 1971, 1974; Durand et al., 1972; Espitalié et al., 1973) indicate a marine origin for the organic content of lower Toarcian shales from the Paris Basin. Recent analyses of lower Toarcian deposits (Serpentinus/Falciferum zone) from northern Italy, Austria, and Greece (Jenkyns, 1985, 1988; Jenkyns and Clayton, 1986, Farrimond et al., 1989) indicate that organic-rich sediments were deposited in the Tethys during this time. The purpose of this study is to complete the characterization of Toarcian organic matter in the Tethyan (Mediterranean) area and the Middle East, as well as to map the quantity and quality of organic matter. Outcrops and well cuttings from Portugal, Spain, Morocco, Italy, Greece, Saudi Arabia, the Persian Gulf, and Iran have been analyzed (Figure 1). Paleontological control based on ammonite fauna is good in most outcrops, except in Greece, but calcareous nannoplankton can be used to identify the lower Toarcian. The biostratigraphic correlations used in this study are summarized in Figure 2.

METHODOLOGY

Outcrop samples were collected as deep as possible below the surface in order to minimize the effects of weathering. The following geochemical techniques were used to characterize organic matter. Rock-Eval pyrolysis was run to provide a rapid determination of the quantity, petroleum potential, thermal maturity, and origin of organic matter (Espitalié et al., 1985a, 1985b, 1986). Elemental analyses of kerogen were made in order to categorize the samples as Types I, II, or III organic matter as defined by Tissot et al. (1974). The Types I and II mainly come from aquatic organic matter, and Type III comes from continental organic matter. Reference Type II has been precisely defined from the early Toarcian in the Paris Basin studied by Tissot et al. (1971). In addition to these three types, there is residual organic matter, either oxidized in subaerial environments and/or during sediment transit or recycled from older sediments. This residual organic matter, sometimes called "Type IV," is located below the evolution path of Type III in the H/C-O/C diagram and has no potential for hydrocarbon production (Tissot, 1984). If samples are thermally mature, the total organic content (TOC) is not entirely pyrolyzed by the Rock-Eval method. For these samples, TOC contents were determined by the Leco method. Gas chromatography of chloroform extracts were performed on selected samples to obtain a better definition of the type and origin of the organic matter. Palynological preparations (HCl, HF treatments) were analyzed in some cases to determine the palynofacies and to confirm the origin of the organic matter.

THE TOARCIAN IN THE WESTERN TETHYAN REGION

Portugal

The Rabacal section (RA, Figure 1) is 142 m thick from the Semicelatum to Meneghinii zones (Mouterde, 1967). The lower Toarcian consists of marls containing small pyritic *Dactylioceras*, corresponding to "bituminous marls," (Mouterde, 1967), followed by flaggy limestones. The middle and upper parts of the section consist of marls and marly limestones that contain bioherms and lenticular Spongae beds. The upper part of the Toarcian (Aalensis zone) is present in the Brehna section (BR, Figure 1). The Toarcian of the Rabacal and Brehna sections is generally very poor in organic matter. TOC values fluctuate from 0.25 to 0.5% in two places located at the middle part of the Bifrons zone and the Speciosum zone. The "bituminous marls" of the Serpentinus zone described by Mouterde (1967) are completely devoid of organic matter. In Figure 3, which shows TOC content and hydrogen index-oxygen index diagrams (HI-OI diagrams), the organic matter is the residual type, and its origin is undetermined. Palynological studies show the abundance of amorphous components with some spores and pollen grains such as *Cyathidites minor*, *C. australis*, *Arapiculatisporites spiniger*, *A. telephorus*, *Classopollis classoides*, and *Alisporites robustus*, and a few dinoflagellate cysts belonging to *Mendicodinium reticulatum*, *Eyachia prisca*, and *Sentusidinium asymetricum*.

Spain

During Liassic sedimentation, shallow marine limestones were predominant in the eastern part of Spain. The Rambla del Salto section (RS, Figure 1) is 77 m thick and composed of alternating bioclastic limestones and marls that are well dated by ammonites (Comas-Renfigo and Goy, 1978). The

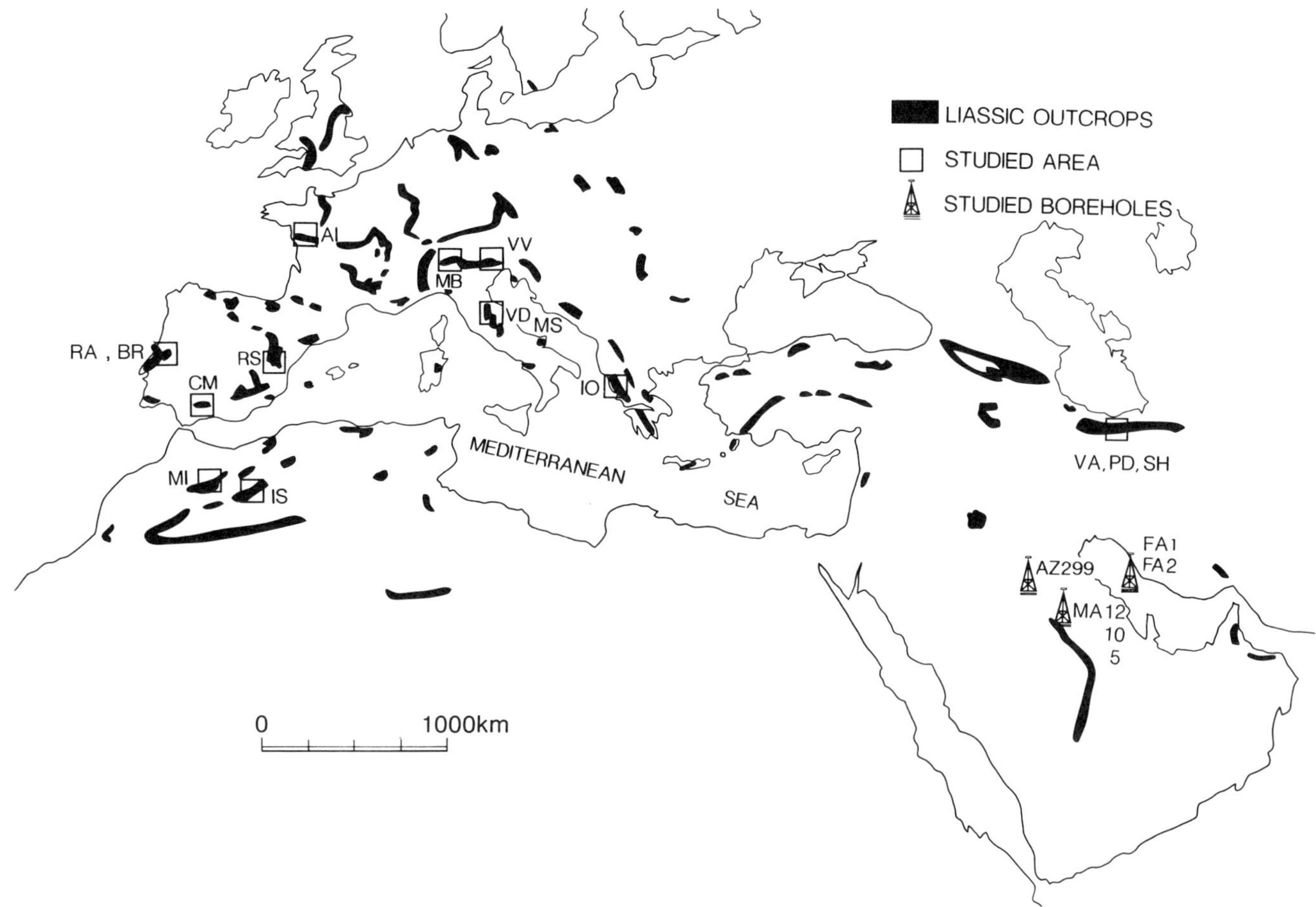

Figure 1. Map of Liassic outcrops in the Mediterranean and Middle East areas (after Gaertner and Walther, 1971) showing locations of sections and boreholes studied.
Sections: AI = Airvault, BR = Brehna, CM = Cerro Mendez, IO = Ionian, IS = Issouka, MB = Monte Brughetto, MI = Moulay Idriss, MS = Monte Serrone, PD = Pol-é-Doktar, Ra = Rabacal, RS = Rambla del Salto, SH = Shemshak, Va = Vaneh, VD = Valdorbia, VV = Val Vajont.
Boreholes: AZ 299 = Az Zabirah; MA 12, 10, 5 = Marrat, FA 1, 2 = Farsi.

entire section is organic poor, with TOC values lower than 0.35%, which corresponds to altered organic matter (Figure 3). Palynological preparations show the predominance of amorphous organic matter with woody debris, pollen, and spores including numerous *Classopollis* and *Chasmatosporites* sp.. Dinoflagellate cysts are rare. In the south of Spain, marly facies are prominent within the Toarcian, and Ammonitico Rosso appears locally in the basin (Azéma et al., 1979). The Cerro Mendez section (CM, Figure 1) is about 150 m thick and is typically yellowish-gray marl with interbedded Ammonitico Rosso facies in the Bifrons zone. All of the Cerro Mendez section is devoid of determinable organic matter with less than 0.25% TOC.

Morocco

Middle Atlas Basin

In the center of the Middle Atlas basin, Toarcian deposits are very thick and are composed of marls with pyritic ammonites (Guex, 1973; Benshili, 1989). On the edge of the basin, this thick series alternates laterally with red bioclastic limestones (Colo, 1961). The Issouka section (IS, Figure 1) has a complete, 150-m thick Toarcian succession composed of well-dated argillites, marls, and marly limestones (Benshili, 1989). These facies correspond to an over-deepened basin that was periodically isolated due to block-faulting subsidence which was greatest at the beginning of the Toarcian Stage (Fedan, 1984, 1985). The Issouka sequence is poor in organic matter. TOC values range from 0.1 to 0.5% (Figure 3). The most carbonate-rich levels have the lowest TOC contents (0.1 to 0.25%). The marly beds are slightly richer (0.25 to 0.5%) but also contain residual organic matter (Figure 3). Palynological studies show an important contribution by amorphous degraded organic matter with 5 to 15% inertinite-vitrinite and a microflora in which *Classopollis* and *Araucariacites australis* are always present. In some levels from the Serpentinus zone, a more representative flora has been found which contains *Verrucosisporites klukiformis*, *Lepto-*

	GREAT BRITAIN 1	FRANCE 2	ITALY 3	MOROCCO 4	PORTUGAL 5	SPAIN 6
UPPER TOARCIAN	LEVESQUEI	AALENSIS	MENEGHINI	AALENSIS	AALENSIS	AALENSIS
		PSEUDORADIOSA		MENEGHINI	MENEGHINI	REYNESI
	THOUARSENSE	INSIGNE	?	INSIGNE	SPECIOSUM	FALLACIOSUM
		THOUARSENSE		RIVIERENSE	BONARELLII	
MIDDLE	VARIABILIS	VARIABILIS	ERBAENSE	GRADATA	GRADATA	GRADATA
	BIFRONS	BIFRONS	BIFRONS	BIFRONS	BIFRONS	BIFRONS
LOWER	FALCIFERUM	SERPENTINUS	FALCIFERUM	SERPENTINUS	SERPENTINUS	SERPENTINUS
	TENUICOSTATUM	TENUICOSTATUM	TENUICOSTATUM	POLYMORPHUM	SEMICELATUM	POLYMORPHUM

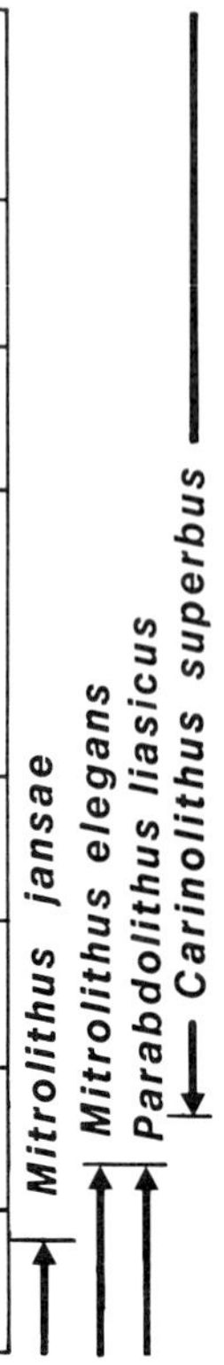

Mitrolithus jansae
Mitrolithus elegans
Parabdolithus liasicus
Carinolithus superbus

Figure 2. Biostratigraphic correlations used in this study for the Toarcian Stage in the European and Tethyan areas. 1- Cope et al. (1980), 2- Gabilly (1976), 3- Gallitelli-Wendt (1969), 4- Benshili (1989), 5-6 Elmi et al. (in press). First and last appearances of some nannofossils after Crux (1987) and Young et al. (1986) are shown.

lepidites equatibossus, *Cyathidites*, and *Vitreisporites pallidus*. A little higher in the section, *Cerebropollenites mesozoicus*, *Apiculatisporites jurassicus*, and *Ischyosporites variegatus* are present. A few levels contain some dinoflagellates such as *Mendicodinium reticulatum*, *Mancodinium semitabulatum*, and rare acritachs (*Micrhystridium lymensis* and *M. fragile*).

South Rifian Ridges

The Moulay Idriss section (MI, Figure 1) is about 150 m thick and has been dated by ammonites by Faugères (1975, 1978). The samples are poor in organic matter, generally lower than 0.5% TOC. Only a few levels from the Serpentinus and upper Bifrons zone contain organic matter. TOC values can reach 1.1%, but hydrogen indices are typically lower than 50 mg HC/g TOC (Figure 3). The elemental analyses of the kerogen in the richest samples show high O/C values, which confirm the oxidation of the organic matter. In a Van Krevelen diagram (Figure 10) they are situated in the residual zone defined by Tissot et al. (1980). The microflora are well preserved and representative in the Serpentinus zone, consisting of miospores, megaspores, and, in some levels, of dinoflagellates. In addition to *Classopollis* and *Araucariacites australis*, *Leptolepidites major*, *L. bossus*, *L. macroverrucosus*, and *Stereisporites stereoides* are present, as well as *Chasmatosporites major* and *C. hians*. Among the megaspores, *Trilites murrayi*, *T. pinguis*, *Horstisporites harrisi*, and *Minerisporites richardsoni* are abundant. Dinoflagellates are numerous higher in the Bifrons and Variabilis zones, consisting mostly of *Mancodinium semitabulatum* and *Mendicodinium reticulatum*.

THE TOARCIAN IN THE APULIAN REGION

During Toarcian times, the Apulian block (Italy, Yugoslavia, and Greece) was a wide continental passive margin bordering the Tethyan ocean. The sedimentation of the Apulian region was characterized by neritic facies, affected locally by intense block faulting, which created deep troughs. These troughs may have been isolated from the ocean or in communication with the Tethys.

Italy

Northern Italy

The Jurassic history of the Alps from northern Italy has been summarized by Aubouin (1963), Bernoulli et al. (1979), and Winterer and Bosselini (1981).

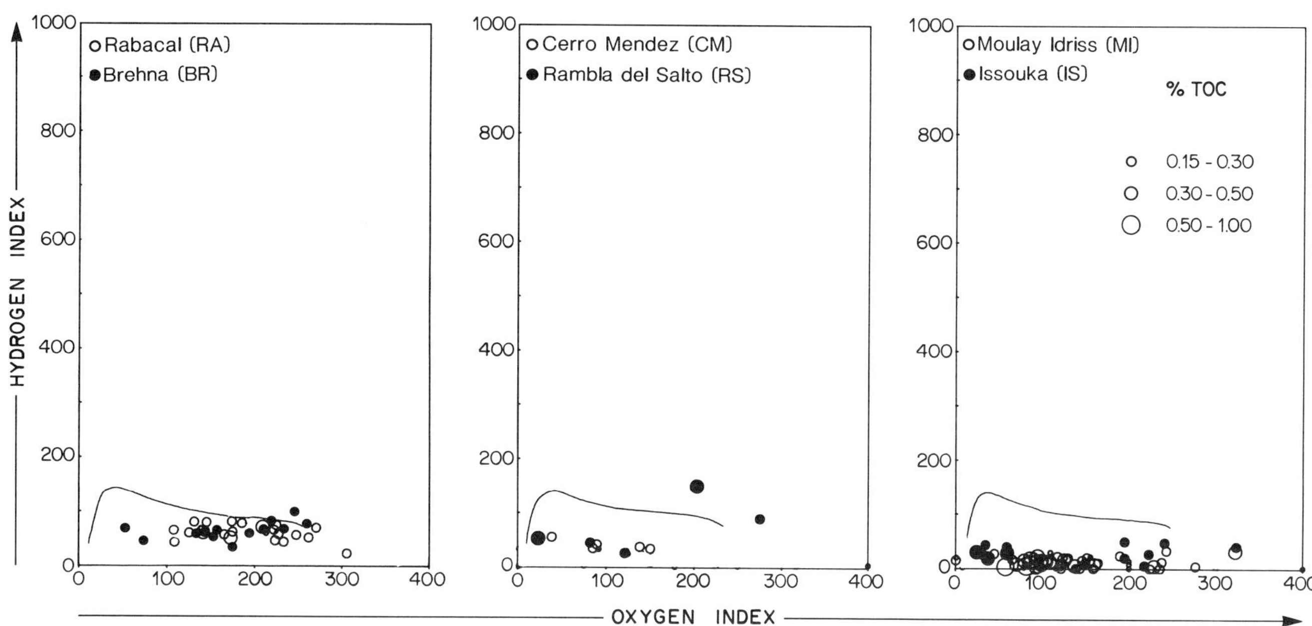

Figure 3. Quantity and characterization of organic matter from Portugal, Spain, and Morocco in HI-OI diagrams. Size of circles is proportional to total organic carbon (TOC) values. This material may either be recycled from older sediments by erosion or oxidized by biological processes.

During the late Lias, sedimentary patterns indicate vertical movement. In the western part of the southern Alps (Lombardian basin) the Ammonitico Rosso facies appears on the flanks of tilted blocks, whereas pelagic mudstone was deposited in the deeper part of the basin. In the eastern part (Belluno trough), sedimentation rates, which had been high since the Triassic, decreased rapidly. Numerous sections with organic-rich mudstones have been described in these two basins, and their organic content has been well documented by Jenkyns (1985, 1988), Jenkyns and Clayton (1986), and Farrimond et al. (1989). Between the Lombardian basin and Belluno trough, the Trento plateau forms a paleogeographic high where Toarcian organic-rich shales occur only on the eastern side. In order to analyze organic matter with the same methodology, two of the sections studied by Jenkyns (1988) were sampled.

The Monte Brughetto section (MB, Figure 1), which is 65-70 m thick, is characterized by shale-limestone cycles (Gaetani and Poliani, 1978). The upper Tenuicostatum or Falciferum zone, in the lower part of this section, is represented by 6 m of brown to black laminated shales rich in fish debris (Tintori, 1977). This level is the only one rich in organic matter (Figure 4), with TOC values ranging from 0.2 to 1.3% and averaging about 0.9%. The hydrogen indices of the richest samples are close to 400 mg HC/g TOC (Figure 5), which indicates a marine origin of the organic matter. However, the less organic-rich levels contain a mixture of Types II and III, or marine oxidized organic matter. The average Tmax is about 437°C and corresponds to the beginning of the "oil window." The other Toarcian samples from this section do not contain any organic matter.

The Val Vajont section (VV, Figure 1) is located in the Belluno area (Jenkyns et al., 1985). Extremely faulted at the base, this section contains 4 m of black mudstone overlain by gray limestones of the Bifrons zone. This black unit contains a maximum of 2.2% TOC with an average of around 0.9% (Figure 4). In the bottom of the section, as well as in the gray limestones from the top of the lower Toarcian, organic matter is absent. The hydrogen index of the richest sample reaches 640 mg HC/g TOC. Mean HI values are around 450 mg HC/g TOC. This organic matter can obviously be related to Type II (Figure 5). The petroleum potential is 4.2 kg HC/t of rock. The degree of maturity is very slightly less than that of the Monte Brughetto section as indicated by average Tmax values of 433 °C.

The elemental analysis of six kerogens coming from these two sections shows high H/C ratios ranging from 1 to 1.3, although the O/C ratios are rather low (0.07). These kerogens are located along the Type II evolution path in a Van Krevelen diagram (Figure 10). The distribution of saturated hydrocarbons (Figure 10) indicates the predominance of even-numbered *n*-alkanes in the C_{17}-C_{19} range together with the high concentration of isoprenoids. These chromatograms confirm the marine origin for the organic matter from northern Italy.

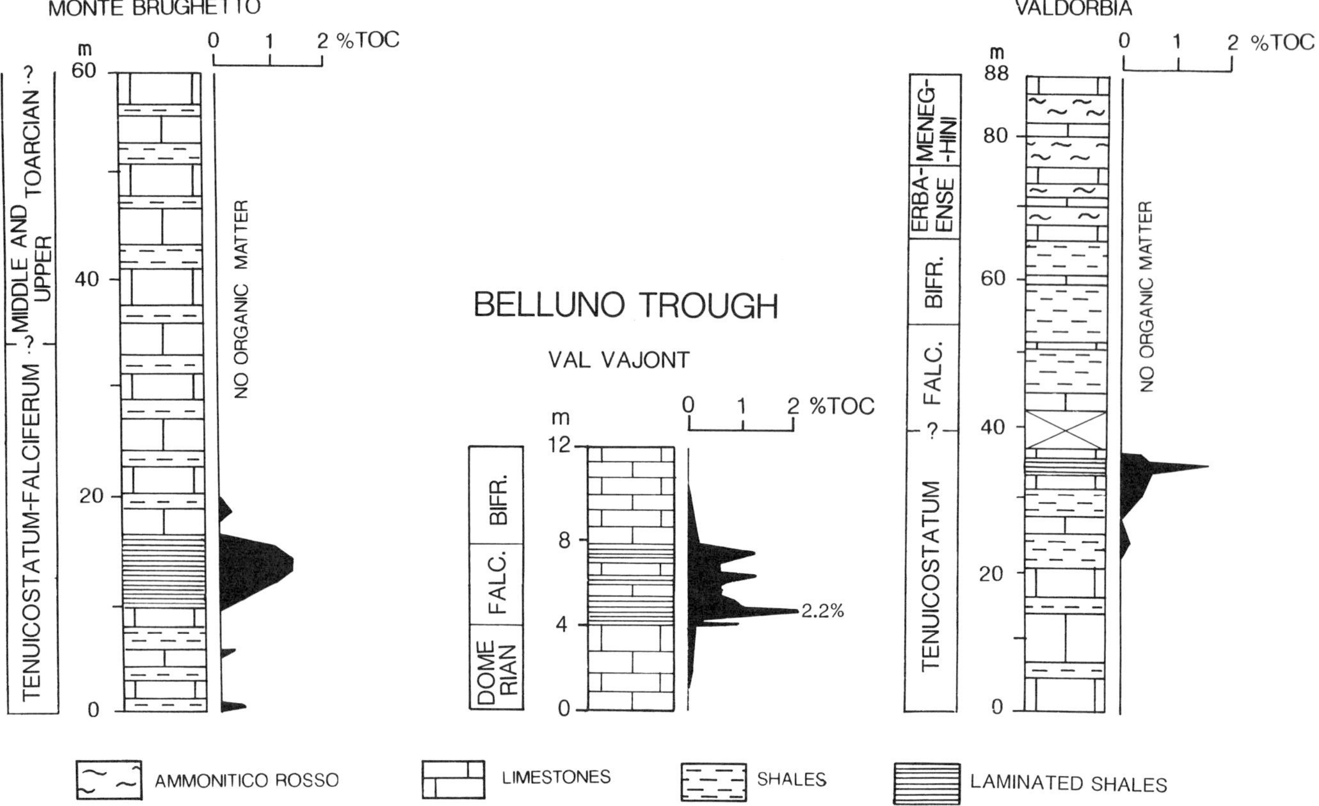

Figure 4. Vertical distribution of organic matter in three sections in Italy (Lombardian, Belluno, and Umbrian basins). The organic content is concentrated within the Falciferum zone in northern Italy and at the top of the Tenuicostatum zone in the Umbrian basin.

Umbrian Basin — Central Italy

The Marches-Umbrian basin belongs to the external zone of the Apennines (central Italy). During the middle Lias, the carbonate platform that had predominated until then, broke up, leading to rapid subsidence and the formation of tilted blocks. The Valdorbia section (VD, Figure 1) is a classic and well-studied outcrop in the Umbrian basin (Donovan, 1958; Elmi, 1981; Jenkyns and Clayton, 1986). The marly facies that developed at the base of the Toarcian is known as "Unita calcareo-marnosa del Sentino" (Centamore et al., 1971) and is transitional with an Ammonitico Rosso within the Erbaense zone. The entire Toarcian interval in the Valdorbia section is 88 m thick. The color of the more argillaceous units is greenish gray and they probably belong to the Tenuicostatum zone. In the upper part of this zone, several units have TOC contents ranging from 0.3 to 1.6% (Figure 4). The other samples from the section are devoid of organic matter, even the argillaceous Falciferum units. The hydrogen index of the organic-rich gray marls reaches 390 mg HC/g TOC (Figure 5) which indicates a marine origin. The average Tmax is close to 427°C, indicating an immature level of thermal maturation for the Valdorbia section. In the southern part of the Umbrian basin, the lower Toarcian crops out as a marly facies known as the "Marne di Monte Serrone" (MS, Figure 1) defined by Pialli (1969). All of the section is devoid of organic matter.

Western Greece — Ionian Zone

At the begining of the Jurassic, the Ionian zone in western Greece was an extending platform. During the late Lias, a linear depression developed in this zone (Aubouin, 1958). The differential subsidence of the basin has caused great facies variations. Whereas hiatuses are located on top of tilted blocks, Ammonitico Rosso facies occur on the slope, and "Schistes à Posidonies Inférieurs" are present in the deeper part of the half-grabens. These siliceous and argillaceous limestones and laminated shale deposits crop out widely in western Greece and on Corfu Island. Some units show reworking of the Ammonitico Rosso facies into the "Schistes à Posidonies Inférieurs." In the 12 sections studied, great variations can be found in thickness and facies, but they are lithologically correlated (Figure 6). The calcareous nannofossil association (*Calyculus* sp.,

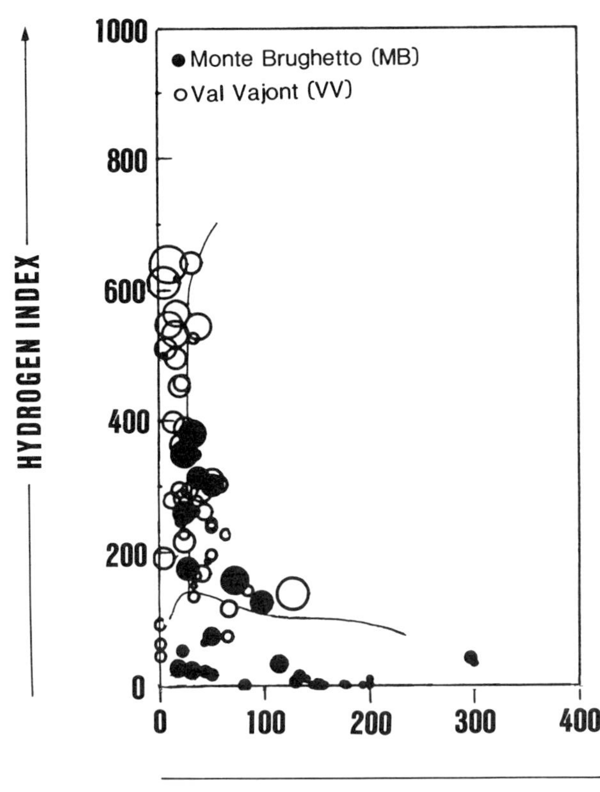

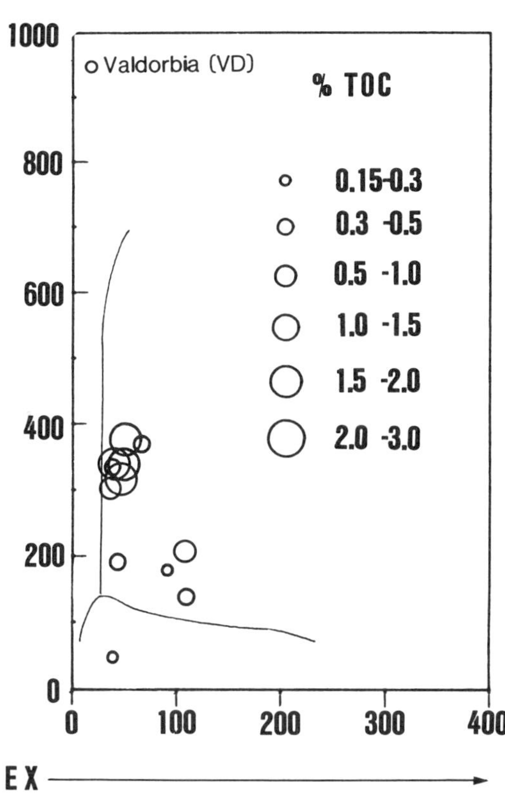

Figure 5. Characterization of the type of organic matter from Italy in HI-OI diagrams. Size of circles is proportional to total organic carbon (TOC). Most of the organic matter is marine, except in the Monte Brughetto section where terrestrial influences may exist in less organic-rich samples.

Carinolithus superbus, *Mitrolithus jansae*, *M. elegans*, *Crepidolithus cavus*, *Lotharingius hauffii*, and *Parabdholithus liasicus*) found in "Schistes à Posidonies Inférieurs" indicates their presence during the early Toarcian (Figure 2). Analysis of organic matter from some of these sections has previously been summarized by Baudin et al. (1988) and Baudin and Lachkar (1990). The formation contains between 0.3 and 5.2% TOC, with an average around 1.2% (Figure 6). The hydrogen index varies greatly but reaches 650 mg HC/g TOC in the richest units (Figure 7). The samples chosen for elemental analysis of their kerogen were selected from several sections. H/C and O/C ratios vary greatly but are located between the Type II and the Type III evolution paths in a Van Krevelen diagram (Figure 10). The saturated hydrocarbons show two modes (Figure 10). The larger one is in the range of C_{17}-C_{19} *n*-alkanes, which is typical of a phytoplanktonic origin for the organic matter. Pristane and phytane are abundant, comprising between 20 and 30% of the saturated fraction. Phytane is slightly more abundant than pristane at the top of the sequence, which might indicate a reducing environment. The smaller mode is in the range of C_{25} and C_{30} and corresponds to a derivative fraction of continental organic matter (Tissot et al., 1975). The Tmax average is 420°C indicating that the thermal evolution of the organic matter is slight. Study of palynological slides also confirms this dual origin. About 90% of the material is amorphous and flocky and contains dinoflagellates and Crassosphaera (Baudin and Lachkar, 1990), which is consistent with the planktonic origin. Rare spores, pollen grains, and woody debris fragments, however, indicate some continental influences. The spores and pollen grains belong mostly to the following taxa: *Chasmatosporites elegans*, *C. hians*, *C. major*, *Deltoidospora toralis*, *D. minor*, *Classopollis* sp., and *Corollina meyeriana*.

THE TOARCIAN OF THE ARABIAN PLATFORM

Saudi Arabia

In central Saudi Arabia, the Marrat formation has been investigated from four boreholes located north

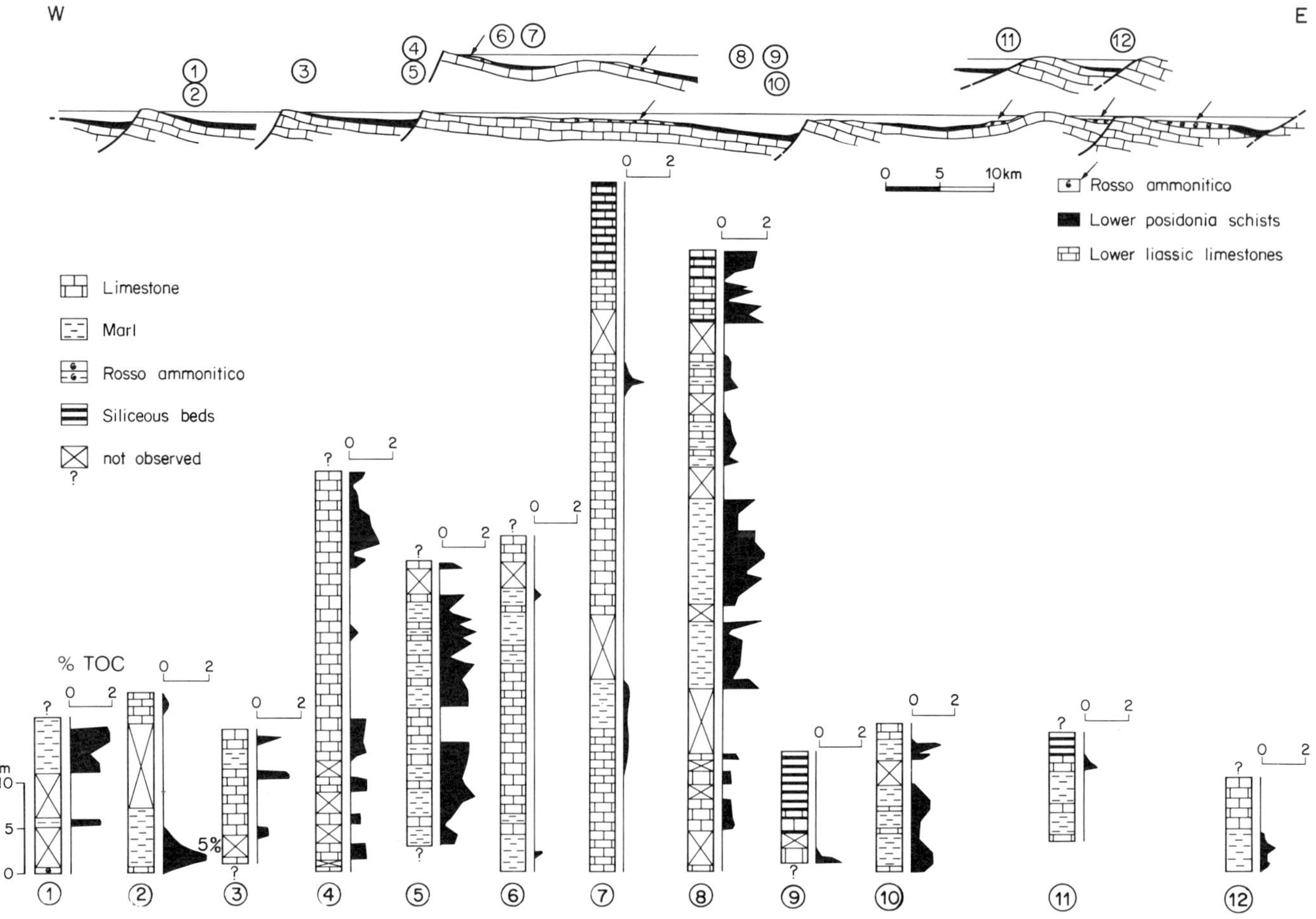

Figure 6. Simplified lithologic sections at the 12 sites studied from the Ionian zone (western Greece), and their approximate locations on a tentative palinspastic profile (modified from IGRS-IFP, 1966). Generally the total organic carbon in these sections ranges from 0.25 to 2%; a maximum value of 5.2% was recorded in section 2, but such an amount is anomalous. The northern basins (sections 6, 7, 11, and 12) were more open to the ocean, and the flank of the tilted block in section 3 contains little organic matter.

of Riyadh. The formation is typically sandy shale, corresponding to a shallow-water marine origin with fluvial influences. At outcrops, two carbonate units contain ammonites which date the Serpentinus zone and the base of the Bifrons zone (Enay et al., 1987). Samples from boreholes (AZ 299 and MA12, MA10 and MA5, Figure 1) are the only ones that show enrichment in organic matter in the lower part of the Serpentinus zone (Figure 8). The TOC values reach 1%, but belong to residual organic matter in an HI-OI diagram (Figure 8). The elemental analyses of the richest samples show a relatively high O/C ratio, implying oxidized organic matter (Figure 10).

Persian Gulf

The Toarcian from the Marrat formation sampled near Riyadh corresponds to littoral facies and is probably not characteristic of the whole Arabian shelf. A few samples from the Persian Gulf have been analyzed to evaluate the organic content in a slightly different environment. Samples from the top of the Marrat formation drilled near Farsi Island (holes FA 1 and FA 2, Figure 1) were analyzed. TOC values are close to 0.5%, and pyrolysis shows very high oxygen indices. However, the top of the Marrat formation in the Persian Gulf seems to be richer in organic matter than the same section at Riyadh.

THE TOARCIAN OF THE NORTH TETHYAN MARGIN

Iran

During the Early Triassic and Lias, the Shemshak formation, which consists of detrital deposits with coal zones, was deposited in what is now the Elburz

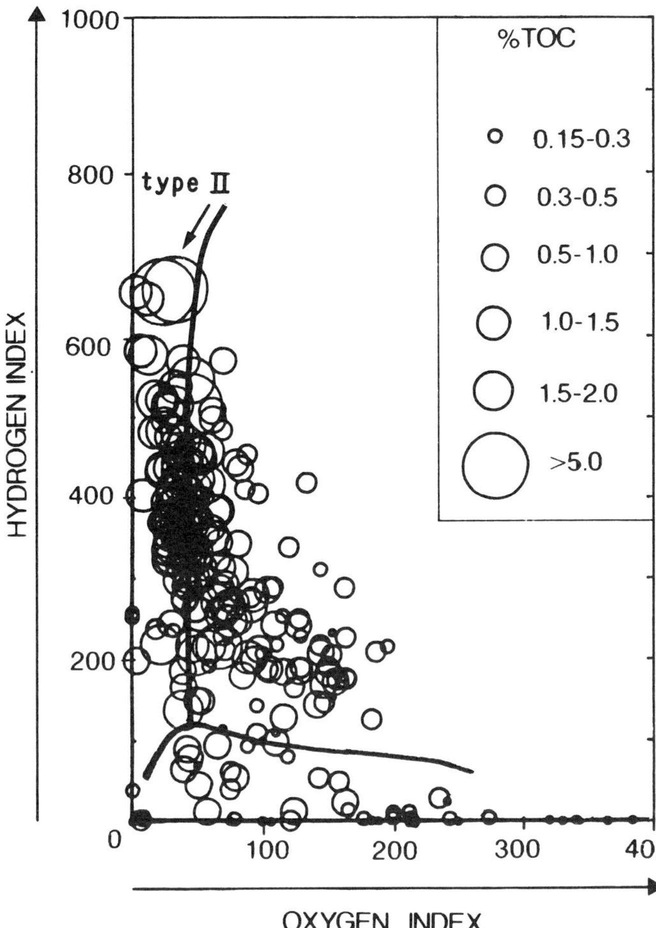

Figure 7. Characterization of the type of organic matter in Ionian Toarcian deposits by HI-OI diagrams. Size of circles is proportional to total organic carbon (TOC). The richest samples are located along a Type II evolution path, suggesting a marine origin.

range (northern Iran) (Assereto, 1966). This formation can be considered as a deltaic complex characterized by its paralic to continental facies and its richness of plant fossils (Stampfli, 1978; Vozennin-Serra and Taugourdeau-Lantz, 1985). During Toarcian time, the depositional environment became more marine. Ammonites from the Serpentinus to the Aalensis zones have been described by Assereto et al. (1968) and Rad (1986). In the Central Elburz, the Toarcian deposits vary between 300 and 450 m in thickness, and consist of alternating sandstones and coaly argillites. In the Vaneh section (VA, Figure 1), the organic matter in the argillaceous units ranges from 0.5 to 7% TOC (Figure 9). The relatively thin coal units have generally high TOC values, some of which reach 60%. This organic matter can be classified as Type III on a HI-OI diagram (Figure 9). Its degree of thermal maturity is very high (Tmax is generally over 485°C), and the entire organic content could be quantified only by the Leco method. The high thermal maturity in this area seems to be due to deep burial and heating by volcanic activity which was vigorous during early Holocene. Elemental analyses were performed on five samples of the kerogen. The H/C and O/C ratios are low, corresponding to the ultimate extent of the path of evolution for Type III in a Van Krevelen diagram (Figure 10). The chromatogram of the saturated hydrocarbons shows the predominance of low carbon numbers ($< C_{20}$) without any carbon preference. The *n*-alkanes distribution decreases abruptly after C_{20}, which is mainly correlated with the catagenesis maturation. The pristane and phytane amounts are low, with a slight predominance of pristane. The Pr/Ph ratios are close to 2, which is usual for Type III organic matter.

DISTRIBUTION OF ORGANIC MATTER DURING THE TOARCIAN

Figure 11 is a compilation of recent paleogeographic maps (Dercourt et al., 1985; Le Nindre et al., 1987; Ziegler, 1988; Dercourt et al., 1990). Paleolatitudinal data for the Toarcian (Honner and Heller, 1983; Channel et al., 1984; Galbrun and Daly, 1987) are in agreement with what is shown on the proposed map. During the Toarcian, the paleogeography of the Mediterranean and Middle East areas was quite different. The oceanic floor of the Tethys widened abruptly to the east. The subduction of the oceanic basement under the Eurasian plate was marked by a volcanic arc, which is especially evident in northern Turkey and Iran. The coal-bearing belt of the temperate humid climate runs approximately along latitude 45°N from Middle Asia to the northern Alps. This setting promoted the formation of deltaic complexes with numerous coal beds, as well as the Shemshak formation. The southern North Sea and northwestern Europe consisted of a wide shelf where emerged lands were bounded by small basins (Ziegler, 1988). Sedimentation within the European shelf was mostly argillaceous, but sandstones and ironstones developed around islands. Toarcian time was also a period of subsidence on the newly formed margins, especially within the Apulian block: Austria, Italy, Yugoslavia, and Greece. The arid tropical zone corresponding to shallow-water carbonate platform facies with reef complexes and, locally, evaporites was located on the northern margin of Africa from Portugal to Saudi Arabia.

All the sections presented here, as well as data published previously (Table 1), have been utilized in the preparation of Figure 11. Geochemical data obtained from the samples have been extrapolated taking into account all available geological and

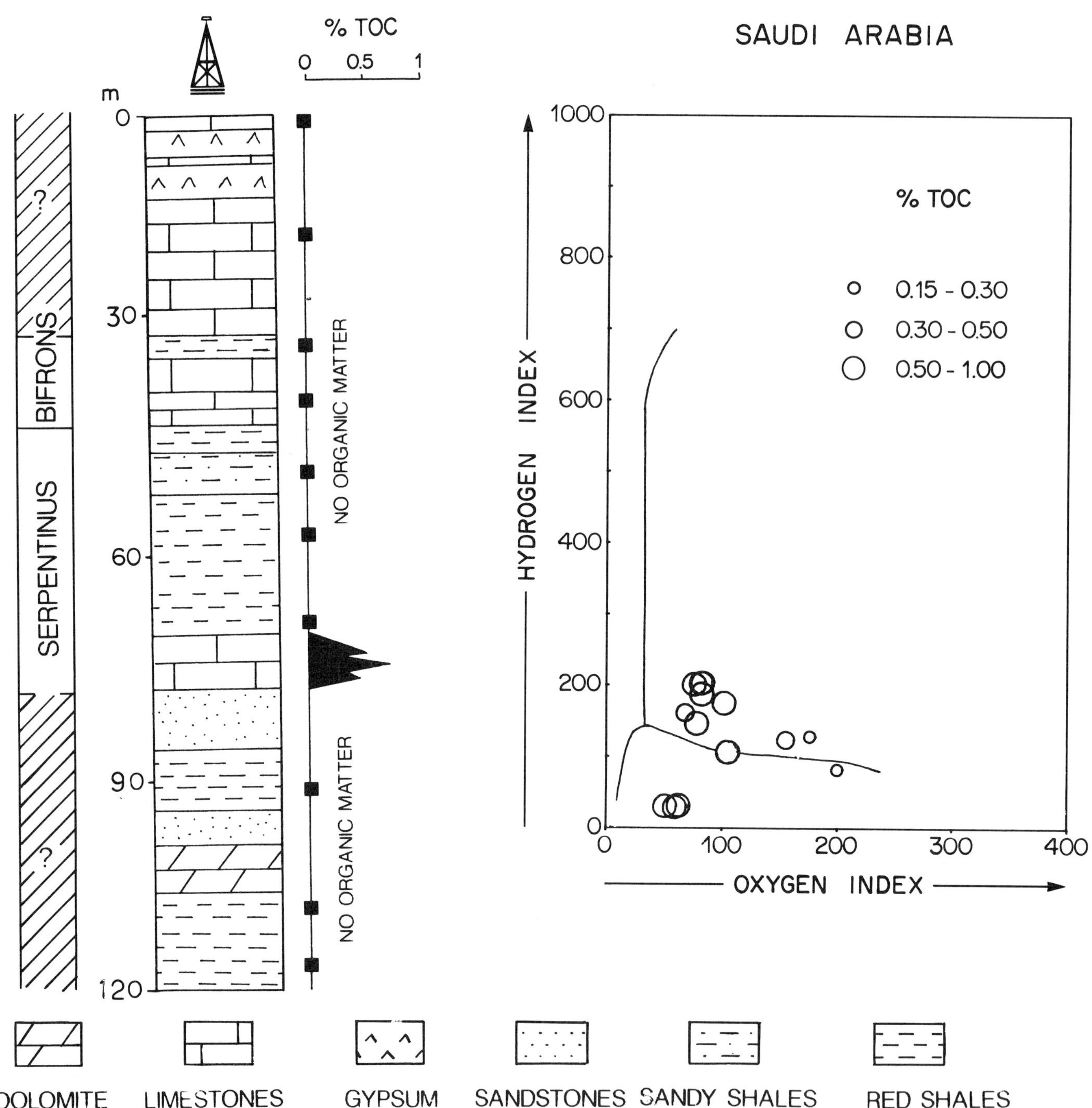

Figure 8. Organic-carbon content of samples from the Marrat formation (Toarcian) from Saudi Arabia. The HI-OI diagram indicates that the organic matter had a residual and/or terrestrial origin.

biostratigraphic data. Previous maps of the distribution of organic matter within the Paris Basin (Espitalié et al., 1987) also have been integrated.

TOTAL ORGANIC CARBON

Figure 12, which is a map of total organic carbon distribution during Toarcian time, clearly shows the differences in amount of organic matter between the boreal realm and the Tethyan domain. The organic-carbon values for the European shelf range from 1 to 20%, with an average of 5%, whereas in the Tethyan margin, organic matter is preserved in small basins. In the section studied, the TOC values fluctuate from 0.1 to 2%, with an average of 1%. A maximum value of 5.2% was recorded in western Greece, but such an amount seems exceptional in Tethyan sections. Generally, organic matter was not preserved in open,

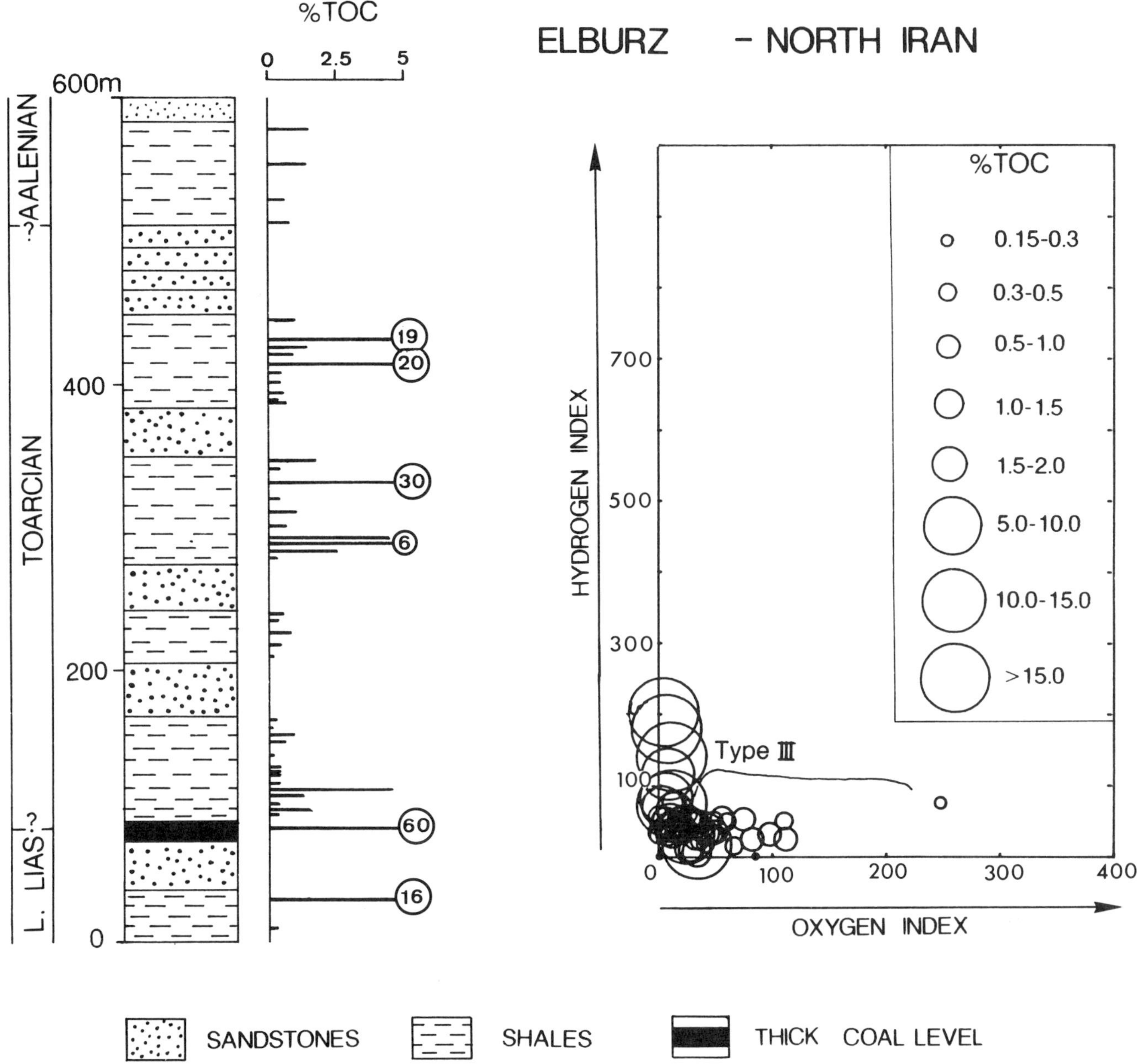

Figure 9. Organic-carbon content of the shaly units in the Vaneh section of the Shemshak formation in northern Iran. Circles indicate amounts of TOC above 5%. In the HI-OI diagram these samples are located near the bottom of the Type III evolution path.

deep basins with oxygenated circulation. However, the Eurasian margin was prone to the deposition of coal. TOC values reach 60% in some units, with a mean value of 2% in the shaly beds. The other types of basins (Portuguese and Spanish margins, Middle Atlas basin and Arabian platform) contain either no organic matter or very little. In all of these basins the organic content is less than 1%, with a mean value of 0.35%.

HYDROGEN INDEX

Figure 13, which shows hydrogen-index distribution for Toarcian rocks, indicates the same HI values for European organic-rich mudstones and the richest samples from small basins within the Tethyan margin. The HI ranges from 400 to 650 mg HC/g TOC in both areas, indicating a marine origin for the organic matter (Type II). However, the less

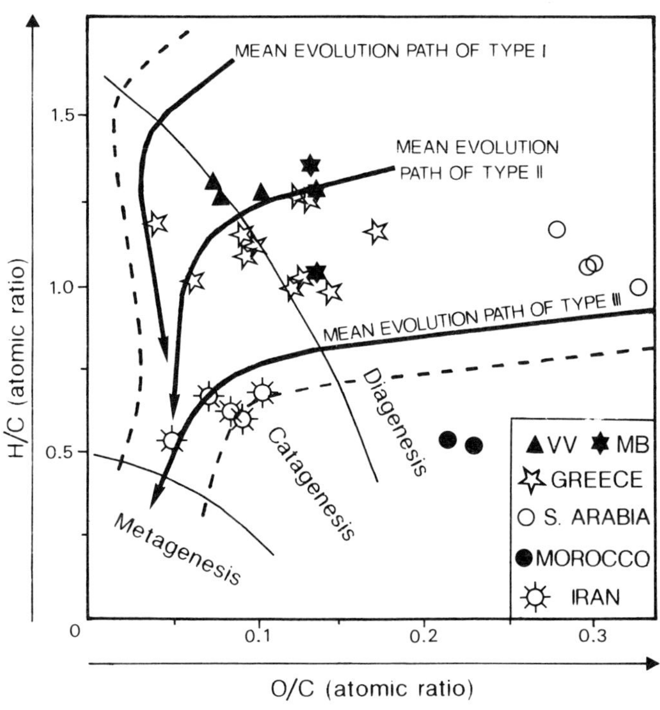

Figure 10. Elemental analysis of kerogen from different Toarcian deposits within the Mediterranean Tethys and Middle East in an H/C-O/C diagram. VV = Val Vajont and MB = Monte Brughetto sections (northern Italy).

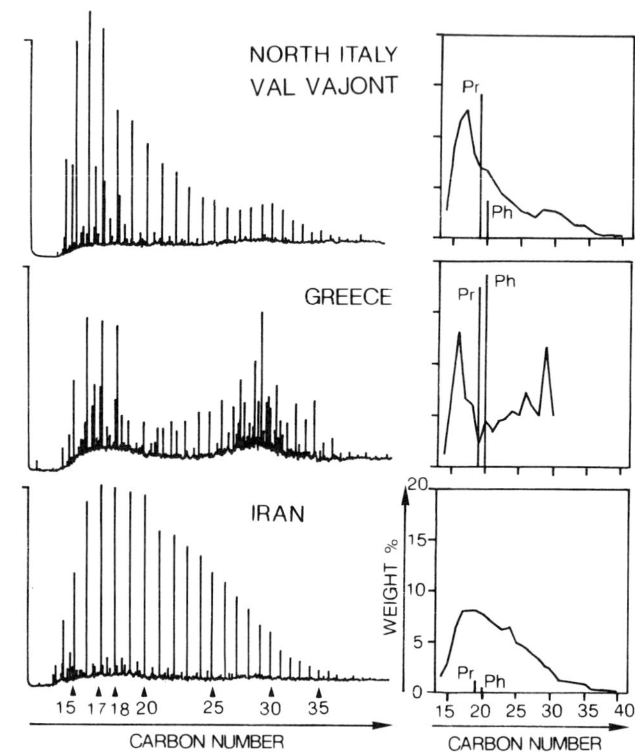

Selected chromatograms of the saturated-hydrocarbon fractions of samples from northern Italy, Ionian zone (Greece), and Iran. Pr = pristane and Ph = phytane.

organic-rich units from Tethyan basins contain a mixture of Type II and III, or marine oxidized organic matter. Elemental analyses of selected kerogen and gas chromatography of the saturated fraction confirm the phytoplanktonic origin. The HI values from Iran are low but correspond to mature continental organic matter (Type III). Low hydrogen indices occur in the Norwegian offshore area where terrestrial organic input has been documented (Nagy et al., 1984; Thomas et al., 1987). The remainder of the Tethyan region has very low hydrogen indices, corresponding to residual organic matter (Type IV).

PETROLEUM POTENTIAL

Figure 14 shows the petroleum potential of Toarcian deposits in kg hydrocarbons/ton of rock. In terms of oil potential, only the small Tethyan basins contain source-rocks of Toarcian age, and these are generally limited to the Falciferum/Serpentinus zone. In any case, their petroleum potential is moderate (1 to 5 kg/t of rock) in comparison to that of northwestern Europe (5 to 20 kg/t). The north Tethyan margin, which is rich in coal, doubtless has a potential for gas that should not be overlooked (2 to 5 kg/t). Rad (1982) concluded that the Iranian coaly formation was the source-rock for western Caspian Sea oils. Other Toarcian deposits in the Mediterranean and Middle East areas have no petroleum potential.

MODEL OF DEPOSITION

The widespread occurrence of lower Toarcian source rocks over European areas and Tethyan basins calls for a global depositional control. The distribution of lower Toarcian organic facies on the flanks of tilted blocks and on isolated pelagic highs in northern Italy led Jenkyns (1985, 1988) to invoke a well-developed oxygen-minimum zone within the midwater.

In light of the present data and maps showing the regional distribution of organic matter, the depositional model can be more precisely developed. Organic-rich facies did not necessarily occur on the flanks of all tilted blocks. For instance, in western Greece, the reworking of Ammonitico Rosso facies within the organic-rich shales suggests that the latter correspond to the deeper part of the basin, and the organic matter does not seem to be well preserved

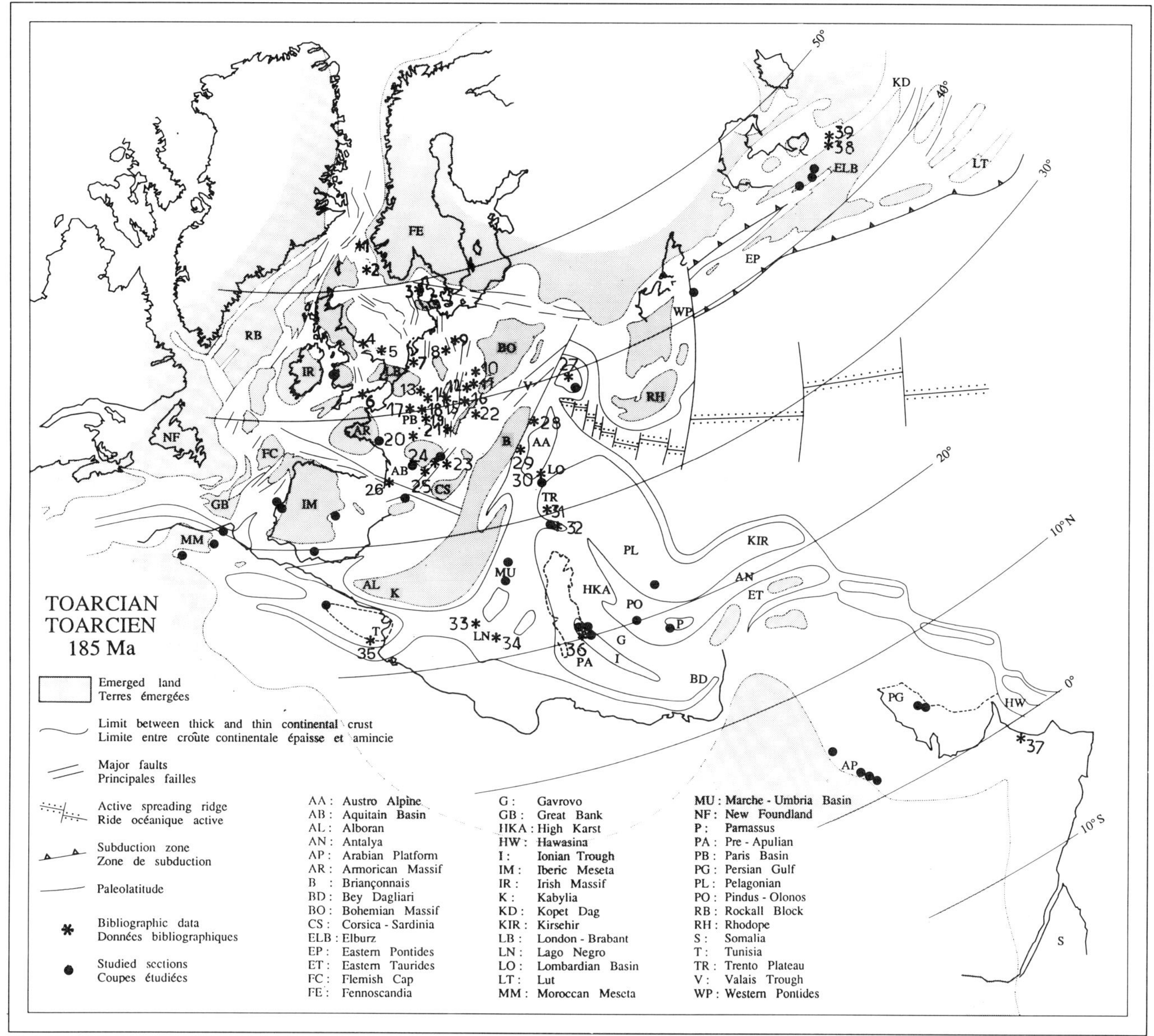

Figure 11. Paleogeographic map of the Mediterranean and Middle East during Toarcian time. Compiled from Dercourt et al. (1985); Le Nindre et al. (1987); Ziegler (1988); and Dercourt et al. (1990). Numbers on the map are keyed to bibliographic references shown on Table 1.

on the flanks of tilted blocks. The relative paleobathymetry within the Apulian area suggests that the pelagic deposits without organic-rich shales (Generoso basin) correspond to the larger and deeper basins. This fact supports an oxygen-minimum zone (Jenkyns, 1985, 1988) but also may be explained by oxidation of the organic matter by deep oceanic currents. We suggest that during Toarcian time, marine organic matter was preserved on the bottoms of half grabens (Lombardian Belluno and Ionian restricted basins), whereas the larger basins were devoid of anoxic conditions. Organic geochemical indicators of paleoenvironment studied in the southern Alps (Farrimond et al., 1989) corroborate this hypothesis.

Besides, the absence of organic matter in western Tethys basins does not agree with high surface productivity coupled with a large oxygen-minimum zone. In a large oxygen-minimum-zone model, organic matter should be preserved everywhere around the "cul de sac" of the Tethys basin.

We propose that oxygen depletion occurred only on the bottoms of restricted newly formed basins within the Tethyan area. The organic-rich Toarcian

Table 1. Supplemental data to accompany Figure 11.

No.	Locality or Basin	Formation	Bibliographic References
1	Horda Platform	?	(Ronnevik et al., 1983)
2	Viking Graben	Dunlin (Drake mbr)	(Nagy et al., 1984; Dahl and Speers, 1987)
3	Jutland	Fjerritslev (FIII)	(Sørensen, 1987)
4	Yorkshire	Jet-Rock	(Morris, 1979a, 1979b; Myers and Wignall,1987; Farrimond et al., 1989)
5	Sole pit trough	Posidonia shales	(Barnard and Cooper, 1983)
6	Dorset	Downcliff shales	(Ebukanson and Kinghorn, 1985, 1986)
7	Rijswijk	Posidonia shales	(Bodenhausen and Ott, 1981)
8	Westphalia	Posidonienschiefer	(Durand, IFP report, 1976)
9	Hils syncline	Posidonienschiefer	(Mann et al., 1986; Rüllkotter et al., 1987; Littke and Rüllkotter, 1987)
10	Franconia	Posidonienschiefer	(Küspert, 1983)
11	Swabian Jura	Posidonienschiefer	(Küspert, 1982, 1983)
12	Dotternhausen	Posidonienschiefer	(Moldovan et al., 1985)
13	Ardennes	Schistes cartons	(Espitalié et al., 1987)
14	Lorraine	Schistes cartons	(Huc, 1976, 1977)
15	Alsace	?	(IFP data, unpublished)
16	Gipft	?	(Küspert, 1983)
17	East of Paris	Schistes cartons	(Espitalié et al., 1987)
18	Champagne-Brie	Schistes cartons	(Espitalié et al., 1987)
19	Bourgogne-Jura	Schistes cartons	(Espitalié and Madec, 1981; Thomas, 1977; Broquet and Thomas, 1979)
20	Couy	Schistes cartons	(Pradier and Gauthier, 1988)
21	Switzerland Jura	Posidonia Schiefer	(Gorin and Feist, 1988, in press)
22	Pre-Alps	Marnes à Ammonites	(Mettraux et al., 1986)
23	Le Brusquet	?	(Tribovillard, 1988)
24	Ardèche	Schistes cartons	(Dromart et al., 1989)
25	Great Causses	Schistes cartons	(Trümpy, 1983)
26	Aquitain	Arg.de Palombières	(Carozzi et al., 1972)
27	Urkùt	?	(Polgari et al., 1989)
28	Sachrang	Allgaü Schichten	(Jenkyns, 1988; Farrimond et al, 1989)
29	Elbigenalp	Allgaü Schichten	(Jenkyns, 1988; Farrimond et al.,1989)
30	Lombardy	Sogno	(Jenkyns, 1988; Farrimond et al.,1989)
31	Val di Vesa	?	(Jenkyns, 1988; Farrimond et al.,1989)
32	Belluno-Julian	Igne	(Jenkyns, 1988; Farrimond et al.,1989)
33	Sclafani	Crisanti	(Jenkyns, 1988)
34	Lago Negro	Sirino	(Jenkyns, 1988)
35	North-South Axis	Nara moyen	(Soussi et al., 1988, 1989)
36	Epirus	Schistes à Posid.	(Jenkyns, 1988; Farrimond et al., 1989)
37	Oman	Mafraq	(Grantham et al., 1988)
38	Jarjan	Shemshak	(Rad, 1982, 1986)
39	Shahpassand	Shemshak	(Rad, 1982, 1986)

units found on isolated paleohighs are assumed to have been formed under local anoxic conditions. In the deeper pelagic basins, the lack of organic matter is interpreted as being due to oxidation caused by oceanic circulation.

CONCLUSIONS

The maps of organic-carbon distribution show the heterogeneity in the quantity and quality of organic matter that was deposited during Toarcian time. During Serpentinus/Falciferum times, the northern European area was the site of deposition of large quantities of marine organic material, whereas lower concentrations occur in the Tethyan region. In the Tethyan area, organic-rich units are preserved only in deep, isolated basins within neritic platforms and contain a more degraded type of marine organic matter than sediments from the European area. The relatively lower organic content, hydrogen indices, and lack of steroids (Farrimond et al., 1989) typical of Tethyan Toarcian shales may indicate that less planktonic material was produced in the surface waters and/or that the water was more oxygenated.

Local environmental factors (e.g., block faulting) apparently controlled the deposition of organic matter on the Tethyan margins during a favorable period of organic productivity promoted by high sea level. The well-oxygenated basins in the western Tethys, Arabian platform, and Persian Gulf were unfavorable environments for the preservation of organic matter, and the northern margin of the Tethys region was prone to coal deposition.

This contrasting distribution of types and quantities of organic matter reveals the existence of several different environments where organic matter was preserved to greater or lesser degrees during the lower Toarcian anoxic event.

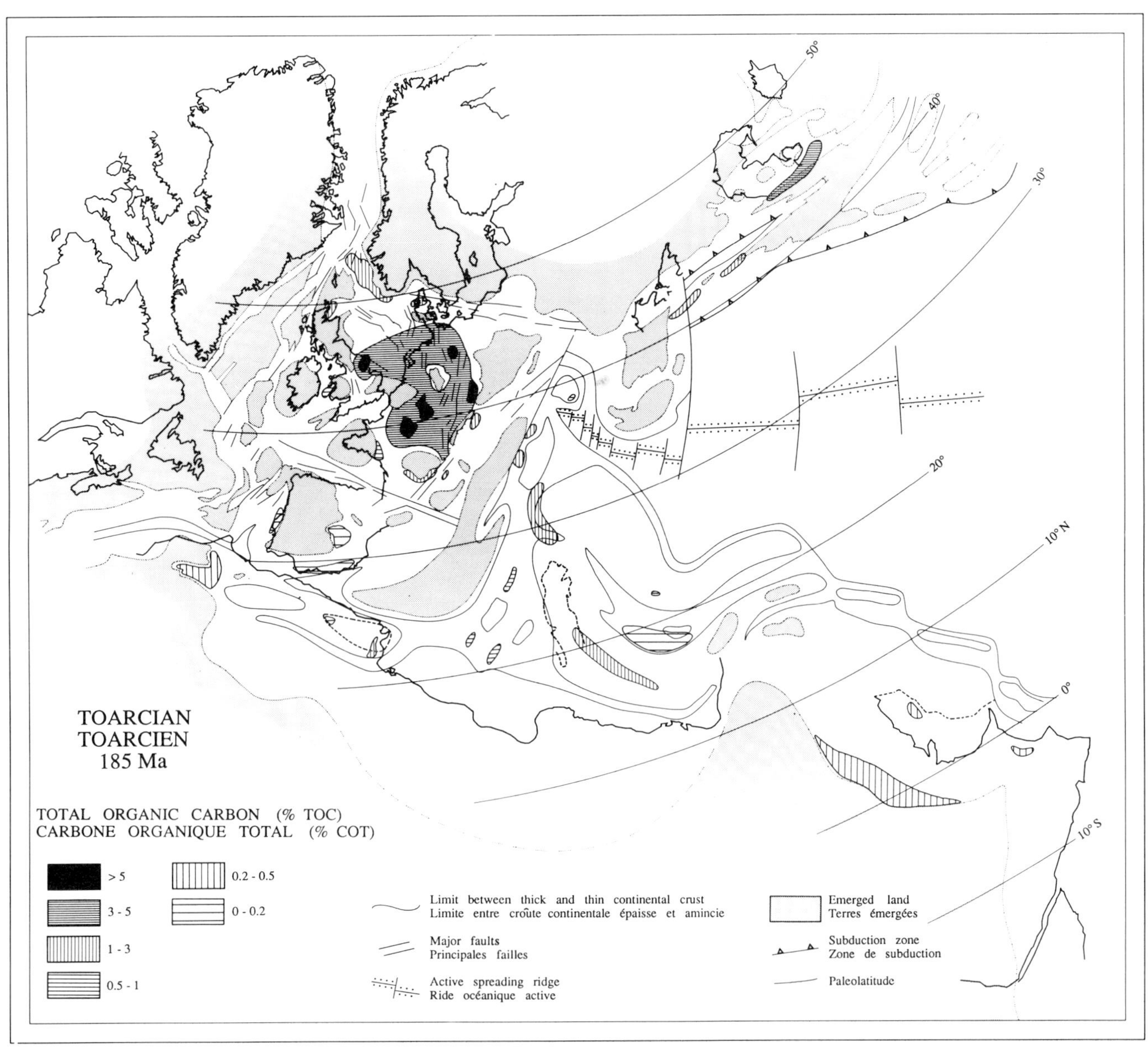

Figure 12. Map showing the total organic carbon (TOC) distribution for Toarcian units.

ACKNOWLEDGMENTS

The authors are grateful to the GS Tethys program for financial aid and scientific facilities. The manuscript has been greatly improved by the suggestions of Drs. S. Herron and R. Stein. We would also like to thank many colleagues who contributed to the study of the Tethyan area, and SNEA(P) and BRGM for the core samples.

REFERENCES

Assereto, R., 1966, The Jurassic Shemshak formation in Central Elburz (Iran): Rivista italiania de Paleontologia, v. 72, p. 1133-1182.

Assereto, R., P. D. W. Barnard, and N. Fantini-Sestini, 1968, Jurassic stratigraphy of Central Elburz (Iran): Rivista Italiana de Paleontologia, v. 74, p. 3-21.

Aubouin, J., 1958, A propos de la série adriatico-ionienne. Essai de corrélation stratigraphique en Epire: Annales Géologique Pays Helleniques, v. IX, p. 171-184, (in French).

Aubouin, J., 1963, Essai sur la paléogéographie post-triasique et l'évolution secondaire et tertiaire du versant sud des Alpes orientales (Alpes méridionales, Lombardie et Vénétie, Italie, Slovénie occidentale, Yougoslavie): Bulletin Société Géologique de France, v. 5, p. 730-766, (in French).

Azéma, J., A. Foucault, and E. Fourcade, 1979, Le Jurassique des cordillères bétiques. Symposium "Sédimentation Jurassique W européenne": Publication Spéciale de l'Association des sédimentologues francais, v. 1, p. 317-333, (in French).

Barnard, P. C., and B. S. Cooper, 1983, Geochemical data related to the Northwest European gas province, *in* Brooks, ed., Petroleum geochemistry and exploration of Europe: Special Publication Geological Society, v. 12, p. 19-33.

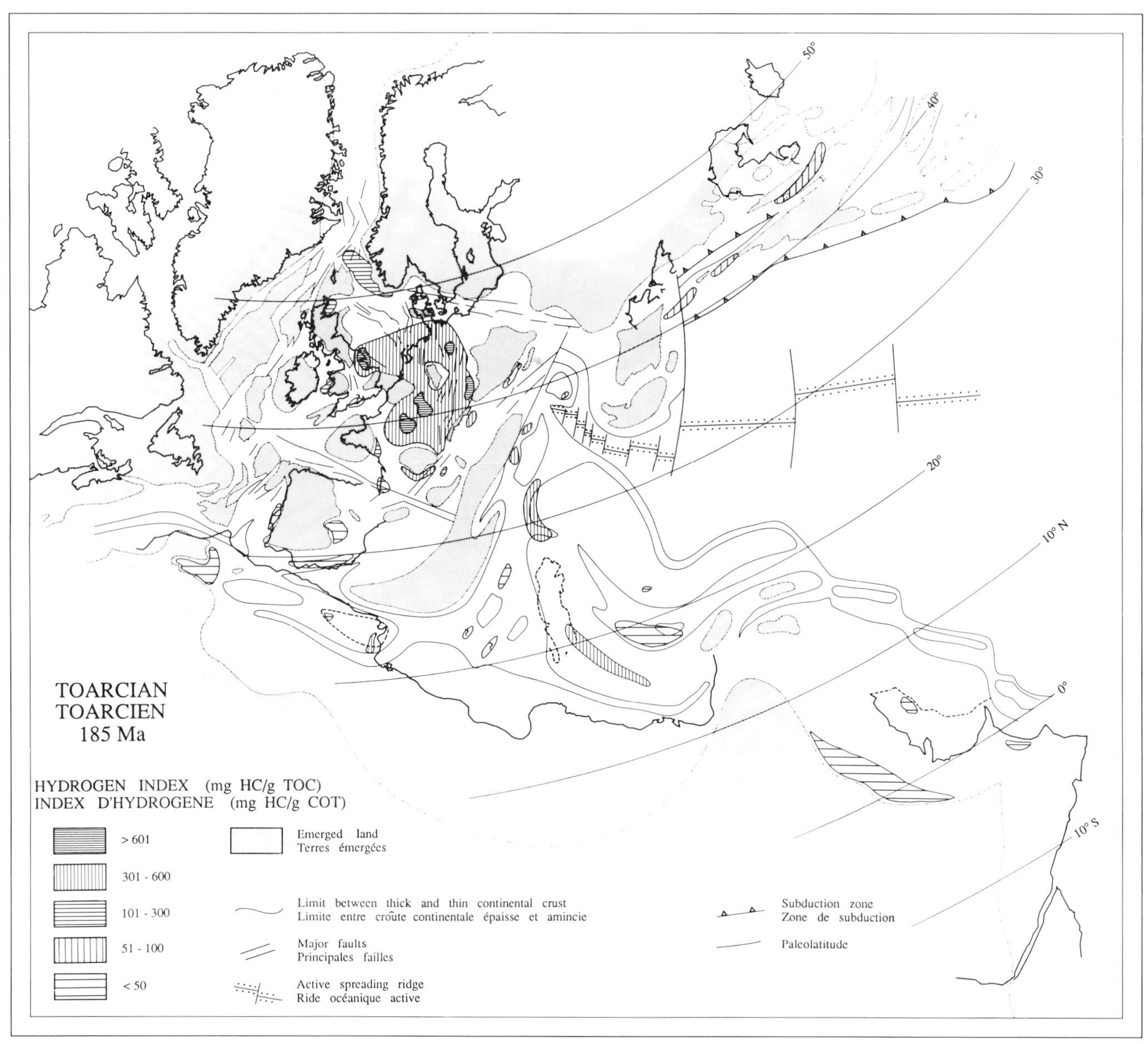

Figure 13. Hydrogen-index (HI) distribution for Toarcian rocks.

Baudin, F., J. Dercourt, J-P. Herbin, and G. Lachkar, 1988, Le Lias supérieur de la zone ionienne (Grèce): une sédimentation riche en carbone organique: Compte Rendu Académie des Sciences de Paris, v. 307, p. 985–990, (in French).

Baudin, F., and G. Lachkar, 1990, Geochimie organique et palynologie du Lias superieur en zone ionienne (Grèce): exemple d'une sédimentation anoxique conservée dans une paléo-marge en distension: Bulletin Société Geologique de France, v. 1, p. 123–132, (in French).

Benshili, Kh., 1989, Lias-Dogger du Moyen-Atlas plissé (Maroc). Sédimentologie, biostratigraphie et évolution paléogéographique: Documents des laboratoires de Géologie de Lyon, v. 106, 285 p., (in French).

Bernoulli, D., C. Caron, P. Homewood, O. Kälin, and J. van Stuijvenberg, 1979, Evolution of continental margins in the Alps: Schweiz Min Pet Mitt., v. 59, p. 165–170.

Bodenhausen, J. W. A., and W. F. Ott, 1981, Habitat of Rijswijk oil province, onshore, The Nethers, *in* Illing and Hobson, eds., Petroleum geology of continental shelf of North-West Europe: Institute of Petroleum, London, p. 301–309.

Broquet, P., and M. Thomas, 1979, Quelques caractères géologiques et géochimiques des schistes bitumineux du Toarcien franc-comtois: Bulletin du Centre de Recherche et d'exploration de Pau, v. 3, p. 265–280, (in French).

Carozzi, A. V., J. Bouroullec, R. Deloffre, and J. L. Rumeau, 1972, Microfaciès du Jurassique d'Aquitaine: Bulletin du Centre de Recherche et d'Exploration de Pau, Special Memoir, v. 1, (in French).

Centamore, E., M. Chiocchini, G. Deiana, A. Micarelli, and U. Pieruccini, 1971, Contributo alla conoscenza del Giurassico dell'Appennino Umbro-Marchigiano: Studi geol. Camerti, v. 1, p. 7–89, (in Italian).

Channel, J. E. T., W. Lowrie, P. Pialli, and F. Venturi, 1984, Jurassic magnetic stratigraphy from Umbrian (Italian) land sections: Earth Planetary Science Letters, v. 68, p. 309–325.

Colo, G., 1961, Contribution à l'étude du Jurassique du moyen-Atlas septentrional: Notes et mémoires du Service géologique du Maroc, v. 139, 226 p., (in French).

Comas-Renfigo, M. J., and A. Goy, 1978, El Pliensbachiense y Toarciense en la Rambla del Salto (Sierra Palomera, Teruel):

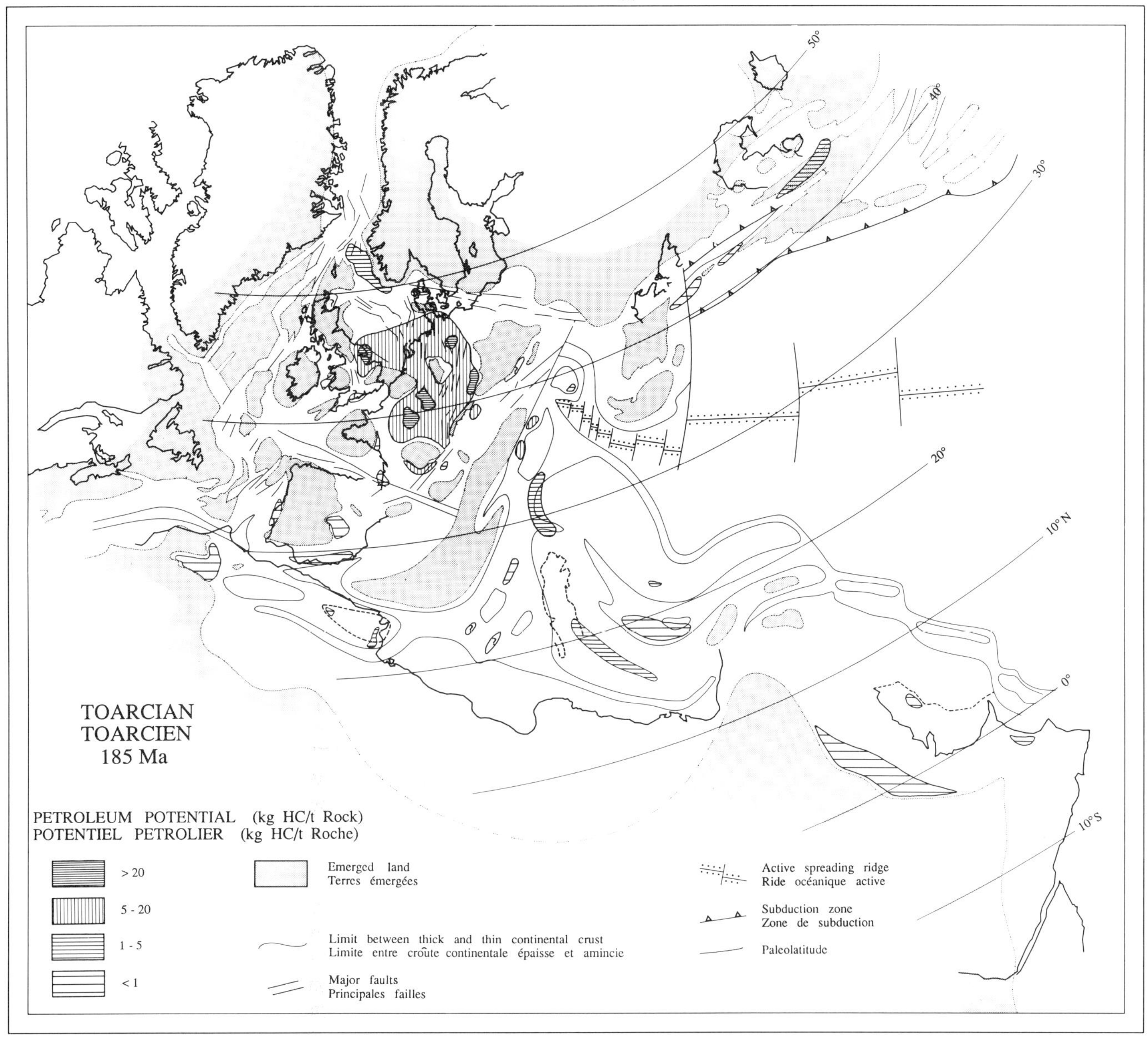

Figure 14. Petroleum potential in kg hydrocarbons/ ton of rock for European, Mediterranean, and Middle East Toarcian deposits.

Guia de las excursiones al Jurasico de la Cordillera iberica, Univ. complutense Madrid, 11 p., (in Spanish).

Cope, J. C., T. A. Getty, M. K. Howarth, N. Morton, and H. S. Torrens, 1980, A correlation of Jurassic rocks in the British Isles. Part I, Introduction and Lower Jurassic: Special Publication Geological Society, v. 14, 73 p.

Crux, J. A., 1987, Early Jurassic calcareous nannofossil biostratigraphic events: Newsletter in Stratigraphy, v. 17, p. 79-100.

Dahl, B., and G. S. Speers, 1987, Organic geochemistry of the Oseberg Field (I), *in* Thomas, ed., Petroleum geochemistry in exploration of the Norwegian shelf: Graham & Trotman, p. 185-196.

Dercourt J., L. V. Zonenshain, L. E. Ricou, V. G. Kazmin, X. Le Pichon, A. L. Knipper, C. Grandjacquet, I. M. Sborshchikov, J. Boulin, O. Sorokhtin, J. Geyssant, C. Lépvrier, B. Biju-Duval, J. C. Sibuet, L. A. Savotin, M. Westphal, and J. P. Lauer, 1985, Geological evolution of the Tethys belt from the Atlantic to the Pamir since the Lias: Tectonophysics, v. 123, p. 241-315.

Dercourt, J., E. Ricou, S. Adamia, G. Czaczar, J. Lefeld, M. Rakus, M. Sandulescu, A. Tollmann, and P. Tchoumatchenko, 1990, Evolution of the northern margin of Tethys from Anisian to Oligocene from Geneva to Baku, *in* Rakus, Dercourt, and Nairn, eds., Mémoire Société géologique de France, v. 154, p. 159-190.

Donovan, D. T., 1958, The ammonite zones of the Toarcian (Ammonitico Rosso facies) of Southern Switzerland and Italy: Eclogae Geologicae Helvetiae, v. 51, p. 33-60.

Dromart, G., J-P. Crumière, S. Elmi, and J. Espitalié, 1989, Géodynamique et potentialité pétrolières d'une marge de bassin: le Jurassique de la bordure ardéchoise (France, Sud-Est): Comptes-Rendus de l'Académie des Sciences, v. 309, p. 1495-1502.

Durand, B., 1976, Résultats d'analyses sur la matière organique des Posidonienschiefer (Toarcien allemand). Comparaison

avec les Schistes-cartons du Toarcien du Bassin de Paris: Rapport interne IFP, v. 23977, 76 p., (in French).
Durand, B., J. Espitalié, G. Nicaise, and A. Combaz, 1972, Etude de la matière organique insoluble (kérogène) des argiles du Toarcien du Bassin de Paris. Première partie : Étude par les procédés optiques, analyse élémentaire, étude en microscopie et diffraction électronique: Revue de l'Institut Francais du Pétrole, v. XXVIII, p. 865-884, (in French).
Ebukanson, E. J., and R.R.F. Kinghorn, 1985, Kerogen facies in the major Jurassic mudrock formations of southern England and the implication on depositional environments of their precursors: Journal of Petroleum Geology, v. 8, p. 435-462.
Ebukanson, E. J., and R. R. F. Kinghorn, 1986, Maturity of organic matter in the Jurassic of southern England and its relation to the burial history of sediments: Journal of Petroleum Geology, v. 9, p. 259-280.
Elmi, S., 1981, Sédimentation rythmique et organisation séquentielle dans les Ammonitico Rosso et les faciès associés du Jurassique de la Méditerranée occidentale. Interprétation des grumeaux et des nodules, *in* Farinacci and Elmi, eds., Proceeding of the Rosso Ammonitico Symposium: Tecnosi, Roma, p. 251-299, (in French).
Elmi, S., A. Goy, R. Mouterde, P. Rivas, and R. Rocha, in press, Correlaciones bioestratigraphicas en el Toarciense de la Peninsula iberica, *in* Actes du Congrès espagnol de Stratigraphie, (in Spanish).
Enay, R., Y. M. Le Nindre, C. Mangold, J. Manivit, and D. Vaslet, 1987, Le Jurassique d'Arabie Saoudite centrale: Nouvelles données sur la lithostratigraphie, les paléoenvironnements, les faunes d'Ammonites, les âges et les corrélations: Géobios, v. 9, p. 13-65, (in French).
Espitalié, J., B. Durand, J-C. Roussel, and C. Souron, 1973, Etude de la matière organique insoluble (kérogène) des argiles du Toarcien du Bassin de Paris. Deuxième partie: Études en spectroscopie infrarouge, en analyse thermique différencielle et en analyse thermogravimétrique: Revue de l'Institut Français du Pétrole, v. XXVIII, p. 37-66, (in French).
Espitalié, J., G. Deroo, and F. Marquis, 1985a, La pyrolyse Rock-Eval et ses applications: Revue Institut Francais du Pétrole, v. 40, p. 563-579, (in French).
Espitalié, J., G. Deroo, and F. Marquis, 1985b, La pyrolyse Rock-Eval et ses applications: Revue Institut Francais du Pétrole, v. 40, p. 755-784, (in French).
Espitalié, J., G. Deroo, and F. Marquis, 1986, La pyrolyse Rock-Eval et ses applications: Revue Institut Francais du Pétrole, v. 41, p. 73-89, (in French).
Espitalié, J., and M. Madec, 1981, Les schistes bitumineux du Toarcien de la bordure orientale du Bassin de Paris: Bulletin du Centre de Recherche et d'exploitation de Pau, v. 5, p. 461-472, (in French).
Espitalié, J., F. Marquis, L. Sage, and I. Barsony, 1987, Organic geochemistry of the Paris Basin, *in* Brooks and Glennie, eds., Petroleum geology of north west Europe: Graham & Trotman, p. 71-87.
Farrimond, P., G. Eglinton, C. Brassell, and H.C. Jenkyns, 1989, Toarcian anoxic event in Europe: An organic geochemical study: Marine and Petroleum Geology, v. 6, p. 136-147.
Faugères, J. C., 1975, Le Toarcien inférieur du Jbel Dehar en N'sour (Rides prérifaines, Maroc): précisions biostratigraphiques et remarques paléontologiques sur les Ammonites g. *Bouleiceras*, Hildaites de la zone à Serpentinus: Bulletin Société Géologique de France, v. 17, p. 116-122, (in French).
Faugères, J. C., 1978, Les rides sudrifaines. Evolution sédimentaire et structurale d'un bassin atlanticomésogéen de la marge africaine: Thesis, University of Bordeaux, 480 p., (in French).
Fedan, B., 1984, Les Formations carbonatées liasiques du plateau de Dwira Enjil Moyen Atlas. Evolution sédimentologique et paléogéographique: Bulletin Institut Sciences, Rabat, v. 8, p. 51-65, (in French).
Fedan, B., 1985, Naissance et évolution d'une plateforme carbonatée: l'exemple du Moyen-Atlas (Maroc) au cours du Lias; comparaisons avec les régions voisines: Bulletin Institut Sciences, Rabat, v. 9, p. 43-65, (in French).
Fleet, A. J., C. J. Clayton, H. C. Jenkyns, and D. N. Parkinson, 1987, Liassic source rock deposition in western Europe, *in* Brooks and Glennie, eds., Petroleum geology of north west Europe: Graham & Trotman, p. 59-70.
Gabilly, J., 1976, Le Toarcien àThouars et dans le Centre-est de la France, biostratigraphie, évolution de la faune (*Harpoceratinae, Hildoceratinae*): CNRS, Les Stratotypes francais, v. 3, 217 p., (in French).
Gaetani, M., and G. Poliani, 1978, Il Toarciano e il giurassico medio in Albenza (Bergamo): Rivista Italiana Paleontologia, v. 84, p. 349-382, (in Italian).
Gaetner, H. R., and H. W. Walther, 1971, International geological map of Europe and Mediterranean region, 1/5,000,000: Bundesantalt für Bodenforshung and UNESCO.
Galbrun, B., and L. Daly, 1987, Etablissement d'un pôle paléomagnétique pour la plate-forme européenne stable à l'époque toarcienne: Compte Rendu de l'Académie des Sciences, v. 305, p. 773-776, (in French).
Gallitelli-Wendt, M. F., 1969, Ammoniti et stratigrafia del Toarciano umbro-marchigiano (Appennino Centrale): Bolletino della Societa Paleontologica italiana, v. 8, p. 11-62, (in Italian).
Gorin, G. E., and S. H. Feist-Burkhardt, 1988, Organic facies of Early Jurassic sediments in the central Jura Mountains, Switzerland (abs.): 7th International Palynological Congress, Brisbane, 1988, p. 58.
Gorin, G. E., and S. H. Feist-Burkhardt, in press, Organic facies of Early Jurassic sediments in the central Jura Mountains, Switzerland: Proc. 7th International Palynological Congress.
Grantham, M., and R. J. Willink, 1988, Origin of crude oils in Oman: Journal of Petroleum Geology, v. 11, p. 61-80.
Guex, J., 1973, Apercu biostratigraphique sur le Toarcien inférieur du Moyen-Atlas marocain et discussion sur la zonation de ce sous étage dans les séries méditerranéennes: Eclogae Geologicae Helvetiae, v. 66, p. 493-523, (in French).
Horner, F., and F. Heller, 1983, Lower Jurassic magnetostratigraphy at the Breggia Gorge (Ticino, Switzerland) and Alpe Turati (Como, Italy): Geophysic Journal research Astronomy Society, v. 73, p. 705-718.
Huc, A. Y., 1976, Mise en évidence de provinces géochimiques dans les schistes bitumineux du Toarcien de l'est du Bassin de Paris: Revue de l'Institut Francais du Pétrole, v. 31, p. 933-953, (in French).
Huc, A. Y., 1977, Contribution de la géochimie organique à une esquisse paléoécologique des schistes bitumineux du Toarcien de l'est du Bassin de Paris: Revue de l'Institut Francais du Pétrole, v. 32, p. 703-718, (in French).
Institute of Geological Research Subsurface and Institut Francais du Pétrole, 1966, Etude géologique de l'Epire (Grèce Nord-occidentale): Edition. Technip, 306 p., (in French).
Jenkyns, H. C., 1985, The early Toarcian and Cenomanian-Turonian anoxic events in Europe: Comparisons and contrasts: Geologische Rundschau, v. 74, p. 505-518.
Jenkyns, H. C., 1988, The early Toarcian (Jurassic) anoxic event: Stratigraphic, sedimentary and geochemical evidence: American Journal of Science, v. 288, p. 101-151.
Jenkyns, H. C., M. Sarti, D. Masetti, and M. K. Howarth, 1985, Ammonites and stratigraphy of Lower Jurassic black shales and pelagic limestones from Belluno trough, Southern Alps, Italy: Eclogae Geologicae Helvetiae, v. 78, p. 299-311.
Jenkyns, H. C., and C. J. Clayton, 1986, Black shales and carbon isotopes in pelagic sediments from the Tethys Lower Jurassic: Sedimentology, v. 33, p. 87-106.
Küspert, W., 1982, Environmental change during oil shale deposition as deduced from stable isotope ratios, *in* Einsele and Seilacher, eds., Cyclic and event stratification, Springer Verlag, p. 482-501.
Küspert, W., 1983, Faziestypen des Posidonienschiefers (Toarcium, Süddeutschland). Eine isotopengeologische, organisch-chemische und petrographische Studie: Ph.D. Dissertation, University of Tübingen, 233 p., (in German).
Le Nindre, Y. M., J. Manivit, and D. Vaslet, 1987, Histoire géologique de la bordure occidentale de la plate-forme arabe du Paléozoique inférieur au Jurassique supérieur. Livre 4: Thèse de doctorat d'état de l'Université Pierre et Marie Curie, v. 4, 111 p., (in French).
Littke, R., and J. Rullkötter, 1987, Mikroskopische und makroskopische Unterschiede zwischen Profilen unreifen und reifen Posidonienschiefer aus der Hilsmulde: Facies, v. 17, p. 171-180, (in German).
Mann, U., D. Leythaeuser, and P. J. Müller, 1986, Relation between source rock properties and wireline log parameters: An example from Lower Jurassic Posidonia shale, NW Germany: Advances in Organic Geochemistry, v. 10, p. 1105-1112.
Mettraux, M., C. Dupasquier, and P. Homewood, 1986, Conditions

de dépôt et diagénèse précoce du Toarcien inférieur des préalpes médianes romandes (Suisse): Documents du Bureau des Recherches Géologiques et Minières, v. 110, p. 231-237, (in French).

Moldowan, J. M., P. Sundararaman, and M. Schoell, 1985, Sensivity of biomarker properties to depositional environment and/or source input in the Lower Toarcian of SW-Germany: Advances in Organic Geochemistry, v. 10, p. 915-926.

Morris, K. A., 1979a, A model for the deposition of bituminous shales in the lower Toarcian, *in* La sédimentation Jurassique W-Européene: Publication Spéciale de l'Association des sédimentologues francais, v. 1, p. 397-406.

Morris, K. A., 1979b, A classification of Jurassic marine shale sequence: An example from the Toarcian (Lower Jurassic) of Great Britain: Paleogeography, Paleoclimatology, Paleoecology, v. 26, p. 117-126.

Mouterde, R., 1967, Le Lias du Portugal. Vue d'ensemble et division en zones: Comm. Serv. Geol. Port., v. 52, p. 209-226, (in French).

Myers, K. J., and P. B. Wignall, 1987, Understanding Jurassic organic-rich mudrocks. New concepts using gamma-ray spectrometry and palaeoecology, *in* Legget and Zuffa, eds., Marine clastic sedimentology: Graham and Trotman, p. 172-189.

Nagy, J., H. Dypvik, and T. Bjaerke, 1984, Sedimentological and paleontological analysis of Jurassic North Sea deposits from deltaic environments: Journal of Petroleum Geology, p. 169-188.

Pialli, G., 1969, Geologia delle Formazioni Giurese dei Monte ad Est di Foligno (Appennino Umbro): Estato da Geologica Romana, v. 9, p. 1-31, (in Italian).

Polgari, M., B. Molak, and E. Surova, 1989, Contribution of the organo-geochemical study of the black shale-Mn-carbonate sequence Urkut (Hungary) and comparison to the Mn-carbonate sequences in Branisko Mts (E. Slovakia) (abs.): Meeting of the Czechoslovakian working group. PICG 254: "Metalliferous black-shales" - 23-25 Mai 1989, Pribram, 18 p.

Pradier, B., and B. Gauthier, 1988, Etude préliminaire de la matière organique sédimentaire, *in* Géologie profonde de la France. Sondage scientifique de Sancerre-Couy (Cher). Rapport d'execution et descriptions préliminaires: Documents du Bureau des Recherches Géologiques et Minières, v. 136, p. 103-118, (in French).

Rad, F. K., 1982, Hydrocarbon potential of eastern Alborz region, NE Iran: Journal of Petroleum Geology, v. 4, p. 419-435.

Rad, F. K., 1986, A Jurassic delta in the eastern Alborz, NE Iran: Journal of Petroleum Geology, v. 9, p. 281-294.

Ronnevik, H., S. Eggen, and J. Vollset, 1983, Exploration of Norwegian shelf, *in* Brooks, ed., Petroleum geochemistry and exploration of Europe: Special Publication Geological Society, v. 12, p. 71-93

Rüllkötter, J., D. Leythaeuser, B. Horsfield, R. Littke, U. Mann, P. J. Müller, M. Radke, R. G. Schaefer, H-J. Shenk, K. Schwochau, E. G. Witte, and D. Welte, 1987, Organic matter maturation under the influence of a deep intrusive heat source : A natural experiment for quantification of hydrocarbon generation and expulsion from a petroleum source rock (Toarcian shale, northern Germany): Advances in Organic Geochemistry, v. 13, p. 847-850.

Sørensen, S., 1987, Basin analysis and maturation modelling onshore Denmark, a case study from Danish first round, *in* Thomas, ed., Petroleum geochemistry in exploration of the Norwegian shelf: Graham & Trotman, p. 153-160.

Soussi, M., M. H. Ben Ismail, A. M'Rabet, and C. Parris, 1988, Les roches mères potentielles des séries jurassiques en Tunisie centrale et méridionale (abs.): 1 Journées tunisiennes de Géologie Appliquée, Sfax, p. 32, (Abstract in French).

Soussi, M., A. M'Rabet, and M. Rabhi, 1989, The Jurassic series of Central Tunisia: An example of shallow to deep marine carbonate shelf (abs.): 10th IAS regional meeting on sedimentology, Budapest, 1 p.

Stampfli, G. M., 1978, Etude géologique générale de l'Elburz oriental au Sud de Gonbad-e-Qabus, NE Iran: Thesis, University of Geneva, 329 p., (in French).

Thomas, B. M., P. Moller Perderren, M. F. Whitaker, and N. D Shaw, 1987, Organic facies and hydrocarbon distribution in the norwegian North Sea, *in* Thomas, ed., Petroleum geochemistry in exploration of the Norwegian shelf: Graham & Trotman, p. 3-26.

Thomas, M., 1977, Les schistes bitumineux du Toarcien de Franche-Comté. Etude géologique, caractérisation géochimique et estimation du potentiel pétrolier: Ph.D. Thesis, Université des Sciences et Techniques de Franche-Comté, Besancon, 145 p., (in French).

Tintori, A., 1977, Toarcian fishes from the Lombardian Basin: Societa Paleontologica italiana bolletino, v. 16, p. 143-152.

Tissot, B., 1984, Recent advances in petroleum geochemistry applied to hydrocarbon exploration: AAPG Bulletin, v. 68, p. 545-563.

Tissot, B., Y. Califet Debyser, G. Deroo, and J. L. Oudin, 1971, Origin and evolution of hydrocarbon in early Toarcian shales, Paris basin, France: AAPG Bulletin, v. 55, p. 2177-2193.

Tissot, B., B. Durand, J. Espitalié, and A. Combaz, 1974, Influence of nature and diagenesis of organic matter in formation of petroleum: AAPG Bulletin, v. 58, p. 499-506.

Tissot, B., G. Demaison, P. Masson, J. R. Delteil, and A. Combaz, 1980, Paleoenvironment and petroleum potential of middle Cretaceous black shales in Atlantic basins: AAPG Bulletin, v. 64, p. 2051-2063.

Tissot, B., R. Pelet, J. Roucaché, and A. Combaz, 1975, Utilisation des alcanes comme fossiles géochimiques indicateurs des environements géologiques: Advances in Organic Geochemistry, v. 7, p. 117-154, (in French).

Tribovillard, N. P., 1988, Contrôle de la sédimentation marneuse en milieu pélagique semi-anoxique. Exemples dans le Mesozöique du Sud-Est de la France et de l'Atlantique: Ph.D. thesis, University of Lyon, 110 p., (in French).

Trümpy, D., 1983, Le Lias moyen et supérieur des Grands Causses: Cahiers de l'université de pau et des Pays d'Adour, v. 19, 276 p., (in French).

Vozenin-Serra, C., and J. Taugourdeau-Lantz, 1985, La flore de la formation de Shemshak (Rhétien à Bajocien - Iran): Rapport avec les flores contemporaines. Implications paléogéographiques: Bulletin Société Géologique de France, v. 8/5, p. 663-678, (in French).

Winterer, E. L., and A. Bosselini, 1981, Subsidence and sedimentation on Jurassic passive continental margin, Southern Alps, Italy: AAPG Bulletin, v. 65, p. 394-421.

Young, J. R., C. Tarquin Teale, and P. R. Brown, 1986, Revision of the stratigraphy of the Longobucco group (Liassic, southern Italy); based on new data from nannofossils and ammonites: Eclogae Geologicae Helvetiae, v. 79, p. 117-135.

Ziegler, P. A., 1988, Evolution of the Arctic-North Atlantic and the western Tethys (with paleogeographic maps): AAPG Memoir, v. 43, p.1-196.

Upper Triassic (Rhaetic) Argillaceous Sequences in Northern Italy: Depositional Dynamics and Source Potential

Marco Stefani
Dipartimento di Scienze Geologiche e Paleontologiche
Universitá di Ferrara, Italy

Mark Burchell
BP Exploration
Glasgow, United Kingdom

Upper Triassic (Rhaetic), thick, argillaceous deposits are the source of important Italian hydrocarbon accumulations. This study integrates sedimentology, paleoecology, and both organic and inorganic geochemistry to determine the genesis and source potential of these marine deposits, which are ascribed to two depositional sequences. The first Rhaetic transgression was associated with an interruption of carbonate production and with a strong increase in the preservation of organic carbon. Deposition of these sequences was controlled by interactions between carbonate and terrigenous sediment sources, under the influence of high-frequency eustatic fluctuations. Most of these organic-rich Rhaetic successions are presently overmature and appear to have entered the oil window during Mesozoic time, prior to trap-forming tectonics. Important exceptions are represented by Mesozoic paleohigh areas, where Triassic deposits reached maturity in the Neogene, or are still immature.

INTRODUCTION

The discovery in 1973 of one of Europe's largest onshore petroleum accumulations, the giant Malossa gas-condensate field near Milan (Erico et al., 1980), and subsequent development of associated satellite fields (Seregna, Pobbiano, Cernusco, and S. Bartolomeo) (Pieri and Mattavelli, 1986) made a large contribution in establishing the Po Plain in northern Italy (Figure 1) as an area of major hydrocarbon production (Mattavelli and Novelli, 1987). Relatively recent work by AGIP workers established Upper Triassic (Rhaetic) shales as the unequivocal source of these large reserves (Riva et al., 1986; Pieri and Mattavelli, 1986; Mattavelli and Novelli, 1987). Outcrops of these Upper Triassic rocks can be examined in Lombardy where they occur in units belonging to the Austroalpine and Southern Alpine orogenic segments (Figure 2). These tectonic domains, which form part of the much larger Alpine belt, developed during deformation of the African-Apulian continental margin of western Tethys (Winterer and Bosellini, 1981).

The thick, generally organic-rich, Rhaetic successions are characterized by cyclic marl-limestone alternations bearing specific faunas that suggest deposition under unfavorable ecological conditions.

The subject of strong environmental biofacies control has produced much debate concerning biostratigraphic definition of the Rhaetian Stage (Ager, 1987). In this article the term "Rhaetic" is primarily used as an informal "lithostratigraphic" term describing distinctive argillaceous units that occur beneath definitively Lower Jurassic (Hettangian) strata.

The major objective of this study is to give a greater insight into Rhaetic source-rock development in northern Italy through investigation of the well-exposed Southern Alpine sections of Lombardy. It combines facies analysis, paleoecological considerations, and both inorganic and organic geochemical

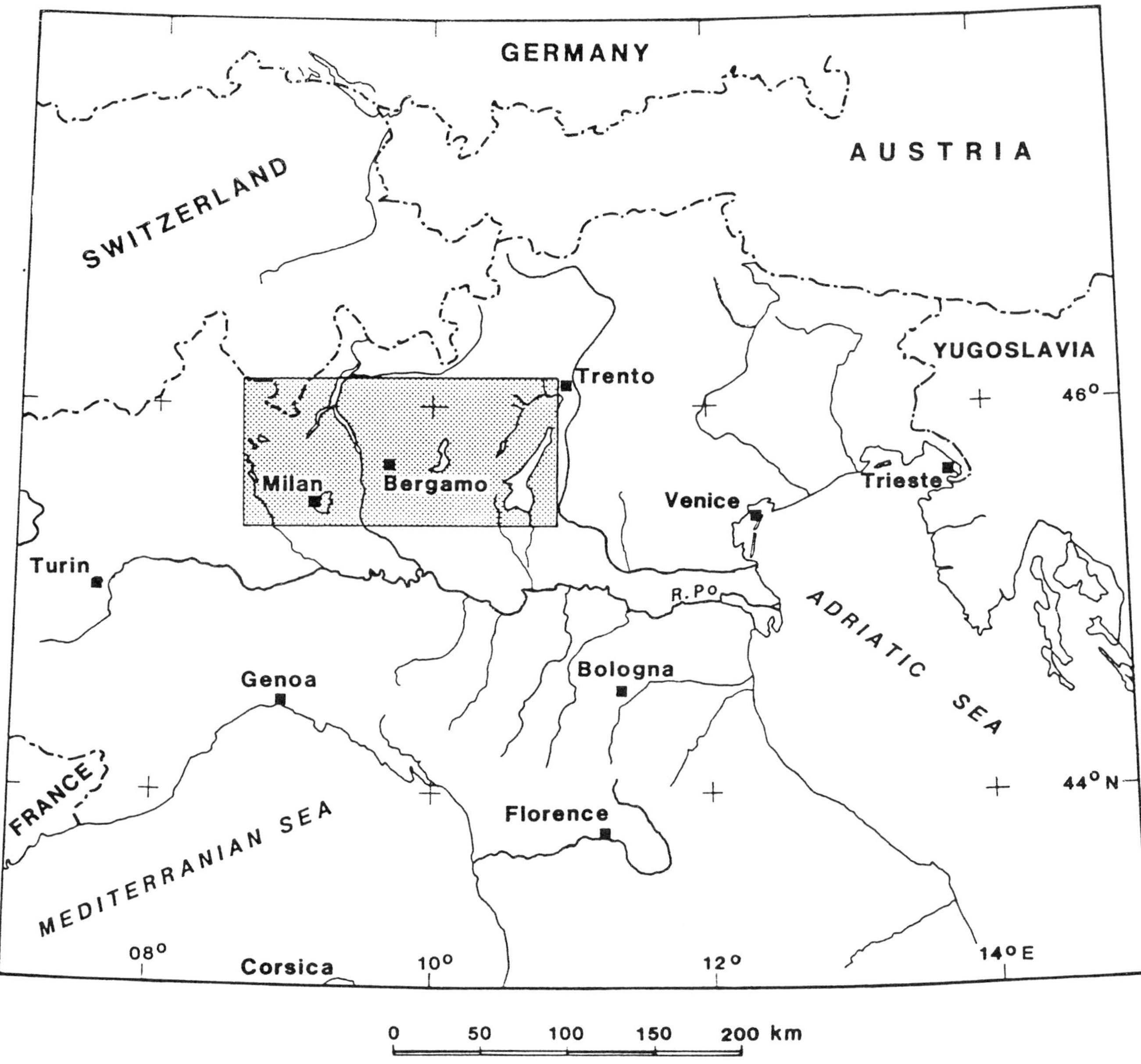

Figure 1. Location of the study area in northern Italy.

research in an integrated study to evaluate source potential by understanding sedimentary dynamics and postdepositional modifications.

STRATIGRAPHIC FRAMEWORK OF RHAETIC SOURCE ROCKS

During the Late Triassic, the western portion of the Southern Alps was characterized by the development of the Rhaetic "Lombardy Basin," the argillaceous deposits of which can be traced from Lake Garda to Lake Lugano (Figure 3). This large basin (more than 1,000 square km) was flanked by shallow-water carbonate platforms (Trento Platform to the east and Lugano Platform to the west) and was characterized by poor communication with open-marine realms.

Stratigraphic Succession in the Lombardy Basin

The initial stages of Late Triassic basin formation in Lombardy commenced when the Norian Dolomia Principale (Figure 4) carbonate platform was dissected by strong synsedimentary tectonics into a series of highs, interspersed with deep localized depressions, the so-called "Aralalta" troughs (Jadoul, 1985). While peritidal deposition continued on the

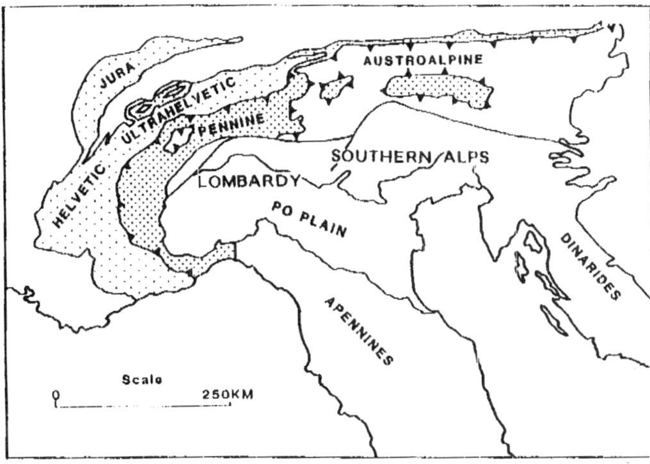

Figure 2. Sketch showing tectonic elements in the Alpine region. The study area is part of the Southern Alpine segment and extends southward into the Po Plain.

highs (upper Lombardian Dolomia Principale), deposition of thick (up to more than 1000 m), locally organic-rich, micritic limestones was initiated in the basins—Calcare di Zorzino (Casati, 1964), part of the Gruppo dell'Aralalta (Jadoul, 1985). These carbonate strata have an extremely sharp contact with overlying argillaceous Rhaetic deposits. This contact is thought to correspond to a generalized environmental deepening and marked interruption of carbonate sedimentation.

The Rhaetic successions of the Lombardy Basin (Figure 4) are subdivided into two main units: (1) a lower shale-dominated formation, the Argillite di Riva di Solto (0–1100 m thick), and (2) an upper, carbonate-dominated formation, the Calcare di Zu (250–1000 m thick) (Gnaccolini, 1965b). The Calcare di Zu is succeeded by shallow-water carbonates—Dolomia a Choncodon (Gnaccolini, 1965a). The two Rhaetic argillaceous formations are normally characterized by repeated asymmetric cycles composed of 20 or 30 marl-limestone couplets that illustrate an upward carbonate enrichment (Masetti et al., 1988; Burchell et al., 1990). Both formations can be further subdivided into a number of constituent members (Stefani, 1989a, b).

The Argillite di Riva di Solto reaches its maximum thickness in the deeper parts of the basin, commonly corresponding to the former depressions ("Aralalta basins"). It consists of two main units—a lower member, almost exclusively argillaceous and commonly organic-rich, passing upward into a second member characterized by a number of superposed limestone-marl asymmetric cycles. In the shallower areas, corresponding to former carbonate highs, the Argillite di Riva di Solto is generally poorly developed, being thin and relatively impoverished in argillaceous facies.

The upper Rhaetic formation, Calcare di Zu, is defined on the basis of increasing carbonate content, which became greater during general environmental shallowing. Three main units can be defined: a lower member characterized by marl-limestone asymmetric cycles; a middle wholly carbonate member, deposited in shallow water; and an upper, deeper water, member largely composed of stacked "asymmetric cycles."

Lateral Variations and Paleotectonic Control

The Lombardian Rhaetic exhibits important lateral thickness and facies variations that are largely related to synsedimentary tectonics (Figure 3). Lateral variations, partially inherited from the previous carbonate paleogeography, were especially important during deposition of the lower portion of the Rhaetic succession (Argillite di Riva di Solto). Overall thickness of these argillaceous units ranges from more than 2 km in the Iseo depocenter to less than 100 m on localized paleohighs. In addition to being relatively thin and containing more carbonates, Rhaetic sequences deposited on these intrabasinal highs (e.g., Botticino, Albenza, Corni di Canzo near Lecco) (Figure 3) also are characterized by features suggesting deposition under much shallower conditions than adjacent shale-dominated depressions (e.g., Lake Iseo Trough).

It is thought that subsidence patterns in the Lombardy Basin were probably controlled by important normal (listric?) synsedimentary faults operating within the crystalline basement (Pieri and Mattavelli, 1986). Such Norian-Rhaetic movements were related to tensional (or transtensional) stress and were probably the forerunners of Liassic tectonics that ultimately led to total foundering of the opposing Tethyan continental margins (Winterer and Bosellini, 1981). The main Rhaetic structural elements were progressively reactivated from the west during the Liassic. The most rapidly subsiding areas of the Rhaetic Lombardy Basin (e.g., west Iseo) were characterized during the Jurassic by high subsidence and sedimentation rates, while the Rhaetic highs remained stable areas, in some cases until the middle Cretaceous (e.g., Corni di Canzo area). This observation has important implications for maturity evolution and present-day source potential of the thicker shale-rich Rhaetic successions, because these sediments underwent much deeper Mesozoic burial than their thinner paleohigh counterparts.

RHAETIC DEPOSITS AS SOURCE ROCKS

Thermal History and Organic Maturity

Reconstructions of source potential in the Lombardian Rhaetic require that various modifications

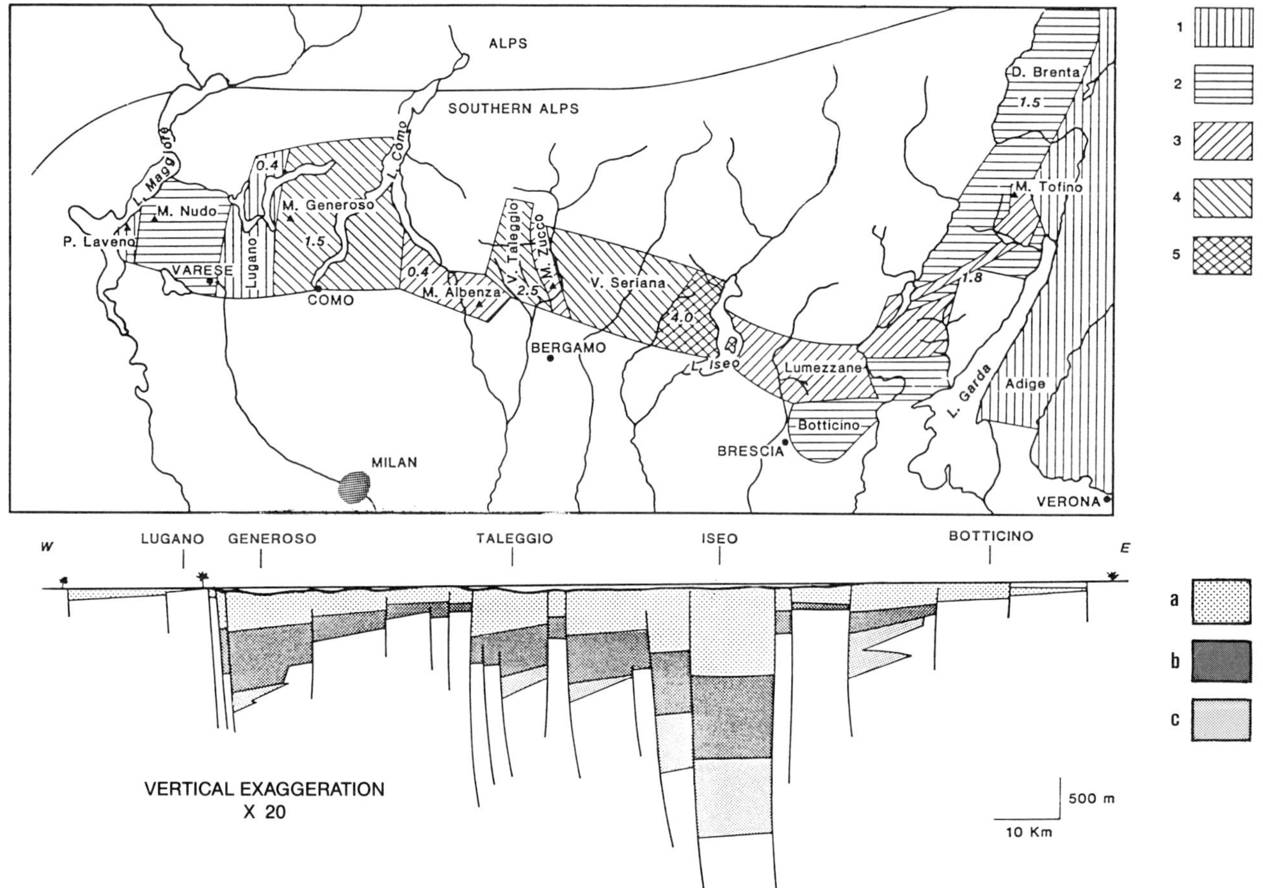

Figure 3. Rhaetic "paleogeographic," nonpalinspastic map of Lombardy and hypothetical cross section across the Lombardy Basin showing conditions at the very close of Triassic time. Numbers indicate the average vitrinite reflectance associated with different areas. Note the good correlation with Rhaetic subsidence trends. Key to patterns: 1 = Peritidal dolomitic successions (confined to the eastern sector), completely lacking argillaceous influx. 2 = Platform successions, dominated by calcareous facies; marly layers are confined to the lower portions. 3 = Relatively thin argillaceous successions, commonly showing shallow-water features. 4 = Thick argillaceous successions, showing basinal features. 5 = Argillaceous successions deposited in the deepest environments. a = Calcare di Zu. b = Argillite di Riva di Solto. c = Aralalta Group.

of organic-matter after burial be taken into account. Commonly, these modifications strongly affect our ability to reconstruct primary organic features.

Clay mineralogy can provide preliminary ideas on the rock thermal history, together with information concerning terrigenous sediment provenance. X-ray diffraction studies show the Lombardian Rhaetic claystones to consist mainly of illite, with varying quantities of mixed-layer illite-smectite and chlorite; kaolinite is generally absent. Although it seems likely that the original detrital assemblages were fairly uniform throughout the basin, diagenetic modifications affected strongly subsiding Mesozoic depressions more severely. While on the Lugano paleohigh (Bernoulli, 1964) (Figure 3) detrital mineralogical associations are fundamentally preserved (Dunoyer de Segonzac and Bernoulli, 1977), the Generoso and Iseo depocenters display much higher illite crystallinity values (Iseo depocenter—ill-xtall index ($\Delta 2\Phi$) approx. 0.5–0.6) and relatively large percentages of diagenetic chlorite (e.g., approx. 25% in the western Iseo Argillite di Riva di Solto). Such modifications suggest that important thermal stresses must have acted on the organic matter deposited in the rapidly subsiding portions of the Lombardy Basin.

More accurate measurements are provided by optical (vitrinite reflectance—%Ro and spore color—SCI) and chemical (organic molecular parameters) maturity-assessment techniques. In the following discussion all data are expressed as equivalent vitrinite reflectance values (%Ro).

As outlined previously, the Lombardian Rhaetic displays marked variations in maturity that are dependent on contrasting burial histories. A range of maturity "zones" situated over discrete Late Triassic paleotectonic blocks can be identified. These blocks are delineated by a set of major extensional faults that played an important role in subsequent

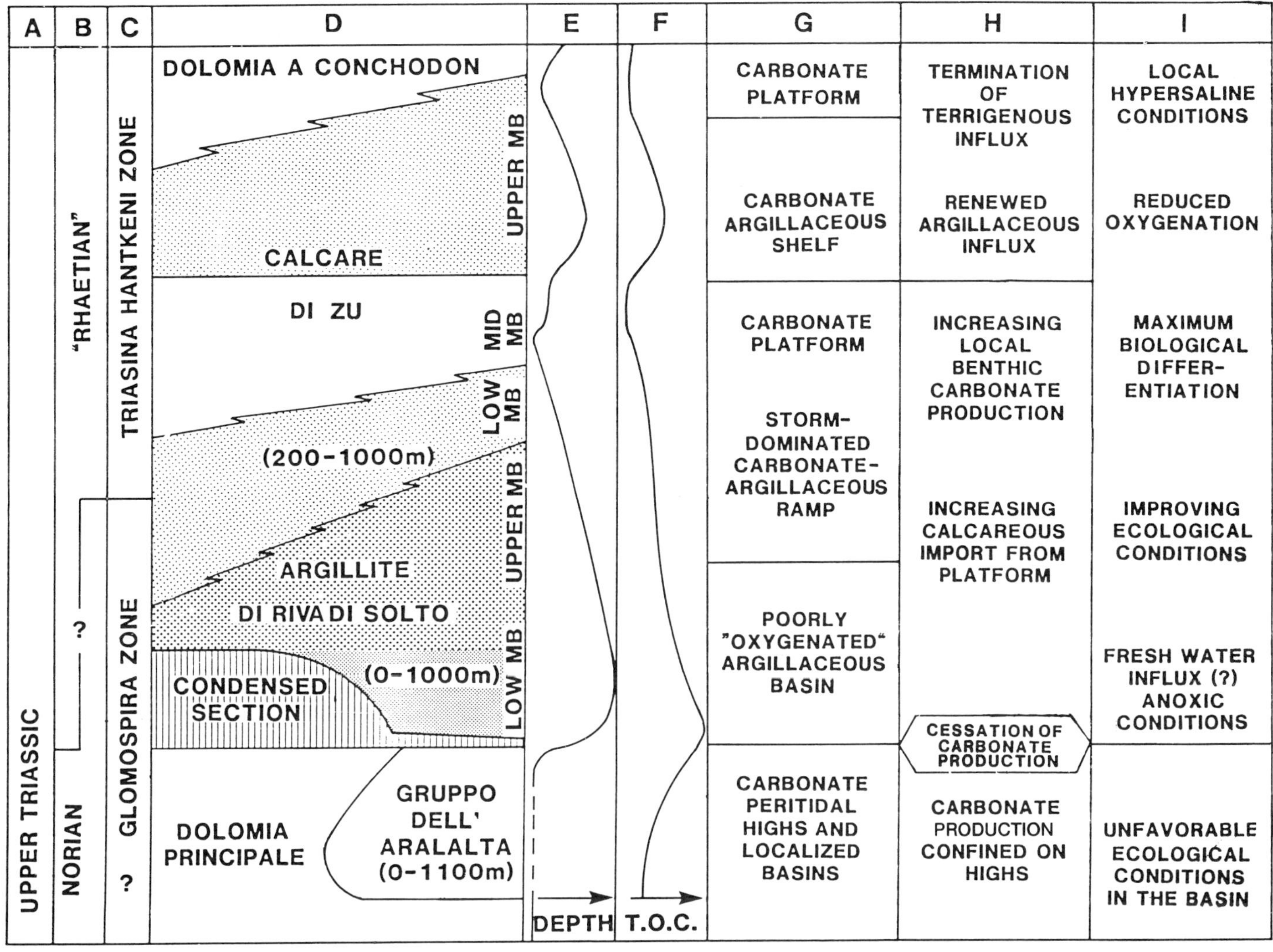

Figure 4. Sketch summarizing Norian-Rhaetic stratigraphy in central Lombardy. Key to columns: A and B—chronostratigraphy (note the uncertainty in the age of the Rhaetian-Norian boundary); C—foraminiferal zones; D—lithostratigraphy (white indicates carbonate units); E—relative depositional depth trends; F—general trends in organic-carbon content (note the correlation with depth trends); G—principal depositional systems; H—evolution of sedimentary sources; I—important environmental features.

Mesozoic subsidence patterns (Figure 3). Average present-day maturity values for the different paleotectonic zones are as follows:

1. Lugano Platform, %Ro = 0.3 (immature)
2. Generoso Trough, %Ro = 1.0–1.5 (overmature for oil generation)
3. Albenza Ridge, %Ro = 0.3–0.5 (immature)
4. Central Lombardy Basin (Imagna to Brembana Valleys), %Ro = 1.5 (west) to 3.2 (east) (overmature)
5. Iseo depocenter, %Ro = 4 (lower Argillite di Riva di Solto); %Ro = 2–3 (upper Calcare di Zu) (highly overmature)
6. Eastern margin of the Lombardy Basin, %Ro = 1.0–2.0 (commonly overmature)

A consideration of the foregoing zonation shows that, with the exception of sediments associated with long-lived Mesozoic paleohighs (Lugano and Albenza), Upper Triassic deposits in Lombardy are predominantly overmature for the generation of liquid hydrocarbons.

These measurements generally agree with "theoretical maturity" calculated using preliminary thermal models which suggest that in central Lombardy the Upper Triassic organic matter commonly reached maturity during the Early Jurassic (Burchell, 1988; Mattavelli and Novelli, 1987). The rapid burial of lower Rhaetic black shales started during the Triassic and continued during the deposition of thick sequences of Liassic sediments (Medolo Group). For example, at the end of the Liassic in the Iseo depocenter, the base of the Rhaetic claystones was buried at more than 5500 m. This deep burial, together with the high geothermal flux related to Late Triassic and Liassic tectonic distention, produced conditions favorable to the complete development of catagenetic modification of the organic matter. Therefore, except for tectonic paleohighs, the generation of liquid hydrocarbons from Rhaetic source rocks took place during the Jurassic.

It should, nevertheless, be stressed that important anomalies between calculated and measured maturities are locally found within the Southern Alps. An interesting example is provided by Rhaetic sediments of the Generoso Trough which are rather less mature than might be expected considering geohistory models. This is by no means an isolated exception since researchers studying subsurface units in the Po Plain area have found immature organic matter over large areas at depths even exceeding 6 km (Riva et al., 1986; Pieri and Mattavelli, 1986). Such amazingly low maturities were related to strong overpressures, probably produced by Alpine tectonic stresses, and are present throughout the Lombardy area (Chiaramonte and Novelli, 1985). However, exact genetic mechanisms for such phenomena are, as yet, unclear.

Sedimentological Features and Organic-Carbon Content

The preceding discussion indicates that the great majority of Lombardy Rhaetic outcrops have an organic content that is highly modified and reduced with respect to amounts present at the time of deposition. Despite these limitations, residual carbon levels (TOC) and Rock-Eval (S1/S2) pyrolysis measurements can identify those levels which were originally richest in organic matter. There is a good correlation between stratigraphic position, sedimentological facies, and total carbon levels, at least as far as samples with similar maturity levels are concerned. Depositional features and organic carbon distribution are compared in Figure 4. The following description is mainly focused on deeper water successions, as they are the ones with the highest original TOC values.

The average TOC content of Norian basinal micritic limestones (Calcare di Zorzino) is not particularly high (0.5–0.6). The highest levels of residual organic carbon are instead found at the base of the Rhaetic argillaceous deposits (Figure 4—uppermost Calcare di Zorzino and lower Argillite di Riva di Solto). The transition between basinal carbonates and Rhaetic argillaceous sediments comprises a distinctive 0.5–5 m interval of finely laminated carbonaceous black shales, some of which are rich in well-preserved fossil fish, reptiles, and crustaceans (e.g., Taleggio and Brembo Valleys, Figure 3, Tintori et al., 1985; Vestino Valley, Pinna, 1976). These black shales are recognizable both in Bergamo and Brescia Provinces where residual TOC contents up to 5% (average 2–3%) are found locally. Such organic-carbon-rich levels testify to a lack of benthic aerobic activity, matched with an interruption of carbonate production.

The lowermost Rhaetic argillaceous unit (lower member of the Argillite di Riva di Solto, Figure 4) continues the pattern of enhanced carbon preservation. These facies appear to have been deposited well below wave base and consist largely of parallel-laminated, black fissile claystones. In basinal settings the unit generally lacks benthic fauna, with its paleontological content commonly restricted to well-preserved fossil fish, crustaceans, insects (Tintori et al., 1985), and medusas (Stefani and Golfieri, 1989). Bioturbation is generally absent in the lowermost portion, while overlying units contain decimeter-scale alternations of bioturbated and unbioturbated beds, probably reflecting fluctuating conditions of anoxia. Present organic content ranges from about 1.5 to 2.5%.

The remaining argillaceous portion of the older Rhaetic terrigenous succession (mid-Argillite di Riva di Solto to mid-Calcare di Zu) shows a progressively greater degree of bioturbation, being typified by *Zoophycus* or *Cruziana* ichnofacies. The gradual improvement in ecological conditions, indicated by an increasingly diversified benthic paleontological association, was matched with a concomitant increase in carbonate and decrease in organic carbon; the average TOC of claystones decreases from 2 to less than 0.5%. Such features are thought to have been caused by progressive shallowing of the depositional environments and by an improvement in bottom water oxygenation.

The central portion of the Rhaetic succession (middle member of the Calcare di Zu) is characterized by shallow-water carbonate facies with relatively diversified coral-bearing faunal associations and a reduced organic-carbon content (<0.1%). Organic-carbon trends in the upper portion of the Rhaetic succession (upper member of the Calcare di Zu, Dolomia a Conchodon) are similar to the previously discussed patterns, although generally at a progressively impoverished level, with the TOC peak corresponding to the second deepening (Figure 4). Consequently, with the exception of localized levels in the transgressive lower portion of the upper member of the Calcare di Zu (TOC contents between 0.5 and 1.5%), the carbonate-enriched upper Rhaetic strata in Lombardy demonstrate significantly lower (<1.0%) residual-carbon levels than those of the Argillite di Riva di Solto.

GEOCHEMICAL STUDY OF SOURCE ROCKS

Iron-Sulfur-Carbon Geochemistry: Salinity Implications

Inorganic geochemistry of Rhaetic deposits, integrated with sedimentological studies, can be useful in reconstructing paleoceanographic conditions that may have facilitated the preservation of organic matter. Study of iron-sulfur-carbon geochemical systematics within deeper water argillaceous Rhaetic sediments has proved useful in understanding the paleosalinity conditions and, therefore, in understanding the paleoceanographic control on oxygen distribution in the Lombardy Basin waters (Figure 5).

In a marine argillaceous mud deposited under reducing conditions, pyrite deposition is controlled by the availability of iron and sulfide ions. The sulfide component derives from anaerobic bacterial reduction of sulfate ions under oxygen-depleted conditions. In a normal-salinity, reducing marine environment there is usually an excess of sulfide ions, and the limiting factor on pyrite formation is the amount of free reactive iron available for combination (Raiswell and Berner, 1985).

Geochemical analyses of lower Rhaetic organic-rich facies (Argillite di Riva di Solto) demonstrate that total pyrite levels are unusually low for sediments which are thought to have developed under normal-marine salinity conditions (cf. Raiswell and Berner, 1987). In addition, the rocks contain an anomalously large percentage of reactive iron (up to 4.7 wt% of bulk rock in the lower Argillite di Riva di Solto in the Iseo depocenter—determined using the method of Berner, 1970). Such evidence suggests a severe reduction in sulfide-ion activity during deposition of the lower portion of the Rhaetic argillaceous sediments. The high organic-carbon levels suggest that the early Rhaetic environmental conditions would almost certainly have favored anaerobic reducing bacteria; consequently, it seems likely that sulfide production was limited by low concentrations of sulfate anions in the water column (e.g., brackish conditions related to fresh-water influx).

Such an observation may provide a mechanism whereby salinity stratification and the production of anoxia at the sediment-water interface may have been induced.

Organic Geochemistry and Source Quality

Detailed geochemical studies of Malossa Group oils or condensates (e.g., Riva et al., 1986; Pieri and Mattavelli, 1986—see Figures 6, 7, 8) suggest derivation from an aluminosilicate-dominated source rock (high abundance of diasteranes) deposited under reducing conditions (pristane/phytane ratio = 1.0–1.2). For correlation purposes, the most abundant peaks in the m/z 191 GCMS (gas chromatography mass spectrometry) trace correspond to the widely occurring 17A (H)-hopanes in the range C27–C35 and to the C29–C30 moretanes. In addition to prominent diasteranes, the m/z 217 fragmentogram displays a relatively high C29 (2OR)/C28 (2OR) ratio. The carbon-isotope composition (delta) of the various fractions ranges from -29.2 to -30.9 for HCS (saturates) and -28.5 to -30.3 for HCA (aromatics).

Although such information is readily available for petroleum products from the Malossa Group, locating sufficiently immature lower Rhaetic sediments on which to perform similar analyses for oil-source correlation is difficult. Nevertheless, Figures 6 and 7 show that lower Rhaetic sample extracts have characteristics that are very similar to Malossa condensate.

Further information concerning original organic composition can be derived from analysis of immature Rhaetic sediments associated with the intrabasinal tectonic highs. Organic-rich facies corresponding to the second deepening episode are locally well developed and provide a partial insight into the more detailed molecular characteristics of Rhaetic source rocks in Lombardy. Finely laminated black shales from the Albenza paleohigh give total organic carbon analyses (TOC) of 0.7–1.5%, Rock-Eval S2 pyrolysis yields of 1–3 kg/tonne, and gas-oil generation indices (GOGI) around 0.45. Saturate gas chromatograms for Rhaetic claystone extracts from this site (Figure 9) show the n-alkane window to be skewed toward the low C-number end of the trace, peaking at C17 and falling smoothly on either side. This, together with a pristane/phytane ratio (Pr/Ph) of around 1.2 and a carbon preference index (CPI) in the region of unity, provides strong evidence against the inclusion of significant quantities of higher land-plant material in the organic fraction. No specific cuticular wax biomarkers (e.g., resin diterpanes) are seen in the GCMS traces (Figure 9), while these samples show a fairly normal C27:C28:C29 marine sterane distribution (Figure 10), and characteristic diasterane compounds are found, suggesting clay-catalyzed diagenesis. A more unusual aspect of the sterane content can be seen in the high total hopane to sterane ratio. This is thought to indicate a relatively small algally derived sterane input; the reasons for this are unclear in what otherwise appears to be an algally dominated source lithology. However, it is possible that the bacterially derived hopane fraction has been enhanced by allochthonous influx of organic debris from surrounding inter/supratidal bacterial ("algal") mats on carbonate platforms. The overall picture for these middle Rhaetic organic-rich zones is generally one of normal marine aluminosilicate source rocks, with negligible inclusion of higher land-plant lipids. It is, however, unlikely that this would have been the case for at least the lower Rhaetic, when rapid influx of terrigenous fines, containing significant quantities of higher land-plant material, was the rule rather than the exception.

As previously discussed, the great majority of lower Rhaetic outcrops of Lombardy are overmature; therefore, it is difficult to determine the original composition of their organic matter. The preferential depletion of the more reactive amorphous and marine organic matter caused the concentration of the more resistant herbaceous and woody land-plant debris that dominates present-day organic composition

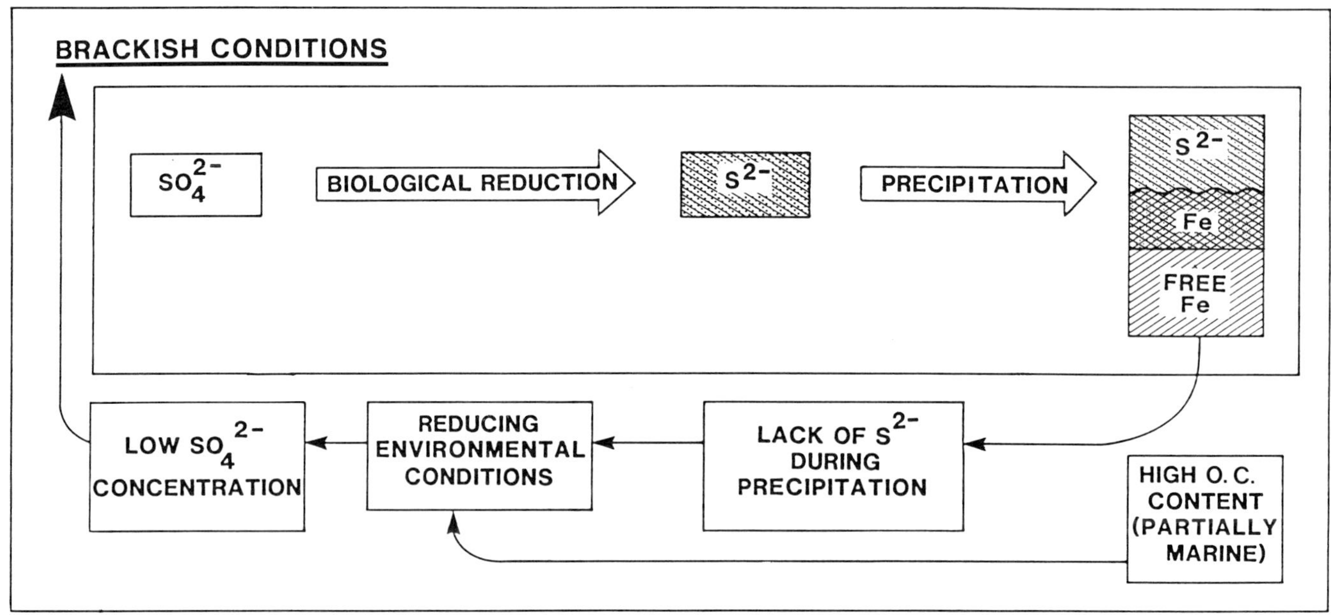

Figure 5. Salinity implications of iron-sulfur-organic carbon geochemistry.

(Figure 11). More labile marine kerogens normally represent less than 10% of the total organic content. Paleohigh successions suffered less severe postdepositional modifications, and therefore they are relatively enriched in marine and amorphous organic matter (up to 25%) compared to their faster subsiding counterparts.

The analysis of middle Rhaetic deposits demonstrates that these facies would be an unlikely source of liquid petroleum, but would probably generate condensate or gas on reaching maturity. It is plausible that due to their high residual TOC values, the lowermost argillaceous Rhaetic sediment (lower Argillite di Riva di Solto) and topmost Norian carbonates (e.g., black shales at the very top of the Calcare di Zorzino) in central Lombardy may have been somewhat richer and more oil prone; however, on the whole, it seems that the Upper Triassic units were relatively lean in organic matter. Nonetheless, the large volume of these formations means that they may have been able to generate great amounts of light oil or condensate.

LATE TRIASSIC SEDIMENTARY EVOLUTION IN LOMBARDY

Carbonate Platforms to Argillaceous Basin: Peak of Organic-Carbon Accumulation

Late Norian evolution in Lombardy (Figure 12) was characterized by the tectonically controlled development (Jadoul, 1985; Picotti and Pini, 1989) of small, relatively deep and commonly poorly oxygenated basins ("Aralalta Basins") where severe ecological conditions prevented indigenous carbonate production. Basinward, redeposition of calcareous oozes from surrounding platforms (Stefani, 1989a), locally associated with subordinate calcarenites, accounted for the development of thick carbonate facies, which were locally enriched in organic carbon (e.g., Calcare di Zorzino), within these depressions.

Both platform and basinal carbonate deposition was halted by events associated with the first Rhaetic deepening phase (top of the Gruppo dell'Aralalta/ Dolomia Principale). Sea-level rise is thought to have "overtaken" sedimentation because a "range" of factors acted coevally to produce unacceptable stress on benthic carbonate productivity. The isolated nature of the small "Aralalta depressions" that prevented deep-water exchange, together with higher fresh-water influx at the surface (see the section on inorganic geochemistry), facilitated the development of strong water-column stratification (Burchell, 1988; Stefani, 1989a). Euxinic conditions, associated with a strong pycnocline, probably developed, allowing strongly enhanced preservation of organic carbon and the exceptional fossilization of nektonic faunas by preventing the reworking of sediments by benthic organisms and aerobic bacteria (black shales at the very top of the Gruppo dell'Aralalta). Short-lived water-mixing episodes, related to stratification instability, may have introduced "jets" of nutrient-rich deep waters into shallow-water platform biotopes. The coeval transgression also may have mobilized a large supply of continental nutrients. The increased nutrient concentration then turned the sea waters eutrophic. It is interesting to note that

Geochemical Parameters – Literature

ARGILLITE DI RIVA DI SOLTO

Bulk

Sample Locality	%Ro	%TOC	S2 (Kg/T)	HI (mgHC/g)	%Organic Matter		
					Am&Mar	Herb	Woody
* Subsurf.-Po Plain	?	1.3	3.1	250	13	59	28
+ Surface-Lombardy	?	0.8	-	-	21	34	45

Molecular

+PRISTANE/PHYTANE : 0.9

+DEL C13: SAC - 29.81, AROM - 28.94, RES - 28.81

MALOSSA CONDENSATE

Bulk

*API : 40-50

*%S : 0.0-0.6

*%RES& ASPH : 0.0-4.0

*V/Ni : 0.3-0.5

Molecular

*PRISTANE/PHYTANE : 1.0-1.2

*C28-AAAR/C29AAAR : 0.2-0.4

*C27S DIA/C27AAAR : 2.7-3.3

*Ts/Tm : 1.0-1.5

+DEL C13: SAC - 30.57, AROM - 28.80, RES - 28.96

* - Riva et al. + - Pieri & Mattavelli (1986)

Figure 6. Summary of published geochemical data for Argillite di Riva di Solto samples and Malossa Group condensates. Direct comparisons between the information shown here and the analyses conducted during this study are hampered by a lack of maturity values for the published data. (Modified after Pieri and Mattavelli, 1986; Riva et al., 1985.)

nutrient excess and sea-water eutrophization are suspected to be major controlling factors in coral-reef and carbonate-platform demise (Hallock and Schlager, 1986). Present knowledge about the ecology of the Norian Lombardy platforms is limited; certainly there is no evidence to suggest domination by hermatypic corals. However, it is possible that a marked increase in nutrient concentration and rapid fresh-water influx could have generated considerable ecological stress within the carbonate-producing communities, which were possibly adapted to oligotrophic conditions (Grigg, 1982; Stefani, 1989a). The resultant reduced level of carbonate production may have been insufficient to counter rapid relative sea-level rise so that large tracts of platform drowned, producing the Lombardy Basin.

Deposition of the main Rhaetic argillaceous succession did not commence until after the initial transgressive phase (Figure 3). It is proposed that, during the lowstand period, terrigenous material was stored close to the source area. Transgression subsequently flooded distal continental areas and allowed an abundant reserve of clay materials, together with continental organic matter, to disperse basinward (Stefani, 1989b). Evolution toward moister conditions also would have influenced terrigenous exportation by increasing soil formation and fluvial delivery of argillaceous sediments.

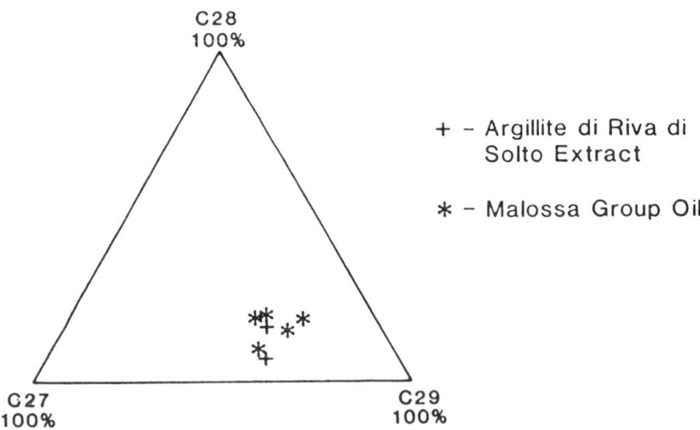

Figure 7. C27:C28:C29 sterane distribution for Malossa oil and extract of the Argillite di Riva di Solto (From Riva et al., 1986).

Sedimentary Evolution of the Rhaetic Lombardy Basin

The strongly inhibited platform communities resulted in the lower part of the Lombardian argillaceous Rhaetic being deposited without carbonate dilution. Furthermore, reduced salinity, seawater stratification, and anoxia are thought to have continued so that the lowest parts of Rhaetic basinal successions accumulated as uniform claystones, originally rich in both marine and continental organic matter (e.g., lower member of the Argillite di Riva di Solto).

Although flooding permanently drowned small carbonate highs in Lombardy, other larger platforms (e.g., Trento and Lugano) survived and, after a possible period of crisis, carbonate production and associated off-platform transport recovered. Detailed analysis (Burchell et al., 1990; Masetti et al., 1988) shows that it was the eustatically and climatically controlled interaction between carbonate platforms and terrigenous basins that gave rise to the high-frequency, marl-limestone, asymmetric cyclicity so evident in the Rhaetic deposits.

As ecological conditions progressively improved, bottom-water oxygenation increased, with an associated reduction in organic-carbon preservation. The greater availability of sediment facilitated a shallowing evolution of the entire basin (upper member of the Argillite di Riva di Solto and lower member of the Calcare di Zu, Figure 3). This evolution eventually caused shallow-water conditions to spread throughout Lombardy (middle member of the Calcare di Zu), with peritidal sedimentation and subaerial exposure dominating the eastern part of the region (Stefani and Golfieri, 1989). During this phase, both terrigenous influx and deposition of organic-rich facies effectively ceased.

Subsequent relative sea-level rise again overcame sedimentation rates. A new deepening evolution took place across Lombardy, together with restoration of terrigenous influx and some reduction of environmental oxygenation (upper member of the Calcare di Zu). After the second deepening, a new shallowing evolution led to the deposition of shallow-water carbonates that were lean in organic matter. This ended the Lombardy Rhaetic successions, and terrigenous input ceased.

Sedimentary Evolution and Sequence Stratigraphy

The Late Triassic evolution of the sedimentary units of Lombardy, together with their geometric organization, suggest that, from a sequence-stratigraphy viewpoint (Vail et al., 1977; Brown and Fisher, 1977; Vail et al., 1987), these units can be ascribed to two superposed depositional sequences. The first of these embraces the upper portion of the Norian carbonates (uppermost Dolomia Principale) and the lower and middle parts of the Rhaetic argillaceous succession. The transgressive surface of this sequence corresponds to a generalized deepening matched with the peak of organic-carbon preservation. The second sequence comprises the remaining portion of the Rhaetic argillaceous succession and part of the following carbonates. In the peripheral areas of the Lombardy Basin, the boundary between the two sequences is marked by subaerial exposure (cf. Stefani and Golfieri, 1989), while in the central portion, the junction is found within the middle member of the Calcare di Zu. A more detailed discussion of Rhaetic sequence stratigraphy was presented by Stefani (1989b).

Regional Context

Much of the foregoing discussion has stressed the link between the upper Norian and Rhaetic and the deposition of organic-rich strata. The reconstruction of depositional dynamics clearly shows that the Rhaetic peaks of organic-carbon preservation are matched with transgressions and deepening episodes (cf. Figure 3 E and F). Viewed in a wider sense, the Late Triassic deposition of organic-rich sediments is seen as a significant regional event since carbon-enriched strata of about this age also are found in many other Italian areas: Brenta Dolomites—"Facies Retiche" (Masetti et al., 1985); Engadine Dolomites—Koessen Formation (Doesseger et al., 1982); Belluno Dolomites—upper Dolomia Principale (Biachin and Masetti, 1985); eastern Southern Alps—Resiutta Shale and Dolomie di Forni (Mattavelli and Rizzini, 1974; Northern Apennines—Formazione della Spezia

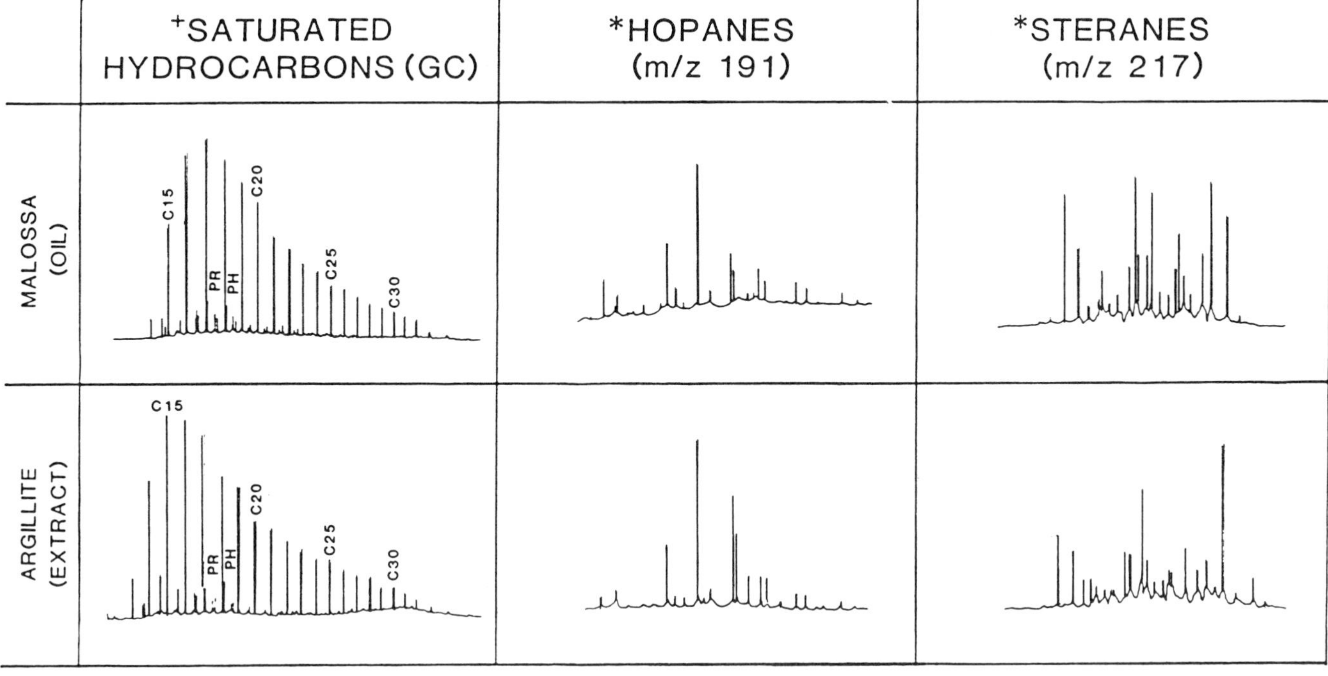

Figure 8. Gas chromatography (GC) and gas-chromatography mass spectrometry (GCMS) traces for Malossa oil and Argillite di Riva di Solto sediment extract (Modified after Pieri and Mattavelli, 1986; Riva et al., 1985).

(La Spezia Gulf) (Ciarapica and Passeri, 1980), Formazione di Monte Cetona (Tuscany/Umbria) (Ciarapica et al., 1987; Stefani and Trombetta, 1989); Central Apennines—"Schisti Bituminosi" (Gran Sasso) (Adamoli et al., 1988), Dolomie di Filettino (Latium) (Pieri and Mattavelli, 1986); Southern Apennines—"Facies Anossiche del Trias Superiore" (Boni et al., 1988); Sicily—Noto/Steppenosa Formations (Patacca et al., 1979); Western Adriatic—upper Anidrite di Burano (Martinis and Pieri, 1962). Although only a few of these formations are important as source rocks, all of them have a strongly increased organic-carbon content compared to adjacent lithologies. Because of the lack of well-documented biostratigraphic markers, it is impossible to reconstruct precise chronological correlations between these formations. Although not demonstrably synchronous in absolute terms, one possibility is that the enhanced carbon preservation recorded by some of these strata was linked to a rapid transgressive phase, as in the Lombardy Basin.

The association of organic-rich intervals with periods of rapid transgression has been well documented from deeper marine facies—Cretaceous (Arthur et al., 1987; Schlanger et al., 1987) and Jurassic (Jenkyns, 1985; 1988) "anoxic events." Recent work by the authors (Burchell, 1988; Stefani, 1989a) suggests that the oceanographic conditions (expanded oxygen minima?) affecting the preservation of organic carbon in the deeper water environments may have "leaked" onto surrounding shelves and platforms causing a similar phenomenon in these areas.

CONCLUSIONS

This study has sought to investigate the widespread Late Triassic development of carbon-rich deposits in Lombardy. It seems likely that, with the exception of the deposits associated with the first Rhaetic deepening (black shales at the base of the Rhaetic argillaceous successions) and perhaps a few zones within the second Rhaetic sequence (lower portion of the upper member of the Calcare di Zu), the original source quality of these facies was only moderate. Nevertheless, because of the exceptional thickness and total volume of these successions, they may have been able to produce exceptionally large quantities of light oil or gas. In most areas, the facies have little remaining potential for liquid petroleum, since the majority of successions appear to have entered the oil-generation window during Jurassic time, prior to the trap-forming compressive tectonics. Important exceptions to this early hydrocarbon generation occur where syndepositional horsts

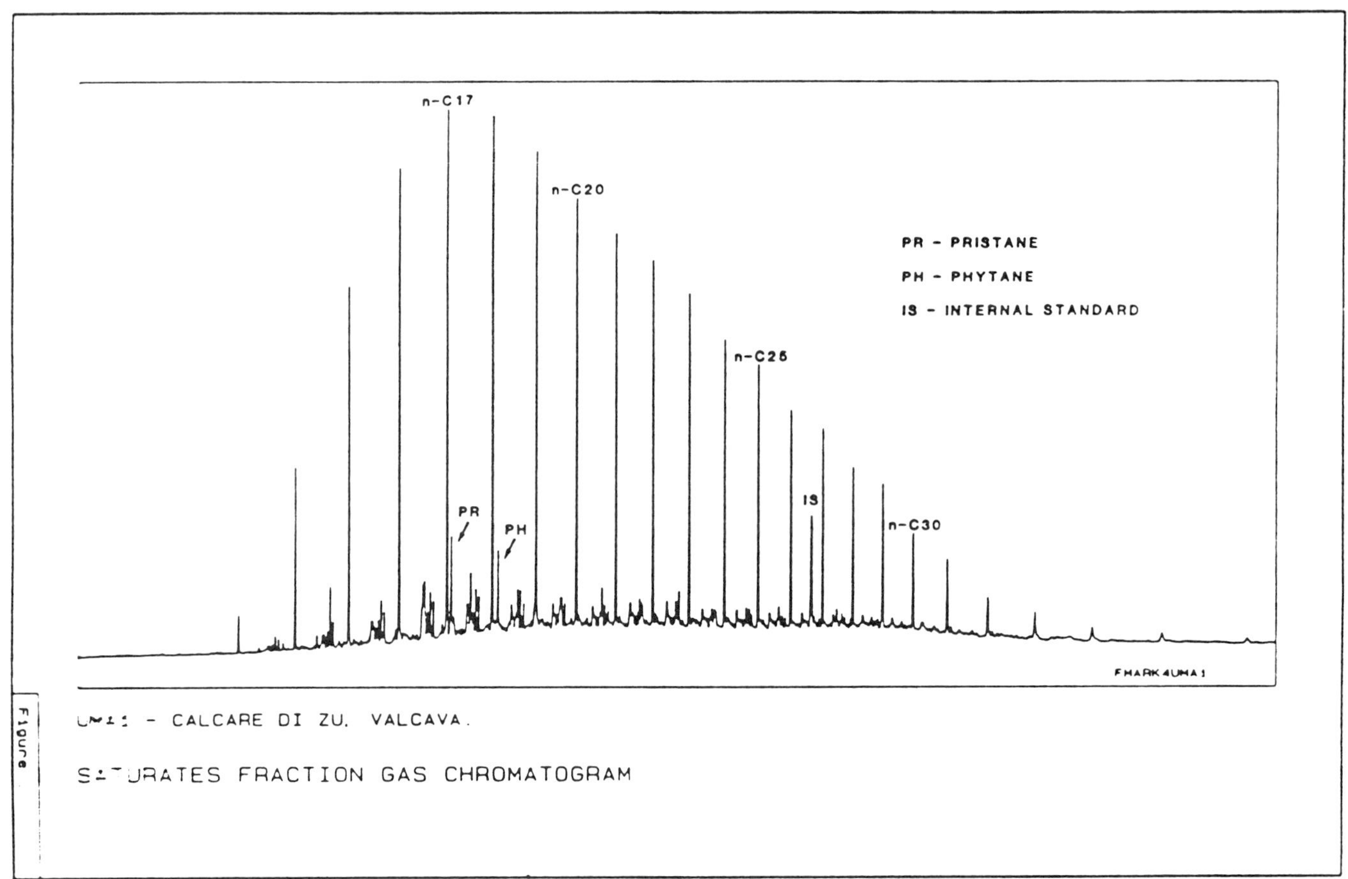

Figure 9. Saturate gas chromatogram of claystones from the Albenza area. Horizontal scale—relative elution time; vertical scale — relative peak intensity. For clarity, all chromatogram plots are normalized to the strongest peak. Axes are linear.

remained as positive features for much of the Mesozoic. With regard to the Rhaetic deposits, the most promising exploration target in the region is the southern extension of the upthrown Mesozoic fault block of the Albenza-Corni di Canzo Ridge (western Bergamo Province, Figure 4). Although immature in outcrop, molasse deposition in the Po Plain is likely to have matured not only the Rhaetic succession, but also the thinner, more oil-prone intervals present in the Toarcian (Jenkyns, 1988).

ACKNOWLEDGMENTS

Thanks must go to Alfonso Bosellini, Maurizio Gaetani, Hugh Jenkyns, Daniele Masetti, and Peter Vail. Work was supported by grants from the Consiglio Nazionale delle Ricerche of Italy (C.N. 890034305), the National Environmental Research Council, and the British Petroleum Company PLC.

REFERENCES CITED

Adamoli, L., A. Bigozzi, G. Ciarapica, S. Cirilli, F. Duranti, L. Passeri, A. Romano, and F. Venturi, 1988, Facies bituminose carniche e facies pelagice hettangiane nella catena del Gran Sasso: Atti 74 Congresso Società Geologica Italiana, p. B 5–8.

Ager, D. V., 1987, A defence of the Rhaetian stage: Albertiana, v. 6, p. 4–21.

Arthur, M. A., S. O. Schlanger, and H. C. Jenkyns, 1987, The Cenomanian-Turonian oceanic anoxic event, II - Palaeoceanographic controls on organic-matter production and preservation, *in* Brooks, J., and A. Fleet, eds., Marine petroleum source-rocks: Geological Society of London special publication, v. 26, p. 401–420.

Berner, R. A., 1970, Sedimentary pyrite formation: American Journal of Science, v. 268, p. 1–23.

Bernoulli, D., 1964, Zur geologie des Monte Generoso (Lombardische Alpen): Beitrag Zeitschrift Geologisch Karte, Zchweiz, N. F., p. 1–34.

Bianchin, G., and D. Masetti, 1985, Carta Geologica del Gruppo della Schiara (Dolomiti Bellunesi): Memorie di Scienze Geologiche-Padova.

Boni, M., A. Lannace, J. Koester, and M. Parente, 1988, Facies anossiche nel Trias superiore dei Monti Picentini (Salerno): Osservazioni preliminari: Atti 74 Congresso Società Geologica Italiana, p. B 28–30.

Brown, L. F., and W. L. Fisher, 1977, Seismic stratigraphy interpretation of depositional surfaces: Examples from the Brazilian rift and pull-apart basins: AAPG Memoir 26, p. 213–248.

Burchell, M. T., 1988, Organic-rich sediments in a rifted basin: The Rhaetic of Western Tethys: Oxford University, Ph.D. Thesis, 254 p.

Burchell, M. T., M. Stefani, and D. Masetti, 1990, Cyclic sedimentation in the Southern Alpine Rhaetic: The importance of climate and eustasy in controlling platform-basin interactions: Sedimentology, v. 37.

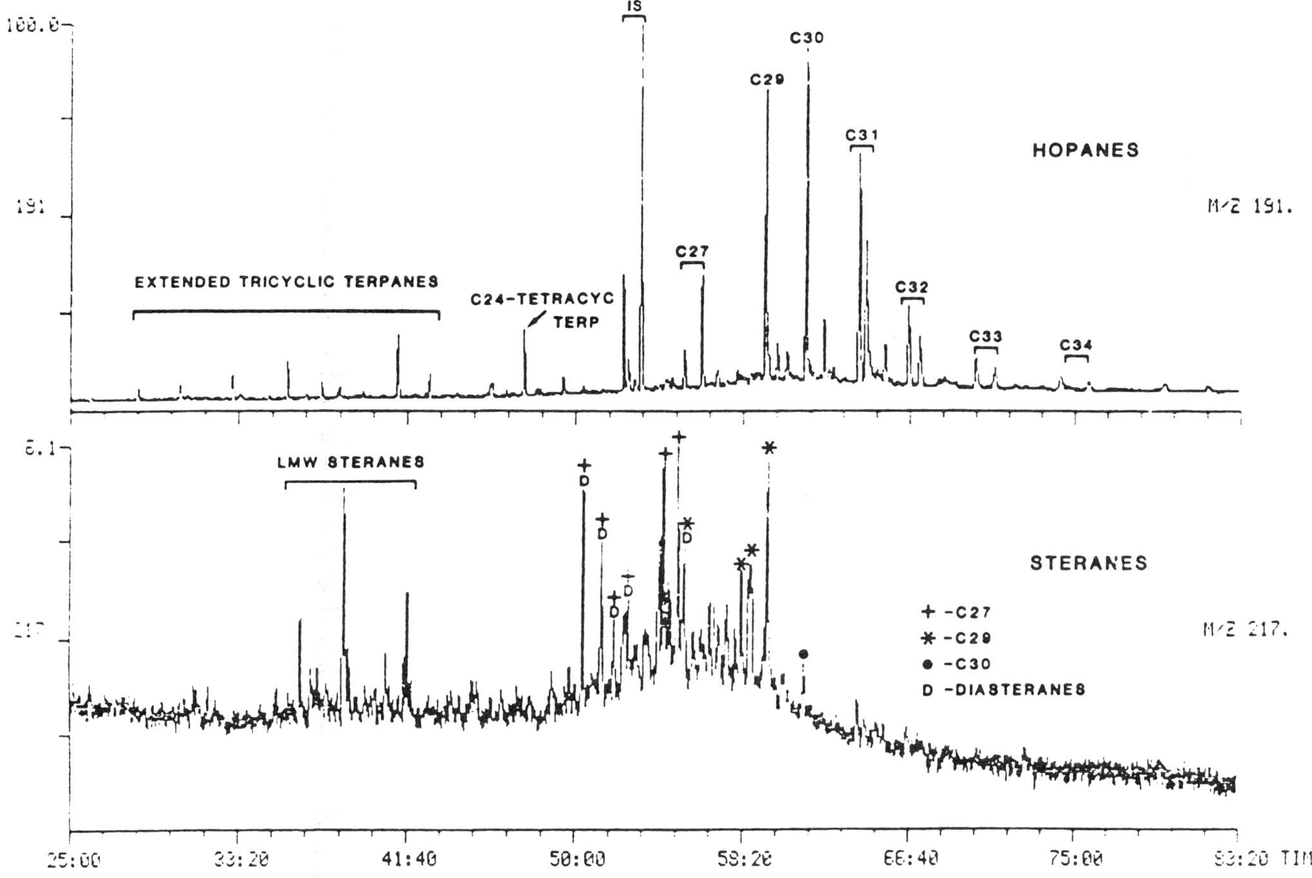

Figure 10. GCMS hopane (m/z 191) and sterane (m/z) 217 traces from Rhaetic claystone extract, Albenza. Note the characteristic diasterane compounds typical of clay-rich, marine shales, and the presence of C24-tetracyclic terpane, abundant in many carbonate-related source intervals.

Casati, P., 1964, Osservazioni stratigrafiche sull'infraretico delle Prealpi Bergamasche: Rivista Italiana di Paleontologia e Stratigrafia, v. 70, p. 447-465.

Chiaramonte, M. A., and L. Novelli, 1985, Organic-matter maturity in Northern Italy: Some determining agents: Organic Geochemistry, v. 10. p. 281-290.

Ciarapica, G., S. Cirilli, L. Passeri, E. Trincianti, and E. Zaninetti, 1987, "Anidriti di Burano" et "Formation du Monte Cetona" (Nouvelle Formation), biostratigraphie du deux series-types du Trias Superieur dans l'Apennin Septentrional: Revue de Paleobiologie, v. 6, p. 341-409.

Ciarapica, G., and L. Passeri, 1980, La lithostratigrafia del promontorio occidentale del Golfo della Spezia: Memorie Società Geologica Italiana, v. 24, p. 193-208.

Doesseger, R., H. Furrer, and W. Muller, 1982, Die Sedimentsarien der Engadiner Dolomiten un ihre lithostratigraphisce Gliederung, Teil 2: Eglogae Geologiscae Helvetiae, v. 75, p. 303-330.

Dunoyer di Segonzac, G., and D. Bernoulli, 1977, Diagenese et Metamorphisme des argiles dans le Rhaetian Sud-Alpin (Lombardie et Grisons): Bull. Soc. Geol. France, v. 18, 1283-1293.

Erico, G., G. Groppi, S. Savelli, and G. C. Vaghi, 1980, Malossa field: A deep discovery in the Po Plain, Italy: AAPG Memoir 30, p. 525-538.

Gnaccolini, M., 1965a, Sul significato stratigrafico della "Dolomia a Conchodon": Rivista Italiana di Paleontologia e Stratigrafia v. 71, p. 155-166.

Gnaccolini, M., 1965b, Il Calcare di Zu e l'Argillite di Riva di Solto, due nuove formazioni del Retico lombardo: Rivista Italiana di Paleontologia e Stratigrafia, v. 71, p. 1099-1121.

Grigg, R. W., 1982, Darwin Point: A threshold for atoll formation: Coral Reefs, v. 1, p. 29-34.

Hallock, P., and W. Schlager, 1986, Nutrient excess and the demise of coral reefs and carbonate platforms: Palaios, v. 1, p. 389-398.

Jadoul, F., 1985, Stratigrafia e paleogeografia del Norico delle Prealpi Bergamasche occidentali: Rivista Italiana di Paleontologia e Stratigrafia, v. 91, p. 479-511.

Jenkyns, H. C., 1985, The Early Toarcian and Cenomanian-Turonian anoxic events in Europe: Comparison and contrasts: Geologische Rundschau, v. 74, p. 505-518.

Jenkyns, H. C., 1988, The Early Toarcian (Jurassic) anoxic event: Stratigraphic, sedimentary, and geochemical evidence: American Journal of Science, v. 288, p. 101-151.

Martinis, B., and M. Pieri, 1962, Alcune notizie sulla formazione evaporitica del Triassico superiore dell'Italia centrale e meridionale: Memorie della Società Geologica Italiana, v. 4, p. 649-678.

Masetti, D., C. Neri, M. Stefani, and R. Zanella, 1985, Cicli e tempestiti nel "Retico" delle Dolomiti di Brenta: Memorie della Società Geologica Italiana, v. 30, p. 267-283.

Masetti, D., M. Stefani, and M. T. Burchell, 1988, Asymmetric cycles in the Rhaetic facies of the Southern Alps: Platform-Basin interactions governed by eustatic and climatic oscillations: Rivista Italiana di Paleontologia e Stratigrafia, v. 94, p. 401-424.

Mattavelli, L., and L. Novelli, 1987, Origin of Po Basin hydrocarbons: Memoire Societe Geologique de France, N.S., n. 151, p. 97-105.

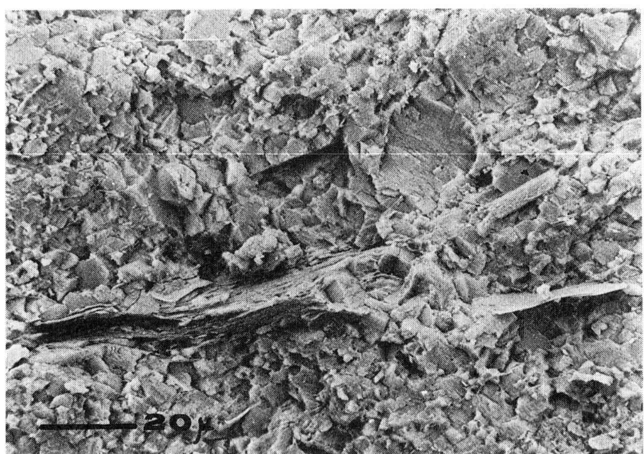

Figure 11. SEM photomicrograph of woody fragment within a deep-water lower Rhaetic carbonate.

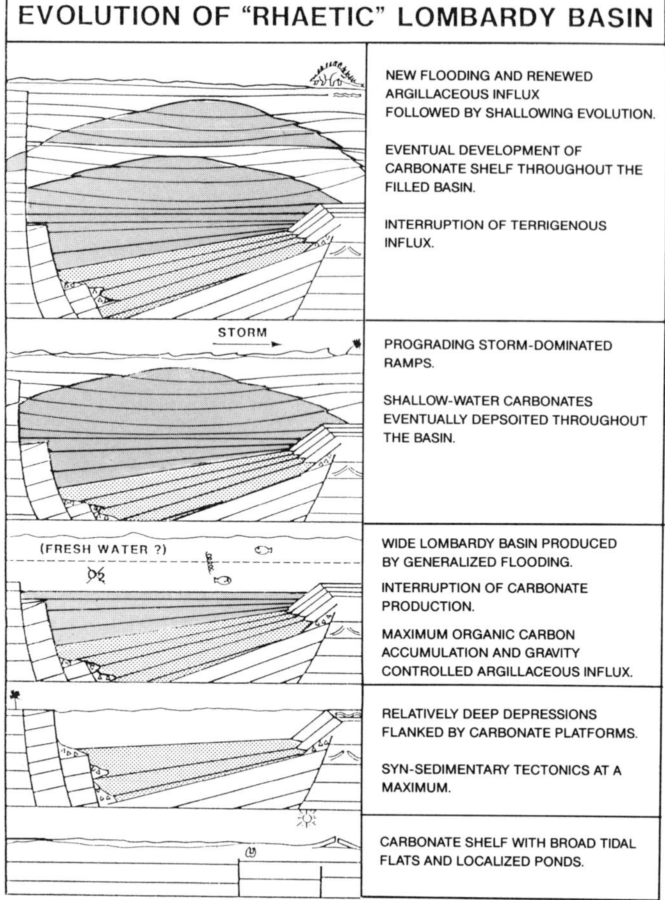

Figure 12. Diagrammatic sketch showing evolution of the Norian-Rhaetic Lombardy Basin. White = shallow-water carbonates; light-gray = basinal carbonates; dark-gray = argillaceous deposits.

Mattavelli, L., and A. Rizzini, 1974, Facies euxiniche nelle dolomie noriche dell'Ampezzano (Udine): Petrografia e sedimentologia: Rivista Italiana di Paleontologia e Stratigrafia, v. 14, p. 111-138.

Patacca, E., P. Scandone, G. Giunta, and A. Liguori, 1979, Mesozoic palaeotectonic evolution of the Ragusa Zone (Southern Sicily): Geologica Romana, v. 18, p. 331-369.

Picotti, V., and G. A. Pini, 1989, Tettonica sinsedimentaria norica nel settore compreso fra il Lago d'Idro e il Lago di Garda: Rendiconti della Società Geologica Italiana, v. 11, p. 225-230.

Pieri, M., and L. Mattavelli, 1986, A geologic framework of Italian petroleum resources: AAPG Bulletin, v. 70, p. 103-130.

Pinna, G., 1976, I crostacei triassici dell'alta Valvastino (Brescia): Natura Bresciana, v. 13, p. 33-42.

Raiswell, R., and R. A. Berner, 1985, Pyrite formation in anoxic and semi-euxinic sediments: American Journal of Science, v. 285, p. 710-724.

Raiswell, R., and R. A. Berner, 1987, Organic-carbon losses during burial and thermal maturation of normal marine shales: Geology, v. 15, p. 853-856.

Riva, A., T. Salvadori, R. Cavaliere, T. Ricchiuto, and L. Novelli, 1986, Origin of oils in the Po Basin (Northern Italy) (abs.): 12th International Meeting on Organic Geochemistry.

Schlanger, S. O., M. A. Arthur, H. C. Jenkyns, and P. A. Scholle, 1987, The Cenomanian-Turonian oceanic anoxic event, I. Stratigraphy and distribution of organic-rich beds and the marine delta C13 excursion, *in* Brooks, J., and A. Fleet, eds., Marine petroleum source-rocks: Geological Society of London special publication, v. 26, p. 323-341.

Stefani, M., 1989a, Stratigrafia e sedimentologia del "Retico marnoso-calcareo" della Lombardia e dell'Appennino Settentrionale: Cicli sedimentari in un sitema deposizionale di rampa, paleogeografia e stratigraifa sequenziale: Tesi di Dottorato Ferrara, p. 331.

Stefani, M., 1989b, Rhaetic marly successions of the Southern Alps: Sedimentary Evolution and Sequence Stratigraphy, Annali dell'Universita di Ferrara, v. 10, p. 21.

Stefani, M., and A. Golfieri, 1989, Sedimentologia e stratigrafia delle successioni Retiche al confine fra Lombardia e Trentino: Rivista Italiana di Paleontologia e Stratigrafia, v. 94, p. 29-54.

Stefani, M., and G. L. Trombetta, 1989, Le successioni retiche della Toscana orientale e dell'Umbria: Cicli sedimentari asimmetrici in un bacino marnoso-calcareo dominato dalle tempeste: Bollettino della Società Geologica Italiana, v. 108, in press.

Tintori, A., G. Muscio, and S. Nardon, 1985, Triassic fossil fish localities in Italy: Rivista Italiana di Paleontologia e Stratigrafia, v. 91, p. 197-210.

Vail, P. R., J. P. Colin, R. J. Du Chene, J. Kuchly, F. Mediavilla, and Trifilieff, 1987, La stratigraphie sequentialle et son application aux corerelations chronostratigraphiques dans le Jurassique du Basin di Paris: Bullettin di la Societe Geologique di France, v. 8, p. 1301-1321.

Vail, P. R., R. M. Mitchum, and S. Thompson, 1977, Seismic stratigraphy and global changes of sea-level, part 2: The depositional sequence as a basic unit for stratigraphic analysis, *in* C. E. Payton, ed., Seismic stratigraphy: applications to hydrocarbon exploration: AAPG Bulletin, v. 65, p. 394-421.

Winterer, E., and A. Bosellini, 1981, Subsidence and sedimentation on a Jurassic passive continental margin, Southern Alps, Italy: AAPG Bulletin, v. 65, p. 393-421.

Global and Regional Controls on Potential Source-Rock Deposition and Preservation: The Cenomanian-Turonian Oceanic Anoxic Event (CTOAE) on the European Tethyan Margin (Southeastern France)

Jean-Pierre Crumière
Université Claude Bernard
Villeurbanne, France

Christine Crumière-Airaud
Université de Provence
Marseille, France

Jean Espitalié
Institut Francais du Pétrole
Rueil-Malmaison, France

Pierre Cotillon
Université Claude Bernard
Villeurbanne, France

On the European margin of the Tethys Ocean, late Cenomanian to early Turonian sea-level rises resulted in a progressive drowning of the Provence rudist-bearing platform, successive breaks in marine sedimentation, and the initiation of deep-water disaerobic conditions in the whole pelagic domain of the Vocontian Basin. Paleogeographic and structural features may have controlled the regional extent of oxygen-minimum layers, the distribution of laminated and organic-rich sediments, and the regional duration of such a worldwide Oceanic Anoxic Event. Sporadic modifications in the water stratification, perhaps due to climatic changes, induced high-frequency changes of the deep-water redox conditions. These changes probably caused short-term fluctuations in the thickness of the oxygen-minimum layers and the ecological perturbations recorded in the deep-water-dwelling planktonic and benthic biota. Periodic disaerobic conditions, which protected the sediments from oxidizing benthic activity, are thought to be the main cause of relative concentrations of hydrocarbon-rich amorphous organic matter found in these potential source rocks.

INTRODUCTION

Major sedimentary and biological crises are recorded in many epicontinental basins and on margins of the Atlantic, Tethyan, or Pacific oceans near the Cenomanian-Turonian boundary (Herbin et al., 1985; Kuhnt et al., 1986; Thurow and Kuhnt, 1986; Schlanger et al., 1987; Arthur et al., 1987). Deposits at this boundary are generally marked by a major discontinuity, a sedimentary break ("Evénement E_2," Graciansky et al., 1984) or low sedimentation rate. The event has been regionally recorded in several successive laminated layers rich in marine organic matter (OM)—the black shale horizons (BSH) of the CTOAE (Schlanger and Jenkyns, 1976) which have been referred to as the Balhoul horizon in Tunisia (Burollet et al., 1978), the Livello Bonarelli in Italy (Jenkyns, 1980; Arthurand Premoli Silva, 1982), the Black Band in the Anglo-Germanic basins (Hart and Bigg, 1981), or the Thomel level in the southern Subalpine domain (Crumière, 1989).

During the past 10 years, numerous studies of middle Cretaceous Oceanic Anoxic Events (OAE) have shown that development of the BSH would have required an increase in the organic productivity by the phytoplankton in the photic zone and good protection of organic matter from oxidation, due to the initiation of deep-water anoxic conditions. Worldwide oceanographic perturbations, related to a major sea-level rise of eustatic origin, have been proposed to explain the wide distribution of Cenomanian-Turonian deep-water anoxic layers. The downward movement of dense epicontinental waters, with a slight enrichment in dissolved salts, might cause a pause in vertical exchanges, water stratification, and the initiation of deep-water, oxygen-minimum layers (Brass et al., 1982; Breheret et al., 1986; and Busson, 1984, 1988). The rhythmic fluctuations observed in the sediments (BSH generally alternate with homogeneous, bioturbated, and organic-depleted levels) are probably due to short-term variations in sedimentary inputs (organic, carbonate, and terrigenous), together with changes in the redox conditions in deep parts of the water column (review in Jacquin and Graciansky, 1988; Crumière et al., 1990).

Our purpose is to characterize both the successive episodes of the sedimentary history and the main oceanographic perturbations recorded throughout the CTOAE, in both deep- and shallow-water environments on the northwestern Tethyan margin. The consequences of major paleoceanographic changes and regional paleogeographic controls on the cyclic deposition and preservation of the potential source rocks will be discussed.

GEOLOGICAL SETTING

During middle Cretaceous time, the kinetics of the Iberian and Apulo-Adriatic plates changed with the beginning of ocean-floor spreading in the northern Atlantic realm (Dercourt et al., 1985). From middle Cenomanian onward, the northwestern margin of the Tethys Ocean was affected by the progressive closure of marine communication between two sedimentary basins (Figures 1, 2) which gradually filled up through Late Cretaceous time (review in Philip et al., 1984). The Pyrenean-Provencal Trough, a narrow, subsident furrow, was connected to the Northern Atlantic Ocean (Figure 1), and its eastern end was bordered by rudistid and carbonate-rich facies—the southern Provence platform (Figure 2). The Vocontian Basin was a western gulf in the European Alpine realm of the Tethys s.l. (Figure 1). In the deeper central and eastern parts of this basin (Figure 2), where lower Cenomanian clayey sediments constitute the uppermost part of middle Cretaceous BSH (Bréhéret and Crumière, 1989), the Cenomanian and Turonian sediments are marked by a cyclic increase in carbonate-rich and well-bioturbated sediments (Figure 3).

Near the Cenomanian-Turonian boundary, a major discontinuity exists in the two paleogeographic domains. It is expressed as an abrupt break in the development of the southern Provence platform (Phillip et al., 1989), and as a depositional hiatus within the carbonate-rich, subpelagic sedimentation of the Vocontian Basin (Ferry et al., 1988). In the deeper parts of this basin, a regional stratigraphic unit composed of 10-m-thick clay-rich sediments (Thomel, 1969) overlies the discontinuity. The several successive layers of 10-cm-thick BSH, constituting the "Thomel level" (Crumière, 1989), represent the latest organic-rich potential source rocks (Figure 3) deposited in that Cretaceous basin (Bréhéret and Crumière, 1989).

THE THOMEL STRATIGRAPHIC LEVEL: A REGIONAL SEDIMENTARY RECORD OF THE CTOAE

The black shale horizons (BSH) of the Thomel level (Figure 2, G_1) are unknown on the margins of the basin. This is perhaps due to a depositional hiatus, together with local submarine erosion. In the central part of the basin, the Thomel level does not exceed a few centimeters in thickness, whereas it may be up to a few meters thick in the deeper eastern domain connected with the Alpine Sea.

The Rock-Eval pyrolysis method (Espitalié et al., 1977, 1985) has been applied to numerous Cenomanian to lower Turonian subpelagic sediments sampled in the Vergons area (type section in the southeastern Vocontian Basin, Figures 2, 3). Total Organic Carbon (TOC) and Hydrogen Index (HI) values increase markedly in the laminated BSH of the Thomel level, whereas TOC and HI values remain low in the well-bioturbated sediments (Figure 3).

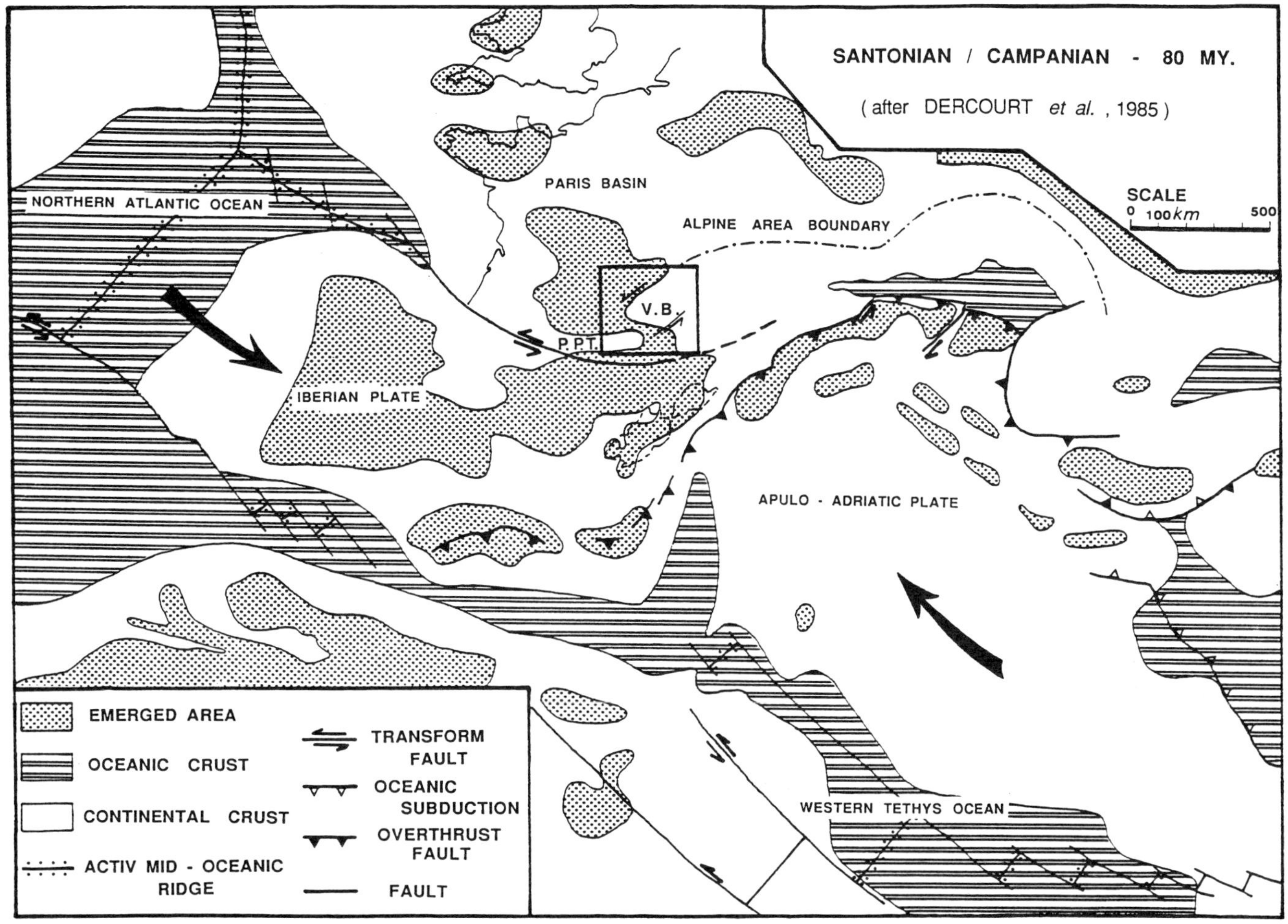

Figure 1. Palinspastic map of the Western Tethys and the Northern Atlantic oceans during Santonian-Campanian time (after Dercourt et al., 1985, pl. 5). V.B. = Vocontian Basin; P.P.T. = Pyrenean-Provencal Trough.

Modifications in the biotic associations occur in the Thomel level (Figure 4a); benthic microfauna and ichnofauna consist only of scolecodont remains of annelid jaws (Courtinat et al., 1990) in the BSH, while the interbedded bioturbated marls contain a poorly diversified benthic microfauna. Throughout deposition of the BSH, the calcareous planktonic microfauna also was decreasing (Figure 4a); only some small globular morphotypes of surface-dwelling foraminifera persisted, while the large keeled morphotypes disappeared. This selective preservation of the smaller forms may rule out the possibility of a drastic carbonate dissolution after sedimentation. According to Hart and Ball (1986), Jarvis et al. (1988), and Hart and Leary (1989), the worldwide disappearance of *Rotalipora* genus (Brotzen) was the result of a rising oxygen-minimum zone arresting the process of gametogenesis in the deeper-water planktonic foraminifera. The abrupt extinction of *R. cushmani* (Morrow) occurred above the sedimentary break at the base of the first BSH of the Thomel level (Figure 4a).

The abrupt initiation of a deep-water oxygen-minimum layer in the Vocontian Basin induced an ecological crisis, throughout the *Archaeocretacea* Zone, in connection with the widespread CTOAE. Biological perturbations, generally found in the Cenomanian-Turonian sediments (review in Jarvis et al., 1988), affected the deep-water basinal environment, resulting in drastic reduction of the planktonic biota at the top of the *Cushmani* Zone followed by a decrease in the benthic activity (Figure 4b).

The following section considers in detail the regional features and the paleoceanographic controls that might have caused these biological perturbations and led to periodic preservation of the organic matter in the sediments.

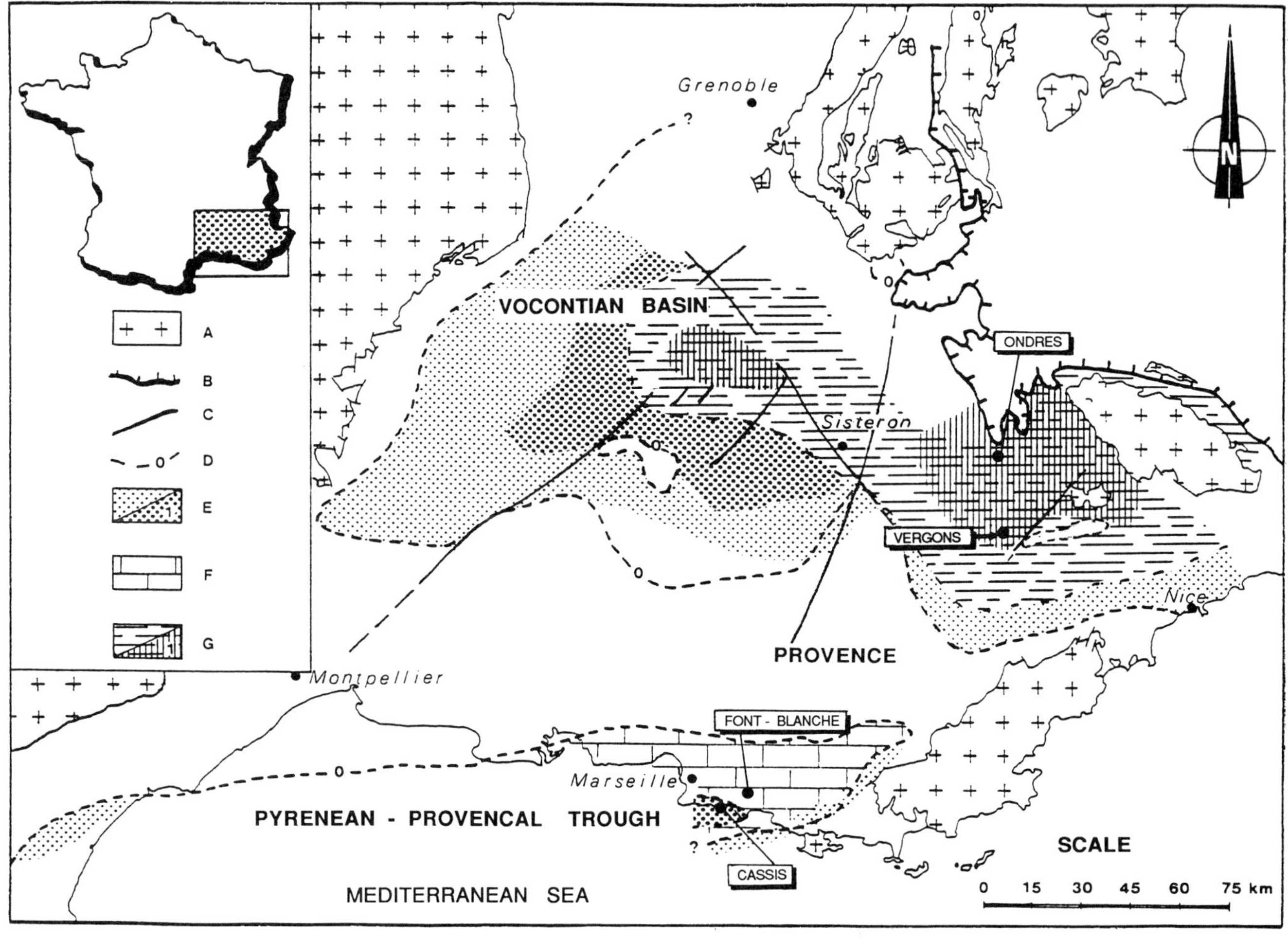

Figure 2. Cenomanian facies in the Vocontian Basin and the Pyrenean-Provencal Trough (after Philip et al., 1984). A = basement; B = Penninic thrust front; C = Cretaceous fault; D = 0-meter isopach; E = channelized sands, sandstones, and sandy bioclastic limestones with shallow-water fauna; E_1 = sandy glauconitic facies; F = rudist-bearing limestones of the southern Provence platform; G = marls and marly limestones with ammonites and planktonic foraminifera; G_1 = black shales (Thomel level).

THE DEEP-WATER OXYGEN-MINIMUM ZONE IN THE VOCONTIAN BASIN

Paleogeographic Setting

In the eastern part of the Vocontian Basin (Figures 2, 5), the Thomel level has been sampled at numerous outcrops between the Digne-Vergons overthrusts and the present-day erosional limit of the Inner Alpine Overnappes. Rock-Eval pyrolysis analyses have been run on about ten black shale samples from each section. The iso-TOC contours based on these analyses (Figure 5) reflect the Vocontian deep-water paleogeochemistry during the CTOAE. The TOC-1% contour delineates the paleogeographic extent of the oxygen-minimum zone on the basin floor. Figure 5 shows a northern deep anoxic area of black shale deposition that was regionally affected by a later thermal maturation of the organic matter (T_{max} > 460°C and a gas window of type II OM) due to the Alpine Overthrusting (Crumière et al., 1988). A southern neritic area unaffected by the deep-water anoxia is shown on Figure 5 at the southern margin of the basin where tilting of faulted blocks is inferred from sedimentary unconformities, indicated by thin or missing units, that have developed since Neocomian times (Cotillon, 1971). Some shoals were affected by bioerosion and mineralization (iron, glauconite, phosphate) until late Albian time (Cotillon, 1985). While sedimentation was more continuous from latest Albian (Vraconian) onward (Cotillon, 1971; Conard-Noireau, 1983), these structural features persisted through Cenomanian-Turonian times (Figure 5), as indicated by missing units and slump features on the sloping transitional area between the basin and the northeastern edge of Provence platform (Crumière, 1989). A large neritic zone (the northern Provence platform) was the site

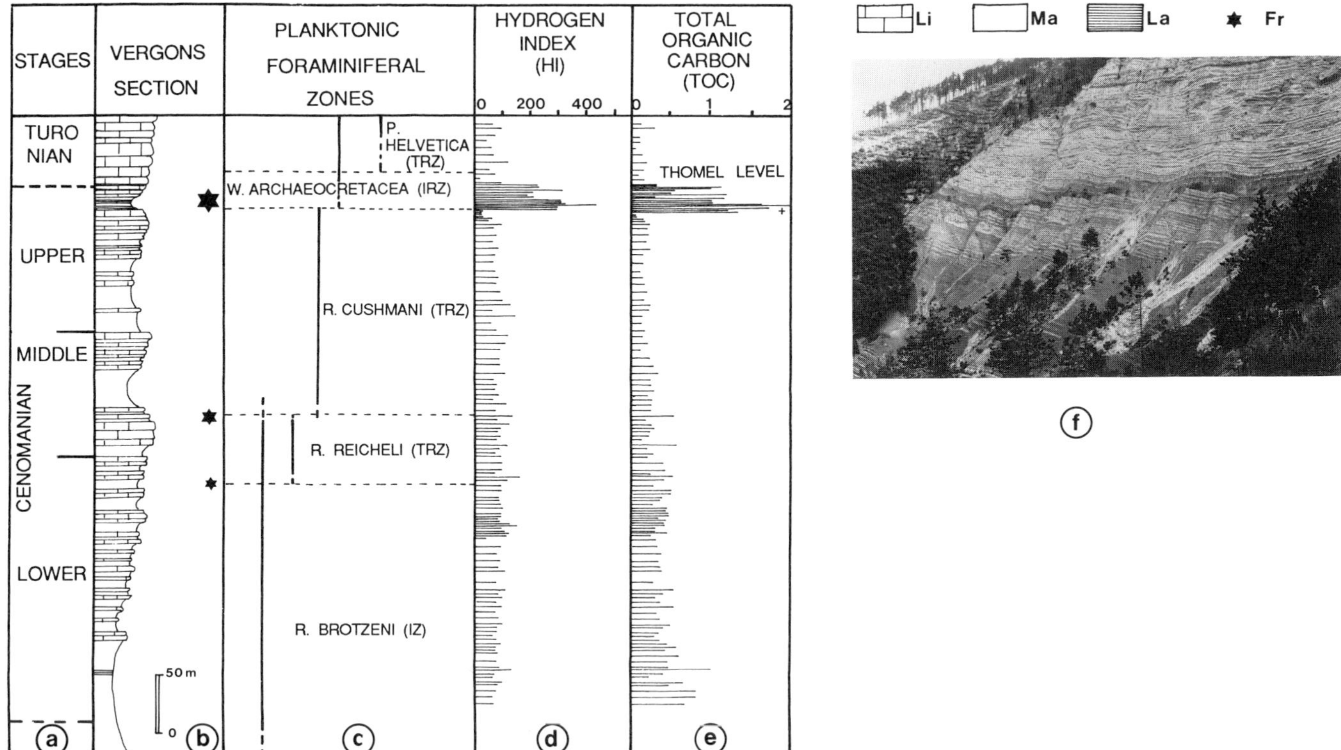

Figure 3. Vergons type section. (a) stages; (b) lithology: Li = limestone, Ma = marls, La = laminites (black-shales); (c) planktonic foraminiferal zones showing vertical extensions of species, Fr = faunistic renewals; geochemical logs of Rock-Eval pyrolysis data : (d) hydrogen index (in mg HC / g TOC), (e) total organic carbon (in %); (f) the Thomel stratigraphic level at the upper part of the outcrop.

of deposition of marly and silty sediments that contain a few glauconite-rich horizons (Thomel, 1969; Donze and Thomel, 1972; Conard-Noireau, 1986). Paleodepths and structural features of the basin and its margins (Figures 2, 5) were major factors in the regional distribution of the BSH.

Short-Term Variations in Thickness of the Oxygen-Minimum Zone

On a basinwide scale, the distribution of the Thomel level deposits suggests that the deep-water stratification was not stable throughout the CTOAE. In the whole central and southern parts of the Eastern Vocontian Basin (type section in Vergons, Figure 5), where subpelagic sedimentation produced alternations of laminated BSH and well-bioturbated organic-depleted marls (Figure 4a), the oxygen-minimum zone periodically reached the basin floor (Figure 4b). This rhythmicity was abruptly interrupted by the slumping of BSH (Figure 4a) coming from the shallower southern slope where euxinic conditions prevailed almost continuously (Figure 4b). In the northern part of the basin (type section in Ondres, Figure 5), the BSH were more numerous, thicker (Figure 4), and more organic rich than in the deeper areas.

The rhythmic variations in the deep-water redox conditions were mainly controlled by short-term fluctuations in the thickness of the oxygen-minimum zone, the causes of which (possibly climatic changes due to the earth's orbital periodicities) are still not understood. Periodic changes in the deep-water paleo-oxygenation of the Vocontian Basin affected a substantial portion of the biota, resulting in the decrease of aerobic benthic activity when the oxygen-minimum zone reached the basin floor and perturbation of gametogenis of some large globular and keeled morphotypes of planktonic foraminifera (Figure 4b). Such deep-water redox cyclicity should have caused periodic preservation of labile OM in sediments.

Cyclic Preservation of the "Amorphous" Organic Content of Sediments

In marine sedimentary rocks, the relationship between TOC values and intensity of bioturbation has already been noticed. For example, in the underlying middle Cretaceous BSH, Bréhéret et al. (1986) point out that when both benthic microfauna

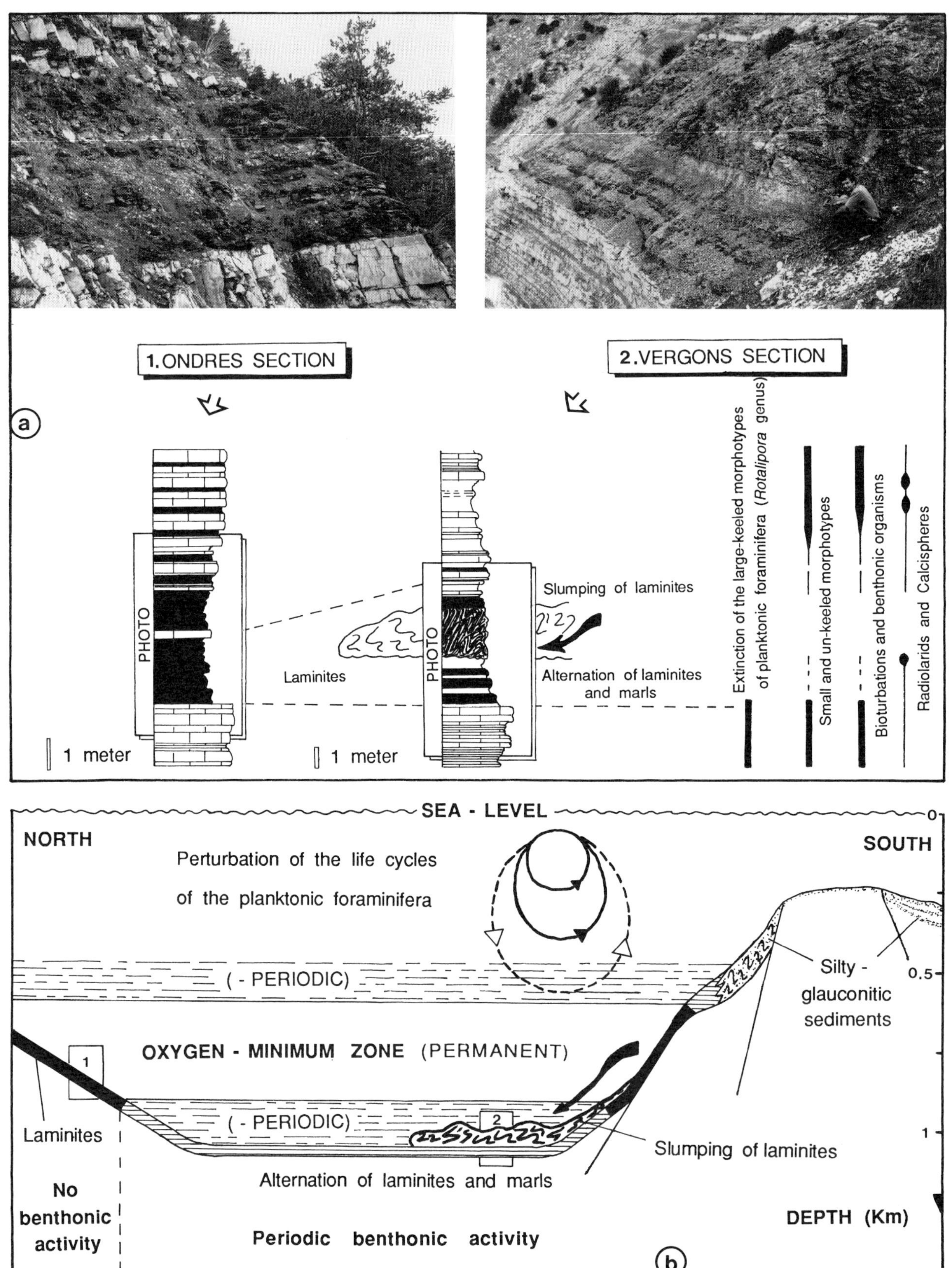

Figure 4. (a) Lithologic correlations between the two type sections (Ondres and Vergons) of the Thomel level in the Eastern Vocontian Basin, showing major changes in the biota. (b) North-south reconstruction of the Eastern Vocontian Basin and its southern margin. Periodic variations in thickness of the deep-water oxygen-minimum layer resulted in regional distribution of the BSH and produced marked effects on the biota.

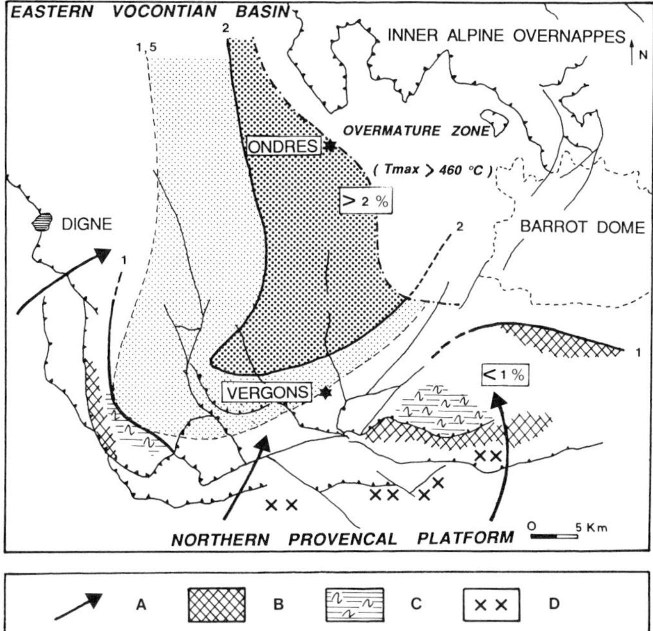

Figure 5. Iso-TOC (in %) contours showing the extent of the deep-water oxygen-minimum zone on the Eastern Vocontian Basin floor during Cenomanian-Turonian times. Locations of the two type sections (Ondres and Vergons) are indicated. Main features of the northern Provence platform : (A) Direction of lower Cenomanian clastic transport; (B) Hauterivian-late Albian shoals (missing section, bioerosion, and mineralization); (C) Cenomanian-Turonian slope (missing section and slump features); (D) Upper Cenomanian glauconite-rich horizons.

and ichnofauna are at a minimum, the TOC values are at a maximum, and the OM contains more hydrocarbons.

In order to determine the effects of such periodic variations in deep-water paleo-oxygenation on organic fractions of sediments, we analyzed the OM contained in several rock samples from the Thomel level through successive geochemical and optical methods (Table 1). We selected six samples from the Vergons section where the OM had remained immature ($T_{max} < 425°C$), as shown by their specific TOC values (Figure 6a).

Rock-Eval pyrolysis data from crude rocks (Table 1) show a decrease in TOC values, mainly expressed by a decrease in residual petroleum potential of the OM (decrease of the S2 peak, Figure 6b). A progressive drift (Figure 7a) of Hydrogen Index (HI) and Oxygen Index (OI) values may indicate an increase in the OM alteration (Espitalié et al., 1977). The laminated BSH (samples 1-4) contain type II OM, characterized by both high HI values (> 200 mg HC/g TOC) and low OI values (< 60 mg CO_2/g TOC). The more homogenous and bioturbated rocks (samples 5-6) contain small amounts of type III OM, characterized by low HI values (< 100 mg HC/g TOC) and very high OI values (> 100 mg CO_2/g TOC).

Such a trend is enhanced by an increase in mineral-matrix effects (Espitalié et al., 1980) due to the very low organic content of the well-bioturbated rocks. In order to minimize the mineral-matrix effects, the kerogen (insoluble OM) was extracted from rocks. The results (Table 1) of both elementary analysis (illustrated in a H/C vs O/C diagram, Figure 7b) and kerogen pyrolysis (Figure 7c) show similar drifts toward the reference lines of type III OM—a decrease in both the H/C and HI values, together with an increase in both the O/C and OI values. Such geochemical data suggest that this organic matter was of continental origin and/or mostly altered.

Optical investigations in transmitted and fluorescent light (Table 1 and Figure 6c, d) reveal that the laminated BSH contain mainly sapropelic-amorphous OM (60–90%) of marine origin and a small amount of humic-terrigenous OM (10-40%) of ligneous origin generally hydrogen depleted. Organic matter of such a maceral composition has geochemical characteristics of a mixture of types II and III kerogen (Crumière and Espitalié, 1989). The well-bioturbated rocks do not contain any amorphous OM, and the small amount of OM present is mostly of terrestrial origin, as ligneous tissues, spores, or pollen.

In summary, the trend of geochemical data from mixed types II and III toward type III (Figure 7) and the decrease in petroleum potential of sediments (Figure 6b) are related to a progressive transformation of the maceral composition of the OM (Figure 6d). The decrease (from 90% to 0) in the amorphous organic content caused the decrease (from 2.1% to 0.1%) in the TOC values, while the quantities of ligneous and coal debris remained constant in the organic fraction of sediments (Figure 6c). Amorphous organic matter, which is usually altered in the water column and in the sediments under oxidizing conditions, has been preserved from oxidation under periodic deep-water disaerobic conditions. Short-term absence of any aerobic benthic activity when the oxygen-minimum layer reached the basin floor should explain the periodic preservation of the initial petroleum potential of the BSH.

MAJOR SEDIMENTARY EPISODES OF THE CTOAE ON THE EUROPEAN TETHYAN MARGIN

We established lithostratigraphic and biostratigraphic correlations (Figure 8) between the Vergons section and shallower environments of the Provence domain unaffected by disaerobic conditions (Figure 2)—Fontblanche, referred to as the type section of the southern Provence platform, and Cassis, the type section of the northern slope of the Provence Trough. Several successive episodes in the sedimen-

	CRUDE ROCK ANALYSIS			KEROGEN ANALYSIS												
	ROCK - EVAL PYROLYSIS					ELEMENTARY ANALYSIS						OPTICAL ANALYSIS : Palynofacies				
									Residue	Atomic ratio						
	TOC	HI	OI	HI	OI	C	H	O	N, S, Fe	H/C	O/C	MOS + MOA	MOX	MOB₁	MOB₂	Others
n° 1	2.1	436	25	508	26	68.92	6.28	9.30	6.00	1.09	0.101	90	-	5	-	5
n° 2	1.3	328	28	430	24	56.05	4.83	7.41	16.80	1.03	0.099	80	10		-	10
n° 3	1.1	307	39	437	29	61.49	5.31	9.09	10.40	1.04	0.111	75	15		-	10
n° 4	1.1	198	66	409	33	37.82	3.01	7.92	33.10	0.96	0.157	60	10	10	10	10
n° 5	0.2	75	100	225	42	38.67	2.72	6.06	34.70	0.84	0.118	5	35	30	10	20
n° 6	0.1	14	130	129	42	27.03	1.43	5.70	45.30	0.63	0.158	-	40	20	20	20

Table 1. Organic data for the six rock samples from the Vergons section (see also Figure 6). Rock-Eval pyrolysis data: TOC, total organic carbon (in %); HI, hydrogen index (in mg HC / g TOC); OI, oxygen index (in mg CO_2 / g TOC). Elementary composition of kerogens in carbon, hydrogen, and oxygen (in %); atomic ratio, H / C and O / C. Distinguishable organic matter in transmitted and fluorescent light (in %) : MOS, amorphous-sapropelic; MOA, amorphous; MOX, opaque coal; MOB_1, opaque ligneous remains; MOB_2, translucent ligneous remains; Others, organic microfossils, spores, and pollen.

tary history of each paleogeographic domain are recognized (Figure 8).

Throughout the *Cushmani* Zone, subpelagic to hemipelagic sediments were deposited in the Vergons and Cassis areas, while shallow-water, rudist-bearing facies developed in the Fontblanche area (Figure 8, unit A). A sedimentary break is present in both domains at the top of this zone.

After this marine hiatus, the *Archaeocretacea* Zone began with the abrupt extinction of *R. cushmani* and with the deposition of the main BSH of the Thomel level (Figure 8, unit C) in the Vocontian Basin only. When sedimentation resumed throughout the area, it was represented by mainly bioclastic deposits, small rudist banks on the Provence platform, calcisphere-rich hemipelagic limestones on its southern flank, and alternation of marls and subpelagic limestones with silty turbidites at the tops of beds (Figure 8, unit D) in the Vocontian Basin. Following this, another sedimentary break occurred, as evidenced by a regional hard-ground on the Provence platform and a soft-ground on the slope. The top of the *Archaeocretacea* Zone is recorded in Vergons by the last four BSH of the Thomel level (Figure 8, unit F), while equivalent deposits are lacking in Provence.

The *Helvetica* Zone corresponds to the deposition of a subpelagic carbonate-rich series in the Vocontian Basin, of marls that contain slump features on the slope, and of glauconite-rich and marly hemipelagic limestones on the whole Provence platform (Figure 8, unit G).

According to numerous authors, the deposition of sediments during Cenomanian-Turonian time took place in a global transgressive context, during the third-order eustatic cycle referred to as UZA-2.5 by Haq et al. (1987). As shown on Figure 8, the sequence boundary and the UZA-2.5 shelf margin wedge systems tract are not clearly defined in carbonate-rich, siliciclastic-depleted units. Two main transgressive pulsations (referred to as UZA-2.5 a and b), at the base and the top of the *Archaeocretacea* Zone, are represented by nondeposition in the shallower Provence environments, and by deposition of BSH in the Vocontian Basin. Two periods of highstand in sea level resulted in the development of numerous slump features and led to the progressive recovery of mainly bioclastic basinal sedimentation.

In summary, two main periods of deposition of BSH in the deeper areas of the basin are contemporaneous with two major discontinuities in the shallower environments. The first discontinuity marked a landward shift of rudistid reefal patches as the result of a first step in the Cenomanian-Turonian sea-level rise (in late Cenomanian; top of the *Cushmani* Zone). The second one corresponded to the complete drowning of the Provence platform (in early Turonian; top of the *Archaeocretacea* Zone). The worldwide CTOAE began on the European Tethyan margin during the late Cenomanian (base of the

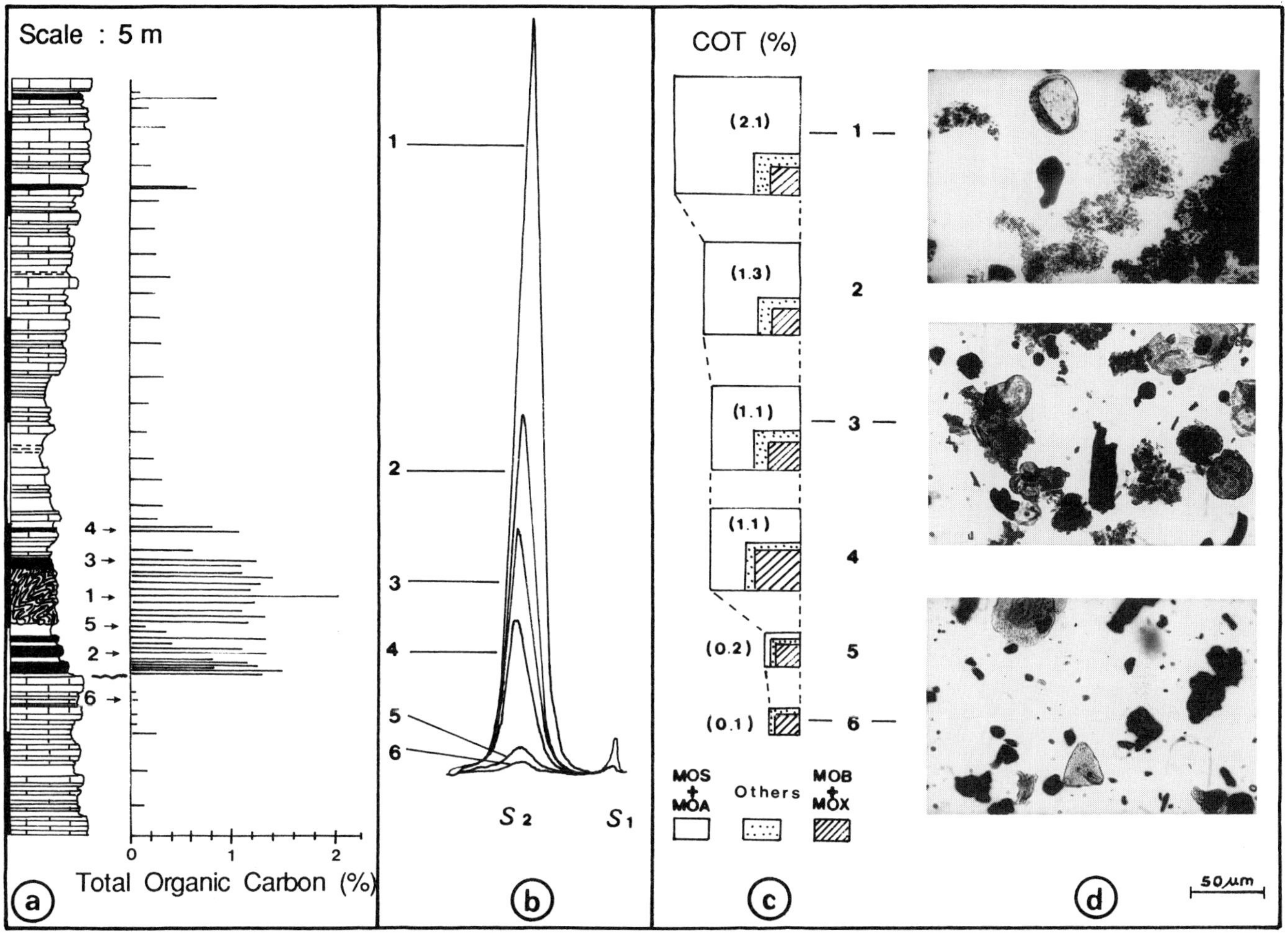

Figure 6. Organic content of the uppermost Cenomanian and lower Turonian rocks in the Vergons section. (a) Log of TOC values and stratigraphic position of the six rock samples; (b) peaks of Rock-Eval pyrolysis (S1, amount of free hydrocarbons; S2, amount of potential hydrocarbons); (c) evolution of maceral compositions: MOS + MOA, amorphous organic matter; Others, organic microfossils, spores, and pollen; MOB + MOX, ligneous and opaque coal remains; (d) palynofacies.

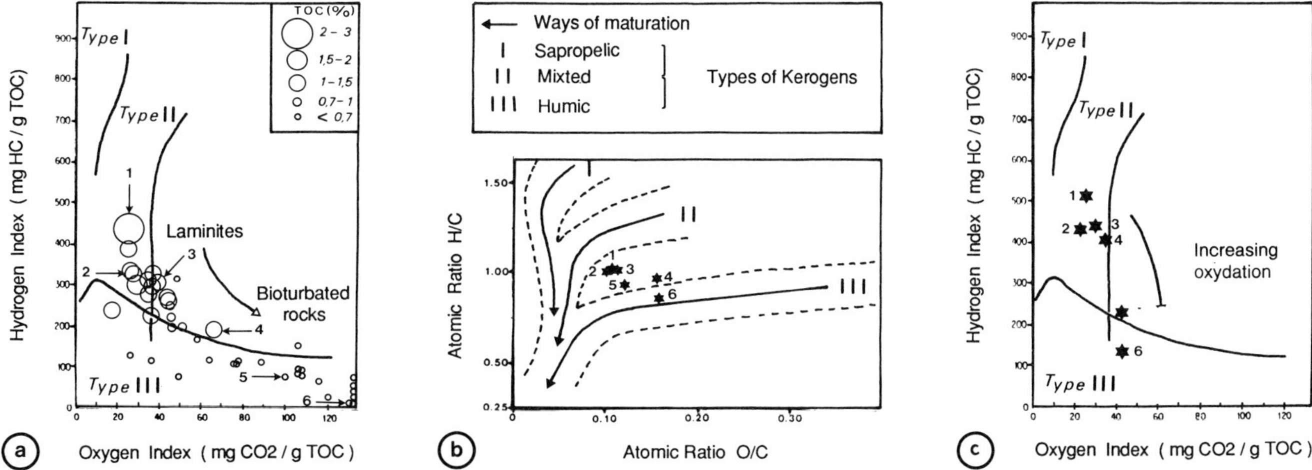

Figure 7. Geochemical data on the organic content of the six rock samples from the Vergons section (see also Figure 6). (a) diagram of HI vs OI (pyrolysis data from the crude rocks); (b) Van Krevelen H/C vs O/C diagram; (c) HI vs OI diagram (pyrolysis data from the kerogens).

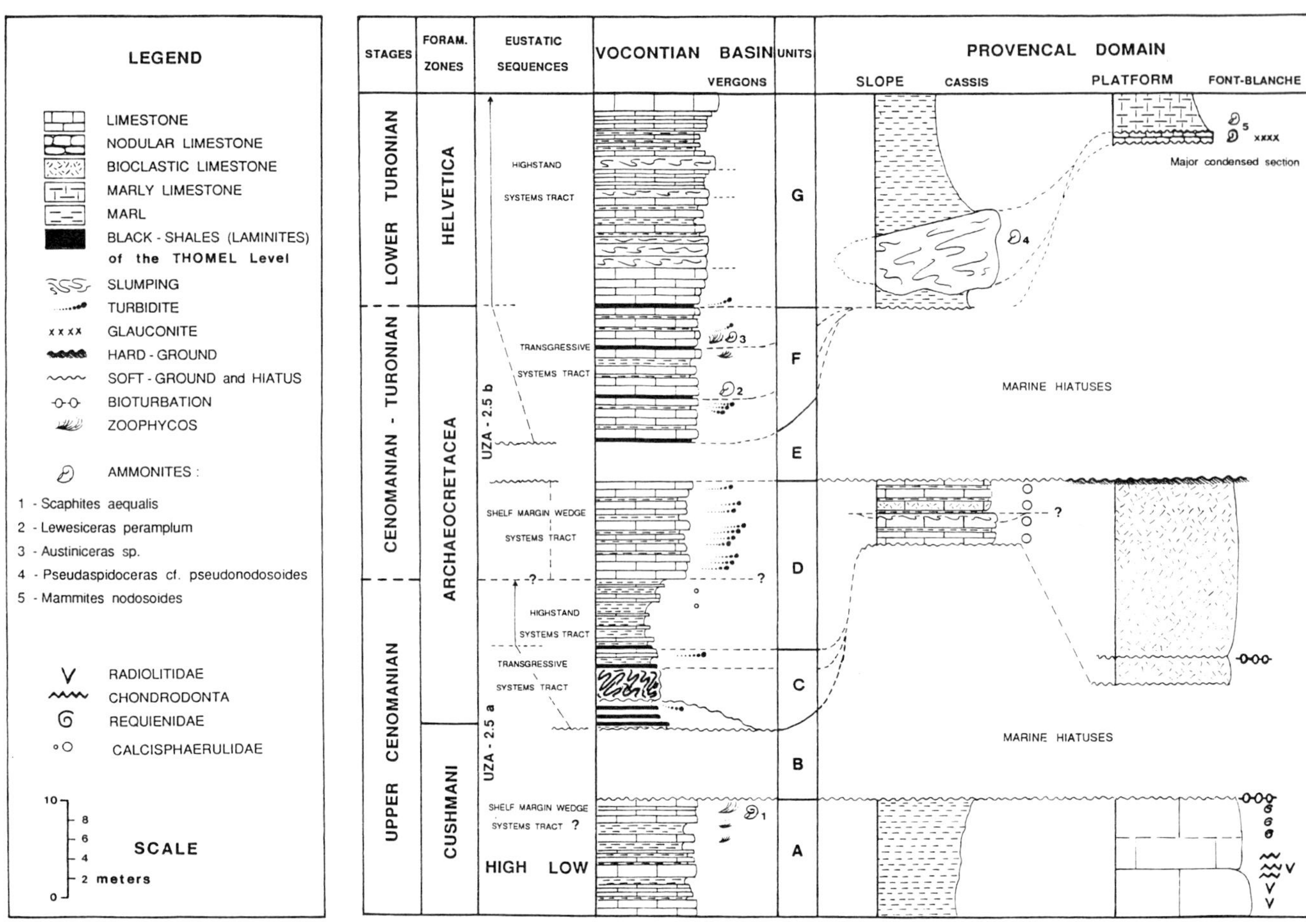

Figure 8. Lithostratigraphic and biostratigraphic correlations between the Provence domain and the Vocontian Basin, during Cenomanian-Turonian times, and interpretation in terms of systems tracts.

Archaeocretacea Zone) and ended during early Turonian (base of the *Helvetica* and *Nodosoides* Zones), when the global sea level reached its highest Mesozoic-Cenozoic stand top of the UZA-2.5 eustatic cycle).

CONCLUSIONS

According to existing paleoceanographic models (Tissot, 1979; Arthur and Schlanger, 1979; Demaison and Moore, 1980), a global eustatic control, represented here as two transgressive pulsations, could have initiated oxygen-minimum layers in deeper-water marine environments. On the European Tethyan margin, latest Cenomanian and earliest Turonian sea-level rises could have caused the abrupt drowning of the Provence platform; the deposition of black shale horizons in the Vocontian Basin; and periods of nondeposition in the shallower environments of longer duration than in the deeper ones. Throughout the *Archaeocretacea* Zone, the increase in oceanic circulation between platforms, epicontinental basins, and oceans might have promoted the connection between different density water masses, thus causing a deep-water stratification of regional extent.

Paleogeographic and structural features (depth of basins, surface of their margins, etc.) might have controlled the connection between shallow environments (source-salt areas) and oceanic water masses; the regional location of oxygen-minimum layers and of BSH distributions; and the regional duration of such deep-water Oceanic Anoxic Events.

Climatic changes, perhaps due to earth orbital periodicities, might have caused high-frequency changes in the deep-water redox conditions. Short-term fluctuations in the thickness of oxygen-minimum layers could have controlled the petroleum potential of the sediments.

The relative concentrations of hydrocarbon-rich organic matter observed in such marine potential source rocks could be due to:

1. an increase in the phytoplankton productivity;
2. a weaker dilution of the organic fraction of marine origin, caused by periodic pauses in basinal sedimentary inputs (planktonic carbonates, outer-platform muds, and siliciclastics); and
3. periodic preservation of the amorphous OM from alteration in the deep-water column and from

oxidizing benthic activity, when the oxygen-minimum layer reached the basin floor.

This last factor should be the major one because the worldwide, abrupt disappearance of the deeper-water-dwelling planktonic biota should have caused a substantial decrease in the basinal sedimentation of sapropelic OM, while the organoclastic inputs remained constant, and while the calcisphere blooms that have been recorded, mainly occurred after the episodes of black shale deposition.

ACKNOWLEDGMENTS

Sincere thanks are due to J. F. Raynaud and the group of the "Service de Biostratigraphie" (Société Nationale Elf Aquitaine (Production), Pau, France) and M. Caron (University of Fribourg, Switzerland) for providing technical support. Thanks are extended to M. P. Aubry-Berggren, J. G. Bréhéret, B. Courtinat, G. Dromart, S. Ferry, P. C. de Graciansky, J. P. Herbin, T. Jacquin, E. Jautée, H. Méon, J. Philip, and G. Tronchetti for helpful discussions and suggestions throughout the course of this study, and to N. Podevigne for photographic enlargements.

We are particularly indebted to M. B. Hart of Plymouth Polytecnic (UK) and to S. O. Schlanger of Northwestern University (Evanston, USA) for their reviews.

REFERENCES CITED

Arthur, M. A., and I. Premoli Silva, 1982, Development of widespread carbon-rich strata in the Mediterranean Tethys, *in* S. O. Schlanger and M. B. Cita, eds., Nature and origin of Cretaceous carbon-rich facies: Academic Press, p. 7-54.

Arthur, M. A., and S. O. Schlanger, 1979, Cretaceous "oceanic anoxic events" as causal factors in development of reef-reservoired giant oil fields: American Association of Petroleum Geologists Bulletin, v. 63, no. 6, p. 870-885.

Arthur, M. A., S. O. Schlanger, and H. C. Jenkyns, 1987, The Cenomanian-Turonian Oceanic Anoxic Event, II. Palaeoceanographic controls on organic-matter production and preservation, *in* J. Brooks and A. J. Fleet, eds., Marine petroleum source rocks: Geological Society Special Publication, v. 26, p. 401-420.

Brass, G. W., J. R. Southam, and W. H. Peterson, 1982, Warm saline bottom water in the ancient ocean: Nature, London, v. 296, p. 620-623.

Bréhéret, J. G., M. Caron, and M. Delamette, 1986, Niveaux riches en matière organique dans l'Albien vocontien; quelques caractères du paléoenvironnement; essai d'interprétation génétique: Documents du Bureau de Recherches Géologiques et Minières, Paris, v. 110, p. 141-191.

Bréhéret, J. G., and J. P. Crumière, 1989, Organic-rich episodes in the Mid-Cretaceous (Aptian to Turonian) pelagic realm of the Vocontian Basin (SE France): Geobios, Lyon, numéro special 11, p. 205-210.

Burollet, P. F., J. M. Mugniot, and P. Sweeney, 1978, The geology of the Pelagian Block: The margins and basins of southern Tunisia and Tripolitania, *in* A. E. M. Nairn, W. H. Kanes, and F. G. Stehli, eds., The ocean basins and margins: Plenum Press, New York, v. 4B, p. 331-359.

Busson, G., 1984, La sédimentation épicontinentale des bassins accomplis et des bassins évaporitiques avortés: effet sur la sédimentation océanique contemporaine et contigue dans le cas du Crétacé saharien et nord atlantique: Comptes Rendus de l'Académie des Sciences de Paris, s. II, v. 299, no. 5, p. 213-216.

Busson, G., 1988, Relations entre les types de dépôts évaporitiques et la présence de couches riches en matière organique (roches-mères potentielles): Revue de l'Institut Francais du Pétrole, v. 43, no. 2, p. 181-215.

Conard-Noireau, M., 1983, La dynamique des dépôts cénomaniens de Haute-Provence: observations nouvelles et implications paléogéographiques: Bulletin de la Société géologique de France, T. XXV, no. 2, p. 239-246.

Conard-Noireau, M., 1986, Le Crétacé supérieur en domaine subalpin méridional; biostratigraphie par les Globotruncanidae et paléogéographie: Thèse d'Etat, Université du Mans, France, 281 p., (unpublished).

Cotillon, P., 1971, Le Crétacé inférieur de l'arc subalpin de Castellane entre l'Asse et le Var. Stratigraphie, sédimentologie: Mémoires du Bureau de Recherches Géologiques et Minières, Paris, no. 68, 313 p.

Cotillon, P., 1985, Hauts-fonds de la marge nord-est provencale au Crétacé inférieur. Un exemple de contrôles tectonique et bathymétrique: Bulletin de la Section des Sciences, Comité des travaux historiques et scientifiques, Paris, T. IX, p. 25-38.

Courtinat, B., J. P. Crumière, and A. M. Bodergal, 1990, Les scolécodontes, reliques des annélides errantes, témoins de la suboxie plutôt que l'anoxie dans des black shales mésozoiques: Comptes Rendus de l'Académie des Sciences de Paris, T. 310, s. II, p. 1089-1093.

Crumière, J. P., 1989, Crise anoxique à la limite Cénomanien-Turonien dans le bassin subalpin oriental (Sud-Est de la France). Relation avec l'eustatisme: Geobios, Lyon, numéro spécial 11, p. 189-203.

Crumière, J. P., C. Crumière-Airaud, and J. Espitalié, 1990, Préservation cyclique de la matière organique amorphe des sédiments au passage Cénomanian-Turonien dans le bassin vocontien (Sud-Est France). Contrôles paléo-océanographiques: Bulletin de la Societé geologique de France, Paris, s. 8, T. VI, no. 3, p. 469-478.

Crumière, J. P., and J. Espitalié, 1989, Caractérisation des types de dépôts organiques par analyse de la forme du pic S2 de pyrolyse Rock-Eval. Comparaison avec des analyses optiques: Comptes Rendus de l'Académie des Sciences de Paris, T. 309, s. II, p. 1413-1417.

Crumière, J. P., F. Pascal, and J. Espitalié, 1988, Evolutions diagénétiques comparées de la matière organique et des argiles. Influence de l'enfouissement normal et d'une anomalie thermique par surcharge des nappes alpines (Crétacé subalpin de Haute-Provence, France): Comptes Rendus de l'Académie des Sciences de Paris, T. 306, s. II, p. 493-498.

Demaison, G. J., and G. T. Moore, 1980, Anoxic environments of oil source bed genesis: American Association of Petroleum Geologists Bulletin., v. 64, no. 8, p. 1179-1209.

Dercourt, J., et al., 1985, Presentation de 9 cartes paléogéographiques au 1/20.000.000° s'étendant de l'Atlantique au Pamir pour la période du Lias a l'Actuel: Bulletin de la Société géologique de France, Paris, T. I, no. 5, p. 637-652, 10 inset maps.

Donze, P., and G. Thomel, 1972, Le Cénomanien de La Foux (Alpes de Haute-Provence). Biostratigraphie et faunes nouvelles d'Ostracodes: Eclogae geologicae Helvetiae, Bâle, v. 65, no. 2, p. 369-389.

Espitalié, J., et al., 1977, Méthode rapide de caractérisation des roches mères, de leur potentiel pétrolier et de leur degré d'évolution: Revue de l'Institut Francais du Pétrole, v. 32, no. 1, p. 23-43.

Espitalié, J., M. Madec, and B. Tissot, 1980, Role of mineral matrix in kerogen pyrolysis: influence on petroleum generation and migration: American Association of Petroleum Geologists Bulletin. v. 64, p. 59-66.

Espitalié, J., G. Deroo, and F. Marquis, 1985, La pyrolyse Rock-Eval et ses applications, Parts 1 and 2: Revue de l'Institut Francais du Pétrole. v. 40, no. 5 and 6, p. 563-784. Part 3 (1986): Revue de l'Institut Francais du Pétrole, v. 41, no. 1, p. 73-89.

Ferry S., J. L. Rubino, and J. P. Crumière, 1988, La regression fini-cénomanienne dans le Sud-Est de la France, *in* S. Ferry and J. L. Rubino, eds., Eustatisme et séquences de dépôt dans le Crétacé du Sud-Est de la France: Géotrope, Lyon, v. 1, p. 95-105.

Graciansky de, P. C., et al., 1984, Ocean-wide stagnation in the late Cretaceous: Nature, v. 5957, p. 346-349.

Haq, B. U., J. Hardenbol, and P. R. Vail, 1987, Chronology of fluctuating sea levels since the Triassic: Science, v. 235, p. 1156-1166.

Hart, M. B., and K. C. Ball, 1986, Late Cretaceous anoxic events, sea-level changes and the evolution of the planktonic foraminifera, *in* C. P. Summerhayes and N. J. Shackleton, eds., North Atlantic palaeoceanography: Geological Society Special Publication, v. 21, p. 67-78.

Hart, M. B., and P. J. Bigg, 1981, Anoxic events in the Late Cretaceous chalk seas of North West Europe, *in* J. W. Neale and M. D. Brasier, eds., Microfossils from recent and fossil shelf seas: The British Micropaleontology Society. v. 14, p. 178-185.

Hart, M. B., and P. N. Leary, 1989, The stratigraphic and paleogeographic setting of the late Cenomanian "anoxic" event: Journal of the Geological Society, London, v. 146, p. 305-310.

Herbin, J. P., et al., 1985, Organic rich sedimentation at the Cenomanian-Turonian boundary in oceanic and coastal basins in the North Atlantic and Tethys, *in* C. P. Summerhayes and N. J. Shackleton, eds., North Atlantic palaeoceanography: Geological Society Special Publication, v. 21, p. 389-422.

Jacquin, T., and P. C. de Graciansky, 1988, Cyclic fluctuations of anoxia during Cretaceous time in the South Atlantic Ocean: Marine and Petroleum Geology, v. 5, p. 359-369.

Jarvis, I., et al., 1988, Microfossil assemblages and the Cenomanian-Turonian (late Cretaceous) Oceanic Anoxic Event: Cretaceous Research, v. 9, p. 9-103.

Jenkyns, H. C., 1980, Cretaceous anoxic events: From continents to oceans: Journal of the Geological Society, London, v. 137, p. 171-188.

Kuhnt, W., J. Thurow, J. Wiedmann, and J. P. Herbin, 1986, Oceanic anoxic conditions around the Cenomanian-Turonian boundary and the response of the biota: Mitt. Geol.-Palaont. Inst., Hamburg, v. 60, p. 205-246.

Philip, J., (coord.) et al., 1984, Crétacé supérieur, *in* S. Debrand-Passart, et al., eds., Synthèse géologique du Sud-Est de la France: Mémoires du Bureau de Recherches Géologiques et Minières, Paris. v. 125, p. 339-387.

Philip, J., C. Airaud, and G. Tronchetti, 1989, Evénements paléogéographiques en Provence (S.-E. France) au passage Cénomanien-Turonien. Modifications biosédimentaires - Causes géodynamiques: Geobios, Lyon, numéro spécial 11, p. 107-117.

Schlanger, S. O., M. A. Arthur, H. C. Jenkyns, and P. A. Scholle, 1987, The Cenomanian-Turonian Oceanic Anoxic Event, I. Stratigraphy and distribution of organic carbon-rich beds and the marine "delta 13 C" excursion, *in* J. Brooks and A. J. Fleet, eds., Marine petroleum source rocks: Geological Society Special Publication, v. 26, p. 371-399.

Schlanger, S. O., and H. C. Jenkyns, 1976, Cretaceous oceanic anoxic events: Causes and consequences: Geologie en Mijnbouw, v. 55. p. 179-184.

Thomel, G., 1969, Etudes stratigraphiques et paléontologiques du Cénomanien subalpin entre Digne et Menton: Thèse d'Etat, Université de Nice, France, 466 p., (unpublished).

Thurow, J., and W. Kuhnt, 1986, Mid-Cretaceous of the Gibraltar Arch area, *in* C. P. Summerhayes and N. J. Shackleton, eds., North Atlantic palaeoceanography: Geological Society Special Publication, v. 21, p. 423-445.

Tissot, B., 1979, La répartition mondiale des combustibles fossiles: La Recherche, Paris, v. 104, no. 10, p. 984-991.

Organic Matter and Evaporites in the Paleogene West European Rift: The Bresse and Valence Salt Basins (France)

Alain Curial
Robert Moretto
Diastrata Sarl
Lyon, France

Daniel Dumas
Groupe J Sarl
Villeurbanne, France

Gilles Dromart
Université Claude-Bernard
Villeurbanne, France

The Bresse and Valence salt basins are in two adjacent segments of the West European Rift that cuts across the eastern part of France. They contain thick Paleogene halite sequences including intercalated and interfingering siliciclastic beds and carbonate and sulfate deposits. Source-rock samples were taken mainly from the depocenters because of maximum sampling coverage. Organic matter is generally immature and occurs primarily within intercalated nonhalitic beds. The amount of organic matter varies considerably with stratigraphic location and lithology. The Bresse basin contains organic matter in (1) the Intermediate Salt Fm (Priabonian), composed of alternating laminated carbonate and primary halite beds; (2) the upper part of the upper Salt Fm (brecciated clayey carbonate beds (Rupelian), affected by multistage halite solution penecontemporaneous with deposition; and (3) the solution breccia that immediately overlies the salt sequence. In the Valence basin, the layers richest in organic matter are in the Subsalt Fm (carbonate beds that are commonly laminated; Priabonian) and the upper part of the Lower Salt Fm (laminites; Rupelian). In both basins, type III kerogen is closely related to terrigenous material. Type I organic matter is abundant in the Valence basin (laminites). Type II seems to be more abundant in the Bresse basin, but its diagnosis remains uncertain. The main depositional environment favoring the preservation of organic matter is likely related to deposition beneath perennially stratified layers of water. Major salt-solution phenomena, which are evident, are supposed to be responsible for concentration of significant amounts of organic matter concurrently with the accumulation of nonsoluble, brecciated residues.

INTRODUCTION

The association of evaporites and sediments rich in deposited organic matter (OM) has been observed in numerous basins of various ages (Kirkland and Evans, 1981; Sonnenfeld, 1985; Warren, 1986; Busson, 1988a, b; Rouchy, 1988; Schreiber, 1988; Evans and Kirkland, 1988). This association is so common that it may be considered as a real succession of genetically linked facies, the organic-rich sediments commonly underlying salt rocks (Busson, 1978, 1979). Great accumulations of salt deposits in some basins of the West European Rift are, from this point of view, of investigative interest.

In this paper, we examine the organic matter of the Paleogene salt basins of Bresse and Valence, located in the French part of the West European Rift system. After giving general information about this rift and the aforementioned basins, we provide quantitative and qualitative data about the organic matter present in these series, as it relates to the lithology and petrographic facies of the organic-matter-bearing sediments and, consequently, to their depositional environments. We summarize and discuss the environments of deposition and the sedimentary events that favored the preservation of organic matter in these two rift basins.

GEOGRAPHICAL AND GEOLOGICAL SETTING

The West European Rift system is composed of a series of grabens that extend along an approximate north-south direction from the North Sea to the Mediterranean Sea. This rift system began to form in the Eocene, experienced most of its tectonic activity throughout late Eocene and Oligocene, and remains somewhat active at present. In France, the main segments of this system are, from north to south, the southern part of the Upper Rhine Graben, the Bresse Graben (or Saône Graben), the Limagnes Grabens, which cut across the French Massif Central, the Valence Graben, and a number of scattered small grabens in southern France (Figure 1). During part of the Paleogene, climatic conditions were favorable for evaporite deposition, allowing thick accumulations of halite and related sediments (up to 2,200 m of deposits) in the depocenter areas of the more actively subsiding troughs (e.g., the salt basins of Mulhouse, Bresse, Valence, Figure 1). In adjacent platform areas deposits are thinner and mainly consist of limestone and calcium sulfate interfingered with terrigenous material.

Until the last few decades, the best known of these troughs was the Upper Rhine Graben, because of the co-occurrence of potash mining in the salt basin of Mulhouse and the ancient petroleum field of Pechelbronn farther to the north. The reservoir formations of this oil field and most of the source rocks are the same age as the salt formations of the Mulhouse basin (see Blanc-Valleron, 1986a). Studies of organic matter in the salt series itself have shown the presence of argillaceous source rocks, having a good petroleum potential, intercalated within halite and potash deposits (Blanc-Valleron, 1986a, b).

In the Bresse and Valence salt basins, the development of artificial salt cavities for underground storage of gas provided a great amount of data (seismic lines, well logs, and numerous cores) which considerably increased our geological knowledge of these basins (Curial, 1986b; Moretto, 1987; Dumas, 1986, 1988). More recently, projects for nuclear waste disposal in the Bresse basin have given rise to additional studies, in particular about organic matter, which have completed previous works (Curial and Moretto, 1988; Baranger, 1989).

The Bresse and Valence Grabens are bounded on the west by crystalline and Mesozoic rocks of the Massif Central (Figure 1). To the east, they are limited by the Mesozoic formations of the Jura and Vercors Chains, which are overthrust upon the eastern part of the Cenozoic basins. During the Paleogene, sedimentation was controlled by two main structural trends. One direction, 020°, is represented by basement highs (Limonest and Montmiral horsts; Figures 1, 2) and by a subsiding axis along which halite deposits are the thickest. Another direction, 050°, is exemplified by the horst of Cormoz (Bresse basin) and the Vienne-Chamagnieu basement high which separates the two grabens.

In east-west cross-sections, the two grabens display an opposite asymmetry (Figure 2). In the Bresse Graben a sub-basin, mainly filled by terrigenous deposits, developed in the west and was separated from the salt basin itself by the Limonest horst. The section through the Bresse graben crosses the region of Etrez, which acted as a depocenter area throughout most of the period of halite sedimentation. This is where a field for underground storage of gas has been developed (17 boreholes). In this area, the Paleogene section is up to 1,600 m thick, and the halite formations account for 1,400 m of this thickness. In the Valence Graben, the east-west asymmetry is reversed. A sub-basin that developed to the east also is filled by terrigenous deposits. The salt basin is confined between the horst of Montmiral and the crystalline basement of the Massif Central. In the northern depocenter of the basin (well TE 1 sector; Figure 1), the Paleogene sequence is up to 2,200 m thick, and includes more than 700 m of halite formations.

In the Bresse basin the origin (marine, non-marine) of the source-brines that led to evaporite formation has not clearly been demonstrated (Curial, 1986b; Moretto, 1987). In the Valence basin, nonreworked nannoflora (coccoliths) have been found in a few laminated horizons, indicating marine influences, at least at certain periods (possibly caused by connection with the sea or seeding from marine aerosols?). In some other levels, presence of glauberite associated

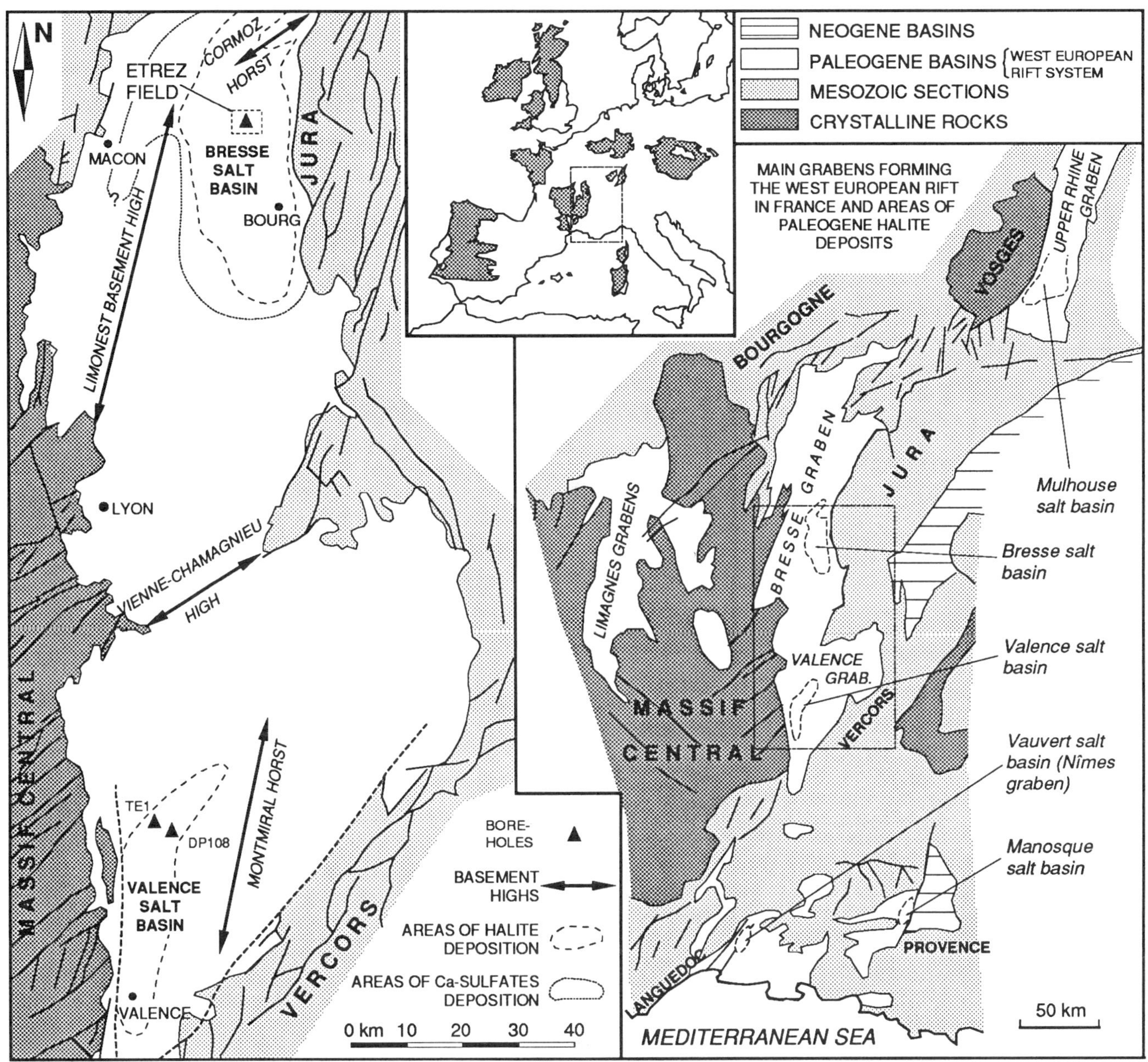

Figure 1. Location map of the Bresse and Valence salt basins in the West European Rift system.

with halite suggests a continental origin for the mother-brines (Dumas, 1988).

ANALYTICAL METHODS

Samples were processed for organic matter at the Institut Francais du Pétrole, using the Rock-Eval pyrolysis method (Rock-Eval II with a carbon module; Espitalié et al., 1985, 1986). Commonly, evaporitic salts themselves are devoid of any organic matter. The analyses have been carried out on associated rocks such as claystone, limestone (some including anhydrite) and mixed, intermediate lithologies. A substantial volume of rock was sampled in order to benefit from representative results. The few samples of small amounts of sediment (ponctual analysis, e.g., single lamina) are specified in the text. The core samples of the Bresse basin were from several boreholes of the Etrez storage field (157 samples). The Valence basin samples were from two boreholes (wells DP 108 and TE 1; see Figure 1) located near the northern depocenter area (140 samples).

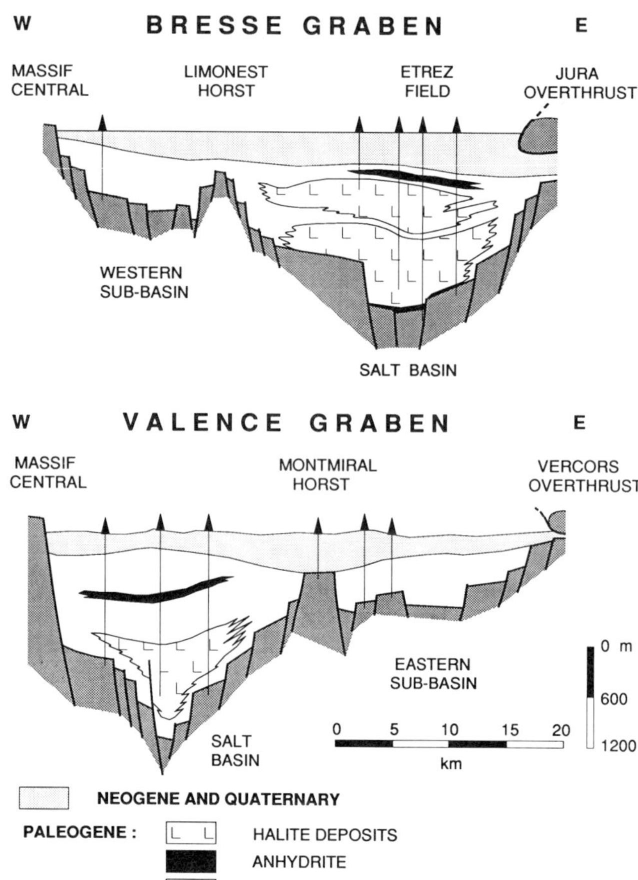

Figure 2. East-west cross sections through the Bresse and Valence grabens showing the asymmetry of each trough (a deep salt basin separated from a shallow sub-basin by a horst), and opposing asymmetry between the two troughs (salt basin to the east in the Bresse Graben, to the west in the Valence Graben).

ORGANIC MATTER IN THE BRESSE SALT BASIN

Overview

In the Bresse basin, halite deposits have been divided into three main formations: the Lower Salt Fm (Unit E5), the Intermediate Salt Fm (Unit E4), both of Priabonian age, and the Upper Salt Fm (Units E3 to E1) of Rupelian age (Figure 3; age assignments are based upon palynology; determined by M. Schüler, in Moretto, 1987). The lower part of the Lower Salt Fm and the Subsalt Fm have never been cored, and their organic content is unknown.

Rock-Eval analyses indicate that the organic matter is immature to low-mature (412°C $< T_{max} <$ 443°C; mean: 430°C). This low maturity is confirmed by the richness in resins and asphaltenes (70% to 95%) analyzed from chloroform extracts (Baranger, 1989). T_{max} values do not disclose any particular evolution versus depth. Samples do not contain any gas, and the generated oil does not exceed 1 mg/g of rock, except for a few beds with accumulations of oil that presumably comes from reworking of older rocks (these levels are marked by stars on the oil log of Figure 3; oil content may reach 2.8 mg HC/g of rock). The Total Organic Carbon (TOC) and the Residual Petroleum Potential (S2) fluctuate according to the lithology and petrographic facies of the OM-bearing horizons, as well as their stratigraphic location. TOC varies from 0.1 to 1.7% (mean of 0.67%), and the residual petroleum potential ranges from 0.25 to 10 mg HC/g of rock (mean 2.4)

On the Hydrogen Index (HI) vs T_{max} diagram (Figure 4), claystone samples display the lowest HI values, indicating a type III kerogen of terrigenous origin. This is consistent with the occurrence of plant debris seen in some beds of silty claystone. Within other lithologies (slighty argillaceous limestone variably rich in anhydrite, and in brecciated deposits), the kerogen is essentially of type II and mixed types II and III. Type II kerogen is commonly considered as derived from marine plankton (Tissot et al., 1974), but in this evaporitic environment, the same faunal and floral assemblages may have developed in brines of continental origin as well as in marine-source brines. We believe that the presence of type II OM does not imply a solely marine origin. As an alternative, a mixture of kerogens of types I and III may have HI values similar to those of type II OM, and the diagnosis of type II may thus be questionable. Another possibility is the oxidation of type I OM. It would lower the HI and put the sample in the type II field on diagrams. A few samples have disclosed high HI values which suggest the actual presence of type I kerogen, usually considered to be of lacustrine origin. Note that the effects of the associated mineral matrix during pyrolysis tend to decrease the HI values, and the samples with low TOC are more sensitive to these effects (Espitalié et al., 1985, 1986). Thus, the occurrence of type I OM and the possibility of its alteration to type II OM in some samples is very likely.

Lower Salt Formation

The upper part of the Lower Salt Fm is composed mainly of clear, displacive halite, well documented in other salt basins (Smith, 1971; Holdoway, 1978; Handford, 1981; Gornitz and Schreiber, 1981). This secondary halite occurs as subhedral to euhedral crystals that developed displacively from interstitial brines within a soft, clayey to carbonate matrix, and which altered the original bedding. Such deposits are commonly related to a shallow-water depositional environment, presumably with episodic subaerial exposures. This may explain the low organic content observed in most of the samples. The TOC amounts and the Residual Petroleum Potential (S2) values are low (except in rare samples which increase the range domains and the mean values), and vary respectively

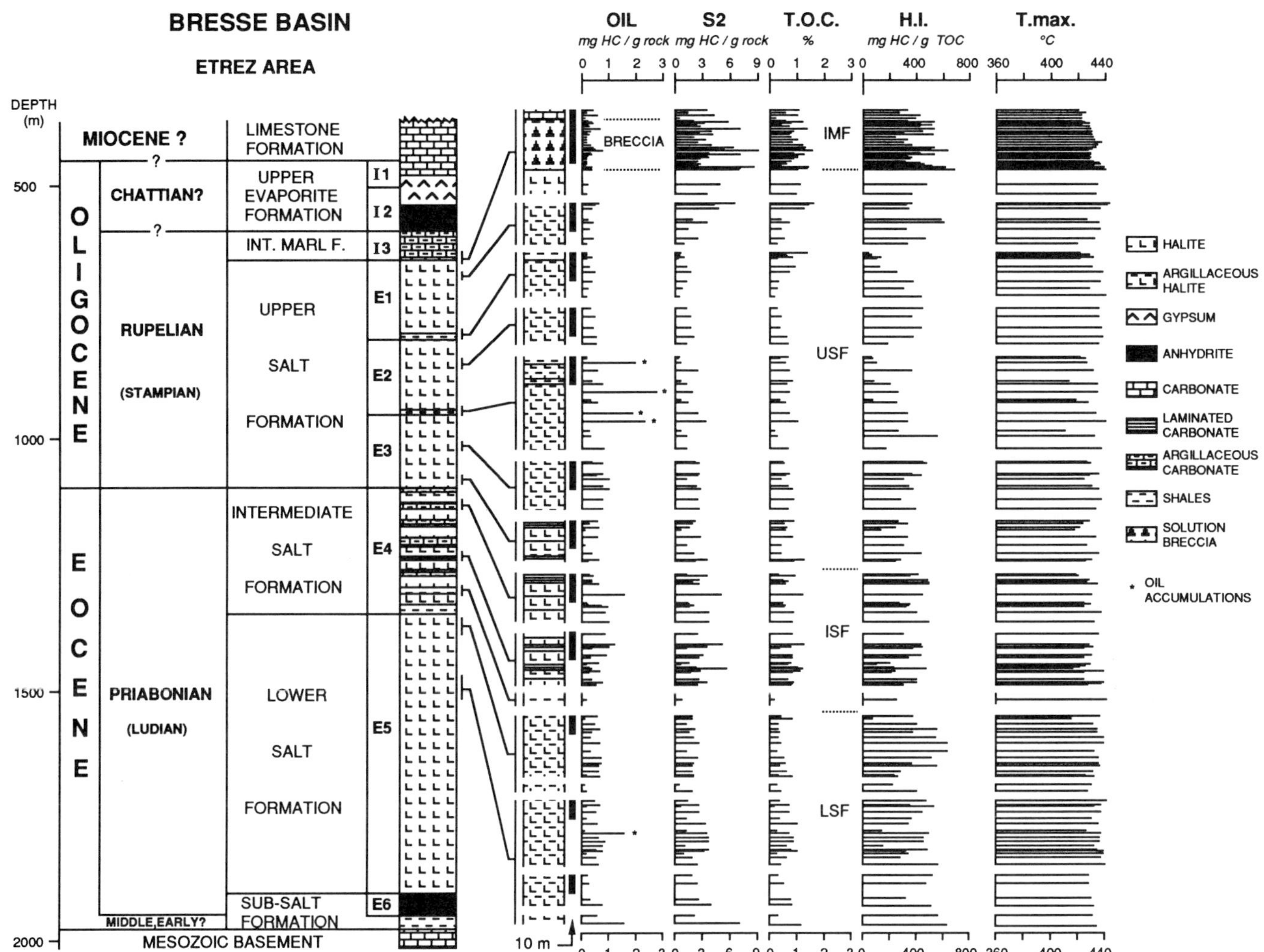

Figure 3. Geochemical logs of the salt formations in the depocenter area of the Bresse basin (gas-storage field of Etrez). The values of oil, S2 (Residual Petroleum Potential), TOC, HI (Hydrogen Index) and T_{max} (Temperature recorded for the maximum kerogen pyrolysis) are measured by Rock-Eval pyrolysis method. Horizons marked by stars (*) refer to accumulation of oil presumably coming from older and mature source rocks.

from 0.1 to 1.2% (mean 0.5), and from 0.5 to 7.1 mg HC/g rock (mean 2.1) (Figure 3).

Intermediate Salt Formation

The Intermediate Salt Fm is composed of syndepositional (primary) halite, gray to white colored and rich in fluid inclusions, enclosing beds of laminated carbonate (millimeters to tens of meters thick). The exception to this morphology is in the lowermost part where thick claystone units are present. The primary halite formed within a brine body; some crystals (pyramidal hoppers) initiated at the air-brine interface before sinking to the bottom, and other crystal types (chevrons) grew on the floor of the brine pool (see Dellwig, 1955; Shearman, 1970; Arthurton, 1973). These depositional layers of primary halite are no more than tens of centimeters in thickness. The top of each layer is affected by a solution surface that truncates the crystals, and may be marked by a film or a bed of nonhalitic sediment, or may pass directly to the next halite layer (Moretto, 1987). Such well-bedded deposits suggest water columns somewhat deeper than those prevailing during the deposition of the displacive halite. Therefore, the limited thickness of each halite bed suggests that the brine pool was affected by rapid oscillations of salinity, and, thus, that the water body was not deep enough to be unaffected by changes in the external conditions (climatic variations and, correlatively, changes in water-supply intensity).

Most of the carbonate beds consist of alternating thin micrite laminae, clayey carbonate enriched in organic matter, and, in some cases, anhydrite. The anhydrite laminae result from the coalescence of very small nodules wrapped in a network of OM. These nodules derive from gypsum-crystal precursors, which appear to have developed within algal mats

(Moretto, 1987). In these facies, the TOC content is somewhat higher than in the previous formation, and ranges from 0.3 to 1.2% (mean 0.7). The Residual Petroleum Potential (S2) varies between 0.7 and 5.5 mg HC/g rock (mean 2.5) (Figure 3). The regular rhythmical pattern of the laminar sequences, the absence of bioturbation traces, the widespread occurrence of the main laminated beds throughout the basin, and the apparent excellence of preservation of organic matter are facts that suggest deposition beneath a stratified water body, under suboxic to anoxic conditions (such stratified water systems in evaporitic contexts have been intensively discussed by Busson, 1979, 1988a). The reducing character of the depositional environment is clearly indicated by pristane/phytane ratios commonly less than 1 (Baranger, 1989).

Upper Salt Formation and Overlying Breccia

The lower part of Unit E3 (Figure 3) exhibits similar lithologies and organic content to the Intermediate Salt Fm (alternation of primary halite layers and beds of laminated carbonate). The upper part of Unit E3 and Units E2 and E1 are mainly made up of secondary halite (equivalent to that of the Lower Salt Fm), suggesting a shallow-water depositional environment. In this interval, the organic content is commonly low. Some thick claystone beds, containing type III kerogen of continental origin (Figure 4), mark the boundaries E2/E3 and E1/E2.

The upper section of the Upper Salt Fm includes intercalations of brecciated material. These deposits and the breccia (tens of meters thick over the Etrez area) that immediately overlies the salt sequence are composed of carbonate matrix, variably argillaceous, enclosing some micritic clasts and anhydrite nodules. They are related to major, multistage halite dissolution, penecontemporaneous with deposition, for which evidence has been found over the northern half of the salt basin (Kiener, 1984; Moretto, 1985; Curial and Moretto, 1985; Curial, 1986a). Successive cycles of deposition/solution of halite occurred in response to tilting movements of the graben floor (uplift of the Cormoz basement high and adjacent areas).

The brecciated deposits contain the most substantial amounts of organic matter, and significant amounts of uranium have been recorded (up to 12 ppm from gamma-ray spectrometry logs), that are quantitatively related to the carbon organic content (Curial and Moretto, 1988). TOC is commonly higher than 1%, with a maximum of 1.7% (Figure 3). The Residual Petroleum Potential (S2) is generally higher than 3, and is up to 9 mg HC/g of rock. On HI vs T_{max} diagrams, the presence of either type I or type III kerogen is seen for some samples, but most occur in the field of type II OM (Figure 4). Baranger (1989) found that the organic fraction of terrestrial origin (plants debris and inertinite) occurs in greater quantity than in the other formations, but the greater abundance of fluorescent compounds, of presumably algal or bacterial origin, would explain the high HI values.

During each dissolution phase, the halite layer previously deposited was partially or completely dissolved, and nonsoluble residues including existing organic matter (algal) were concentrated and mixed with allochthonous material (continental OM). The uranium richness and the apparently good preservation of the organic matter indicates a subaqueous, low-oxygen environment that seems inconsistent with active salt dissolution, suggesting a significant hydrological circulation. Indeed, dissolution beneath a stagnant water body would imply the formation of a saturated-brine layer covering the bottom salt that would rapidly stop the dissolution process. In order to explain the concurrent preservation of organic matter and halite dissolution, it is necessary to invoke a transport of reworked organic matter from the area of active dissolution (north part of the basin) southward toward sectors of lower energy level that were covered by a stratified water system (Curial, 1986a). This interpretation is consistent with the northern origin of continental water supply during this period (Curial, 1987).

Petroleum Potential

An estimate of the Total Petroleum Potential (TPP = oil + S2) has been computed section by section, based on the thickness of the OM-bearing horizons, using a mean rock density of 2.4 g/cm^3 (Figure 5). Organic matter is most abundant in: (1) the breccia which directly overlies the salt sequence (mean TPP = 4.2 mg HC/g rock; 0.12 Mt HC/km^2), (2) the upper part of the Upper Salt Fm (mean TPP = 3.6 mg HC/g rock; 0.15 Mt HC/km^2), and (3) the Intermediate Salt Fm (mean TPP = 3.0 mg HC/g rock; 0.47 Mt HC/km^2).

ORGANIC MATTER IN THE VALENCE SALT BASIN

The Paleogene series of the Valence basin has been subdivided into five formations (Figure 6). Only the uppermost part of the Subsalt Fm is known. It is formed by limestone, claystone, and anhydrite horizons. The Lower Salt Fm is made up of halite (commonly of primary origin, rich in fluid inclusions, and showing pyramidal hoppers and cornet and chevron crystals) enclosing thick claystone and limestone beds. The Intermediate Marl Fm consists of shales and sandy shales, with some thin layers of limestone. In the northern part of the basin, the Upper Evaporite Fm is mainly made up of dolomite, limestone, and anhydrite layers (DP 108 well). Farther to the south, a halite series developed that

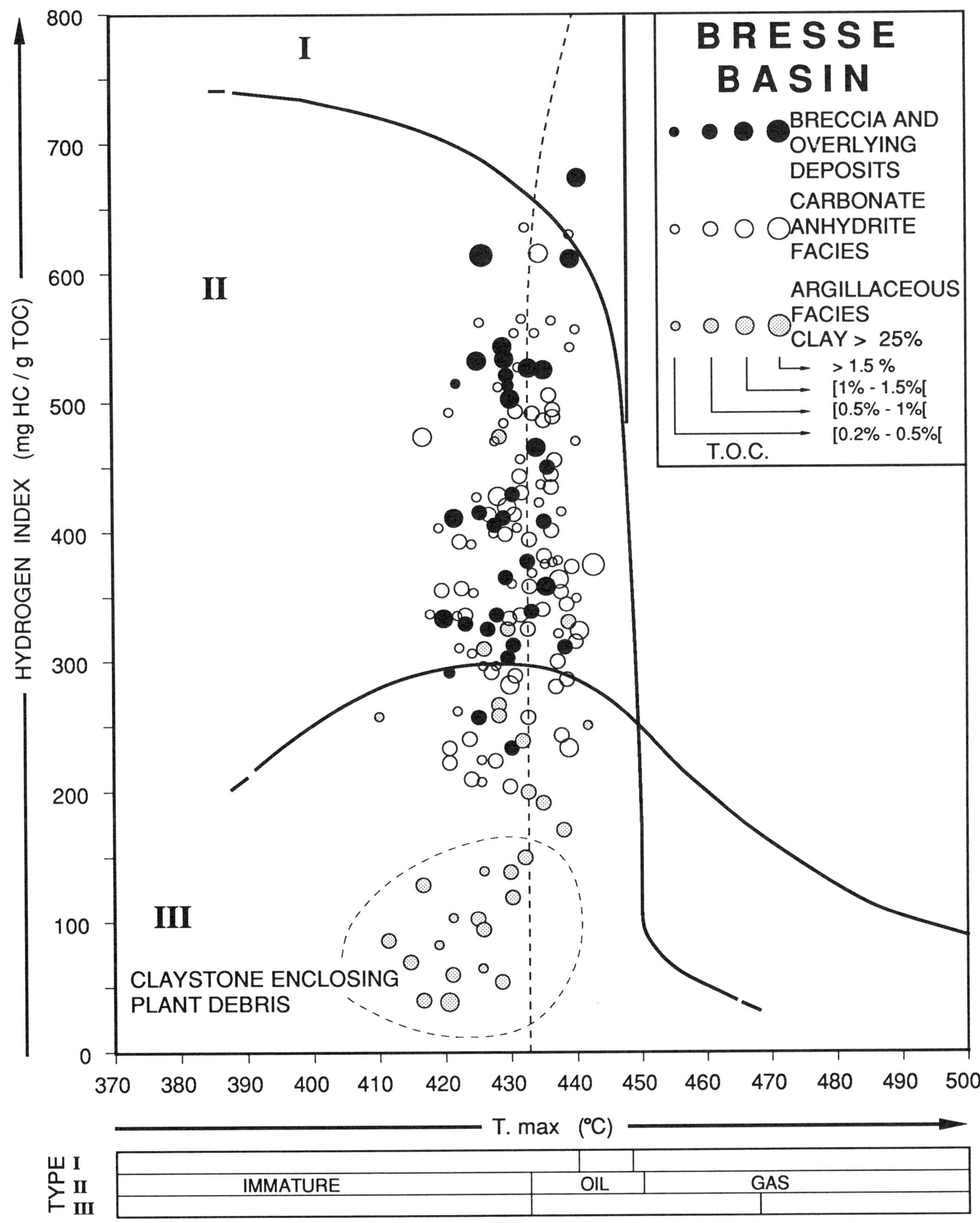

Figure 4. HI vs T_{max} diagram for samples taken in the depocenter area of the Bresse basin, according to the main lithologies of the organic-matter-bearing horizons. Size of circles refers to TOC amounts.

BRESSE BASIN — UNITS Thickness (m)	T.O.C. (%) mean *min. max.*	TOT. PETROLEUM POTENTIAL (S1+S2) (kg HC / ton rock) mean *min. max.*	CUMULATED THICKNESS OF SOURCE ROCKS (m)	APPROXIMATE TOTAL PETROLEUM POTENTIAL (10^6 t HC / km^2)
BRECCIA (12)	0.90 *0.2 - 1.6*	4.20 *0.6 - 10.7*	12 Solution breccia	0.121
E1 (140)	0.84 *0.3 - 1.7*	3.63 *1.1 - 6.9*	17 Argillaceous limestone	0.148
CLAYSTONE (2)	0.57 *0.1 - 1.3*	1.04 *0.3 - 2.1*	2 Claystone plant debris	0.005
E2 (150)	0.44 *0.3 - 0.5*	1.97 *1.5 - 2.5*	10 Limestone	0.047
CLAYSTONE (9)	0.58 *0.4 - 0.8*	1.86 *0.4 - 4.6*	4 Claystone plant debris	0.018
E3 (140)	0.62 *0.2 - 1.2*	2.60 *0.8 - 5.4*	25 Limestone laminites	0.156
E4 upper part (170)	0.71 *0.3 - 1.2*	3.00 *0.8 - 6.3*	65 Laminites	0.468
E4 lower part (45)	1 sample 0.10	0.33	15 Claystone plant debris	underestimated 0.011
E5 upper part (225)	0.53 *0.1 - 1.2*	2.60 *1.4 - 8.7*	33 Argillaceous limestone	0.206
893 m	**APPROX. TOTAL PETROLEUM POTENTIAL (10^6 tons HC / km^2)**			1.2

Figure 5. Approximate Total Petroleum Potential (TPP) values computed for various sections forming the salt series of the Bresse basin. Intervals of most interest are the breccia overlying the salt sequences, the upper section of the Upper Salt Fm (unit E1), and the Intermediate Salt Fm (unit E4).

also includes anhydrite and glauberite. These salts presumably formed from continental-source brines (Dumas, 1988). The Terminal Paleogene Fm consists of shales and sands, with some gypsum occurrences in the lowermost part. Time correlations proposed between the Bresse and Valence basins (Curial et al., 1988) are given in Figure 7.

T_{max} values obtained by Rock-Eval pyrolysis are commonly lower than 435°C, indicating that the OM is immature (Figure 6). Samples do not contain any

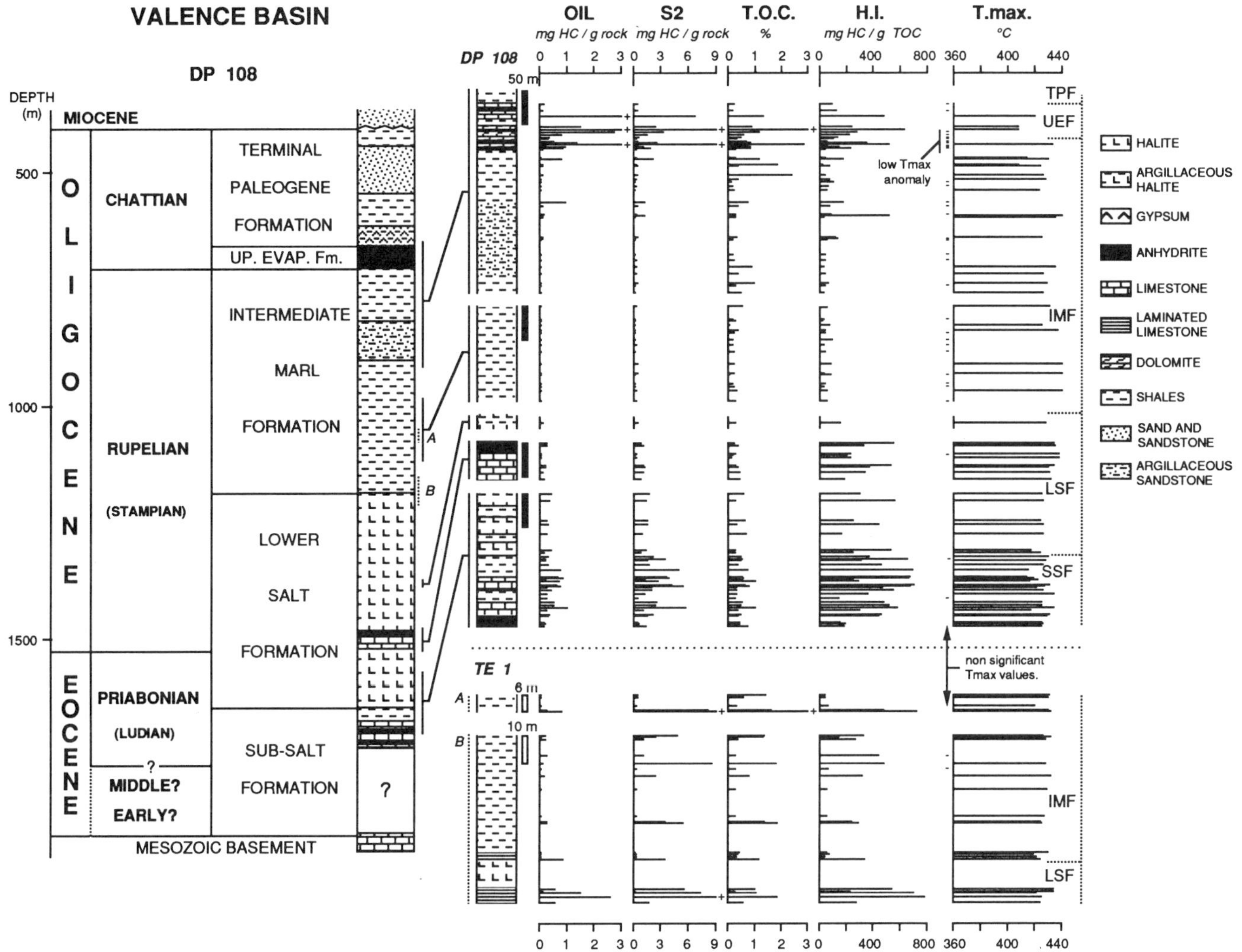

Figure 6. Geochemical logs of the salt formations in the northern depocenter area of the Valence basin (wells TE 1 and DP 108). Nonsignificant T_{max} values are not shown.

gas, and oil amounts rarely exceed 1 mg HC/g of rock, except for the particular case of the Upper Evaporite Fm and the immediately underlying deposits. The TOC content and the Residual Petroleum Potential (S2) are usually low in argillaceous units, and more substantial in carbonate-rich facies.

In the Subsalt Fm, TOC amounts range from 0.3 to 1% (mean 0.5%) (Figure 6). The Residual Petroleum Potential (S2) varies from 0.4 to 5.8 mg HC/g of rock (mean 2.4). The highest values were recorded in finely laminated and undisturbed carbonate. The organic matter of these laminites yields high HI values, which suggest the presence of type I kerogen (Figure 8), as confirmed by the S2 peak morphology on pyrograms (S2 peaks for type I kerogen are narrow and symmetric; Espitalié et al., 1985, 1986). As with the laminated carbonate of the Bresse basin, deposition of such laminated facies presumably took place beneath stratified water bodies (Dumas, 1988). Other samples are fine-grained limestone that are variably rich in clays and anhydrite and contain type II kerogen and mixed types II and III.

The uppermost part of the Lower Salt Fm includes thin source rocks showing a good Residual Petroleum Potential (S2) ranging from 1 to 14.3 mg HC/g of rock (mean 6.1) (TE 1 well; Figure 6). However, their cumulative thickness does not exceed 1.5 m. These layers consist of laminated carbonate similar to that of the Subsalt Fm. HI values are greater than 700 mg HC/g of TOC (Figure 8), probably indicating type I kerogen. The chloroform extract is composed of resins (73%), saturated hydrocarbons (22%), and aromatic hydrocarbons (5%). The pristane/phytane ratio is low (0.12) and reflects an anoxic environment related to deposition beneath stratified water bodies (Dumas, 1988).

The Intermediate Marl Fm contains very small amounts of organic matter, except within rare thin beds of fine-grained limestone and laminites (Figure

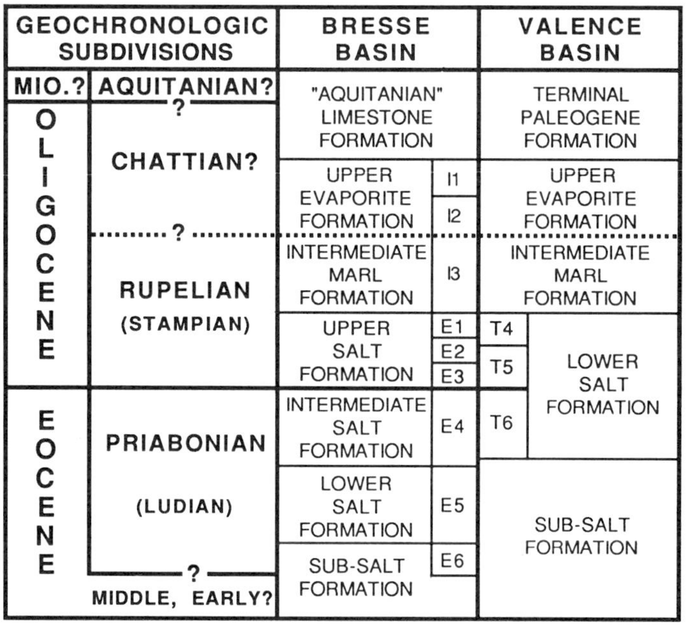

GEOCHRONOLOGIC SUBDIVISIONS		BRESSE BASIN		VALENCE BASIN	
MIO.?	AQUITANIAN?	"AQUITANIAN" LIMESTONE FORMATION			TERMINAL PALEOGENE FORMATION
OLIGOCENE	CHATTIAN?	UPPER EVAPORITE FORMATION	I1, I2		UPPER EVAPORITE FORMATION
OLIGOCENE	? RUPELIAN (STAMPIAN)	INTERMEDIATE MARL FORMATION	I3		INTERMEDIATE MARL FORMATION
OLIGOCENE	RUPELIAN (STAMPIAN)	UPPER SALT FORMATION	E1, E2, E3	T4, T5	LOWER SALT FORMATION
EOCENE	PRIABONIAN (LUDIAN)	INTERMEDIATE SALT FORMATION	E4	T6	LOWER SALT FORMATION
EOCENE	PRIABONIAN (LUDIAN)	LOWER SALT FORMATION	E5		SUB-SALT FORMATION
EOCENE	? MIDDLE, EARLY?	SUB-SALT FORMATION	E6		SUB-SALT FORMATION

Figure 7. Stratigraphic correlations between Paleogene formations in the Bresse and Valence salt basins (after Curial et al., 1988).

6). In claystones, HI values are generally low, indicating a type III kerogen of continental origin which is consistent with the common presence of vegetal remains. Samples having the lowest HI values (lower than 50 mg HC/g TOC) contain type III OM, presumably altered during transport and deposition. The beds of fine-grained limestone seem to contain type II kerogen and mixed types II and III. Type I OM has clearly been detected in a sample from a thin bed of laminated carbonate.

High values of oil, S2, and TOC (respectively up to 22 mg HC/g rock, 49 mg HC/g rock, and 7.9%; Figure 6) have been measured from several thin intercalations rich in organic matter, but these are not characteristic of the upper Evaporite Fm. These OM-rich layers are only a few millimeters thick, and their cumulative thickness does not exceed 20 cm. The dolomite, limestone, and anhydrite layers which compose most of the Upper Evaporite Fm, and the immediately underlying deposits, are basically poor in organic matter.

The large amount of heavy compounds in the chloroform extract (95.6% resins, 0.9% asphaltenes) indicates immature organic matter. In addition, that may explain the very low T_{max} values, lower than 410°C ("Low T_{max} Anomaly" zone; Figure 8), measured from most of the Upper Evaporite Fm samples. Indeed, these compounds are cracked at temperatures ranging from 300 to 400°C. If they occur in high proportions, T_{max} values tend to decrease independently of the kerogen state of maturation. These thin layers presumably correspond to impregnated source-rocks, but additional studies are needed to confirm this interpretation.

Considering Total Petroleum Potential (TPP; Figure 9), and apart from the Upper Evaporite Fm, organic content is most significant in the upper part of the Lower Salt Fm (mean TPP = 6.9 mg HC/g rock; 0.025 Mt HC/km^2), and in the known part of the Subsalt Fm (mean TPP = 2.8 mg HC/g rock; 0.56 Mt HC/km^2).

SUMMARY AND CONCLUSIONS

Lithology and Its Relationship to Organic Matter

In both the Bresse and Valence basins, type III kerogen is commonly associated with argillaceous facies (exclusive in claystone beds; mixed with other types in laminated limestones of the Bresse basin). This kind of organic matter entered the basins conjointly with terrigenous influx, and is of little interest from the point of view of its petroleum potential.

Type I kerogen is characterized by the highest HI values, and consequently shows the best petroleum potential. Type I OM has clearly been found in the Valence basin. It seems to be dominant in laminite horizons, and is independent of their stratigraphic location. Type I OM also appears to be present in the brecciated deposits of the Bresse basin, but it has not been detected in the laminated limestone beds of the Intermediate Salt Fm. These laminated limestones, as well as most of the weakly argillaceous facies (fine-grained limestone), presumably contain a mixture of types III and II. As previously mentioned, the origin of type II remains somewhat uncertain. A mixture of types III and I kerogens is a possible explanation; Baranger (1989) has shown that bacterial activity can alter the original biologic signal, and he considers this organic material to be a mixture of terrestrial (higher vegetals) and algal origins.

Commonly, TOC content is higher in the carbonate-rich facies than in argillaceous deposits. Where a significant amount of sulfate (nodular anhydrite) exists in the sediment (micrite), the organic content is generally low, presumably because the conditions associated with the intrasedimentary growth of calcium sulfate from interstitial brines involved oxidation of the organic materials.

Environments of Deposition and Their Relationship to Organic-Matter Preservation

The main conditions required for the formation of a source rock are the production of a significant biomass (OM) during sedimentation, and its preser-

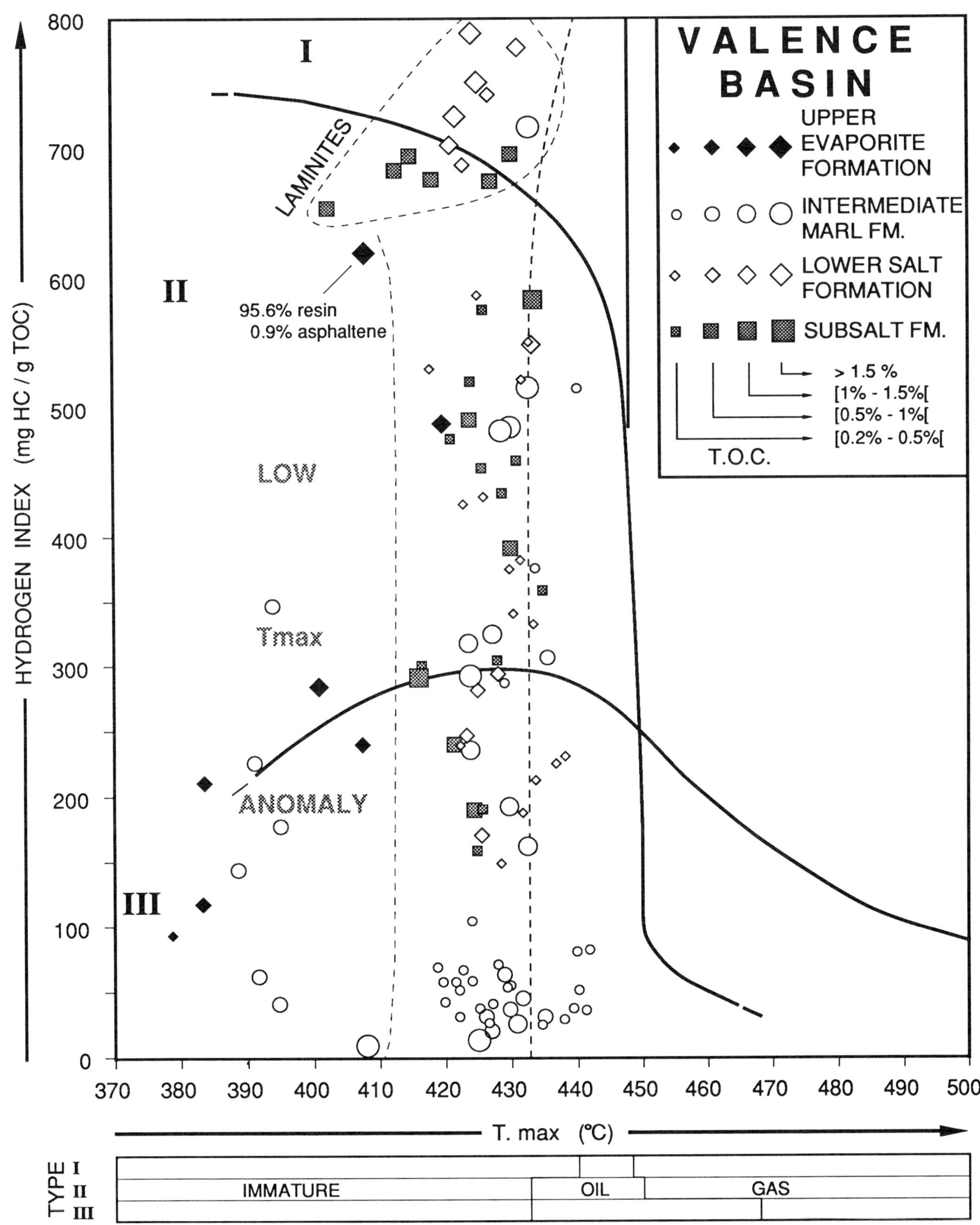

Figure 8. HI vs T_{max} diagram for samples taken in the northern depocenter area of the Valence basin. Size of circles, squares and lozenges refers to TOC amounts. The Low T_{max} Anomaly zone corresponds to samples containing great amounts of heavy compounds (resins and asphaltenes that crack between 300 and 400°C) which tend to decrease T_{max} values independently of the real maturation state of the kerogen.

VALENCE BASIN FORMATIONS Thickness (m)	T.O.C. (%) mean *min. max.*	TOT. PETROLEUM POTENTIAL (S1+S2) (kg HC / t rock) mean *min. max.*	CUMULATED THICKNESS OF SOURCE ROCKS (m)	APPROXIMATE TOTAL PETROLEUM POTENTIAL (10^6 t HC / km^2)
UPPER EVAPORITE F. (50)	**1.62** *0.1 - 7.9*	**11.77** *0.1 - 70.8*	**0.2** Thin beds mm's thick	**0.005**
INTERMEDIATE MARL F. (500)	**0.49** *0.0 - 2.4* *(0.8 - 3.3)* *	**0.56** *0.0 - 5.6* *(1.5 - 24.2)* *	**500** Shales and sandy shales	**0.672**
LOWER SALT uppermost part (15)	**0.90** *0.3 - 1.9*	**6.88** *0.7 - 17.3*	**1.5** Laminites	**0.025**
LOWER SALT lower part (460)	**0.32** *0.1 - 0.7*	**1.19** *0.2 - 1.8*	**70** Limestone and claystone	**0.200**
SUB-SALT F. uppermost part (100)	**0.51** *0.3 - 1.0*	**2.80** *0.7- 6.8*	**85** Laminites carbonate	**0.564**
SUB-SALT F. lower part (approx. 350)	U N K N O W N			?
1125 m	**APPROX. TOTAL PETROLEUM POTENTIAL (10^6 tons HC / km^2)**			**1.5**

* Values corresponding to rare thin beds rich in O.M.

Figure 9. Approximate Total Petroleum Potential (TPP) values computed for various sections of the salt series of the Valence basin. Intervals of most interest are the uppermost part of the Lower Salt Fm and the upper part of the Subsalt Fm. Presumably impregnated source rocks occur as very thin layers (millimeters thick) in the upper Evaporite Fm.

vation. It is now well established that evaporitic environments are able to generate large quantities of organic matter (Evans and Kirkland, 1988), and studies made on solar ponds exemplify the high productivity of mesosaline to hypersaline brines (see review of Cornée, 1988). Since biologic productivity is very great irrespective of the evaporitic environment, the fundamental problem that remains is the degree of organic-matter preservation. This varies considerably from one environment to another. Preservation of organic matter implies suboxic to anoxic conditions during and after deposition, and rapid burial without any major reworking or dissolution phenomena, except those which might act as an enhancing factor.

Based on the lithologies of the salt formations in the Bresse and Valence basins, and their organic contents, we are suggesting two main, but opposite in relation to water depth, depositional environments which variously favor organic-matter preservation. The first is a shallow-water environment with sporadic subaerial exposure. The corresponding salt deposits are characterized by a typical rubbly appearance, due to extensive intrasedimentary growth of halite crystals from interstitial brines. This early diagenetic halite facies is encountered throughout the Lower Salt Fm and most of the Upper Salt Fm in the Bresse basin. Some occurrences also are evident in the upper part of the Lower Salt Fm in the Valence basin, as infilling of solution-related cavities (Dumas, 1988). Successive and superposed episodes of solution/reprecipitation of halite are common in such a context, sometimes acting until complete alteration of the original texture of the host sediment occurs ("haloturbation"; Smith, 1971). Consequently this type of shallow-water environment appears to be unfavorable for good preservation of organic matter, as shown by sample analyses.

The second environment is presumed to be a significantly deeper water body (in the order of several meters to several tens of meters) that is persistent through time. Influx of continental water during wetter periods (and/or marine water in environments connected with the sea) into a preexisting saturated brine pool may induce the formation of a stratified water body, thus allowing the deposition of laminites such as those described in both the Bresse and Valence basins. Such stratified systems are known for their ability to produce and preserve organic matter (Busson, 1979, 1988a). Planktonic organisms develop in the upper water layer, that is both oxygenated and of a lower salinity, and are preserved on the bottom because of the lower oxygenation level of the deeper, more concentrated brine. Corresponding sediments are finely laminated and not bioturbated because the anoxic conditions prevent any benthonic life. The organic content of such deposits may be substantial. Consistent with this model of density stratification, lamina rich in coccoliths were described in some laminites of the Valence basin (Guillevin in Goy and Busson, 1980; Dumas, 1988). Some remains of fresh-water teleostean fishes (euryhaline) also were documented (Gaudant and Guillevin, 1979). In the Bresse basin, the beds of laminated carbonate (Figure 3, Units E4 and E3) contain fresh-water planktonic organisms that appear as very fine (5 to 10 μm), spherical particles (Baranger, 1989).

A thick laminite bed may develop only if the stratified water system remains constant for a long time. Progressive concentration of the upper water layer will force a mixing of waters and a return to evaporite deposition. If the water supply diminishes, the basin will tend toward desiccation. The depositional environment will become similar to that previously described (shallow water). Reworking and dissolution of previously deposited sediments may occur, resulting in the destruction of organic matter. However, if the water remains sufficiently deep because of an adequate water supply, the deposits will be preserved and progressively buried. The petrographic and sedimentological characteristics of the Intermediate Salt Fm in the Bresse basin suggest that it is an example of a stratified depositional environment that remained for a long time, demonstrating the establishment of a perennially stratified water system.

In both the Bresse and Valence salt basins, evidence of dissolution has been recognized at various scales (crystal to basinal scales), whatever the depositional environment. The most spectacular form of halite dissolution, and the most interesting from the point of view of organic-matter concentration and preservation, is seen in the upper section of the Upper Salt Fm in the Bresse basin. Successive salt-solution phases took place under subaqueous conditions and affected a considerable volume of salt (see Curial, 1986a, 1986b), resulting in the accumulation of both nonsoluble residues substantially enriched in organic matter and allochthonous material. This example points out that, under some circumstances, halite dissolution processes may act as a concentrating mechanism for organic matter.

ACKNOWLEDGMENTS

We thank Gaz de France Company and the Mines des Potasses d'Alsace for permission to sample their cores. We extend our appreciation to the Agence Nationale pour la Gestion des Déchets Radioactifs (CEA-ANDRA-DESI) for permission to publish this paper. We are greatly indebted to J. Espitalié (Institut Francais du Pétrole) for sample analyses using Rock-Eval pyrolysis method. Our thanks also go to B. C. Schreiber, J.-M. Rouchy and M. L. Helman for critical review of the manuscript.

REFERENCES

Arthurton, R. S., 1973, Experimentally produced halite compared with Triassic layered halite rock from Cheshire, England: Sedimentology, v. 20, p. 145–160.

Baranger, P., 1989, Origine et diagenèse des constituants organiques associés aux niveaux salifères tertiaires de Bresse (France). Etude expérimentale de la réduction d'espèces métalliques par la matière organique sédimentaire: Thesis, University of Orléans (France), 325 p.

Blanc-Valleron, M.-M., 1986a, La sédimentation salifère Paléogène du bassin potassique de Mulhouse, premiers apports de l'étude géochimique de la matière organique contenue dans les passées argileuses: Documents du Bureau de Recherches Géologiques et Minières, v. 110, p. 111–126.

Blanc-Valleron, M.-M., 1986b, Enseignements géologiques tirés de l'étude systématique de la matière organique d'un bassin salifère. Le bassin potassique de Mulhouse (Alsace, France): Comptes-Rendus de l'Académie des Sciences de Paris, v. 302, p. 825–830.

Busson, 1978, L'unité des faciès confinés en milieu de plate-forme carbonatée: Documents des Laboratoires de la Faculté des Sciences de Lyon, HS v. 4, p. 87–112.

Busson, G., 1979, Couches laminées riches en matière organique et précédant les roches salines: les enseignements d'un enchaînement de faciès: Documents des Laboratoires de la Faculté des Sciences de Lyon, v. 75, p. 5–18.

Busson, G., 1988a, Relations entre les types de dépôts évaporitiques et la présence de couches riches en matière organique (roches-mères potentielles): Revue de l'Institut Francais du Pétrole, v. 43, p. 181–215.

Busson, G., 1988b, Introduction, *in* G. Busson, ed., Evaporites et hydrocarbures: Mémoires du Muséum national d'Histoire naturelle, (C), v. 55, p. 11–13.

Cornée, A., 1988, Productivité microbienne, benthique et planctonique, dans les milieux hypersalés actuels: une revue, *in* G. Busson, ed., Evaporites et hydrocarbures: Mémoires du

Muséum national d'Histoire naturelle, (C), v. 55, p. 19-41.
Curial, A., 1986a, Les dissolutions dans la partie supérieure du Saliféré paléogène de Bresse (SE de la France), chronologie, extension et mécanismes: Revue de Géologie dynamique et de Géographie physique, v. 27, p. 225-235.
Curial, A., 1986b, La sédimentation salifère et suprasalifère du Paléogène bressan (France): comparaison entre les données diagraphiques et lithologiques. Etude diagraphique du champ d'Etrez et synthèse du bassin: Documents de l'UA 1209 (CNRS) (1986), v. 11, 251 p., and Documents des Laboratoires de Géologie de Lyon (1987), v. 100, 192 p.
Curial, A., 1987, Le détritisme durant la sédimentation salifère paléogène en Bresse méridionale: Géologie Alpine, HS, v. 13, p. 391-395.
Curial, A., and R. Moretto, 1985, Le Paléogène salifère de Bresse (département de l'Ain, SE de la France). Etude diagraphique et sédimentologique: Documents du GRECO 52 (CNRS), v. 5, p. 93-153.
Curial, A., and R. Moretto, 1988, La matière organique du bassin salifère paléogène de Bresse: caractérisation par pyrolyse Rock-Eval, relation avec l'uranium et détection par spectrométrie du rayonnement gamma naturel, *in* G. Busson, ed., Evaporites et hydrocarbures: Mémoires du Muséum national d'Histoire naturelle, (C), v. 55, p. 71-94.
Curial, A., D. Dumas, and R. Moretto, 1988, Evolution comparée et corrélation entre deux bassins salifères paléogènes du rift ouest européen: les fossés de Bresse et de Valence (France, Sud-Est): Comptes-Rendus de l'Académie des Sciences de Paris, v. 306, p. 655-661.
Dellwig, L. F., 1955, Origin of the Salina salt of Michigan: Journal of Sedimentary Petrology, v. 25, p. 83-93, p. 95-102, p.107-110.
Dumas, D., 1986, La sédimentation paléogène dans le fossé salifère de Valence: géométrie des dépôts et apercu des mécanismes sédimentaires: Géologie de la France, v. 4, p. 399-416.
Dumas, D., 1988, Le Paléogène salifère du bassin de Valence (Sud-Est de la France) : Géométrie et sédimentologie des dépôts, synthèse de bassin: Thèse University Claude Bernard, Lyon, 293 p.
Espitalié, J., G. Deroo, and F. Marquis, 1985-86, La pyrolyse Rock-Eval et ses applications: Revue de l'Institut Francais du Pétrole, part I and part II, v. 40, p. 563-579 and 755-784.
Espitalié, J., G. Deroo, and F. Marquis, 1986, La pyrolyse Rock-Eval et ses applications: Revue de l'Institut Francais du Pétrole, part III, v. 41, p. 73-89.
Evans, R., and D. W. Kirkland, 1988, Evaporitic environments as a source of petroleum, *in* B. C. Schreiber, ed., Evaporites and Hydrocarbons: New York, Columbia University Press, p. 256-299.
Gaudant, J., and Y. Guillevin, 1979, Découverte de Dapalis (Smerdis) macrurus (poisson téléostéen) dans l'Oligocène du bassin de Valence (Drôme) : implications stratigraphiques et paléogéographiques: Bulletin du Muséum national d'Histoire naturelle, (C), v. 4, p. 25-33.
Gornitz, M. V., and B. C. Schreiber, 1981, Displacive halite hoppers from the Dead Sea: Some implications for ancient evaporite deposits: Journal of Sedimentary Petrology, v. 51, p. 787-794.
Goy, G., and G. Busson, 1980, Inventaire des évaporites du sous-sol: les bassins profonds du Trias du Sud-Est Francais, du Paléogène de Bresse et de Valence: Compte-rendu d'étude DGRST, action concertée "valorisation des ressources du sous-sol", 169 p.
Handford, C. R., 1981, Coastal sabkha and salt pan deposition of the lower Clear Fork Formation (Permian), Texas: Journal of Sedimentary Petrology, v. 51, p. 761-778.
Holdoway, K. A., 1978, Deposition of evaporites and redbeds of the Nippewalla Group, Permian, western Kansas: Kansas Geological Survey Bulletin 215, 43 p.
Kiener, M., 1984, Etudes sédimentologiques et corrélations par diagraphies: le sommet de la série saliféré et sa couverture dans le sondage EZ 08 (Paléogène de Bresse) : Compte-rendu de la journée sur le sel, Ecole Polytechnique de Palaiseau, p. 207-231.
Kirkland, D. W., and R. Evans, 1981, Source-rock potential of evaporitic environments: AAPG Bulletin, v. 65, p. 181-190.
Moretto, R., 1985, Sédimentologie de la série salifère du Paléogène de Bresse (France): Bulletin de la Société Géologique de France, v. 1, p. 849-855.
Moretto, R., 1987, Etude sédimentologique et géochimique des dépôts de la série salifère paléogène du bassin de Bourg-en-Bresse (France): Documents de l'UA 1209 (CNRS), v. 12, 252 p. and Mémoires des Sciences de la Terre, v. 50, 252 p.
Rouchy, J.-M., 1988, Relations évaporites-hydrocarbures: l'association laminites-récifs-évaporites dans le Messinien de Méditerranée et ses enseignements, *in* G. Busson, ed., Evaporites et hydrocarbures: Mémoires du Muséum national d'Histoire naturelle, (C), v. 55, p. 43-69.
Schreiber, B. C.,1988, Relations entre évaporites et accumulations d'hydrocarbures, *in* G. Busson, ed., Evaporites et hydrocarbures: Mémoires du Muséum national d'Histoire naturelle, (C), v. 55, p. 15-18.
Shearman, D. J., 1970, Recent halite rock, Baja California, Mexico: Institute of Mining and Metallurgy Trans., v. 79, p. 155-162.
Smith, D. B., 1971, Possible displacive halite in the Permian Upper Evaporite Group of northeast Yorkshire: Sedimentology, v. 17, p. 221-232.
Sonnenfeld, P., 1985, Evaporites as oil and gas source rocks: Journal of Petroleum Geology, v. 8, p. 253-271.
Tissot, B., B. Durand, J. Espitalié, and A. Combaz, 1974, Influence of nature and diagenesis of organic matter in formation of petroleum: AAPG Bulletin, v. 58, p. 499-506.
Warren, J. K., 1986, Shallow-water evaporitic environments and their source-rock potential: Journal of Sedimentary Petrology, v. 56, p. 442-454.

Distribution of Cenomanian-Turonian Organic Facies in the Western Mediterranean and Along the Adjacent Atlantic Margin

Wolfgang Kuhnt
Universität Tübingen
Tübingen, Federal Republic of Germany

Jean Paul Herbin
Institut Francais du Pétrole
Rueil-Malmaison, France

Jurgen Thurow
Universität Tübingen
Tübingen, Federal Republic of Germany

Jost Wiedmann
Universität Tübingen
Tübingen, Federal Republic of Germany

Cenomanian/Turonian organic-rich sediments are widespread in the western Mediterranean and Atlantic and have been studied in outcrops, DSDP/ODP sites, and petroleum exploration wells. This study includes sections from Italy (Apennines, Southern Alps), Tunisia (Bahloul), Algeria, Morocco (Rif Mountains, Atlas Mountains, Tarfaya), Gibraltar Arch, Spain (Celtiberic Ranges, Betics, Bay of Biscay, Galicia Margin), Senegal (Cape Verde Basin, Casamance), and Nigeria (Benue, Calabar Flank).

Detailed data from thick pelagic sequences (e.g., Tarfaya basin) made it possible to construct a detailed stratigraphic framework using planktonic foraminiferal, radiolarian, and cephalopod zonations. We calculated sedimentation rates and organic-matter-accumulation rates based on this stratigraphic framework. The results are compared to paleogeography and paleobathymetry, in order to recognize variations of middle Cretaceous organic matter accumulation in time and space. The following general trends in organic-matter accumulation were observed. Cenomanian/Turonian organic-rich sediments occur in a wide range of bathymetric settings, except for coastal and shallow-marine shelf areas. Although total organic carbon content (TOC) can be high in the deep sea, highest accumulation rates of organic matter occur in outer-shelf environments of low-latitude areas (e.g., Tarfaya and Senegal shelf basins).

Pre-Cenomanian sediments of the Western Mediterranean and adjacent Atlantic margin are characterized by low TOC with an important terrestrial component. During the Cenomanian, TOC increased, and the marine component became dominant, culminating around the Cenomanian/Turonian boundary with TOC up to 40%. Since the Turonian, organic-rich sediments have progressively declined, being replaced by more oxygenated sediments.

Local subsidence history and tectonic stress led to different stages of maturation of the organic matter. In areas of thick accumulation (especially the shelf environment) and favorable thermal history, deposits around the Cenomanian/Turonian boundary can be of economic interest as petroleum source rocks.

INTRODUCTION

Distinct carbonaceous clay or marlstone deposits (some of which are famous for fish fossils) of Cenomanian/Turonian age have been recognized in the Western Mediterranean area for a long time. Schlanger and Jenkyns (1976) and Jenkyns (1980) claimed that these organic-carbon-rich zones represent time-consistent, geographically widespread units that were deposited in shallow-water and pelagic environments ("oceanic anoxic event"). De Graciansky et al. (1982) and Müller et al. (1984) referred to coeval organic-rich facies in the North Atlantic as "Evénement 2." The term "Cenomanian/Turonian boundary event" (CTBE) has been used recently to describe not only the occurrence of organic-rich sediments but also particular biofacies and geochemical features at the C/T boundary (Thurow and Kuhnt, 1986; Herbin et al., 1986b; Kuhnt et al., 1986; Brumsack and Thurow, 1986; Arthur et al., 1987; Schlanger et al., 1987). Detailed local descriptions of Cenomanian/Turonian organic facies in the Western Mediterranean and along adjacent Atlantic margins were given by Arthur and Premole Silva, 1982; Einsele and Wiedmann, 1982; Thurow et al., 1982; Wiedmann et al., 1982; Herbin et al., 1986b; Kuhnt et al., 1986; Thurow and Kuhnt, 1986; Arthur et al., 1987; Schlanger et al., 1987; Thurow et al., 1988. In contrast to the well-established geochemical and sedimentological features of the C/T organic-rich sediments, little is known about changes of organic-matter-accumulation rates with time and paleogeographic position. The approach of our study is mapping and quantification of the accumulation of Cenomanian/Turonian organic-rich sediments in as many sites as possible and thus evaluating organic-matter accumulation in time and space.

Our data base is material from more than 40 Cenomanian/Turonian outcrop sections, exploration wells, and DSDP/ODP sites in the Western Mediterranean region and along adjacent Atlantic margins. Data have been evaluated from eight study areas, some of which provide transects that reach from the middle Cretaceous coastline down to the deep sea. The study areas (Figure 1) from south to north are: (1) Iberian transect (six outcrop sections and ODP Site 641); (2) Alboran/Gibraltar seaway transect (eight outcrop sections); (3) Italian transect (six outcrop sections in four principal areas from northern Italy to Sicily); (4) eastern North African margin transect of the Algerian Tell and the Tunisian basin (two outcrop sections); (5) western North African margin transect of the Rif Mountains (northern Morocco) (seven outcrop sections); (6) Tarfaya/Agadir transect (three outcrop sections, one exploration well in the Moroccan coastal basins, and DSDP sites 135, 137, and 138); (7) Casamance transect offshore Senegal (six exploration wells and DSDP Site 367); (8) Benue transect of Nigeria and Cameroon (eight outcrop sections).

More than 2000 samples of late Albian to Turonian sediments from these locations have been studied using sedimentological, paleontological, and Rock-Eval techniques. The data obtained allow at least approximate calculations of organic-matter-accumulation rates for each site.

CENOMANIAN/TURONIAN ORGANIC-RICH SEDIMENTS IN THE STUDIED AREA—SHELF TO BASIN TRANSECTS

Iberia (Celtiberic Ranges, Betics, Bay of Biscay, Galicia Margin)

Along the continental margins of the late Mesozoic Iberian microplate, Cenomanian and Turonian sediments have been studied within three different paleogeographic provinces.

1. Celtiberic ranges of northeastern Spain: shallow neritic shelf deposits transgressive on the Iberian Meseta.
2. Subbetic and Penibetic Zones of the Betic Cordillera (southern Spain): slope areas and continental margin highs and basins along the southern margin of the Iberian microcontinent. Paleoenvironmental settings include half-graben structures of tilted blocks, marly slope environments, and marginal highs with condensed sequences.
3. Galicia Margin (DSDP Site 641A) offshore northwestern Spain: deeply subsided part of Iberian Atlantic continental margin. Middle Cretaceous deposition below the CCD in a sediment-starved abyssal environment.

Four different paleoenvironments can be discriminated, which exhibit characteristic sedimentary features across the Cenomanian/Turonian boundary (Figure 2).

Neritic shelf deposits of the Celtiberic Ranges (Picofrentes section near Soria) comprise reduced sequences and a hardground near the Cenomanian/Turonian boundary (Wiedmann, 1975). This hardground is overlain by 30 cm of phosphatic marly limestones with flaser bedding which belong to the *Fallotites subconciliatus* ammonite zone that is included in the topmost Cenomanian, and corresponds to the lower part of the *W. archaeocretacea* foraminiferal zone. The Cenomanian/Turonian boundary is placed 1.5 m above the hardground at the base of the *Paramammites (?) saenzi* zone. The total organic carbon content (TOC) is very low (0 to 0.9%) in this sequence, indicating a well-oxygenated environment.

Margin basins in a bathyal position are observed in the Penibetic Zone of the southern Iberian margin. Cenomanian sediments of this zone include pelagic

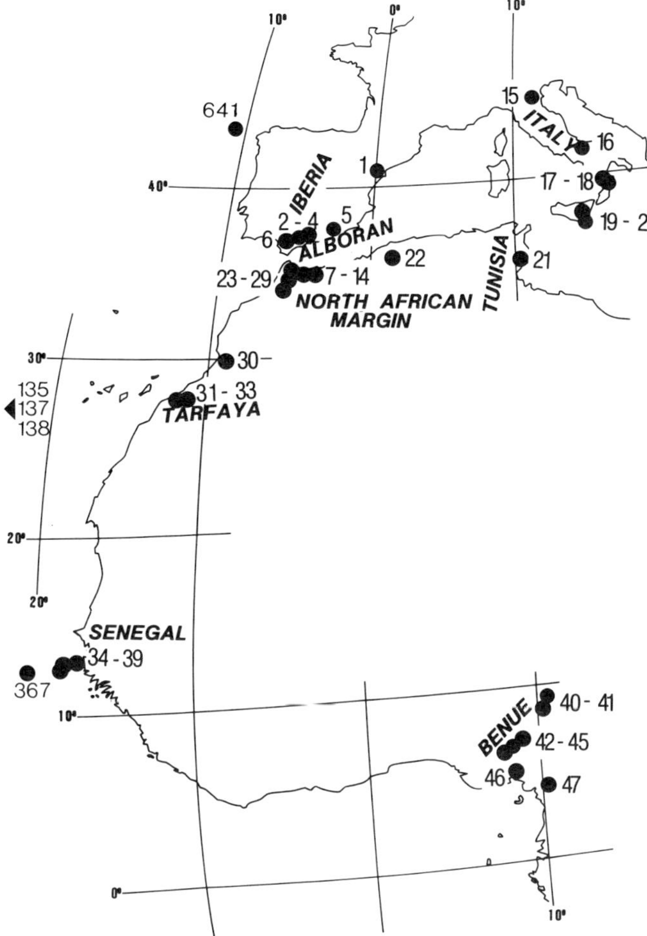

Figure 1. Locality map of the main study areas. The following sections were studied. Iberia: (1) Picofrentes section near Soria, Celtiberic Ranges; (2) Alpandeire section, Penibetic Zone; (3) Baños Sulfurosos section, Sierra de Los Canutos north of Manilva, southwestern part of the Penibetic Zone; (4) Sierra Blanquilla section, main road Ronda-El Burgo, east of the Mirador, Penibetic Zone; (5) Rio Fardes section, east of Granada. Subbetic Zone; (6) Sierra de Las Cabras section, Subbetic Zone south of Arcos de La Frontera; (641) Site 641A of ODP Leg 103, Galicia margin (Boillot et al., 1987). Alboran margin transect: (7) Dorsale of the Moroccan Rif, Chaine de Haouz, north of Tetouan; (8) Predorsalien of the Moroccan Rif, Chrafate section F15/1, southeast of Chechaouen; (9) Predorsalien of the Moroccan Rif, Chrafate section F15/2; (10) Mauretanien Flysch Units of the Moroccan Rif, Beni Ider section M39; (11) Mauretanien Flysch Units of the Moroccan Rif, Beni Ider section M13; (12) Mauretanien Flysch Units of the Moroccan Rif, Beni Ider Section M18, north of Chechaouen (Thutow and Kuhnt, 1986); (13) Massylian Flysch Units of the Moroccan Rif, Melloussa Section D96; (14) Massylian Flysch Units of the Moroccan Rif, Melloussa sections M7 and M62 (Thurow and Kuhnt, 1986). Italy: (15) Euganean Hills, near Padua; (16) Umbrian Apennines, Gubbio sections in the Bottaccione and Contessa valleys and Piobbico section; (17) Calabria, Oriolo, western section (proximal sequence); (18) Calabria, Oriolo, eastern section (distal sequence); (19) Sicily, Malvagna section; (20) Sicily, Floresta Section. Tunisia and Algeria: (21) Oued Bahloul section, east of Makhtar and Djebel Mrhila section near Sbeitla (Tunisia); (22) Monts de Tessala, western part of the Tell Atlas south of Oran (Algeria). North African Margin Transect in the Moroccan Rif: (23) Djebel Tselfat area, external Prerif Zone, east of Sidi Kacem; (24) Djebel Mguedrouz, internal Prerif Zone; (25) Kelaat Boukkorra section L30, southern part of the Loukkos Zone; (26) Pont du Loukkos section L1, northern part of the Loukkos Zone; (27) section M8, internal Tanger Unit, south of Tangier; (28) section M90, internal Tanger Unit, west of Tetouan; (29) Beni Hessane section M61, north of Chefchaouen. Moroccan Coastal Basins: (30) Agadir coastal section (Wiedmann et al., 1982); (31) BRPM Exploration Well S13, western (distal) part of the Tarfaya basin; (32-33) Sebkha Tazra and Sebkha Houiselgua sections, eastern (proximal) part of the Tarfaya basin; (135,137,138) Sites 135, 137 and 138 of DSDP Leg 14 (Hayes et al., 1972). Senegal Transect: (34-39) Casamance exploration wells CM10, CM2, CM4, CM1, NC1, and DM1. offshore Senegal; (367) Site 367 of DSDP Leg 41 (Lancelot et al., 1977). Benue Transect, Nigeria: (40) Ashaka quarry, northern Nigeria; (41) Pindiga section, northern Nigeria; (42) Ngeleg-bokwa Section, central Nigeria; (43) Ngbanocha section, central Nigeria; (44) Lokpanta section, central Nigeria; (45, 46) Odukpani section, Calabar Flank, southern Nigeria; (47) Mungo River section, Cameroon.

marls and limestones with rare chertification. TOC is low (0 to 0.3%); a few darker layers exhibit up to 1% TOC of a prevailing terrestrial kerogen type. Sedimentation of a distinct black-shale type (fish shales) with intercalations of radiolarian sands starts at the top of the *R. cushmani* zone. This sequence, which is 120-130 cm thick, is characterized by extremely high (up to 33%) TOC values of marine kerogen type I/II. These black shales are overlain by cherty pelagic limestones. In the lowermost limestone bed planktonic foraminifers of the *W. archaeocretacea* zone have been identified. The first occurrence of *P. helvetica* is observed a few centimeters above the last black shale layer. In several sections, which were probably situated on the elevated parts of the tilted blocks, these black shales are missing. A hiatus is observed here between marly limestones with *R. cushmani* and chertified limestones with rich radiolarian assemblages and *P. helvetica*.

Slope sequences in a bathyal position characterize middle Cretaceous sedimentation in the Subbetic Zone (Comas et al., 1982). Channel sequences and proximal fan deposits with intense slumping are locally observed (e.g., the middle Cretaceous sediments of the Fardes Formation along the Rio Fardes

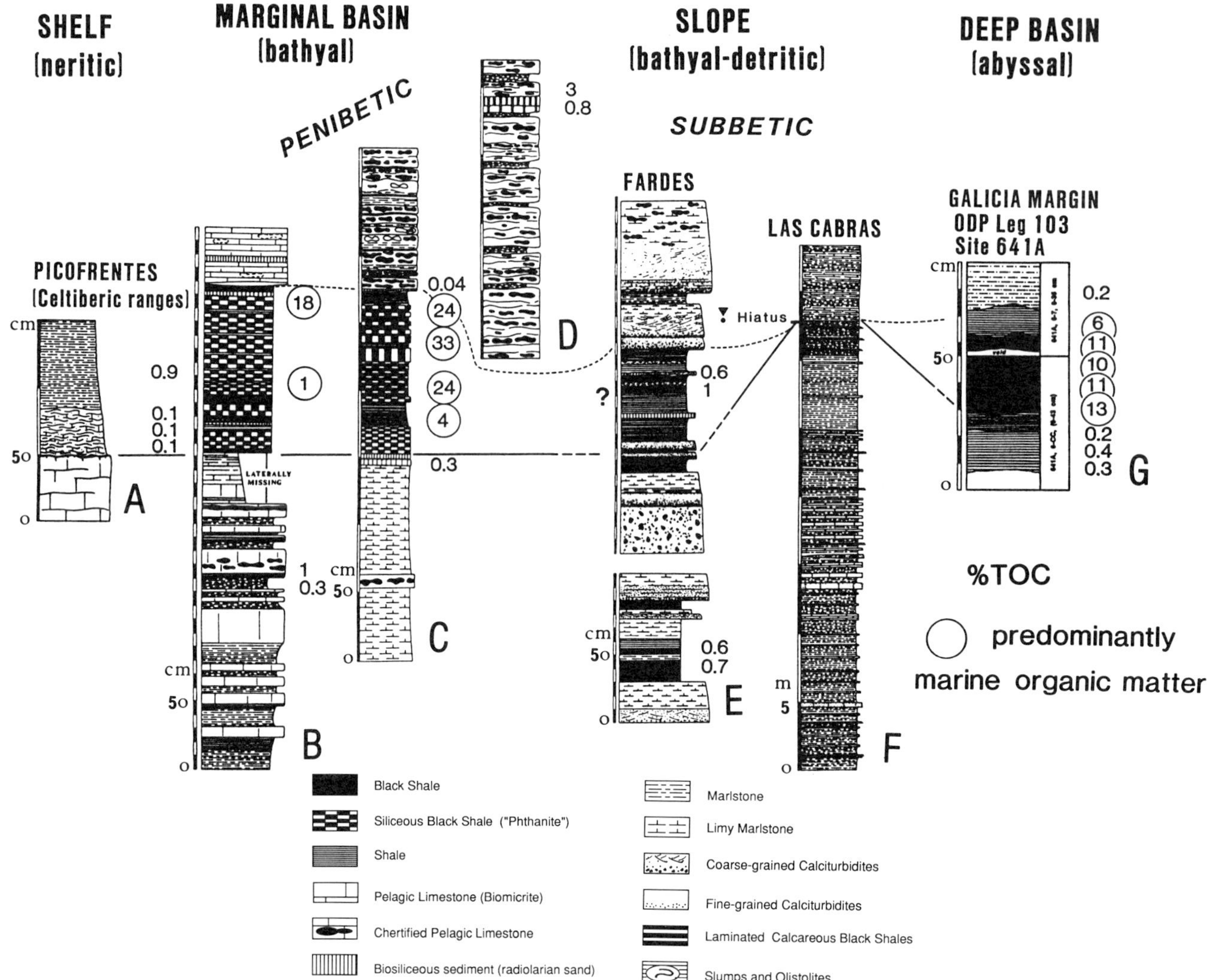

Figure 2. Sections studied along the margins of the Iberian microplate. (A) Picofrentes section near Soria; (B) Alpandeire section; (C) Baños Sulfurosos section, Sierra de Los Canutos north of Manilva (from Kuhnt et al., 1986); (D) Sierra Blanquilla section, main road Ronda-El Burgo, east of the Mirador; (E) Rio Fardes section, east of Granada; (F) Sierra de Las Cabras section; (G) Site 641A of ODP Leg 103, Galicia Margin (modified after Thurow et al., 1988). Lines indicate the beginning and end of organic-rich sedimentation corresponding to the *W. archaeocretacea* foraminiferal zone.

northeast of Granada). At the Cenomanian/Turonian boundary the clastic input ceased, and a thin sequence of fine-grained biogenic turbidites, radiolarian-rich calcilutites, and black shales developed. However, in the analyzed section TOC values remain low (up to 1.2%) and no marine signal was observed in the Rock-Eval data (the Hydrogen Index (HI) is below 30, whereas the Oxygen Index (OI) generally is above 100). In the westernmost part of the Subbetic Zone (Sierra de Las Cabras) the late Cenomanian/Turonian is locally characterized by an important hiatus which covers several foraminiferal zones (e.g., the *R. cushmani, W. archaeocretacea*, and *P. helvetica* zones).

Deep basinal sequences of the western Iberian continental margin (DSDP Site 641A) exhibit an extremely thin (30 cm) intercalation of black organic-rich clay at the Cenomanian/Turonian boundary (Thurow et al., 1988). TOC values of this black band average 11%, and kerogen is of type II. This black-shale event covers the Cenomanian/Turonian boundary interval (which corresponds more or less to the *W. archaeocretacea* foraminiferal zone) dated by radiolarians, palynoflora, and agglutinated foraminifera (Thurow et al., 1988). Sedimentation rate, and consequently accumulation rate of organic matter, is low in this abyssal site.

Italy (Southern Alps, Apennines, Calabria, Sicilia)

Cenomanian/Turonian organic-rich sediments have been studied in four paleogeographically different settings of Italy (Figure 3).

1. The Euganian Hills, an uplifted part of the southern Alpine realm within the Po Plain (Padua area of the River Po depression): Organic-carbon-rich shales that locally contain rich fish faunas (SORBINI, 1974) are intercalated in a pelagic limestone sequence.
2. The Central Apennines of the Gubbio area (Piobbico and Gubbio sections): A thin layer of organic-rich sediments at the Cenomanian/Turonian boundary ("Bonarelli horizon") is intercalated in Cretaceous pelagic limestones ("Scaglia facies"). A detailed description is given by Arthur and Premoli Silva (1982).
3. Turbidite sequences of the Calabrian Arch (Oriolo sections). These sequences form the eastward prolongation of the North African Mauretanian and Massylian flysch nappes. Biosiliceous sediments dominate at the C/T boundary; organic-rich layers are thin (few centimeters), laminated, black claystones with very high TOC (up to 32% of marine-kerogen type).
4. "Argille Scagliose" facies of northeastern Sicily (Floresta section), which can be compared to the deep-basinal (Massylian) sequences in North Africa: Black, carbonaceous, finely laminated shales, occasionally rich in fish remains or radiolarians ("phthanites"), are intercalated in carbonate-free, deep-water claystone.

Comparison of these different settings is problematic because they cover not only a present day north-south distance of more than 1000 km (and even a broader paleolatitudinal range) but also include paleobathymetric settings from the upper bathyal (Euganean Hills) down to abyssal ("Argille Scagliose") facies deposited below the CCD. Moreover these areas most probably belonged to different Mesozoic continental margins. However, organic-rich sedimentation at the Cenomanian/Turonian boundary exhibits surprising similarities to corresponding paleogeographic settings along the Iberian continental margin. The most striking features are: (1) extremely high TOC values (up to 32%) of marine kerogen restricted to a thin layer at the Cenomanian/Turonian boundary; (2) low sedimentation rate of biogenic organic-rich sediments; and (3) an important dilution of organic-rich shale by biosiliceous intercalations (radiolarian sands). These factors result in comparatively low organic matter (OM) accumulation rate in all Italian sections studied, although a distinct peak of high TOC content is observed at the Cenomanian/Turonian boundary.

Gibraltar Arch (Flysch Zones and Continental Margin of the Alboran Microcontinent)

The Flysch Zones of the Gibraltar Arch area (Figure 4) comprise sediments deposited in the Atlantic-Tethys seaway between the late Mesozoic North African margin, the Iberian margin, and the former Alboran microcontinent (Thurow and Kuhnt, 1986; Thurow, 1987). Cretaceous deep-sea fans and basin plain associations have been deposited in a deep, but rather narrow, elongated basin with detrital input from the different borderlands. A depositional environment below the calcium carbonate compensation depth (CCD) can be concluded for all these sediments due to the absence of calcium carbonate and calcareous microorganisms. Typical "flysch-type" or "abyssal-type" (Kuhnt and Kaminski, 1989) associations of agglutinated deep-water benthic foraminifera are characteristic (Kuhnt, 1987; Thurow, 1987). Sedimentation rates are comparatively low (up to 10 m/m.y., Thurow, 1987).

Black shales with high TOC values of marine kerogen sharply mark a very distinct event at the Cenomanian/Turonian boundary, which is defined by the last occurrence of *Rotalipora cushmani* in calciturbidites (Thurow and Kuhnt, 1986) and the first occurrence of the radiolarian species *Crucella cachensis* in biosiliceous layers (Thurow, 1988). Organic-carbon-rich layers are generally associated with radiolarian enrichment. The most substantial accumulations of organic carbon occur in the dark carbonaceous biosiliceous shales ("phthanites") of the most basinal ("Massylian") sequences, which may reach 10 m in thickness. These sequences comprise gray-green intensively silicified shales and thin, very fine-grained turbidites with up to 0.9% TOC, laminated black shales (up to 13% TOC), and radiolarites (up to 2% TOC). Generally the highest enrichment of organic matter is observed in the lowermost 1-3 m of the succession, with a remarkable decrease toward the top. The kerogen is type II but is commonly degraded by slight catagenesis (Herbin et al., 1986b; Thurow and Kuhnt, 1986).

Northern Morocco (Thrusted North African Margin in the Rif Mountains)

Cenomanian/Turonian sediments have been studied along a shelf-to-basin transect south of Tetouan from the external Prerif Zone in the south to the internal part of the Tanger Unit/Beni Derkoul Unit in the north (Figure 5). Distribution of benthic foraminiferal assemblages along this transect (Kuhnt, 1988) gives evidence for a more or less continuous deepening from south (upper bathyal) to north (abyssal environment).

Marly sediments with an abundant planktonic microfauna and distinct slumps prevail in the upper part of the slope. Calciturbidites and hemipelagic sediments with slumping mainly in the Cenomanian part occur in the lower part of the slope. Depauperate, partly dissolved, calcareous microfossils, silification, chert nodules, and dominating radiolarians in the microfossil assemblages are common in the Turonian. Redeposited organic matter in fine-grained turbidites probably makes up the major part of the organic-matter accumulation, especially in these

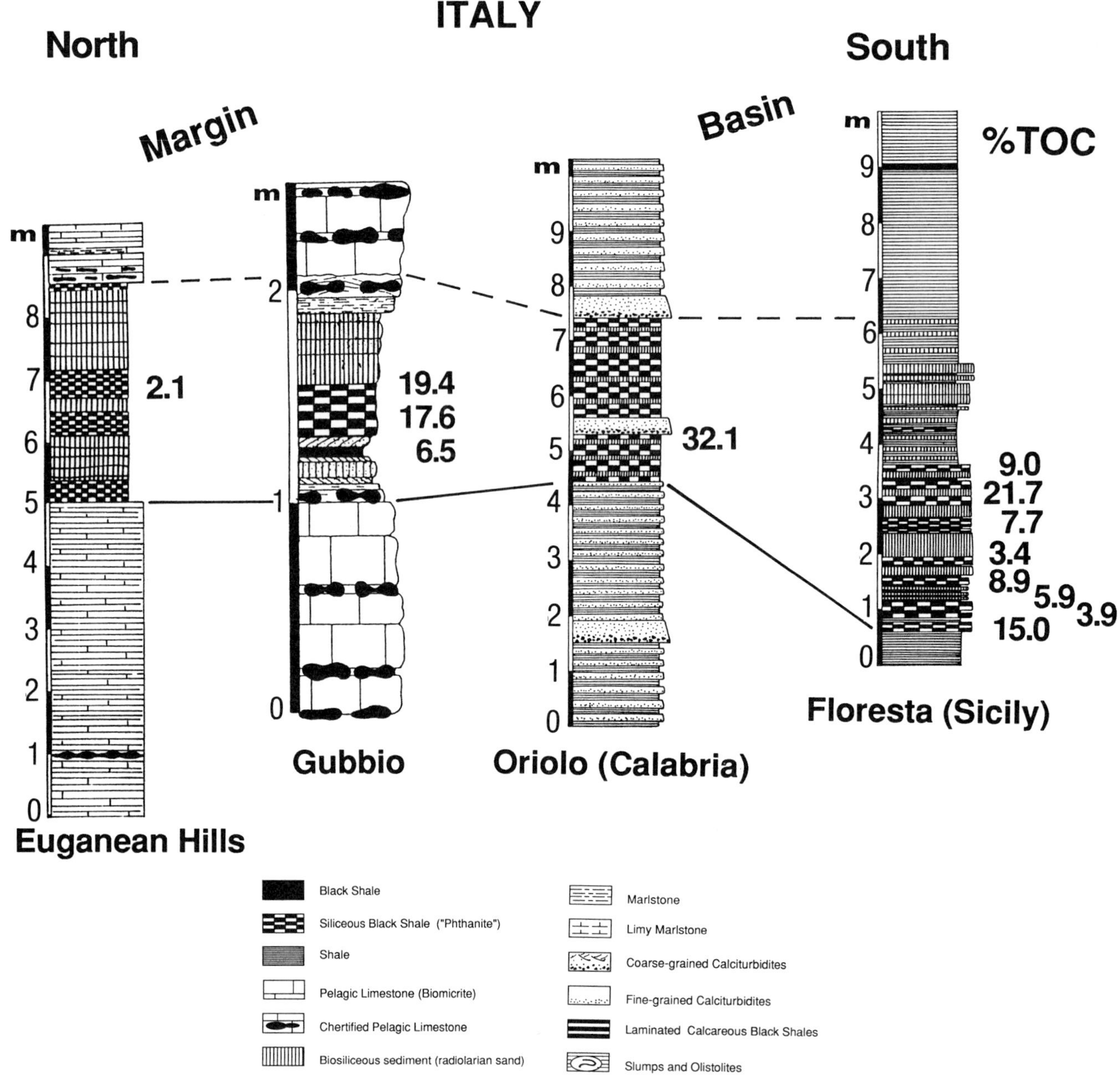

Figure 3. Sections studied along a north-south transect through Italy. The Gubbio section is modified from Arthur and Premoli Silva (1982). Lines indicate beginning and end of organic-rich sedimentation corresponding to the *W. archaeocretacea* foraminiferal zone ("Bonarelli horizon").

distal slope sequences. Erosional features and hiatuses due to gravitational redeposition occur in both environments, but are more common in the upper-slope sequences.

Earlier reconstructions of this transect (deGraciansky et al., 1986) concluded that a continuous gradation with depth occurs from deep-water, carbonaceous (anoxic), carbonate-depleted phthanites in the north to the aerobic neritic limestones which transgress onto the Moroccan Meseta to the south. However, this model did not take into account field data from the external Prerif zone, where Cenomanian bituminous, fish-bearing shales occur (Daguin, 1927; Arambourg, 1954), which have TOC values of more than 30% (Figure 5). Also in the more basinward succession of the internal Prerif zone organic-rich sediments (TOC up to 2.4%) are observed in a slumped marly sequence of late Cenomanian age (Kuhnt et al., 1986).

Tunisia (Bahloul) and Algeria (North African Margin)

The inner-neritic-facies belt observed in Spain and Morocco, with an oyster, echinid, and vascoceratid

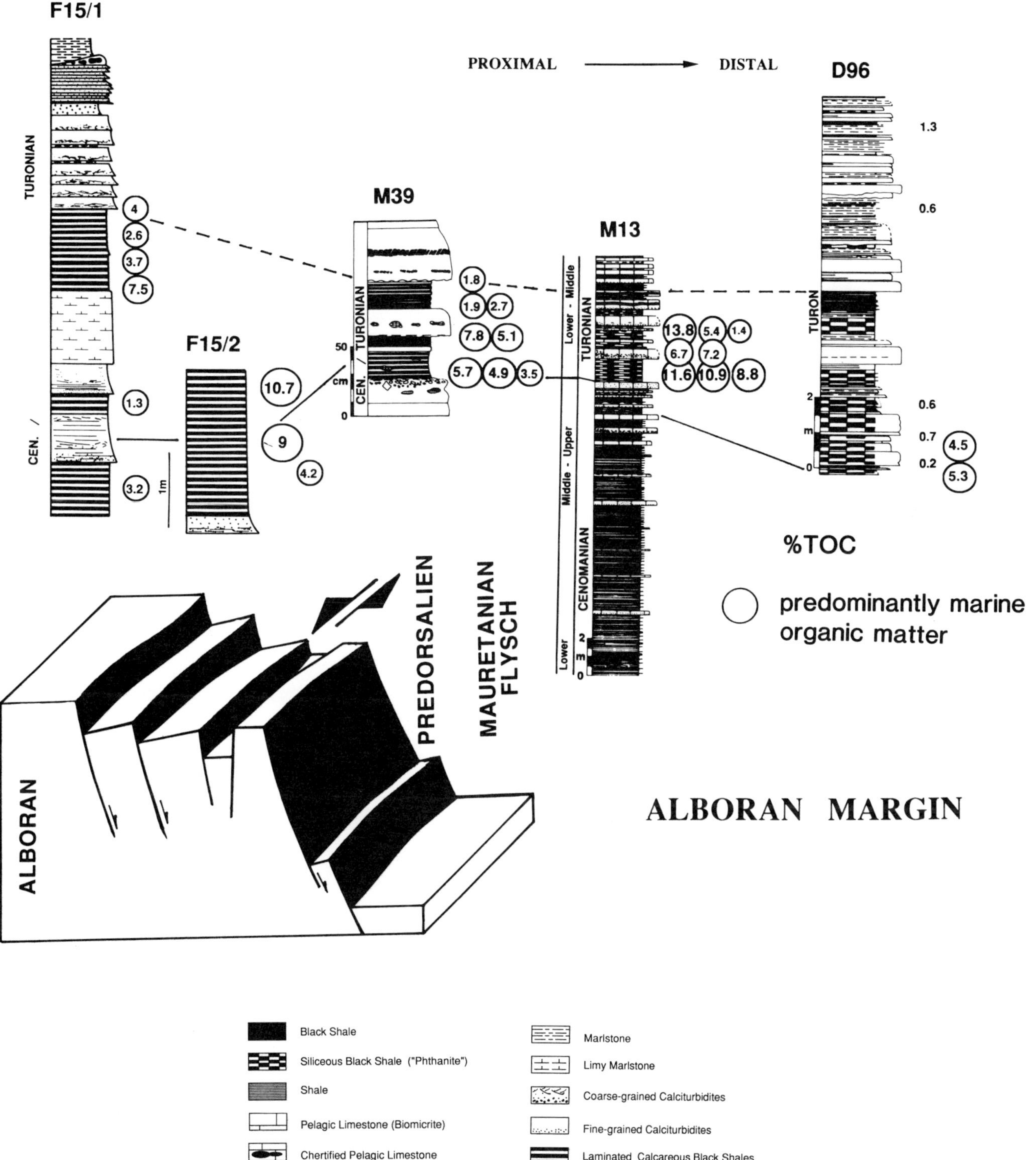

Figure 4. Sections studied along the Alboran transect (Alboran margin and adjacent flysch trough in the Gibraltar Arch area). (F15/1) western Chrafate section, southeast of Chechaouen; (F15/2) eastern Chrafate section; (M39) Cluse des Beni Ider; (M13) Cap Malabata, east of Tangier; (D96) distal part of the Mauretanian or proximal part of the Massylian Flysch Units between Tangier and Tetouan. Sections M39, M13, and D96 are from Herbin et al. (1986b), and Thurow (1987). Lines indicate beginning and end of organic-rich sedimentation corresponding to the *W. archaeocretacea* foraminiferal zone.

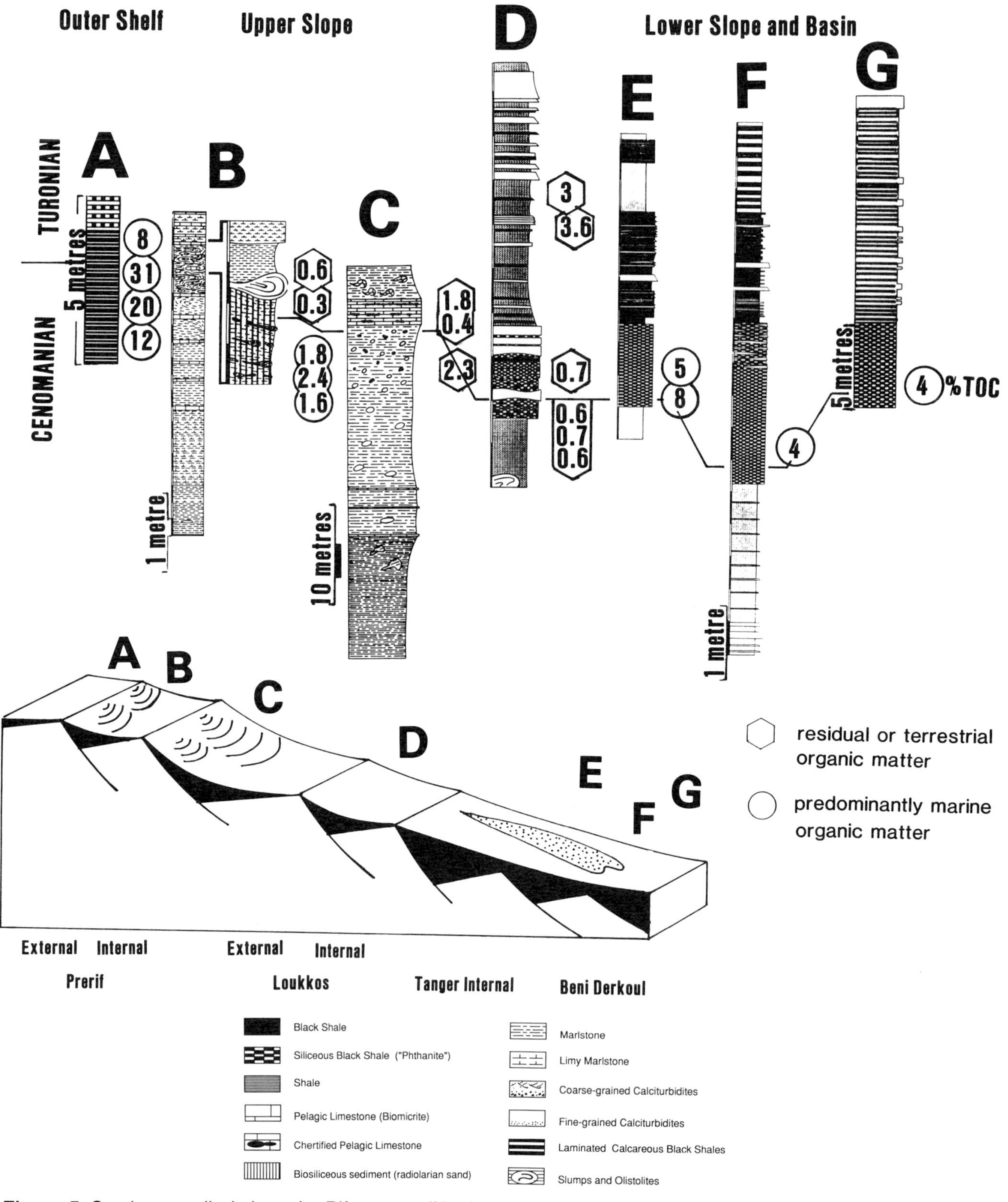

Figure 5. Sections studied along the Rif transect (North African margin). (A) Djebel Tselfat area, external Prerif Zone, east of Sidi Kacem; (B) Djebel Mguedrouz, internal Prerif Zone (Kuhnt et al., 1986); (C) Kelaat Boukkorra section L30, southern part of the Loukkos Zone; (D) Pont du Loukkos section L1, northern part of the Loukkos Zone; (E) section M8, internal Tanger Unit, south of Tangier; (F) section M90, internal Tanger Unit, west of Tetouan; (G) Beni Hessane section M61, Beni Derkoul Unit of Lespinasse (1975) and Kuhnt (1987), north of Chechaouen.

biofacies, is also present in Tunisia and Algeria. The most interesting locality where the Cenomanian/Turonian boundary is exposed is that of the Djebel Mrhila near Sbeitla (Pervinquiere, 1903, 1907). Here, the upper Cenomanian is represented by bluish-gray claystones and marls of the *Metoicoceras geslinianum* ammonite zone and indicator fossils of shallow neritic environments such as *Exogyra, Neithea*, and *Hemiaster*. The basal Turonian is marly and characterized by a more pelagic cephalopod-echinid fauna (Salaj, 1978). The composition and diversity of these macrofaunal assemblages are largely comparable with those of the Picofrentes section in the Celtiberic Ranges, Spain.

The Cenomanian/Turonian organic-rich facies of the Bahloul Formation (Burollet, 1956) in northern Tunisia was deposited in a deeper shelf depression of the late Mesozoic North African continental margin. The paleogeographic distribution of these organic-rich limestones was rather widespread along the northwestern margin of the subsiding Pelagian platform. About 30 to 50 m of organic-rich laminated marls and limestones were deposited in an open-marine shelf environment (Herbin et al., 1986b; Kuhnt et al., 1986; Arthur et al., 1987). The TOC of the bituminous limestones is relatively high (up to 4.7%), and the hydrogen indices reach 670 mg HC/g TOC. Their maturity is moderate with an average T_{max} of 431°C, indicating low post-Turonian burial rates. This is consistent with the end of subsidence within these basins due to the first compressive phases along the North African margin (Burollet and Ellouz, 1986).

Farther to the west (Algeria), late Cenomanian to early Turonian bituminous marls in an outer-neritic paleoenvironmental setting are reported in the Bou Saada area (Emberger, 1960; Kieken, 1962), in the autochthonous massifs of the Ouarsensis (Mattauer, 1958), and in the Titteri Mountains (Guillemot, 1952). Fish skeletons and biosiliceous intercalations (radiolarian sands) were commonly observed within the dark laminated sediments (Mattauer, 1958). This organic-rich facies very likely formed a continuous band along the middle Cretaceous North African shelf edge from the Moroccan Prerif Zone in the west to Tunisia in the east.

Along the deeper parts of the Tellian margin in northern Algeria, middle Cretaceous organic-rich sediments have been encountered as large olistolites in a slope sequence of the Monts de Tessala (Fenet, 1975), which corresponds to the middle part of the North African continental margin (Wildi, 1983). The largest olistolites comprise laminated black shales, more than 2 m thick, with TOC values up to 4.7%. A degragation of the organic matter in this area, where Late Cretaceous and early Cenozoic subsidence was important and Miocene tectonism led to nappe structures, is indicated by comparatively low HI (about 100 mg HC/g TOC) as well as very low OI (7 to 24) and a high T_{max} of about 460°C.

Moroccan Epicontinental Seas and Atlantic Coastal Basins (Atlas Mountains, Tarfaya)

The shallow-marine environments of the transgressive sequences of the Moroccan Meseta are characterized by a typical biodetrital dolomitic limestone formation of Turonian age (Busson, 1984; Wiedmann et al., 1978; Thein, 1988). The neritic carbonates were deposited in oxygenated water and contain the typical "Tethyan" shallow-marine oyster, echinid, and vascoceratid biofacies. The macrobiota commonly show high diversification. Coquinites, tempestites, and a few oolite bars indicate strong water agitation (Thein, 1988). No concentrations of organic matter have been observed in this environment.

The Turonian bituminous marls of the Tarfaya coastal basin were deposited in an open-shelf sea where water depth was controlled chiefly by the rise of sea level during Cenomanian time and the difference between accumulation rate and subsidence (Einsele and Wiedmann, 1982). A combination of both these mechanisms resulted in a Turonian water depth on the order of 200–300 m.

The development of organic-matter content and type is studied in detail in well S13 of the Moroccan BRPM (Figures 6, 7). Lowermost Cenomanian sediments of the Tarfaya coastal basin are characterized by low TOC content with mixed terrestrial and marine kerogen types. During the middle and late Cenomanian, TOC increased, and the marine component became dominant. Nonlaminated, bioturbated, olive-green marls and clays prevail or occur as intercalations in the lower/middle Cenomanian part of the sequence. Since late Cenomanian time finely laminated marlstones prevailed, and bioturbated layers are very rare; the sediment consists of dark-brown, laminated kerogenous chalks, alternating with nonlaminated lighter colored limestones containing a lower kerogen content. The sedimentation rate can be surprisingly high (up to 100 m/m.y. for the *W. archaeocretacea* zone). Sediments consist almost exclusively of organic matter and biogenic carbonate. TOC is very high in the laminated strata (up to 20%). The biogenic-carbonate content reaches up to 90% and is composed mainly of planktonic foraminifera and calcareous nannoplankton. Rare thin-bedded shell layers, probably storm generated, are present in post-lower/middle Turonian sediments. Since late Turonian time, organic-rich sediments progressively decreased and were replaced by more oxygenated sediments.

Senegal (Cape Verde Basin, Casamance; Site 367 DSDP)

The Senegal basin (Figure 8), which is the largest Mesozoic basin along the West African margin of the

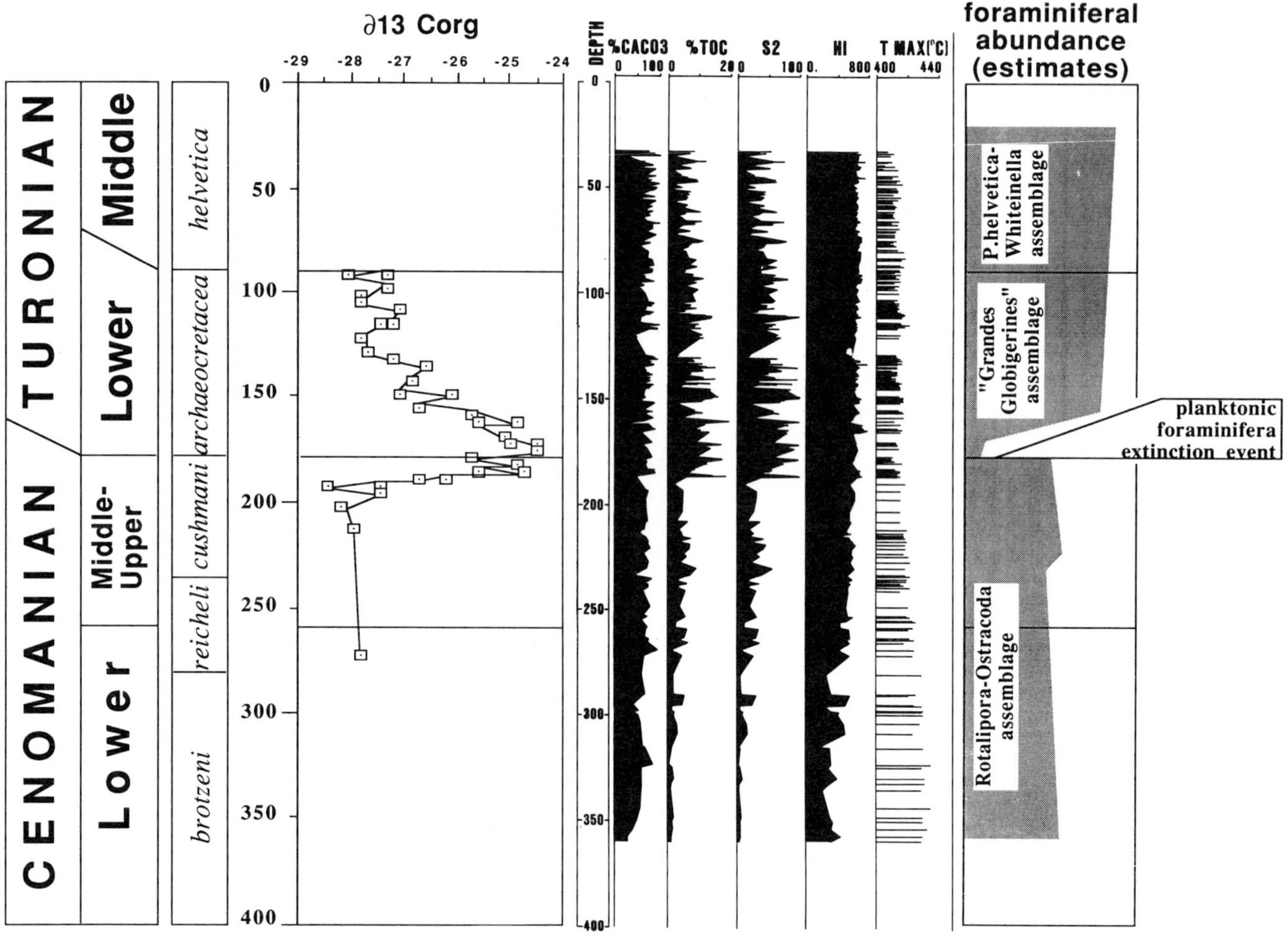

Figure 6. Cenomanian/Turonian organic-rich sedimentation and chemostratigraphy and biostratigraphy of the S13 well in the Tarfaya coastal basin. $\delta^{13}C$ data are from Thurow et al. (1988).

North Atlantic (340,000 km^2), was extensively explored 20 years ago. The results of this exploration activity were somewhat disappointing, showing only huge amounts of heavy oils sealed in Tertiary reservoirs.

Although very few organic-geochemistry studies have been carried out, particularily in order to solve the problem of the origin of these altered oils, it is generally assumed that the bituminous clays of the Brikama Formation of early Turonian age are the main source rocks in this area. However, their maturity is too low—above the oil window—in the wells studied. The effective source rock would be consequently located in deeper parts along the continental margin.

The data so far published mainly pertain to the Casamance area and onshore wells (HERBIN et al.,1986b). An east-west section across the margin shows that the total organic carbon content increases from the continent toward the open oceanic basin, i.e., from about 1% TOC within the detrital zone in the eastern part (well DM1) to about 10% TOC in the more pelagic area (CM10). The same east-west trend can be observed with the type of organic matter; the preservation of organic matter of planktonic origin (type II) is apparently higher toward the open ocean (well CM10).

About 330 km west of the Casamance area, at the base of the continental rise in the Cape Verde Basin, DSDP Site 367 shows a similar enrichment in organic matter at the Cenomanian-Turonian boundary (CTB). The sediments are characterized by exceptionally high concentrations of organic matter; the average TOC is about 30%, and even reaches 40%. Paleodepth reconstructions place the black shale sedimentation of Site 367 at a paleodepth of about 3700 m (Chenet and Francheteau, 1979). The occurrence of both deposits on the shelf and in the deep sea implies, that paleoceanographic conditions at the CTB were favorable for the preservation of planktonic matter within a large part of the oceanic water column.

Nigeria (Benue Trough, Calabar Flank)

Cenomanian/Turonian sediments have been studied in various bathymetric and depositional

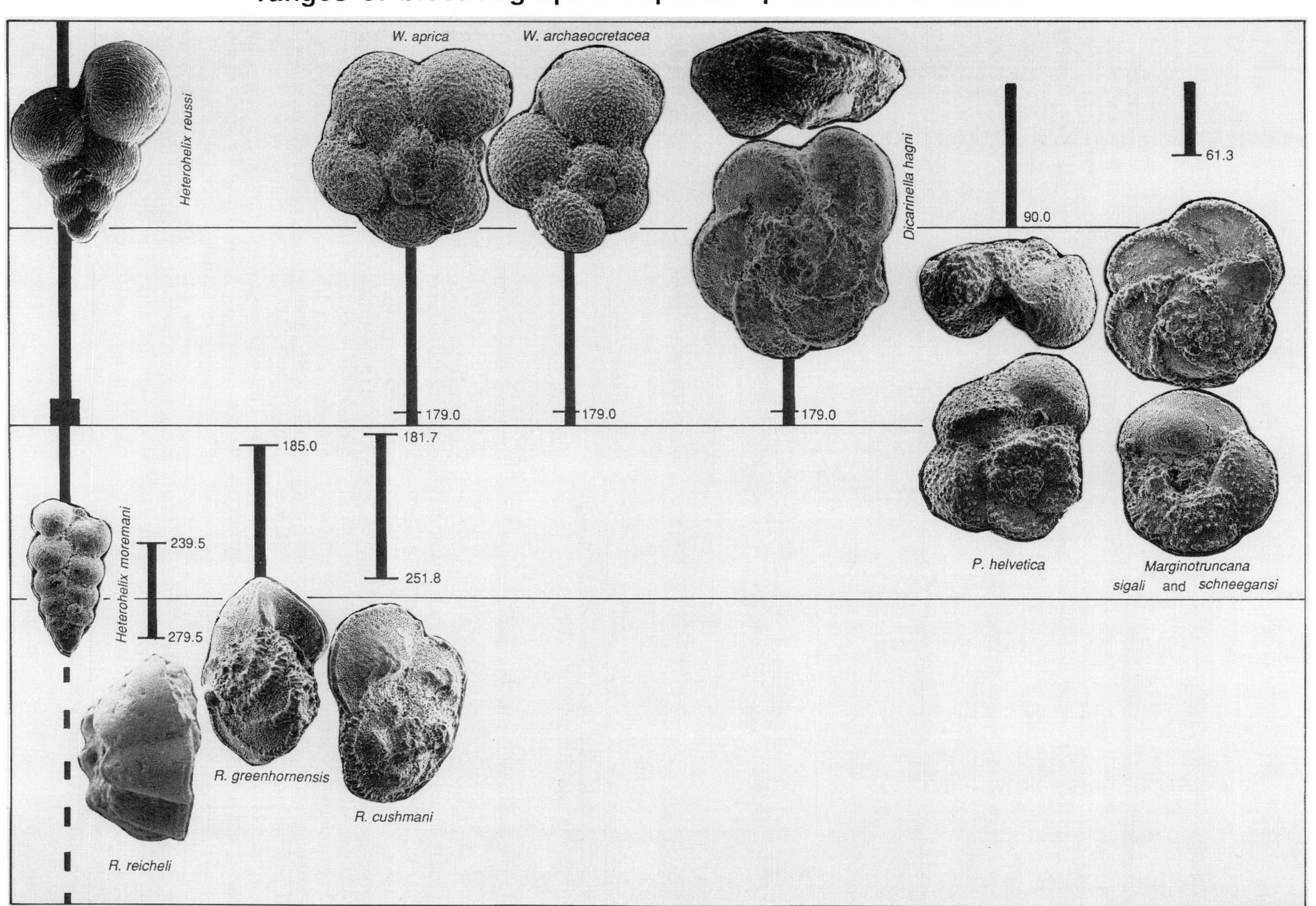

Figure 7. Planktonic foraminiferal biostratigraphy of the S13 well in the Tarfaya coastal basin.

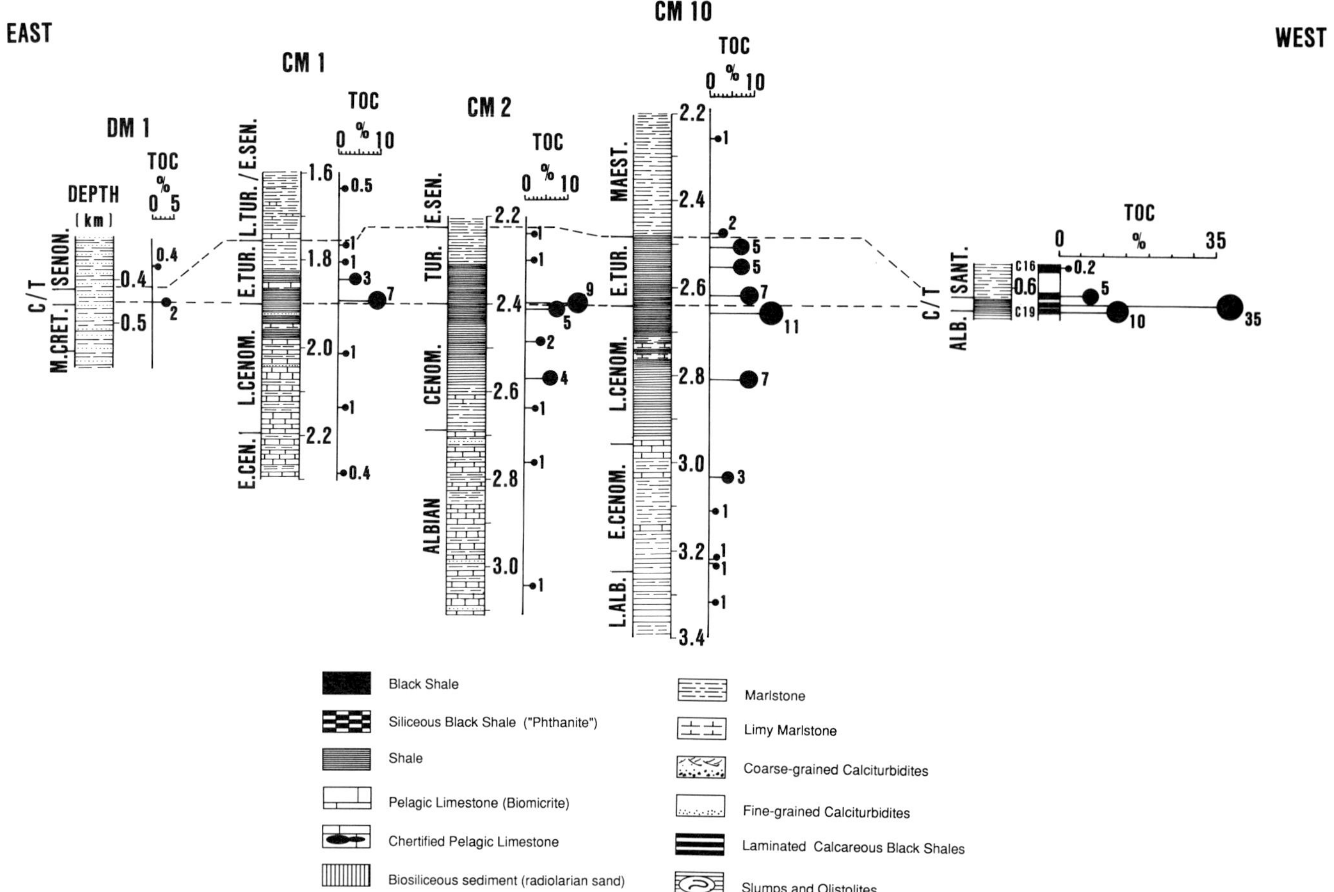

Figure 8. Sections studied along the Senegal margin transect (modified after Herbin et al., 1986b).

settings in the Nigerian Benue trough (Petters, 1978; Petters and Ekweozor, 1982) and adjacent areas (Calabar Flank, southern Nigeria, and Mungo River section, southeastern Cameroon). Four principal environments can be discriminated, ranging from coastal areas in the north to a deep outer shelf basin and shelf/slope transition in the south (Figures 9, 10): (1) shallow-water Gongila (Ashaka quarry) and Pindiga Formations of northern Nigeria; (2) shallow-water sequences with enhanced clastic input in a marginal position in the central part of the Benue Rift (Ngelegbokwa section); (3) deep-shelf sequences in the central and southern parts of the Benue trough (Ngbanocha and Lokpanta sections); and (4) continental-margin sequences, close to the shelf edge along the Calabar Flank (southeastern Nigeria). The biostratigraphic framework is compiled in Figure 11.

The Pindiga and Gongila Formations of northern Nigeria (Popoff et al., 1986; Allix, 1987; Meister, 1989) have alternating units of marine limestone and shale/mudstone about 30 m (Ashaka section) to 60 m (Pindiga section) thick. The lower parts of these formations consist mainly of biogenic shallow-water limestones that have been dated as upper Cenomanian to lower Turonian on the basis of the rich ammonite content. Sediments of the uppermost Cenomanian/lower Turonian *Vascoceras bulbosum* and *Paravascoceras costatum* ammonite zones are strongly condensed, include hardgrounds, and presumably contain hiatuses near the Cenomanian/Turonian boundary. The upper parts of these formations (dated as middle Turonian by ammonites) are characterized by a sequence of dark-gray fissile shales with intercalated, probably storm-generated, lumachelle beds. Shale samples across the Cenomanian/Turonian-boundary show surprisingly low TOC values (generally below 0.1%) in spite of the dark-gray color of the shale.

The Cenomanian to Turonian sedimentary sequence of the central and southern Benue trough is strongly influenced by the global eustatic sea-level rise that marked the Cenomanian/Turonian boundary. This resulted in the buildup of a wide shelf platform on the western flank of the basin, where condensed sections, tempestites, and tidalites can be observed (Ojoh, 1988). Black, laminated, organic-rich calcareous marls and limestones, with abundant early Turonian inoceramids, including biostratigra-

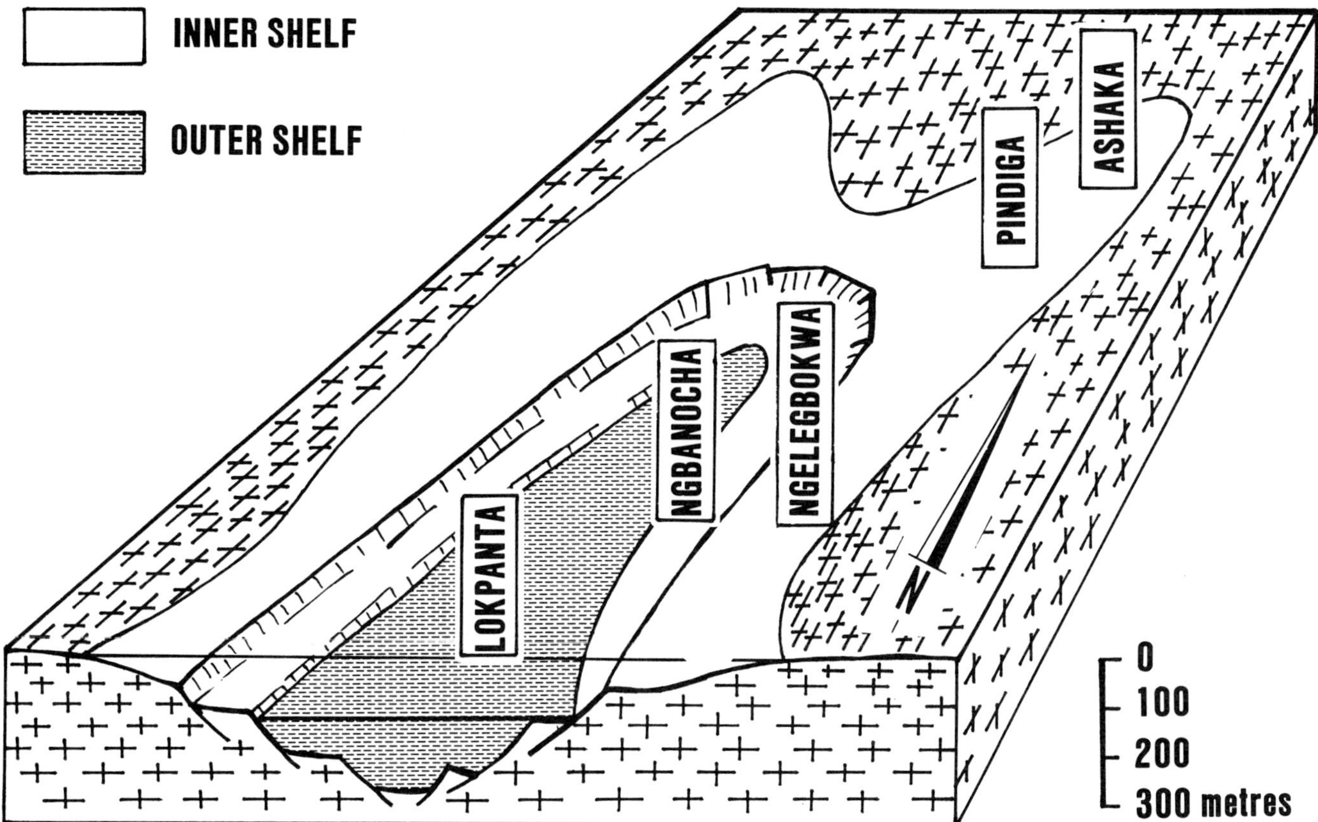

Figure 9. Simplified paleogeographic sketch of the middle Cretaceous Benue trough. Paleogeographic positions of the sections studied are indicated.

phically important forms like *Mytiloides opalensis* and *Sergipia* sp. occur in the central part of the basin. The microfauna is characterized by rich, exclusively planktonic, foraminiferal assemblages of the *Whiteinella archaeocretacea* zone. TOC is high (up to 5%), and the kerogen is mainly of type II. This local expression of the CTBE is followed by a sequence of fine-grained sandstones, lumachelle beds, tempestites, limestones, and hardgrounds, which commonly exhibits characteristic features of condensed sequences.

The Calabar Flank, located at the eastern part of the Gulf of Guinea, west of the Cameroon volcanic trend, has been considered as part of the South Atlantic continental margin southeast of the Benue Rift (Nyong and Ramanathan, 1985). The paleogeographic evolution of this area was characterized by a tilted-block structure, resulting in a horst-and-graben morphology trending northwest-southeast (Buser, 1966). The stratigraphic development along the Calabar Flank was dominated by vertical movements of these fault blocks and by eustatic sea-level changes. A significant unconformity separates gently folded late Albian to Cenomanian strata from the overlying monoclinally dipping Turonian sediments (Nyong and Ramanathan, 1985). Sediments of the Cenomanian/Turonian boundary seem to be missing due to this regional unconformity. The lower part of the Turonian sequence is composed of a thickly bedded, intensely bioturbated (Akpan, 1985) limestone unit associated with pelecypods, gastropods, ammonites, and echinoids (Dessauvagie, 1972). Planktonic foraminifera dominate the microfossil assemblages. TOC content of dark marly intercalations in this sequence is up to 1.3%; kerogen is mainly of marine type II (HI up to 356, OI 8–20).

VARIATION OF ORGANIC MATTER WITH TIME

Stratigraphic Framework

Biostratigraphy

In most of the studied transects, planktonic foraminifera have proved to be the most useful for biostratigraphic correlation and calculation of

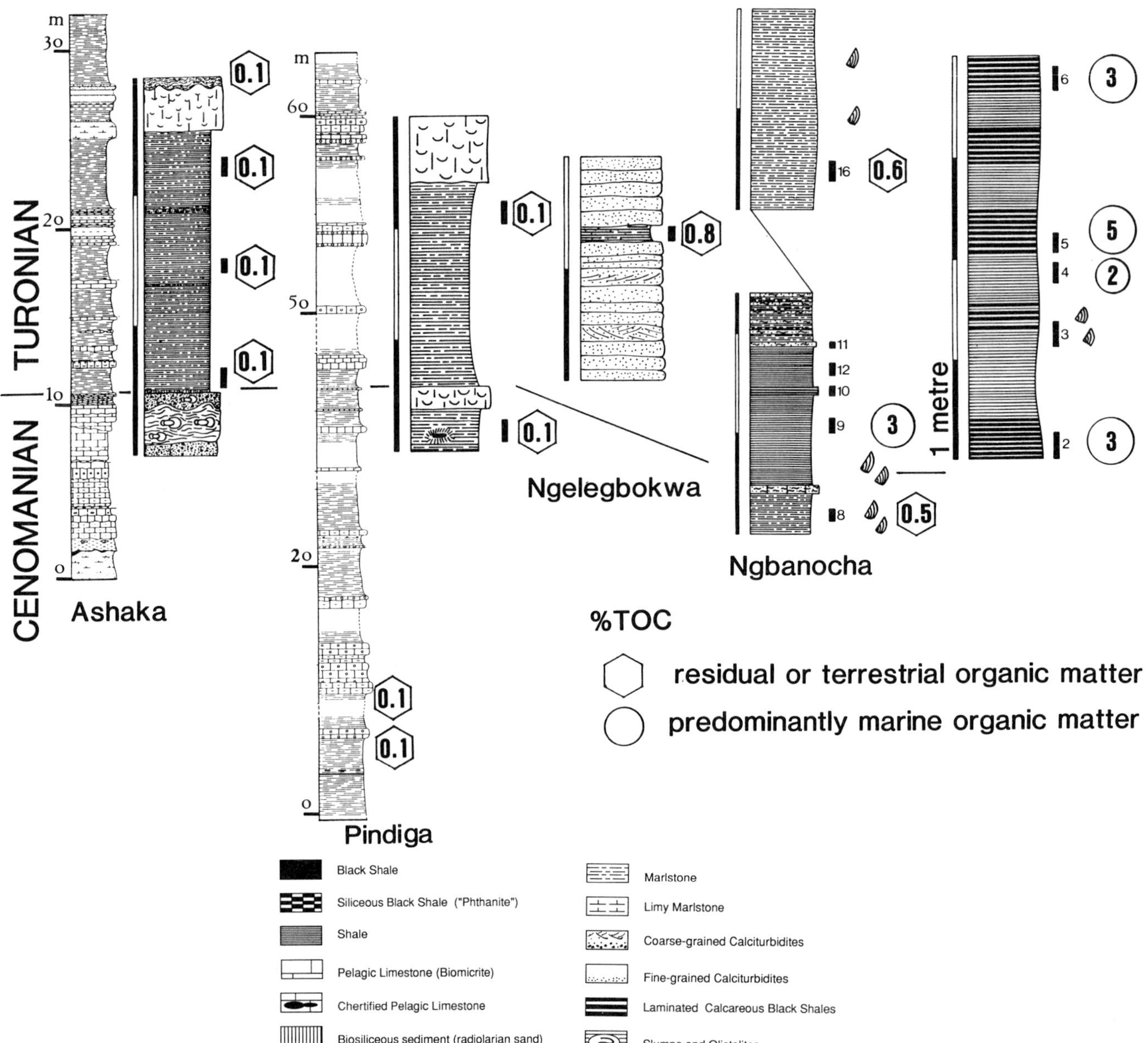

Figure 10. Sections studied along the Benue transect, Nigeria. The overview logs for the Pindiga and Ashaka sections are from Popoff et al. (1986). Scale bar for the detailed sections is 1 m.

sedimentation and accumulation rates. The following zonal scheme (modified after Robaszynki and Caron, 1979a,b and Caron, 1985) is the basis of our biostratigraphic framework.

R. globotruncanoides zone: Interval from the appearance of the marker to the entry of *Rotalipora reicheli*. Additionally, *R. brotzeni* and *Rotalipora montsalvensis* are present.

R. reicheli TRZ: Defined by the co-occurrence of the marker species with the planktonic foraminiferal assemblage of the *R. globotruncanoides* zone.

R. cushmani TRZ: The lower boundary of this zone is defined by the FAD of *Rotalipora cushmani*, which is nearly coeval with the extinction of *Rotalipora reicheli*. The upper boundary is defined by the extinction of the genus *Rotalipora*. Characteristic associated faunal elements in the *R. cushmani* zone are *Rotalipora greenhornensis, Praeglobotruncana stephani*, and *Hedbergella delrioensis-Whiteinella baltica*. The following important "events" have been observed at the top of the *R. cushmani* TRZ and the base of the *W. archaeocretacea* zone which allow a detailed stratigraphic subdivision of this highly interesting interval which is generally referred to as Cenomanian/Turonian boundary event (CTBE): (1)

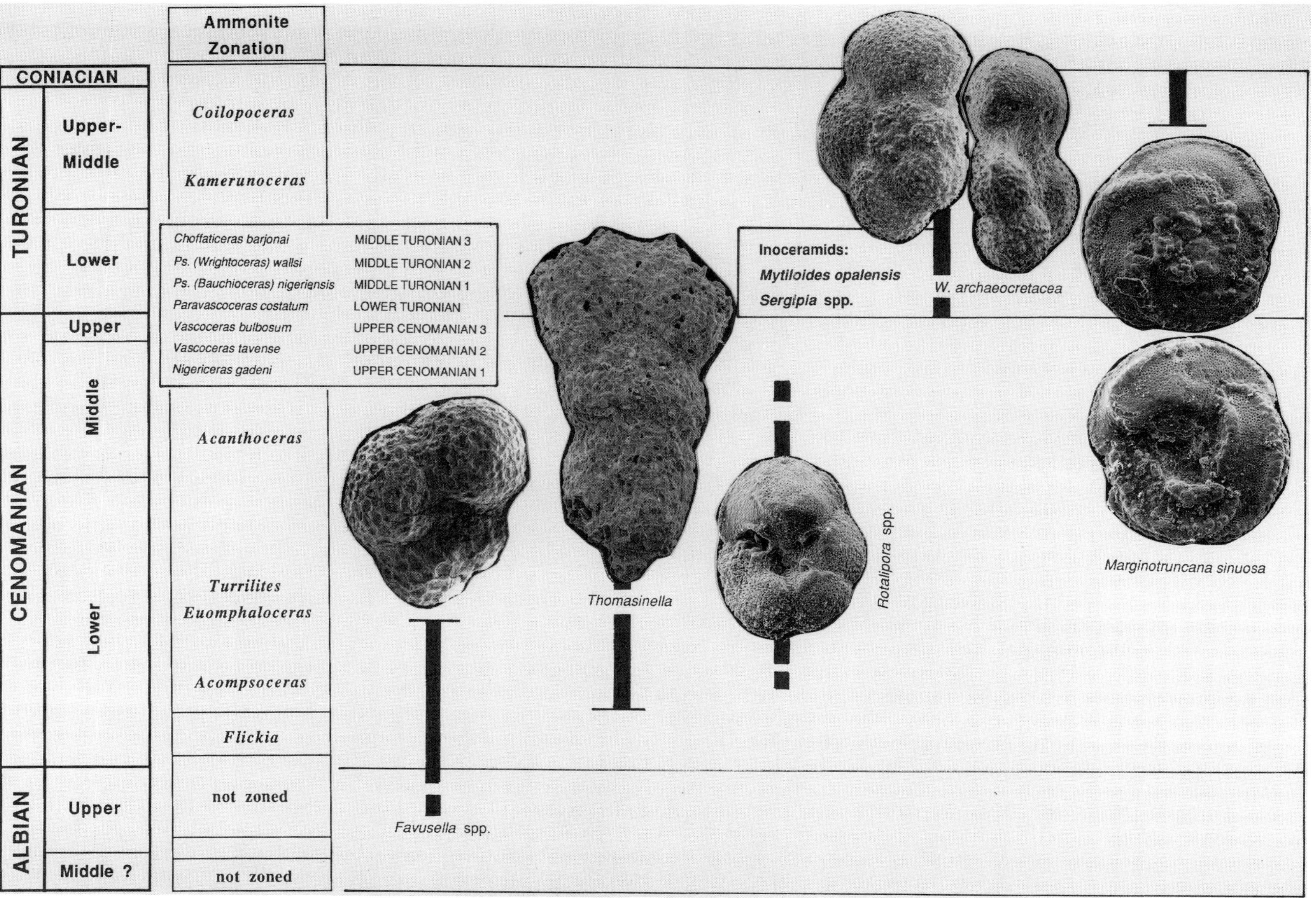

Figure 11. Biostratigraphic framework for the middle Cretaceous in the Benue trough, based on ammonites, inoceramids, and foraminifera.

extinction of *Rotalipora greenhornensis*; (2) strong increase of radiolarian abundance; (3) occurrence of "atypical" specimens of *Rotalipora cushmani* with high trochospire, dorsoconvex outline, and weakly developed keels (these forms may be compared to the North American *Anaticinella* stock or the *Rotalipora turonica-Rotalipora thomei* morphogroup of the Tethyan Realm); (4) positive $\delta^{13}C$ excursion; (5) extinction of *Rotalipora cushmani*; and (6) occurrence of heterohelicid/hedbergellid-dominated planktonic foraminiferal assemblages.

W. archaeocretacea IZ ("Zone à grandes Globigérines"): According to Wonders (1980) defined as interval from the extinction of *Rotalipora cushmani* to the first occurrence of *Praeglobotruncana helvetica*. An assemblage of unkeeled planktonic foraminifera with the guide species *Whiteinella aprica* and *Whiteinella archaeocretacea* is characteristic. The first keeled forms of the genus *Dicarinella* appear within this zone. Quite common in the Tarfaya basin are *Dicarinella imbricata* and *Dicarinella hagni*.

P. helvetica TRZ: The base of this zone is generally defined with the first appearance of *Praeglobotruncana helvetica*. In the Tarfaya sequence a transitional zone is observed at the base of the *P. helvetica* TRZ; it is transitional in two aspects: (1) taxonomic—transitional forms between *P. praehelvetica* and *P. helvetica* and (2) statistical—extremely rare occurrence of *P. helvetica* in the lowermost part of the zone may bias the definition of the lower zonal boundary (e.g., addicted by the richness and preservation of the faunal assemblages). A subdivision of the *P. helvetica* zone is possible by the FADs of different *Marginotruncana* species: (1) FAD of *M.* aff. *renzi* at the base of the *P. helvetica* TRZ (*M.* aff. *renzi* is possibly identical with the North American species *Marginotruncana roddai* which has a FAD between the FAD of *P. praehelvetica* and *P. helvetica* in the Greenhorn Formation (Eicher and Diner, 1985); (2) FAD of *M. schneegansi* and *M. sigali* in the lower part of the *P. helvetica* zone; (3) FAD of *M. paraconcavata, M. marianosi* and *M. pseudolinneiana* in the middle/upper part of the *P. helvetica* zone (exact positions of these datum levels need to be determined precisely).

M. schneegansi IZ: The lower boundary of this zone is defined by the LAD of *Praeglobotruncana helvetica*. Within the zone, the following datum levels are useful for further subdivision: (1) FAD of *Marginotruncana coronata* and *M. marginata* and (2) FAD of *M. sinuosa* and *M. undulata* in the upper part of the *Schneegansi* zone.

D. primitiva TRZ: Interval from the FAD of *D. primitiva* to the appearance of *Dicarinella concavata*. The zonal marker seems to be an intermediate form between *Marginotruncana schneegansi* and *Dicarinella concavata*, and the transitions to both species are gradual (compare Wonders, 1980). Especially the lower boundary of the *D. primitiva* zone is not easy to define because of the very gradual transition from *Marginotruncana* of the *paraconcavata-schneegansi* group to typical *Dicarinella primitiva*. FAD of *Hedbergella flandrini* seems to be ± coeval with the base of the *D. primitiva* TRZ.

D. concavata PRZ: The lower boundary is well defined by the FAD of *D. concavata*, which is an easily recognized datum level (for example in the Tarfaya sections); the top is defined by the FAD of *Dicarinella asymetrica*.

D. asymetrica TRZ: The definition of the FAD of *D. asymetrica* is difficult because of a very continuous evolution from *D. concavata* to *D. asymetrica*. The definition of the base of the *D. asymetrica* zone therefore is very arbitrary, somewhere in the evolutionary lineage of the *concavata-asymetrica* group and needs calibration by other biostratigraphic markers. The upper limit of this zone (i.e., late Santonian) is not reached in any of the studied sections.

In addition to planktonic foraminifera, the following fossil groups provide reliable datum horizons for chronostratigraphic correlation of the studied sequences: calcareous nannoplankton (Bralower, 1988); radiolarians (Pessagno, 1976; Thurow et al., 1982; Thurow and Kuhnt, 1986; Thurow, 1987; Kuhnt et al., 1986; Thurow, 1988); ammonites (Wiedmann et al., 1978; Kuhnt et al., 1986); inoceramids (Wiedmann and Kauffman, 1978, Kuhnt et al., 1986); dinoflagellates (Herbin et al., 1987a; Thurow et al., 1988); deep-water benthic agglutinated foraminifera (Moullade et al., 1988; Kuhnt et al., 1989). The zonal schemes applied for chronostratigraphic correlation in this study, and calibration to the time-scale of Haq et al. (1987) are compiled in Figure 12.

Stratigraphic Importance of the Late Cenomanian Carbon-Isotope Shift

Scholle and Arthur (1980) documented a prominent positive excursion of 3-4 ‰ in $\delta^{13}C$ from pelagic limestones deposited in the area of the North Atlantic and its marginal basins during the middle Cretaceous (upper Cenomanian). The main reasons for excursions in $\delta^{13}C$ values have been seen in significant changes in the ratio of oxidized (carbonate) and reduced (organic) carbon deposited in sedimentary strata (Pratt, 1983; Schlanger et al., 1987). Generally parallel trends in $\delta^{13}C$ values for carbonate and organic carbon were noted by Fischer and Arthur (1977) for Tertiary to Recent marine rocks. Pratt (1983), Thurow et al. (1988), and Arthur et al. (1988) documented a pronounced positive shift in $\delta^{13}C$ values for organic carbon and carbonate in deposits spanning the Cenomanian/Turonian boundary. The global character of this positive excursion has been postulated by Scholle and Arthur (1980) and Schlanger et al. (1987). Enhanced burial of large amounts of organic matter, probably enabled by at

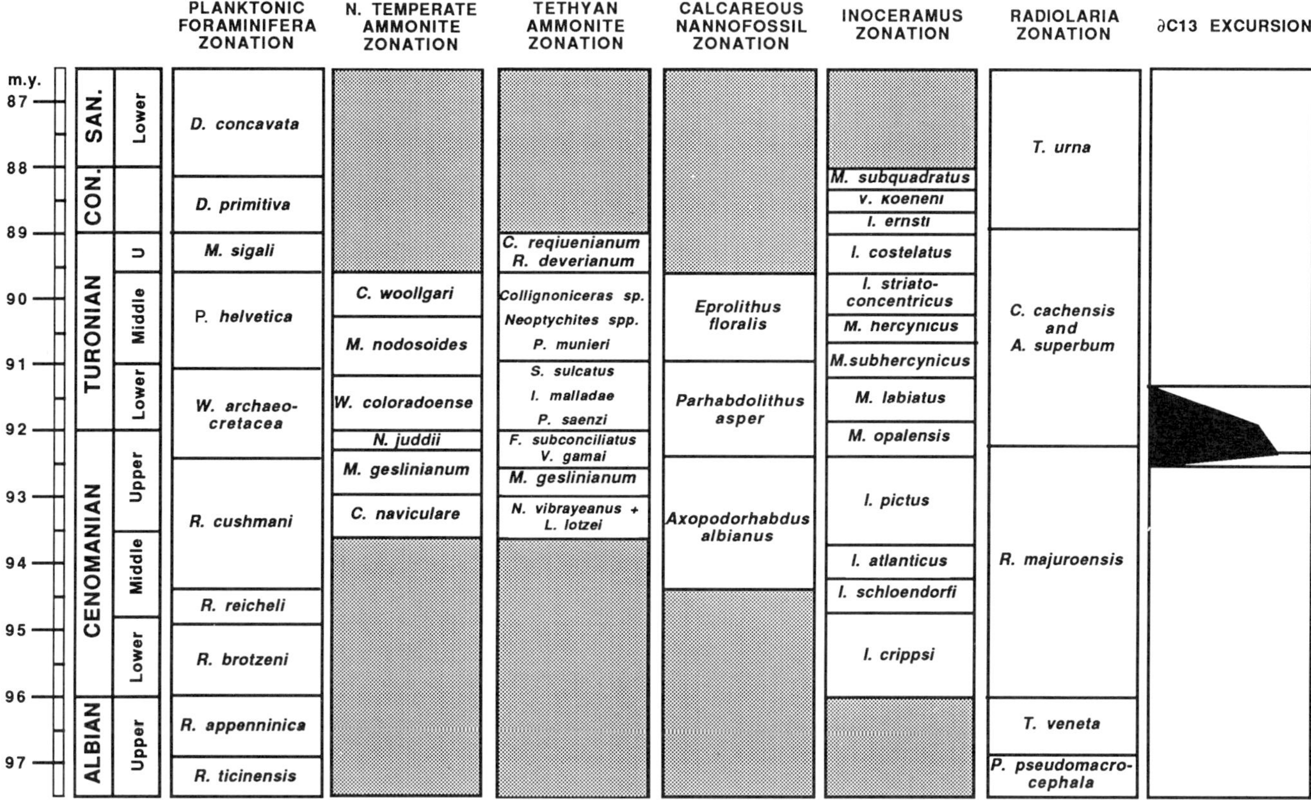

Figure 12. Chronostratigraphic framework and correlation of zonal schemes for the different faunal groups used for biostratigraphic correlation.

least locally enhanced productivity of marine plankton, is the most likely cause of a geologically brief increase in ^{13}C in the global ocean (Summerhayes, 1987). The onset of the positive $\delta^{13}C$ excursion, therefore, is an interesting tool for stratigraphic correlation (Thurow et al., 1988).

A precise correlation of this isotope shift to the planktonic foraminiferal zonation is possible in the Tarfaya sequence (well S13). The first unusually high $\delta^{13}C$ values have been observed a few decimeters below the extinction of *Rotalipora cushmani* at 183.2 m, and the re-occurrence of "normal" carbon-isotope compositions occurs at 160.5 m in the middle part of the *W. archaeocretacea* zone. Between these two events the maximum values in TOC also can be observed (Figure 6).

Variation of Total Organic Carbon Content (TOC) and Kerogen Type with Time

Examples of the variation of organic-matter content and type with time are given from the Moroccan Tarfaya coastal basin (Figure 13) and from the Nigerian Benue trough (Figure 14).

In the Tarfaya basin the total organic carbon content exhibits an overall increase with time from the lower Cenomanian to the lower Turonian (Figure 7). Lower Cenomanian samples have average TOC values of about 3%, and the hydrogen indices are close to 400, except in a few richer levels (TOC about 5%) where HI reaches 500 mg HC/g TOC. A slight increase of TOC content is found in middle and upper Cenomanian samples, with an average TOC of 5%, and hydrogen indices range from 500 to 700 mg HC/g TOC. A significant shift in TOC values is observed at the top of the Cenomanian (upper 2 m of the *Rotalipora cushmani* zone). The deposits of the uppermost Cenomanian and basal Turonian show a sharp enrichment in organic matter, with an average TOC of 10% and maximum values up to 20%. This coincides with the positive excursion in $\delta^{13}C$. Above this zone of highly enriched organic matter, continuously decreasing TOC values are observed.

With the exception of the lower Cenomanian, where the organic material is mixed between type II and type III, sediments from middle Cenomanian to middle Turonian contain mainly type II organic matter of planktonic origin. The hydrogen indices are high and constant, fluctuating only within a small range between 600 and 700 mg HC/g TOC (Figure 13).

In the Benue trough, sediments of Cenomanian and middle Turonian to Coniacian age are characterized by low TOC content and low hydrogen indices (Figure 14). The highest content of organic matter occurs

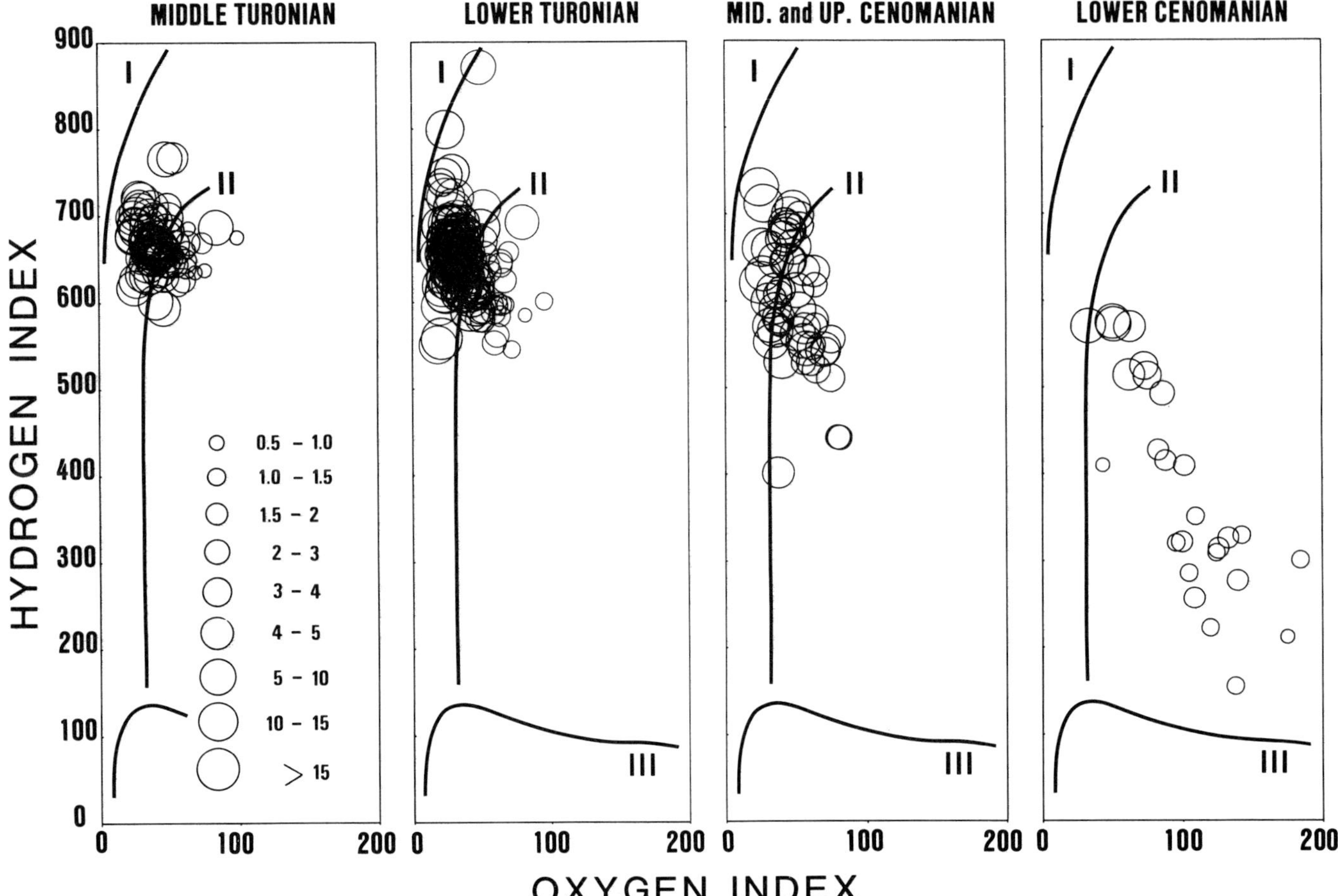

Figure 13. Evolution of organic matter with time in the S13 well of the Moroccan Tarfaya coastal basin. Kerogen types for four selected time periods.

in lower Turonian sediments and corresponds to the highest hydrogen indices, i.e., to the maximum of preserved autochthonous organic matter.

These trends of organic-matter enrichment and kerogen type during the Cenomanian/Turonian in the Tarfaya basin and the Benue trough also can be observed in the intermediate Senegal basin (Figure 8) or farther north in the Penibetic area (Figure 2). The most significant enrichment in organic matter occurs near the boundary between upper Cenomanian and lower Turonian and has been observed in all areas studied. It can be interpreted as the overprint of a global paleoceanographic event on the local depositional history of organic matter.

Sedimentation Rates and Organic-Matter Accumulation

Organic matter (OM) accumulation rates have been calculated for upper Cenomanian to lower Turonian (*W. archaeocretacea* zone, CTBE) sediments from various environments in the Nigerian Benue trough, Northwest African Atlantic coastal basins, western Mediterranean, and Iberian transects. For the determination of the organic-matter accumulation rates, 21 outcrop sections covering the *W. archaeocretacea* zone have been measured. The percentage of the thickness of each section having a TOC content above 0.5% is calculated (Table 1). Other sediment types such as terrigenous turbidites, calciturbidites, biogenic limestones, and radiolarian sands with TOC contents of 0 to 0.3% have not been taken into account for the calculation of OM accumulation rates. The organic-carbon accumulation rates were calculated by multiplying the average weight fraction (weight percent divided by 100) of this component by the total mass accumulation (see equation in Table 1). Sedimentation rates have been calculated for an interval covering the *W. archaeocretacea* foraminiferal zone (corresponding to 1.3 m.y. according to the time scale of Haq et al., 1986).

The organic-carbon content (weight percent) was determined from bulk samples using a Rock-Eval analyzer. For the calculation of OM accumulation rates, mean values of marine organic carbon were

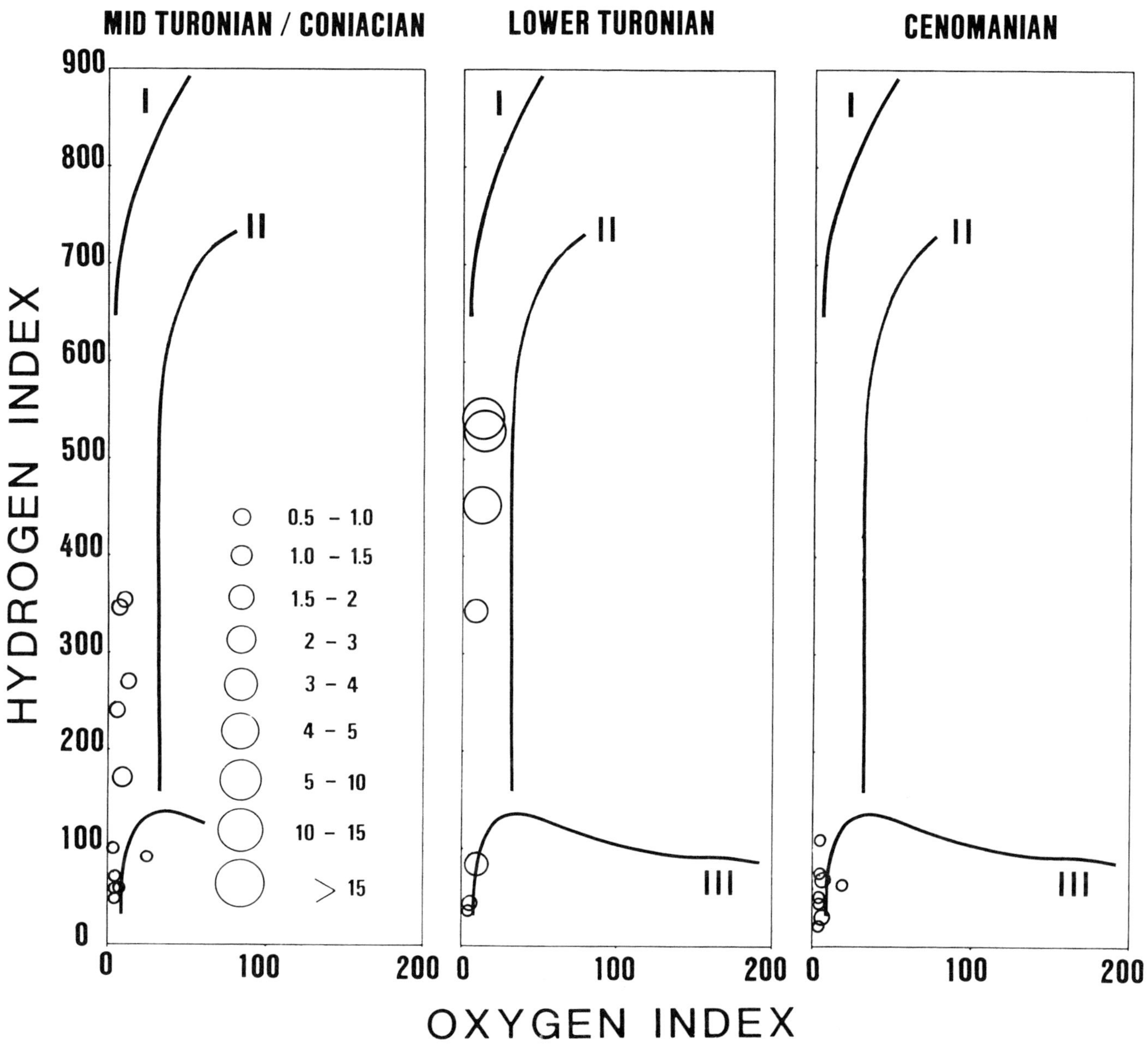

Figure 14. Kerogen types in the Nigerian Benue trough for three selected time periods.

used, i.e., TOC values were corrected for their content of terrigenous organic matter. Using a specific gravity of 1 g/cm^3 for organic matter and 2.7 g/cm^3 for nonorganic matter, the average grain density of organic-rich shales ranges from 2.2 to 2.6 g/cm^3 (compare Leine, 1986).

Calculated rates have been plotted on a paleogeographic map (Figure 15) to obtain geographic variations in intensity of OM accumulation. From this map it can be seen that highest accumulation rates of organic matter occur in low-latitude areas.

The organic-carbon/sedimentation-rate relationship (Figure 16) has been used to compare depositional environments of the middle Cretaceous Western Mediterranean and Atlantic margins with modern and Quaternary data (Stein, 1986). Under oxic deep-water conditions, TOC is positively correlated with sedimentation rate and negatively correlated with water depth (Figure 16A, field A). Modern high-productivity (e.g., upwelling) areas are characterized by high sedimentation rates and high TOC values (Figure 16A, dotted field A′). No correlation between sedimentation rates and TOC values is obvious in anoxic deep-water environments as, for example, the modern Black Sea (Figure 16A, field B; Degens and Ross, 1974; Stein, 1986). Figure

Table 1. Data base for calculation of OM-accumulation rates in sediments represented by the *W. archaeocretacea* zone for 21 sections studied.

AREA	SITE	PALEO LATITUDE	PALEO WATER DEPTH	THICKNESS OF ARCHAEO-CRETACEA ZONE (1.3 my)	PERCENTAGES OF LITHOLOGY TYPES organic rich shale	detrital sediment	radiolarian sands	biogenic limestones	SEDIMENTATION RATE* (cm/1000y)	NUMBER OF ROCK-EVAL ANALYSES	AVERAGE TOC OF OM-RICH STRATA***	**OM-ACCUMULATION RATE****
Benue	Lokpanta	5-10 S	300 m	4 m	100%	-	-	-	0,3	5	4%	**0,3**
Benue	Ngbanocha	5-10 S	200 m	8 (±3) m	25%	75%	-	-	0,6	6	2%	**0,2**
Senegal	DM1	5 N	50 m	30 (±10) m	averaged	-	-	-	2,3	2	1%	**0,5**
Senegal	CM1	5 N	150 m	150 m	averaged	-	-	-	11,5	8	4%	**10,6**
Senegal	CM2	5 N	200 m	180 m	averaged	-	-	-	13,9	9	4%	**12,8**
Senegal	CM10	5 N	300 m	150 m	averaged	-	-	-	11,5	12	7%	**18,5**
Senegal	Site 367	5 N	>3000 m	4 m	100%	-	-	-	0,3	>40	20%	**1,4**
Tarfaya	S13	15 N	300 m	80 (±10) m	98%	-	-	2%	6.2 (6.0)	153	8%	**11**
N-Africa	Ext. Prerif	20 N	300 m	5 m	90%	-	10%	-	0.4 (0.36)	4	20%	**1,7**
Tunisia	Bahloul S	20 N	200 m	30 m	60%	40%	-	-	2.3 (1.5)	Arthur et al.	3%	**1**
Tunisia	Bahloul N	20 N	200 m	40 (±10) m	90%	-	-	10%	3.1 (2.8)	5	4%	**2,6**
Alboran	Predorsal	25 N	2000 m	6 m	30%	70%	-	-	0.5 (0.2)	12	5%	**0,2**
Alboran	Mauretan	25 N	2500 m	3 m	20%	80%	-	-	0.2 (0.04)	55	5%	**0,1**
Alboran	Massylian	25 N	>3000 m	10 m	30%	-	70%	-	0.8(0.3)	26	5%	**0,3**
Iberia	S25	25 N	1000 m	1.3 m	50%	-	50%	-	0.1(0.05)	6	12%	**0,1**
Iberia	S17	25 N	1000 m	1.2 m	50%	-	50%	-	0.1(0.05)	16	12%	**0,1**
Iberia	Hole 641A	30 N	4000 m	0.4 m	100%	-	-	-	0,03	Thurow et al.	10%	**0,1**
Italy	Euganean	28 N	1000 m	3.5 m	20%	-	80%	-	0.3(0.06)	2	2%	**0**
Italy	Gubbio	26 N	1500 m	1 m	40%	-	60%	-	0.08 (0.03)	9	12%	**0,1**
Italy	Oriolo	23 N	3000 m	3 m	5%	90%	5%		0.23(0.01)	2	20%	**0,1**
Italy	Floresta	22 N	>3000 m	6 m	40%	-	60%	-	0.5(0.2)	10	12%	**0,6**

* value in brackets is sedimentation rate only for OM-rich sediments

**in g / square m x y

***only marine organic matter

For calculation of OM-accumulation rates the following formula has been used: OM-Acc = 0.1 TOC x SR x D

Average dry density D has been calculated with 2.3 g / ccm (grain density)

16B shows the organic-carbon/sedimentation-rate relationship for characteristic sections in the Western Mediterranean and adjacent Atlantic margins for the time span of the *W. archaeocretacea* foraminiferal zone. Terrigenous detrital (turbiditic) layers have been deducted from the calculated bulk thickness. The following types of environments can be clearly identified in this plot.

1. Equatorial zones of enhanced OM accumulation, mainly situated in outer-shelf environments (e.g., Tarfaya basin, Casamance margin) have medium to high TOC values and comparatively high sedimentation rates.
2. Slope environments with a remarkable amount of fine-grained detrital input have comparatively low TOC values, probably due to some dilution, but enhanced sedimentation rates (North African margin).
3. Continental margin basin and deep-sea environments are generally characterized by very high TOC values but extremely low sedimentation rates. Here a latitudinal factor is obvious in sedimentation rates with clearly distinguishable equatorial (DSDP Site 367) and north temperate (England) end members.
4. Neritic environments have extremely low sedimentation rates (condensed sequences and hiatuses are common) and low TOC values.

OM-accumulation values have been used to estimate paleoproductivity of surface waters (e.g., Müller and Suess, 1979 and Betzer et al., 1984 in oxic paleoenvironments and the study of Bralower and Thierstein, 1984 in anoxic pelagic environments). According to Müller and Suess (1979), in pelagic to hemipelagic sediments deposited under more or less oxic sea-water conditions, organic-carbon-accumulation rates have positive correlation with sedimentation rates and paleoproductivity. According to Betzer et al. (1984) only a very small part of the primary produced organic matter reaches the sea floor. This flux of organic carbon to the sea floor is a function of water depth and primary production of organic carbon at the surface.

Under anoxic conditions the preservation of organic matter is much higher than in oxic environments (Demaison and Moore, 1980). Bralower and Thierstein (1984) suggest from comparison of accumulation rates of marine organic carbon (CA) and primary production rates in recent anoxic environments that at least 2% of the primary organic carbon is preserved in the sediment. This means for calculating the paleoproductivity from the accumulation rates during the late Cenomanian and early Turonian, where anoxic water conditions prevailed, that paleoproductivity was about 50 times as high as the OM-accumulation rates.

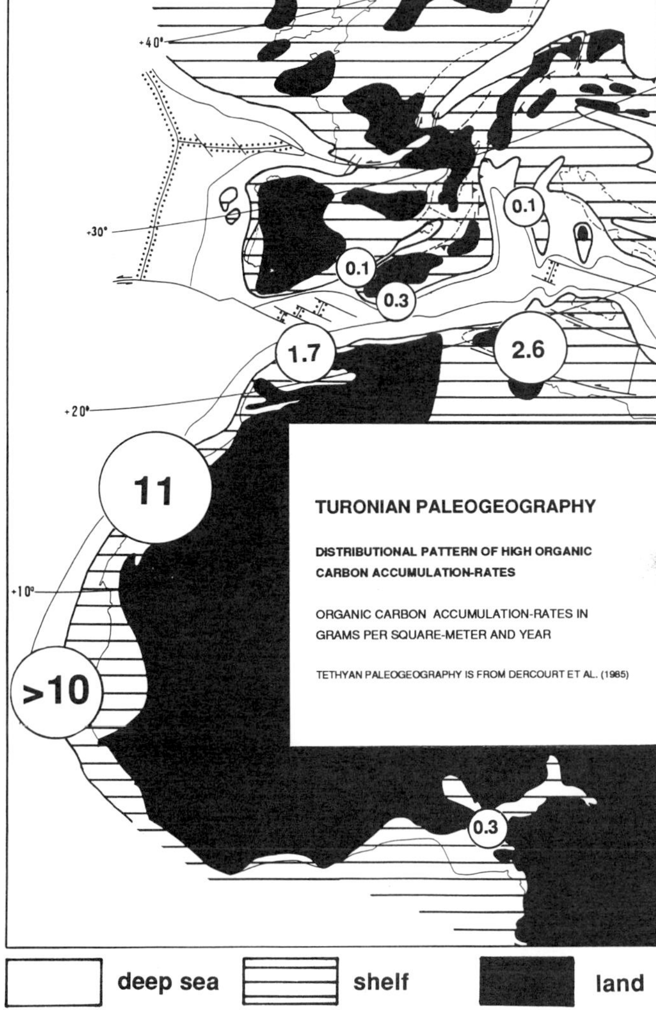

Figure 15. Paleogeographic sketch for the Western Mediterranean and adjacent Atlantic margins during Turonian time (modified and enlarged after Dercourt et al., 1985) and accumulation rates of organic matter.

LATITUDINAL DISTRIBUTION OF CENOMANIAN/TURONIAN ORGANIC FACIES

The organic pulse which characterizes the boundary between the upper Cenomanian and the lower Turonian can be described in terms of wavelength (period of deposition) and amplitude (quantity of the organic matter preserved) (Figure 17). On a south-north transect along the African and European Atlantic margins, maximum organic-matter preservation has been observed in the southern part of the North Atlantic (Cape Verde Basin, DSDP Site 367). Farther to the north, in the Tarfaya coastal basin, the amplitude (in % TOC) is only half that of the Cape Verde Basin, but the organic-rich facies existed over a long period of time; due to the extraordinary sedimentation rates of organic-rich chalks, the accumulation rates are even higher than in the Cape Verde Basin. In the southernmost part of the European margin (Penibetic Zone in southern Spain), the wavelength of the organic signal is strongly reduced, although the maximum TOC contents remain high. Finally, off the Galicia margin (ODP Site 641), the preservation of organic matter is restricted to a very short interval at the Cenomanian/Turonian boundary (short wavelength), and the maximum TOC content is about 10% (low amplitude). This general trend of decreasing quantity of preserved organic matter and shorter duration of the organic pulse from south to north is confirmed by data from the Celtic Basin (DSDP Site 551), where a sequence of green clay with a thin organic-rich layer (about 10% TOC) occurs within pure chalks (Herbin et al., 1986b).

South of the Cape Verde Basin, organic-rich sediments of Cenomanian/Turonian age have been observed in the Nigerian Benue trough. Here duration of the organic-rich pulse is again restricted to the lower Turonian *Mytiloides opalensis* zone, and the measured TOC values do not exceed 5%. Farther to the south, in the Gabon-Congo basins, a period of Cenomanian-Turonian time was highly favorable to the preservation of organic matter, but accumulation rates were comparatively low (Herbin et al., 1986a, 1987b).

The accumulation of organic matter in the western part of the North Atlantic exhibits a similar latitudinal trend. A maximum of organic-matter accumulation occurred at about the time of the Cenomanian/Turonian boundary in the low-latitude areas off Colombia and Venezuela within the so-called La Luna and Querecual formations, which are locally effective source rocks.

In conclusion, highest accumulation rates of organic matter at about the time of the Cenomanian/Turonian boundary have been observed in low-latitude areas, culminating between 5°N and 15°N paleolatitude in the Tarfaya Atlantic coastal basin and offshore Senegal (Figure 17). In addition, the duration of organic-matter accumulation is longest in low-latitude areas and decreases toward the north and the south (Figure 17). This trend can be observed in shelf basins as well as in the deep sea, and thus is independent of the bathymetric setting of the studied sections, but it may indicate a generally enhanced paleoproductivity in tropical waters during Cenomanian/Turonian time.

This latitudinal gradient in the distribution of Cenomanian/Turonian organic-rich facies differs strikingly from organic-matter accumulation and productivity patterns in the upper Lias (Baudin et al., this volume), in the Kimmeridgian, in the late Lower Cretaceous (Aptian/Albian), and in the Recent, where high-latitudinal regions are the preferable areas of marine OM production and accumulation (Herbin et al., 1989). Only an abnormal situation in

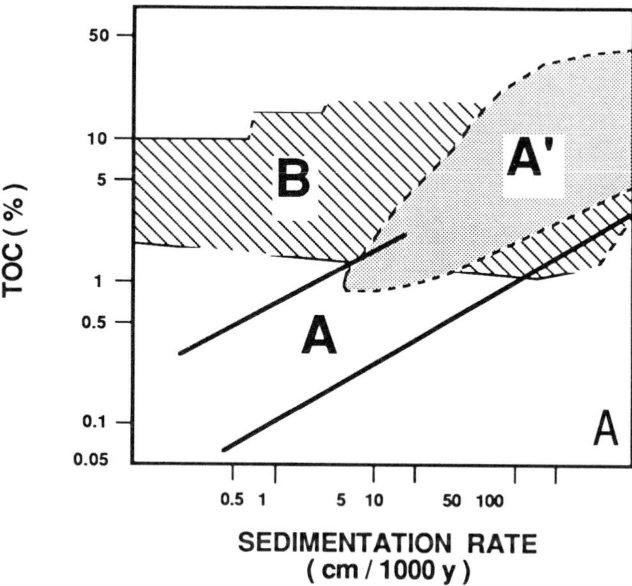

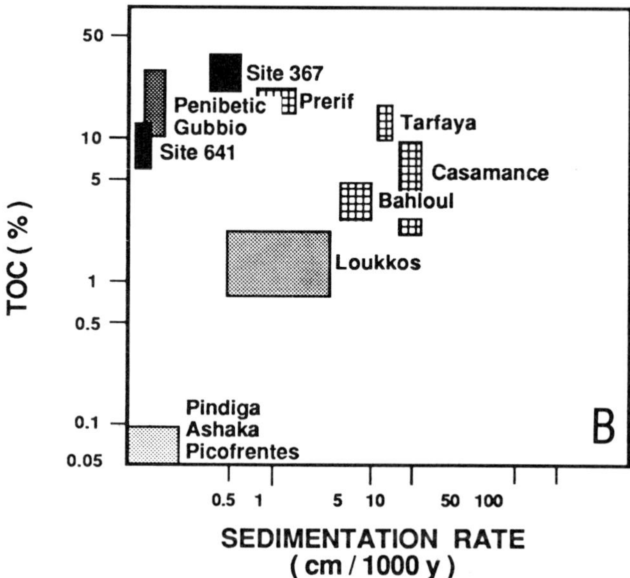

Figure 16. A. Plot of sedimentation rate against TOC values for typical modern environments (from Stein, 1986). The depositional environments, corresponding to the fields A, A′, and B, are defined as follows (Stein, 1986): (A) Oxic deep-water conditions. (A′) Conditions of high organic productivity (e.g., in modern upwelling areas), characterized by high TOC values and high sedimentation rates of pelagic material. (B) Anoxic deep-water conditions. B. Plot of sedimentation rate against TOC values for some of the middle Cretaceous sections studied.

global climate and oceanic circulation could have led to this strange productivity pattern. In this context, it is of interest to note that the Cenomanian/Turonian organic pulse reaches its maximum in wavelength and amplitude during the middle Cretaceous opening of deep marine gateways between the North and the South Atlantic.

VARIATIONS OF ORGANIC-MATTER CONTENT WITH BATHYMETRY AND DEPOSITIONAL ENVIRONMENT

Based on the analysis of more than 40 sections across the Cenomanian/Turonian boundary in the Western Mediterranean area and along the adjacent Atlantic margins, a generalized model of OM accumulation that depends on paleobathymetry and depositional environment has been developed (Figure 18).

The shallow marine tropical seas along the western Tethys shorelines, which were created by the late Cenomanian global transgression, generally lack organic-rich facies.

The laminated bituminous chalks of equatorial deeper shelf to uppermost slope environments exhibit the highest OM-accumulation rates of all studied environments. TOC values range from 5% to about 20%. The kerogen is of type II, with hydrogen index generally above 500, and up to 700. Typical examples for this environment are the distal parts of the Nigerian Benue trough, the Casamance basin offshore Senegal, the Tarfaya and Agadir coastal basins of the Moroccan Atlantic coast, and parts of the Bahloul Formation of the North African margin in Tunisia.

Interestingly enough the geographic distribution of this high-OM-accumulation/high-productivity facies was restricted to an equatorial belt from paleolatitudes of about 10°S (Benue trough of Nigeria) to 20°N (Agadir coastal basin and Tunisian Bahloul Formation). Similar outer-shelf environments farther to the south (e.g., Angola basin) and farther to the north (e.g., Vocontian trough, Northern Iberian margin along the Bay of Biscay, "black band" of the "Pläner" and chalk facies in northern Germany and England) exhibit only comparatively small amounts of organic-matter accumulation.

The bituminous laminated sediments of continental margin basins are characterized in each site by extremely high TOC values (commonly up to 20%; maximum values reach more than 30%). Kerogen is of type II with HI values of 600-700 and OI values below 50. These high organic-matter contents in these entirely biogenic sediments contrast with extremely low sedimentation rates of about 1 m/m.y. and resulting comparatively low OM-accumulation rates. The paleogeographic distribution of this environment appears to be restricted to the European, Iberian, and Apulian continental margins in the Western Mediterranean area, which corresponds to

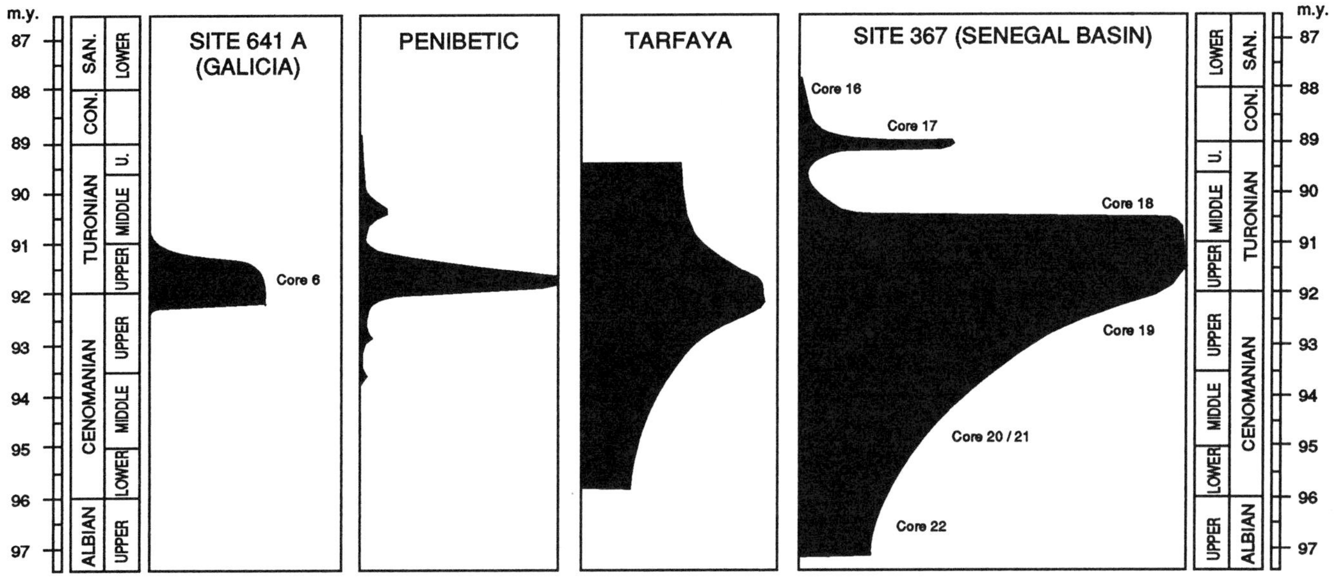

Figure 17. Plot of organic carbon content against time for four selected sections along the eastern margin of the North Atlantic from middle latitude to equatorial paleogeographic settings. Because of major coring gaps (the hole was only spot cored) the plot for Site 367 is discontinuous in the Albian to middle Cenomanian part (19-m coring gap between cores 22 and 21, and 30.5 m between cores 20 and 19) and the upper part (11.5-m coring gap between cores 18 and 17, and 66.5 m between cores 17 and 16) (Lancelot et al., 1977).

a comparatively narrow paleolatitudinal band at about 25°N.

The black siliceous organic-rich sediments of the abyssal realm are intercalated in the deep-sea claystones of the abyssal North Atlantic and the Western Mediterranean "argille varicolore" and turbidite sequences. The TOC content is generally high, and exhibits a tendency to increase toward the tropical sites. Average values are about 10% at the Galicia bank (ODP Hole 641A), up to 20% in the "phthanites" of the Gibraltar seaway, and reach more than 30% at DSDP Site 367. Although peak TOC's tend to be highest in the deep sea, the accumulation of organic matter was rather different. In the shallower environments the organic matter was diluted by carbonate. Most of the studied deep-water sediments were deposited below the CCD, and the diluting carbonate material was eliminated.

PALEOCEANOGRAPHIC IMPLICATIONS: PRODUCTIVITY VERSUS PRESERVATION MODELS

Generally two models have been proposed for depositional environments of organic-rich sediments (compare discussion in Degens et al., 1986).

1. The preservation/restricted-basin model requires restricted vertical circulation, intense dysoxia in deeper parts of the water column, and, consequently, reduction of oxygen, especially at deep-water sites of deposition.
2. The productivity/oxygen-minimum model requires input of organic matter by high surface productivity, leading to a midwater oxygen-minimum layer. Enhanced nutrient supply from upwelling waters is the main process in this model.

In modern oceans these two models appear to be almost exclusive. Both these models have been applied to Cenomanian/Turonian organic-rich depositional environments (e.g., Schlanger and Jenkyns, 1976; De Gracianski et al., 1982; Einsele and Wiedmann, 1982; De Boer, 1982; Thurow et al., 1982; Bralower and Thierstein, 1984; Busson, 1984; Herbin et al., 1986a,b; Kuhnt et al., 1986; Arthur et al., 1987; Summerhayes, 1987; Thurow et al., 1988). Our data base for Cenomanian/Turonian organic-rich deposition in the Western Mediterranean and adjacent Atlantic margins provides arguments for both models. The productivity/oxygen-minimum model is supported by the following observations.

1. The maximum OM-accumulation rates are observed in outer-shelf/upper-slope environments,

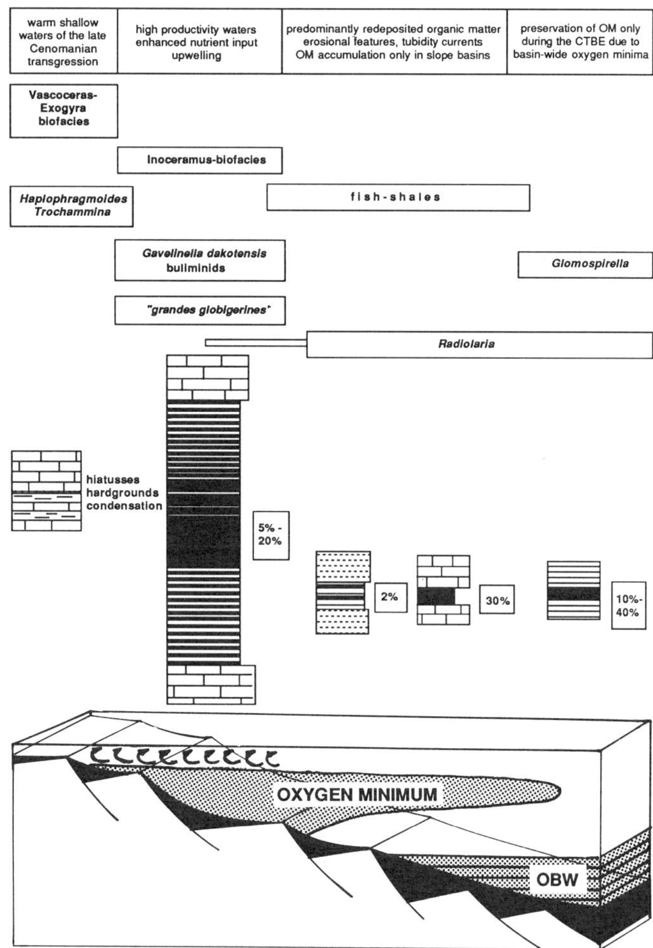

Figure 18. Model for distribution of organic-rich facies and biofacies types at the Cenomanian/Turonian boundary along a hypothetical low-latitude continental margin.

corresponding to modern high-productivity upwelling zones.

2. The fauna of the shelf deposits in Tarfaya and Nigeria include elements of northern temperate zones. Such faunas are interpreted as indicators for the onset of coastal upwelling of cold deep oceanic water along the Northwest African margin as early as middle Cenomanian (Einsele and Wiedmann, 1982) and for the development of an intensified and expanded oxygen-minimum layer (Kuhnt et al., 1986). These faunas include: north temperate ammonite faunas; dominance of inoceramids in contrast to a dominance of subtropical molluscan types (oysters) below the C/T boundary event; high radiolaria/planktonic-foraminifera ratios; planktonic foraminiferal faunas which show evidence for cool-water influence (low diversity, dominance of unkeeled morphotypes); dominance of certain spongy (high-latitude) taxa in radiolarian faunas; heterohelicid-dominated assemblages. Hedbergellid and heterohelicid taxa are commonly the only planktonic species in shallow-water sediments of low-latitude areas (e.g., Nigerian Benue trough, Petters, 1982; Western Interior seaway of the U.S., Eicher and Worstell, 1970). They also are the only planktonic forms in the high-latitude seas, thus indicating a cool-water habitat. Thus heterohelicids probably are indicators of cool, high-productivity (i.e., upwelling) water masses. An interesting observation to solve this apparent conflict in environmental interpretation of unkeeled planktonic assemblages in the Cretaceous is seen in the change from normal diverse planktonic faunas to purely unkeeled assemblages at the maximum of transgression in an open-marine basin in low latitudes, as in Tarfaya.

3. Accumulation rates of biogenic sediments consisting exclusively of calcareous plankton (nannoplankton, calcispheres, and planktonic foraminifers), phosphatic fish remains, and organic matter of marine origin were unusually high. Sediments of this composition are usually interpreted as indicative of high surface productivity.
4. There was a positive excursion of $\delta^{13}C$ values which is interpreted as a signal of high organic productivity (SUMMERHAYES, 1987).

In contrast to these observations are data which favor a preservation/restricted-basin model for Cenomanian/Turonian organic-rich sedimentation.

1. Highest TOC values of up to 40% are observed in the deepest parts of the North Atlantic ocean and Tethys seaway, clearly showing anoxity of bottom waters in the abyssal realm during the time represented by the Cenomanian/Turonian boundary.
2. Abyssal black shales from the Cenomanian/Turonian boundary are characterized by a very special trace-metal chemistry which is best explained by preservation of organic matter in a completely anoxic basin (Brumsack and Thurow, 1986; Thurow et al., 1988).
3. Primary productivity of marine organic matter, which can be deduced from the accumulation rates and the sedimentation rate-TOC relation (Stein, 1986), was generally lower than in modern upwelling zones (Figure 16). The only exception is seen in the Tarfaya basin, where analogies to modern upwelling areas are evident.

From these observations we conclude that during the Cenomanian/Turonian both mechanisms must have led to organic-carbon accumulation on the sea floor. This implies that paleoceanographic conditions were completely different from both the Early Cretaceous oceans with their sluggish circulation and warm saline bottom waters *and* from those we find in oceans of today. These unusual paleoceanographic conditions possibly characterized the transitional stage of the Western Tethys from an east-west orientated, strongly stratified ocean toward a "modern" north-south orientated ocean by opening of a deep-water connection between the South and North Atlantic. It can be speculated that during this transitional stage an old nutrient-rich water mass invaded the North Atlantic through the opened deep-water connection (compare the model of Summerhayes, 1987). As a result, local upwelling with enhanced surface productivity started in the most favorable zones close to the equatorial belt. This concentration of important upwelling zones in the equatorial belt can be deduced from the latitudinal distribution of high organic-matter-accumulation rates. An additional productivity spike in some parts of the ocean may have thus caused short-term severe oxygen depletion of the water mass and the enhanced preservation of organic matter in large parts of the ocean.

MATURATION OF THE ORGANIC MATTER AND PETROLEUM POTENTIAL

During pyrolysis of organic matter by Rock-Eval, the temperatures reached at maximum hydrocarbon release (T_{max}) for peak S_2 depend on both the stage of maturation and the nature of organic matter (Espitalié et al., 1977). If the nature of organic matter remains constant, as at the Cenomanian/Turonian boundary (type II), a maturation scale of the organic matter can be derived from the T_{max} values. This maturation scale derived from the Rock-Eval can be compared with vitrinite reflectance. A reflectance of approximately 0.5% which defines the boundary between immature and mature organic matter roughly corresponds to a T_{max} of 435°C for organic materials of type II. The maximum temperature of pyrolitic hydrocarbon release (T_{max}) in the studied area was about 400° in DSDP/ODP holes, about 405° in the upper part of the Tarfaya section, and more than 470° in strongly tectonized areas of the maghrebidian flysch belt. An overview of T_{max} values within the different basins studied is given in Table 2. T_{max} values below 435°C probably indicate that these sediments have not been buried deep enough or have not been heated sufficiently to reach the oil-generation window (Espitalié et al., 1984; Leythaeuser et al., 1980).

The oil yield of the sediments is generally above 5% at the Cenomanian/Turonian boundary (*W. archaeocretacea* zone). Some of these sediments can be regarded as economically valuable oil shales. Small mines were developed during the 1930s and 1940s in the Betic Cordillera and in the Tangier area to obtain crude oil from these oil shales (Hernandez-Pacheco, 1937).

Table 2. T_{max} values of Rock-Eval analysis for the different sections studied and comparison of the different basin types. Only samples with more than 2% (in some sections more than 5%) TOC content and kerogen type II have been taken into account. The abnormally high T_{max} values for the Loukkos sections in the Rif are probably due to mixed kerogen types and those for the Ngbanocha section in Nigeria are probably due to local volcanic heat flow.

TRANSECT	SITE	Tmax highest	Tmax lowest
SPAIN	S17	424	420
	S25	426	417
	641A	413	405
ITALY	Gubbio	419	416
	Oriolo west	432	432
	Oriolo east	466	464
	Floresta	428	423
ALBORAN	Chrafate	444	434
	Oued M'ter	465	460
	Cluse (M39)	461	451
	Malabata (M13)	426	422
	D96	438	430
RIF	Tselfat	431	426
	Mguedrouz	432	429
	Loukkos south	448	444
	Loukkos north	468	457
	Tanger Intern	449	444
	Beni Derkoul	470	464
TUNISIA	Bahloul	431	429
AGADIR	Coastal Section	417	414
TARFAYA	S10	422	406
SENEGAL	CM10	429	422
	367	403	396
NIGERIA	Ngbanocha	459	457
	Lokpanta	433	428
	Calabar	434	424

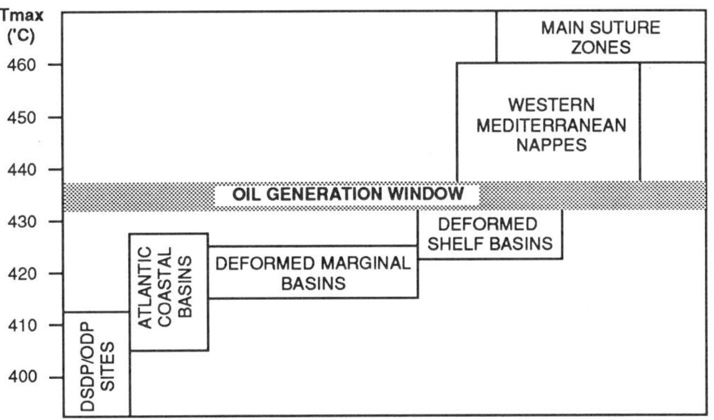

CONCLUSIONS

The temporal and spatial distribution pattern of Cenomanian-Turonian organic facies in the Western Mediterranean and along the adjacent Atlantic margin shows several characteristic features in time and space which can be a key to better understanding of the middle Cretaceous paleoceanography and which can be used to make predictions concerning the source-rock potential of Cenomanian/Turonian strata on a global scale. The following observations appear to be of special importance.

1. Organic-facies distribution in time:

- Pre-Cenomanian sediments are characterized by low organic-matter content with an important terrestrial component.
- During the Cenomanian, TOC increased and the marine component became dominant.
- Culmination of organic-rich deposition occurred at the Cenomanian/Turonian boundary (*W. archaeocretacea* zone).
- Since the early Turonian, organic-rich sediments progressively decreased and were replaced by more oxygenated sediments.
- Onset, culmination, and end of organic-rich sedimentation are correlated with paleobathymetry and paleolatitude.

2. Organic-facies distribution in space:

- Cenomanian/Turonian organic-rich sediments can be observed in all bathymetric settings from the shelf to the deep sea.
- Highest accumulation rates of organic matter occur in outer-shelf environments and in low-latitude areas.
- Accumulation of Cenomanian/Turonian organic-rich sediments began earlier and lasted longer in low latitudes.

From these observations it can be deduced that Cenomanian/Turonian organic facies have high source-rock potential in low-latitude shelf basins along the African continental margin from the North African margin in the north to the Nigerian Benue trough in the south.

ACKNOWLEDGMENTS

W. K. and J. W. thank the Deutsche Forschungsgemeinschaft (Grant Wi 112/21 "Kretazische Küstenbecken und Südatlantik") for financial support of the field work in Nigeria and Cameroon. The BRPM of Morocco supported core material from the Tarfaya exploration well S13. Cordial thanks are due to Drs. I. deKlasz, E. Nyong, Ph. Olivier, K. Ojoh, and M. Popoff for joint field work and additional sample material. The paper benefits from critical reviews of Dr. W. Dean and Dr. J. C. Goff.

REFERENCES CITED

Akpan, E. B., 1985, Ichnology of the Cenomanian-Turonian of the Calabar Flank, southeastern Nigeria: Geologie en Mijnbouw, v. 64, p. 365-372.

Allix, P., 1987, Environnements mesozoiques de la partie Nord-Orientale du fosse de la Benue (Nigeria). Benue Seminar, Bulletin des Centres de Recherches Exploration-Production Elf-Aquitaine, 11, 155-157.

Arambourg, C., 1954, Les Poissons Crétacés du Jebel Tselfat (Maroc): Notes du Service Géolgique du Maroc, v. 118, p. 1-188.

Arthur, M. A., W. E. Dean and L. M. Pratt, 1988, Geochemical and climatic effects of increased marine organic burial at the Cenomanian/Turonian boundary: Nature, v, 335, p. 714-717.

Arthur, M., and Premoli Silva, 1982, Development of widespread organic carbon-rich strata in the Mediterranean Tethys, *in* S. O. Schlanger and M. B. Cita, eds., Nature and origin of Cretaceous carbon-rich facies: London, Academic Press, p. 7-54.

Arthur, M. A., S. O. Schlanger, and H. C. Jenkyns, 1987, The Cenomanian-Turonian Oceanic Anoxic Event, II. Paleoceanographic controls on organic matter production and preservation, *in* J. Brooks, and A. Fleet, eds., Marine petroleum source rocks: Geological Society of London Special Publication 26, p. 401-420.

Betzer, P. R., et al., 1984, Primary productivity and particle fluxes on a transect of the equator at 153°W in the Pacific Ocean: Deep Sea Research, v. 131, p. 1-11.

Boillot, G., et al., 1987: Initial Reports (Part A) of the Ocean Drilling Program, v. 103, 663 p.

Bralower, T. J., 1988, Calcareous nannofossil biostratigraphy and assemblages of the Cenomanian-Turonian boundary interval: implications for the origin and timing of oceanic anoxia: Paleoceanography, v. 3, p. 275-316.

Bralower, T. J., and H. R. Thierstein, 1984, Low productivity and slow deep-water circulation in mid-Cretaceous oceans: Geology, v. 12, p. 614-618.

Brumsack, H.-J., and J. Thurow, 1986, The geochemical facies of black shales from the Cenomanian/Turonian Boundary Event, *in* E. T. Degens, et al., eds., Biogeochemistry of black shales: Mitteilungen aus dem Geologisch-Paläontologischen Institut der Universität Hamburg, v. 60, p. 247-265.

Burollet, F., 1956, Contribution à l'étude stratigraphique de la Tunisie centrale: Annales des Mines et de la Geologie, Tunisia, v. 18, p. 1-350.

Burollet, F. and N. Ellouz, 1986, L'évolution des bassins sédimentaires de la Tunisie centrale et orientale: Bulletin des Centres de Recherches Exploration-Production Elf-Aquitaine, v. 10, p. 49-68.

Buser, H., 1966, Paleostructures of Nigeria and adjacent countries: Geotektonische Forschungen, v. 24, p. 1-90.

Busson, G., 1984, Relations entre la sédimentation de Crétacé moyen et supérieur de la plate-forme du nord-ouest africain et les dépôts contemporain de l'Atlantique centre et nord: Eclogae Geologicae Helvetiae, v. 77, p. 221-235.

Caron, M., 1985, Cretaceous planktic foraminifera, *in* H. M. Bolli, et al., eds., Plankton stratigraphy, New York, Cambridge University Press, p. 17-86.

Chénet, P. Y., and J. Francheteau, 1979, Bathymetric reconstruction method: application to the Central Atlantic Basin between 10°N and 40°N, *in* T. Donelly, et al., compilers: Initial Reports of the Deep Sea Drilling Project: v. 51, 52, 53, p. 1501-1514.

Comas, M. C., M. Ortega-Huertas, A. Lopez-Galindo, and P. Fenoll, 1982, Pelitas turbiditicas y pelitas hemipelagicas en la Formacion Fardes (Albiense-Cretacico superior, Cordilleras Beticas): Cuadernos Geologicos Ibericos, v. 8, p. 483-498.

Daguin, F., 1927, Contribution a l'étude géologique de la région prérifaine (Maroc occidental): Notes du Service Geologique du Maroc, v. 1, p. 1-406.

De Boer, P. L., 1982, Aspects of Middle Cretaceous pelagic sedimentation in southern Europe: Geologica Ultraiectina, v. 31, p. 1-112.

Degens, E. T., and D. A. Ross, 1974, The Black Sea, geology, chemistry, and biology: AAPG Memoir 20, 633 p.

Degens, E. T., P. A. Meyers, and S. C. Brassell, eds., 1986, Biogeochemistry of black shales: Mitteilungen aus dem Geologisch-Paläontologischen Institut der Universität Hamburg, v. 60, (SCOPE/UNEP Sonderband), p. I-XXIV, 1-421.

De Graciansky, P. C., M. Brosse, G. Deroo, J. P. Herbin, L. Montadert, C. Müller, J. Sigal, and A. Schaaf, 1982, Les formations d'age Crétacé de l'Atlantique nord et leur matière organique. Paléogéographie et milieux de depot: Revue de l'Institut Francais du Pétrole, v. 39, p. 275-336.

De Graciansky, P. C., G. Deroo, J. -P. Herbin, T. Jacquin, F. Magni, L. Montadert, L., and C. Müller, 1986, Ocean-wide stagnation episodes in the Late Cretaceous: Geologische Rundschau, v. 75, p. 17-41.

Demaison, G. J., and G. T. Moore, 1980, Anoxic environments and oil source bed genesis: AAPG Bulletin, v. 64, p. 1179-1209.

Dercourt, J., et al., 1985, Présentation de 9 cartes paléogéographiques au 1/20.000.000e s'étendant de l'Atlantique au Pamir pour la période du Lias à l'Actuel: Bulletin de la Societe Géologique de France, v. 8, p. 637-652.

Dessauvagie, T. J. F., 1972, Biostratigraphy of the Odukpani (Cretaceous) type section, Nigeria, *in* T. J. F. Dessauvagie and A. J. Whitman, Proceedings on the Conference on African Geology, University of Ibadan, 1970, p. 207-218.

Eicher, D. L., and P. Worstell, 1970, Cenomanian and Turonian foraminifera from the Great Plains, United States: Micropaleontology, v. 16, p. 269-324.

Eicher, D. L., and R. Diner, 1985, Foraminifera as indicators of water mass in the Cretaceous Greenhorn Sea, Western Interior, *in* L. M. Pratt, et al., eds., Fine-grained deposits and biofacies of the Cretaceous Western Interior seaway: Evidence of cyclic sedimentary processes: SEPM Second Annual Midyear Meeting, Golden, Colorado, Field Trip No. 9, Society of Economic Paleontologists and Mineralogists, p. 60-71.

Einsele, G., and J. Wiedmann, 1982, Turonian black shales in the Moroccan coastal basins: First upwelling in the Atlantic Ocean, *in* U. von Rad, et al., eds., Geology of the Northwest African continental margin: Berlin, Springer, p. 396-414.

Emberger, J., 1960, Esqisse géologique de la partie orientale des monts de Ouled Nails: Bulletin du Service de la Carte Géologique de l'Algérie, Serie 27, p. 1-398.

Espitalié, J., J. L. Laporte, M. Madec, F. Marquis, P. Leplat, J. Palt, and A. Boutefeu, 1977, Méthode rapide de characterisation des roches méres, de ler potentiel pétrolier et de ler degré d'évolution: Revue de l'Institut Francais Petrole, v. 32, p. 23-42.

Espitalié, J., F. Marquis, and I. Barsony, 1984, Geochemical logging, *in* K. J. Voorhees, ed., Analytical pyrolysis, techniques and applications: Butterworths, p. 276-304.

Fenet, B., 1975, Recherches sur l'alpinisation de la bordure septentrionale du bouclier africain à partir de l'étude d'un élément de l'orogénèse nord-maghrébine: les Monts du Djebel Tessala et les Massifs du littoral oranais: Thèse Science, Universite de Nice, 301 p.

Fischer, A. G., and M. A. Arthur, 1977, Secular variations in the pelagic realm, *in* H. E. Cook and P. Enos, eds., Deep-water carbonate environments: SEPM Special Publication 25, 19-50.

Guillemot, J., 1952, La bordure sud-tellienne dans le Titteri: XIXe Congress Géologique International Alger. Monograph Rég. 1re sér., Algérie, no. 9.

Haq, B. U., J. Hardenbohl, and P. R, Vail, 1987, Chronology of fluctuating sea levels since the Triassic (250 million years ago) to present: Science, v. 235, p. 1158-1167.

Hayes, D. E., et al., 1972: Initial Reports of the Deep Sea Drilling Project, v. 14, 975 p.

Herbin, J. P., F. Magniez-Jannin, and C. Mueller, 1986a, Mesozoic organic rich sediments in the South Atlantic: Distribution in time and space, *in* E. T. Degens et al., eds., Biogeochemistry of black shales: Mitteilungen aus dem Geologisch-Paläontologischen Institut der Universität Hamburg, v. 60, p. 71-97.

Herbin, J. P., L. Montadert, C. Mueller, R. Gomez, J. Thurow, and J. Wiedmann, 1986b, Organic-rich sedimentation at the Cenomanian-Turonian boundary in oceanic and coastal basins in the North Atlantic and Tethys, *in* C. P. Summerhayes and N. J. Shackleton, eds., North Atlantic Palaeoceanography: Geological Society of London Special Publication 22, p. 389-422.

Herbin, J. P., E. Masure, and J. Roucaché, 1987a, Cretaceous formations from the lower continental rise off Cape Hatteras: organic geochemistry, dinoflagellate cysts, and the Cenomanian/Turonian boundary event at Sites 603 (Leg 93) and 105 (Leg 11), *in* J. Van Hinte et al., compilers: Initial Reports of the Deep Sea Drilling Project, v. 93, p. 1139-1162.

Herbin, J. P., C. Mueller, P. C. DeGraciansky, T. Jacquin, F. Magniez-Jannin, and P. Unternehr, 1987b, Cretaceous anoxic events in the South Atlantic: Revista Brasileira de Geociencias, v. 17, p. 92-99.

Herbin, J. P., A. Y. Huc, F. Baudin, J. Dercourt, J. R. Geyssant, W. Kuhnt, and J. Thurow, 1989, Toarcian, Kimmeridgian and Cenomano-Turonian organic-rich shales: Similarities and differences (abs.): 28th International Geological Congress, Washington, D.C., v. 2, p. 52.

Hernandez-Pacheco, F., 1937, Los materiales bituminosos de la Serrania de Ronda (Málaga): Boletin de la Sociedad Espanola do Historia Natural, v. 36, p. 245-275.

Jenkyns, H. C., 1980, Cretaceous Anoxic Events: From Continents to Oceans: Journal of the Geology Society of London, v. 137, p. 171-188.

Kieken, M., 1962, Les traits essentiels de la Géologie Algérienne, *in* Livre à la Mémoire du Professeur Paul Fallot: Société Géologique de France, v. 1, p. 545-614.

Kuhnt, W., 1987, Biostratigraphie und Paläoenvironment der externen Kreideserien des westlichen Rif und Betikum - ein Ansatz zur Rekonstruktion der Kreide-Paläogeographie des Gibraltarbogens: Dissertation, Universität Tübingen, 271 p.

Kuhnt, W., 1988, Upper Cretaceous foraminiferal assemblages of the external units of the Rif (Northern Morocco) — a paleobathymetric model of the late Mesozoic North African continental margin: Actes du Xéme Colloque Africain de Micropaléontologie. Revue de Géologie Méditerranéenne (Marseille), v. 14, p. 109-131.

Kuhnt, W., J. Thurow, J. Wiedmann, and J. P. Herbin, 1986, Oceanic anoxic conditions around the Cenomanian/Turonian boundary and the response of the biota, *in* E. T. Degens et al., eds., Biogeochemistry of black shales: Mitteilungen aus dem Geologisch-Paläontologischen Institut der Universität Hamburg, v. 60, p. 205-246.

Kuhnt, W., and M. Kaminski, 1989, Upper Cretaceous deep-water agglutinated benthic foraminiferal assemblages from the western Mediterranean and adjacent areas, *in* J. Wiedmann, ed., Cretaceous of the Western Tethys: Proceedings of the 3rd International Cretaceous Symposium, Tübingen, p. 91-120.

Kuhnt, W., M. A Kaminski and M. Moullade, 1989, Late Cretaceous deep-water agglutinated foraminiferal assemblages from the North Atlantic and its marginal seas. Geologische Rundschau, 78/3, 1121-1140.

Lancelot, Y., et al., 1977: Initial Reports of the Deep Sea Drilling Project, v. 41, p. 1-1259.

Leine, L., 1986, Geology of the Tarfaya oil shale deposit, Morocco: Geologie en Mijnbouw, v. 65, p. 57-74.

Lespinasse, P., 1975, Géologie des zones externes et des flyschs entre Chaouen et Zoumi (centre da la Chaine Rifaine, Maroc): Thèse, Paris, 248 p.

Leythaeuser, D., H. W. Hagemann, A. Hollerbach, and R. G. Schaeffer, 1980, Hydrocarbon generation in source beds as a function of type and maturation of their organic matter: a mass balance approach: Proceedings 10th World Petroleum Congress, v. 2, p. 31-41.

Mattauer, M., 1958, Etude Géologique de l'Ouarsensis oriental (Algérie): Bulletin du Service de la Carte Géologique de l'Algérie (Nouvelle Série), v. 17, p. 1-534.

Meister, C., 1989, Les ammonites du crétacé supérieur d'Ashaka (Nigéria): Bulletin des Centres de Recherches Exploration-Production Elf-Aquitaine, v. 13, Supplement, p. 1-84.

Moullade, M., W. Kuhnt, and J. Thurow, 1988, Agglutinated benthic foraminifers from Upper Cretaceous variegated clays of the North Atlantic Ocean (DSDP Leg 93 and ODP Leg 103), *in* G. Boillot, et al., compilers: Proceedings of the Ocean Drilling Program, Scientific Results, v. 103, p. 349-377.

Müller, C., A. Schaaf, and J. Sigal, 1984, Biochronostratigraphie des formations d'âge Crétacé dans les forages du DSDP dans l'Océan Atlantique Nord: Revue de l'Institut Francais du Petrole, v. 39, p. 3-23.

Müller, P. J., and E. Suess, 1979, Productivity, sedimentation rate, and sedimentary organic matter in the oceans, I., Organic carbon preservation: Deep Sea Research, v. 26, Part A, p. 1347-1362.

Nyong, E., and R. Ramanathan, 1985, A record of oxygen-deficient paleoenvironments in the Cretaceous of the Calabar Flank, southeastern Nigeria: Journal of African Earth Sciences, v. 3, p. 455-460.

Ojoh, K.A., 1988, Evolution geodynamique des bassins albo-santoniens du sud-ouest du fosse de la Benoue (Nigeria): Thèse Aix-Marseille III, 241 p.

Pervinquière, L., 1903, Etude Géologique de la Tunisie Centrale: Direction Géneral Trav. Publ. Carte Géologique Tunisie.

Pervinquière, L., 1907, Etudes de Paléontologie Tunisienne. I. Céphalopodes des terrains secondaires. Géologique de la

Tunisie Centrale. Direction Général Trav. Publ. Carte Géologique Tunisie, 438 p.
Pessagno, E. A., Jr., 1976, Radiolarian zonation and stratigraphy of the Upper Cretaceous portion of the Great Valley Sequence, California Coast Ranges: Micropaleontology, Special Publication 2, 95 p.
Petters, S. W., 1978, Mid-Cretaceous paleoenvironments and biostratigraphy of the Benue Trough, Nigeria: GSA Bulletin, v. 89, p. 151-154.
Petters, S. W., 1982. Central West African Cretaceous-Tertiary benthic foraminifera and stratigraphy: Palaeontographica, Abt. A, v. 179, p. 1-107.
Petters, S. W., and C. M. Ekweozor, 1982, Petroleum geology of Benue trough and southeastern Chad basin, Nigeria: AAPG Bulletin, v. 66, p. 1141-1149.
Popoff, M., J. Wiedmann, and I. DeKlasz, 1986, The Upper Cretaceous Gongila and Pindiga Formations, northern Nigeria: Subdivisions, age, stratigraphic correlations and paleogeographic implications: Eclogae Geologicae Helvetiae, v. 79, p. 343-363.
Pratt, L. M., 1983, Isotopic studies of organic matter and carbonate in mid-Cretaceous strata near Pueblo, Colorado, *in* E. G. Kauffman, ed., Depositional environments and paleoclimates of the Greenhorn tectono-eustatic cycle, Rock Canyon Anticline, Pueblo, Colorado: Penrose Conference on Cretaceous Paleoclimates, Florissant, Colorado, Field Excursion Guidebook, p. 77-98.
Robaszynski, F., and M. Caron, 1979a, Atlas de Foraminiferes planctoniques du Cretace moyen. Part 1: Cahiers de Micropaleontologie, v. 1, p. 1-185.
Robaszynski, F., and M. Caron, 1979b, Atlas de Foraminiferes planctoniques du Cretace moyen. Part 2: Cahiers de Micropaleontologie, v. 2, p. 1-181.
Salaj, J., 1978, The geology of the Pelagian block: The eastern Tunisian Platform, *in* A. E. M. Nairn et. al., eds., The oceans basins and margins. Volume 4B, The Western Mediterranean: New York, Plenum Press, p. 361-416.
Schlanger, S. O., and H. C. Jenkyns, 1976, Cretaceous anoxic events: causes and consequences: Geologie en Mijnbouw, v. 55, p. 179-184.
Schlanger, S. O., M. A. Arthur, H. C. Jenkyns, and P. A. Scholle, 1987, The Cenomanian-Turonian Anoxic Event, I: Stratigraphy and distribution of organic carbon-rich beds and the marine δ13C excursion, *in* J. Brooks, and A. Fleet, eds., Marine petroleum source rocks: Geological Society of London Special Publication 26, p. 371-399.
Scholle, P. A., and M. A. Arthur, 1980, Carbon-isotope fluctuations in Cretaceous pelagic limestones: potential stratigraphic and petroleum exploration tool: AAPG Bulletin, v. 64, p. 67-87.
Sibuet, J.-C., et al., 1979: Initial Reports of the Deep Sea Drilling Project, v. 47, 787 p.
Sorbini, L., 1974, Un nuovo giacimento di pesci fossili cretacei nel territoria di Cinto Euganeo: Bollettino del Museo Civico di Storia Naturale di Verona, v. 1, p. 538-540.
Stein, R., 1986, Surface-water paleo-productivity as inferred from sediments deposited in oxic and anoxic deep-water environments of the Mesozoic Atlantic Ocean, *in* E. T. Degens et al., eds., Biogeochemistry of black shales: Mitteilungen aus dem Geologisch-Paläontologischen Institut der Universität Hamburg, v. 60, p. 55-70.
Summerhayes, C.P., 1987, Organic-rich Cretaceous sediments from the North Atlantic, *in* J. Brooks, and A. Fleet, eds., Marine petroleum source rocks: Geological Society of London Special Publication 26, p. 301-316.
Thein, J., 1988, Turonian paleogeography of the High Atlas Mountains (Morocco) and the North Atlantic: Zeitschrift der Deutschen geologischen Gesellschaft, v. 139, p. 261-287.
Thurow, J., 1987, Die kretazischen Turbiditserien im Gibraltarbogen: Bindeglied zwischen atlantischer und tethyaler Entwicklung: Dissertation, Universität Tübingen, 495 p.
Thurow, J., 1988, Lower and Middle Cretaceous radiolarians of the North Atlantic (ODP Leg 103 and DSDP Sites 398 and 603), *in* G. Boillot, et al., compilers: Proceedings of the Ocean Drilling Program, Scientific Results, v. 103, p. 379-418.
Thurow, J., W. Kuhnt, and J. Wiedmann, 1982, Zeitlicher und paläogeographischer Rahmen der Phtanit- und Black Shale-Sedimentation in Marokko: Neues Jahrbuch fuer Geologie und Paläeontologie, Abhandlungen, v. 165, p. 147-176.
Thurow, J., and W. Kuhnt, 1986, Mid-Cretaceous of the Gibraltar Arch area, *in* C. P. Summerhayes, and N. J. Schackleton, eds., North Atlantic Palaeoceanography: Geological Society of London Special Publication 21, p. 423-445.
Thurow, J., M. Moullade, H. J. Brumsack, E. Masure, J. Taugourdou, and K. Dunham, 1988, The Cenomanian-Turonian Boundary Event (CTBE) at Leg 103/Hole 641A, *in* G. Boillot, et al., compilers: Proceedings of the Ocean Drilling Program, Scientific Results, v. 103, p. 587-634.
Wiedmann, J., 1975, Subdivisiones y precisiones bio-estratigráficas en el Cretácico superior de las Cadenas Celtibéricas: Actas Ier Symposium Cretácico Cordillera Ibérica, Cuenca, 1974, p. 135-153.
Wiedmann, J., and E. G. Kauffman, 1978, Mid-Cretaceous biostratigraphy of northern Spain: Annales de la Museum des Historie Natural, Nice, v. 4 (1976), p. iii1-iii22.
Wiedmann, J., A. Butt, and G. Einsele, 1978, Vergleich von marokkanischen Kreide-Küstenaufschlüssen und Tiefseebohrungen (DSDP): Stratigraphie, Paläoenvironment und Subsidenz an einem passiven Kontinentalrand: Geologische Rundschau, v. 67, p. 454-508.
Wiedmann, J., A. Butt, and G. Einsele, 1982, Cretaceous stratigraphy, environment, and subsidence history at the Moroccan continental margin, *in* U. von Rad, et al., eds., Geology of the Northwest African continental margin: Berlin (Springer), p. 366-395.
Wildi, W., 1983, La chaine tello-rifaine (Algérie, Maroc, Tunisie): structure, stratigraphique et évolution du Trias au Miocène: Revue de Géologie dynamique et de Géographie physique, v. 24, p. 201-297.
Wonders, A. A. H., 1980, Middle and Late Cretaceous planktonic foraminifera of the Western Mediterranean area: Utrecht Micropaleontology Bulletin, v. 24, p. 1-158.

Trends in Organic Geochemistry and Petroleum Exploration in Italy

B. Tissot
Institut Francais du Pétrole
Rueil-Malmaison, France

L. Mattavelli
AGIP
Milano, Italy

E. Brosse
Institut Francais du Pétrole
Rueil-Malmaison, France

Exploration efficiency has increased since 1970 due to improvements in geophysical methods and to the development of organic geochemistry. A further increase in the success ratio of wildcat wells is expected during the 1990s by integrating all of the earth sciences in basin modeling.

Explorationists expect geochemistry to provide a clear answer to several essential questions for prospect evaluation, i.e., the existence and location of areas where hydrocarbons have been generated by the source rocks, the amount and type of hydrocarbons formed, the migration paths, and the quantity and type of petroleum accumulated in traps. With kinetic models of oil generation, such an answer can now be given to all questions related to the formation of hydrocarbons. Regarding migration, both approaches (observations and basin modeling) offer promising trends for individual trap evaluation. However, further work is needed before a quantitative figure of in-place reserves can be given.

In Italy, exploration encounters very different and complex geological situations, all governed by two regional major events: the Tethyan rifting during Mesozoic times, and the Alpine folding during the end of Tertiary times. In a well-known tectonic context, success in exploration may be greatly enhanced by taking into account modern concepts of organic geochemistry and basin modeling. Examples are given of bacterial and thermal gas prospects in the Po Basin (northern Italy), mainly from late Lower Tertiary and early lower Mesozoic sources, respectively. Occurrences of liquid oil at great depth in the same basin are also described, and the influences of kerogen type, burial history, and geothermal gradient are discussed. Immature heavy oils are present in the Adriatic Basin and in Sicily. Their conditions of formation are interpreted with respect to type of kerogen and to burial history. More generally, the important role of late subsidence associated with the Europe-Africa collision is emphasized.

INTRODUCTION

From the outset of the exploration for hydrocarbon resources, the discovery of new fields has become, in progressive stages, a technically more and more difficult task. Two main avenues are being pursued by the petroleum industry to take up this challenge: (1) improving the recovery ratio by suitable production methods, and (2) improving the efficiency of exploration by considering not only the location and the size of the traps but also the entire natural process that leads to the formation and accumulation of hydrocarbons in a given sedimentary basin.

Organic geochemistry is a young science born from the petroleum industry. In the last three decades it has established the concepts and the tools that help answer the following basic questions: where does oil or gas come from? What is the mechanism of its formation? How does it migrate in the subsurface? What amount of hydrocarbons is a given area likely to have generated?

The use of organic geochemistry by oil companies is gaining ground every day and is acknowledged as a highly profitable step in an exploration strategy (Sluijk and Parker, 1986). As a matter of fact, advances in seismic prospecting which serve to delimit traps with better accuracy, together with geochemistry that serves to estimate the generating capacity of source rocks and the behavior of the reservoirs, increased the success ratio of wildcats from one out of ten to one out of seven, between 1970 and 1986. By systematic basin modeling, i.e., a full integration of geophysical, geological, and geochemical data, it is hoped that a further degree will be gained in the 1990s, with a success ratio as high as one out of four.

The usefulness of organic geochemistry in oil and gas assessment stems from the logical position it holds in exploration thinking. One aspect is to locate a trap with a good seal in a favorable drainage area. It then becomes possible to calculate its maximum capacity to hold oil or gas. But this approach never predicts or explains why oil or gas is present. In contrast, for a given trap that has been geologically described and geometrically delineated, organic geochemistry attempts to answer the questions related to actual filling with oil or gas: where are the source rocks? What quantities of hydrocarbons have been formed? When and how did they migrate to the pools? By considering the generating capabilities of the source rocks, the nature and the quantities of the hydrocarbons produced (gas, condensate, light or heavy oil), and the efficiency of migration pathways, organic geochemistry is predictive and greatly enhances the chances of discovery.

FUNDAMENTALS OF PETROLEUM FORMATION

Organic geochemistry is now a coherent framework of well-established deterministic concepts dealing with the formation and occurrence of petroleum (Tissot and Welte, 1984; Tissot et al., 1987). Petroleum is considered to originate from kerogen, the organic matter contained in the sediments. Kerogen is not indefinitely stable. After burial, increasing temperature causes thermal degradation of kerogen and cracking into hydrocarbons. Then the liquid (or gaseous) hydrocarbons escape from the source rock. Since they are lighter than water, they migrate in a separate phase toward reservoirs and traps, along relatively permeable paths, and eventually up to the surface. The basin-modeling approach makes a quantitative estimate of these phenomena, using all available geophysical, geological, and geochemical data. It integrates principles and parameters in numerical software, which tends to become one of the most promising tools to reduce risk in petroleum exploration.

One of the main aims of organic geochemistry is to describe and predict the composition of hydrocarbons formed from kerogen as a function of burial depth (Figure 1). In a zone of incipient diagenesis, the sediment is the center of active microbial reworking. Bacteria are able to transform deposited organic matter to some extent, producing mainly methane gas. If favorable conditions are present, pools containing bacterial gas may be formed as in the Po Basin in northern Italy. The organic residue not destroyed by microbes recombines progressively by polycondensation, polymerization, and loss of functional groups, finally forming kerogen, a macromolecule insoluble in organic solvents. At greater burial depths, the temperature becomes high enough (~60°C) to prevent bacterial life, and the kerogen starts undergoing the cracking process. The first products are carbon dioxide and water, which mark the end of diagenesis. At this stage, the chemical structure and composition of kerogen present specific features depending mainly upon the nature of the biogenic material from which it was derived. Type I kerogen, known from sediments where algal precursors are concentrated, or which have been intensely reworked by microorganisms, is very aliphatic and relatively rich in ether functional groups. Type II kerogen, coming from marine planktonic material, may be characterized by more polyaromatic nuclei than in type I, and by heteroatomic ketone, carboxylic acid, and ester functional groups. Special attention must be paid to sulfur, which may be abundant in both types I and II. The chemical position of sulfur inside the macromolecule, in sulfide bonds or in thiophenic cycles, influences to some extent the cracking behavior (Orr, 1985). But the incorporation of sulfur into organic matter, as a function of depositional or diagenetic environments, is not yet completely understood. Lastly, type III kerogen, formed from continental plant material, is much richer in polyaromatic nuclei and contains numerous oxygenated functional groups: phenols, quinones, and aromatic acids (Tissot and Welte, 1984; Behar and Vandenbroucke, 1987).

Diagenesis is followed by catagenesis, which governs the oil-producing zone. At the beginning, the hydrocarbons formed have a high molecular weight. They combine with the geochemical fossils, preserved without any major chemical change from living precursors, to form heavy "immature" oils. Such oils are known, for instance, in Sicily. As temperature increases, the proportion of light components grows, both from the primary cracking of kerogen and as byproducts from the secondary cracking of oil. This is exemplified by the light oils, condensates, and gases

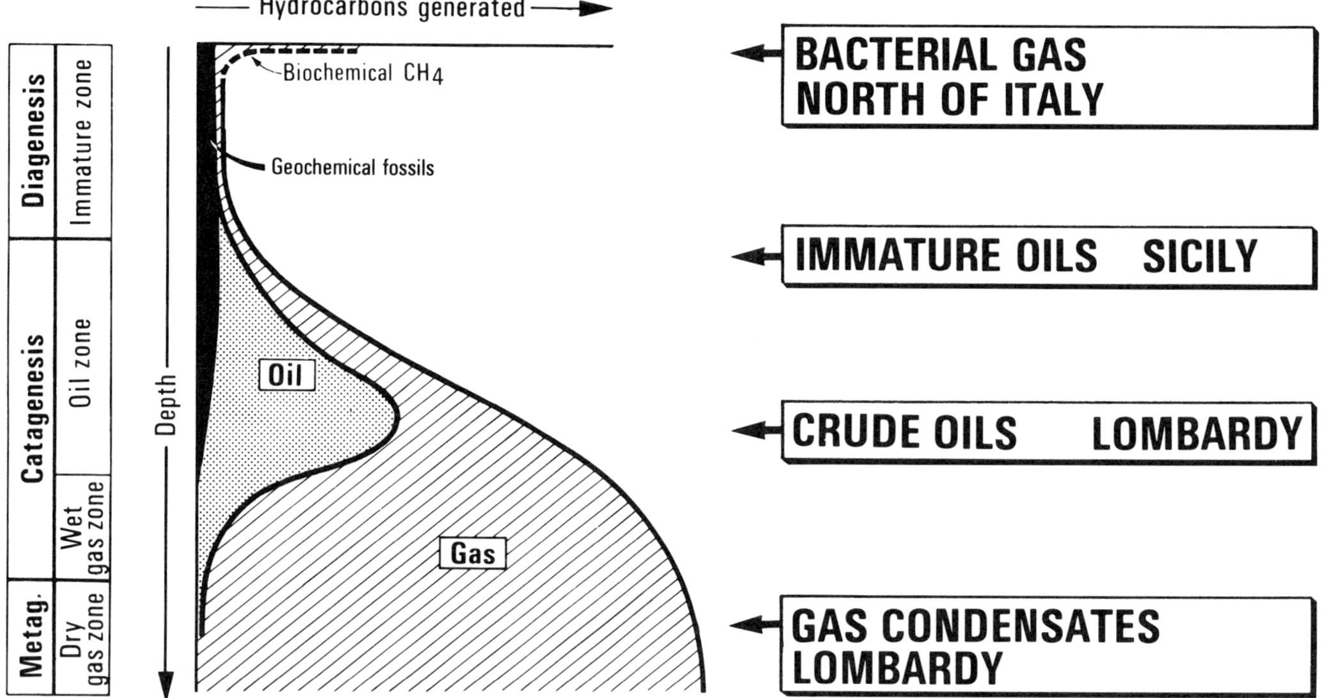

Figure 1. Composition of hydrocarbons generated from kerogen as a function of burial depth. Below the zone of bacterial activity, which may provide commercial pools of biogenic methane if favorable conditions prevail, kerogen undergoes thermal cracking and produces progressively lighter hydrocarbons. Because of its complex geological framework, Italy may serve to illustrate a wide variety of situations.

of Lombardy. Lastly, during metagenesis, only methane is stable, together with a coke residue that becomes more and more condensed, and finally is transformed into graphitic carbon.

All these progressive stages may be illustrated in Italy, where a complex geological history has provided a wide variety of petroleum situations. We shall successively examine examples of biogenic gas and of thermogenic oil and gas from the north of the country, as well as examples of immature oils from Sicily.

In the second half of the 19th century, Italy ranked among the leading oil-producing countries in the world. But, in spite of continued exploration efforts the production did not increase significantly until the middle of the 20th century (Rocco, 1959). The very complicated tectonic framework of the Apennines, which were the main target of exploration because of numerous seepages, explains the low success rate (Pieri and Mattavelli, 1986). After the Second World War, with the renewal of some geological concepts and the use of modern techniques, discoveries increased. Original reserves discovered by the end of 1988 were 690 billion m^3 of gas and 150 MT of oil located in more than 200 fields. The difficulties Italian petroleum geologists have to face has forced them to take an interest in up-to-date tools and ideas developed for exploration. Organic geochemistry, structural surveys with renewed interpretations (leading to recent discoveries of oil beneath the allochthonous nappes in the southern Apennines), and now integrated basin modeling are considered to be the best way to work in this territory. Obviously, the present jump in petroleum exploration in Italy is a consequence of scientific progress.

NORTHERN APENNINES (BIOGENIC GAS)

Geologically, northern Italy is dominated by three large structural domains (d'Argenio et al., 1980; d'Argenio and Alvarez, 1980; Pieri and Mattavelli, 1986; Novelli et al., 1987) (Figure 2). These are from north to south: (1) the southern Alps, outcrops of which end in the Lakes area but which can be traced farther beneath the Po Plain, with a general southern vergence (Castellarin and Vai, 1982; Roure et al., 1989); (2) the Po Plain, from Lombardy (Milano) to Veneto (Venice), a complex basin characterized by a thick upper Tertiary and Quaternary cover, deposited in the foredeep where subsidence was caused by the continental collision between the European and African plates (Pieri and Groppi, 1981); and (3) the northern Apennines, which have a northern vergence (Pieri and Groppi, 1981).

Since the Oligocene, up to 8000 m of sediments have been deposited in the Po Basin, mainly as thick sequences of terrigenous turbidites. For the petro-

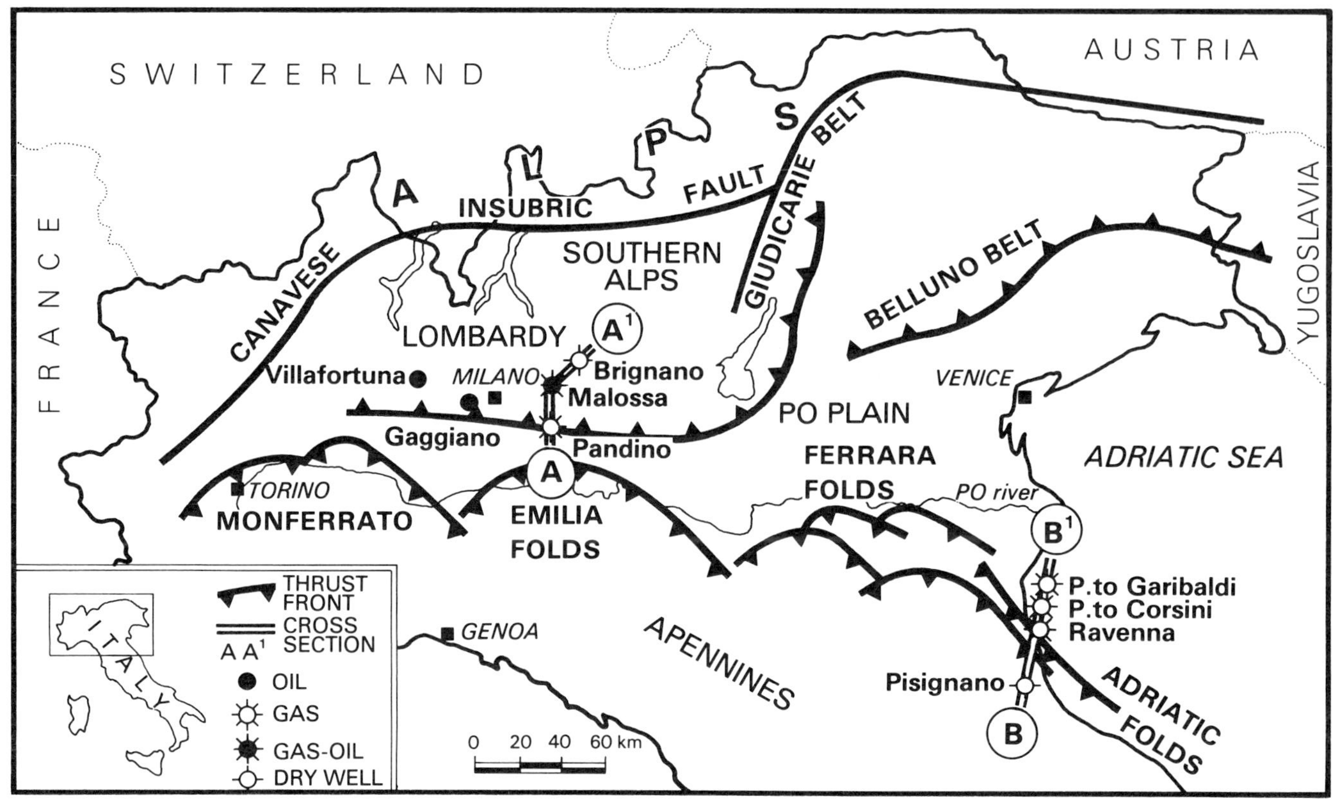

Figure 2. Tectonic map of northern Italy showing locations of cross sections A-A′ and B-B′.

leum industry, the Ravenna area is exceptionally prolific. It contains 65% of the gas produced in the whole Po Basin. Geochemical investigations have proven that all this gas in the Ravenna area is of bacterial origin (Mattavelli et al., 1983).

Because of their simple composition, hydrocarbon gases, and especially the "dry" gas, can be characterized by a very few parameters compared to that used for oils or source-rock organic contents. Two parameters may be used (Bernard et al., 1977; Tissot and Bessereau, 1982; Schoell, 1984). They are the isotopic composition, i.e., the relative abundance of C and H stable isotopes as referred to a standard, and the percentage of hydrocarbons higher than methane (Table 1). As far as methane is concerned, the different possibilities of its origin are clearly discriminated in a diagram where the ^{13}C concentration (shown as $\delta^{13}C_1$) is plotted against the composition of the gas $C_1/\Sigma C_n$, i.e., the ratio of methane to the sum of measured hydrocarbons (Figure 3). Biogenic gas, known for a long time as "swamp gas" in marshes and produced by bacterial activity operating during diagenesis below the sulfate-reduction zone, is characterized by an almost pure methane composition and very low $\delta^{13}C$ values, down to -80‰. This type of gas occurs in the northern Apennines. Other types of gas are the thermogenic dry gas from metagenesis, characterized by relatively high $\delta^{13}C$ values, and the thermogenic or mixed gas associated with liquid hydrocarbons formed in the catagenesis zone with intermediate isotopic compositions.

The coincidence of several favorable phenomena leads to the remarkable situation encountered in the Ravenna area (Figure 4).

1. The rate of sedimentation was high in this subbasin which was actively subsiding during the Pliocene. Although the average total organic carbon content is moderate (0.1 to 0.6%) and the kerogen contains more than 80% woody and herbaceous detritus (Pieri and Mattavelli, 1986; Mattavelli and Novelli, 1987), the high rate of sedimentation prevented organic matter from becoming degraded (Berner, 1974; Curtis, 1977; Pelet, 1984).
2. Synsedimentary tectonic activity, which was widespread in the Pedeapennine Fold Belt from the Miocene to the Pleistocene, created several kinds of structural and stratigraphic traps.
3. The sedimentary sequences were turbiditic and characterized by alternating pervious and impervious layers, resulting in the entrapment of gas in multipay pools, as in the Porto Garibaldi-Agostino field (Figure 5).
4. The "cold" thermal regime, shown by the very low temperature values recorded at depth (around 60°C at 4000 m, see Figure 5), ensured that

Table 1. Types of gases according to their composition and isotopic characters (after Mattavelli, 1987).

GENETIC TYPE OF NATURAL GASES	STABLE ISOTOPES $\delta^{13}C_1$	STABLE ISOTOPES δD	HIGHER HYDROCARB C_{2^+}
BIOGENIC GASES "B"	-75 ÷ -60 ‰	-180 ÷ -210 ‰	< .2 %
MIXED GASES "M"	-60 ÷ -50 ‰	-180 ÷ -210 ‰	.1 ÷ 5 %
THERMOGENIC GASES "T"	-50 ÷ -30 ‰	-150 ÷ -180 ‰	.3 ÷ 10 %

bacterial life, which is no longer possible above 60°C, would develop in a thick zone of the sedimentary cover.

It must be noted that these favorable conditions are more or less linked together, i.e., subsidence and heat flow through geodynamics, temperature and sediment nature through petrophysical properties, etc. Use of a modeling approach for the integration of all the components of such a dynamic system provides obvious advantages.

LOMBARDY (THERMAL OIL AND GAS)

The Lombardy province in the northwestern part of the Po Basin (Figure 2) has a particular complexity with respect to hydrocarbon occurrences because oil, gas, and gas-condensate fields coexist (Novelli et al., 1987). Two different situations will be commented upon. First is the Malossa gas-condensate field which belongs to the south-verging structures of the Alpine domain and contains source rocks of Triassic (especially Rhaetian) age (Novelli et al., 1987; Stefani, this volume). On the other hand, the Gaggiano oil field is situated in a small foreland area that was almost unaffected by Alpine tectonics. Therefore, the tectonic setting of each field is different, and their geological history and the source rocks involved are also distinctly different.

The main reservoirs in the Malossa area (Figure 6) are located in the Triassic Hauptdolomit. The traps are faulted anticlines that occur at depths between 4300 and more than 6000 m in the most distal nappes of the southern Alps. Folding occurred in the late Miocene (Chiaramonte and Novelli, 1986; Pieri and Mattavelli, 1986). The Pliocene/Pleistocene cover is relatively thin. In the well Malossa-3, the condensate is 53°API, the C_{15+} content is about 8.5% (Pieri and Mattavelli, 1986), and $\delta^{13}C_1$ is -36°/oo.

The oils in the Gaggiano (Figure 7) and Villafortuna areas accumulated in Upper and Middle Triassic dolomites. The traps are found in weakly deformed tilted blocks under the major unconformity that marks the beginning of Tethyan history in this region. In this part of the foreland, Pliocene and Quaternary sediments are considerably thicker than in the Malossa area. Oils are 42° API gravity and are encountered at depths up to 6200 m (Pieri and Mattavelli, 1986).

The Mesozoic carbonate sequences of the northwestern Po Basin, where the reservoirs of Malossa, Gaggiano, and Villafortuna are located, are presently characterized by very abnormal hydraulic pressures. It is probable that the compressive tectonic movements disrupted the Mesozoic carbonate aquifers (Novelli et al., 1987).

During Middle and Late Triassic times, before the opening of the Ligurian Tethys between Europe and Africa, the carbonate platform in northern Italy was broken by crustal distension into a series of shelves and more rapidly subsiding troughs. Complex and changing paleogeographic patterns developed, which are not yet completely understood (Gaetani, 1975; Laubscher and Bernoulli, 1977; d'Argenio et al., 1980; Pieri and Mattavelli, 1986; Burchell, this volume). Especially during Ladinian and Rhaetian ages, conditions were locally favorable for the development of anoxic conditions in the basins.

The geochemical analyses carried out on the oil samples and on bitumens extracted from possible source rocks have established a reliable correlation between Malossa hydrocarbons and Riva di Solto shales, which are of Rhaetian age. Gaggiano oils have many chemical features in common with Malossa

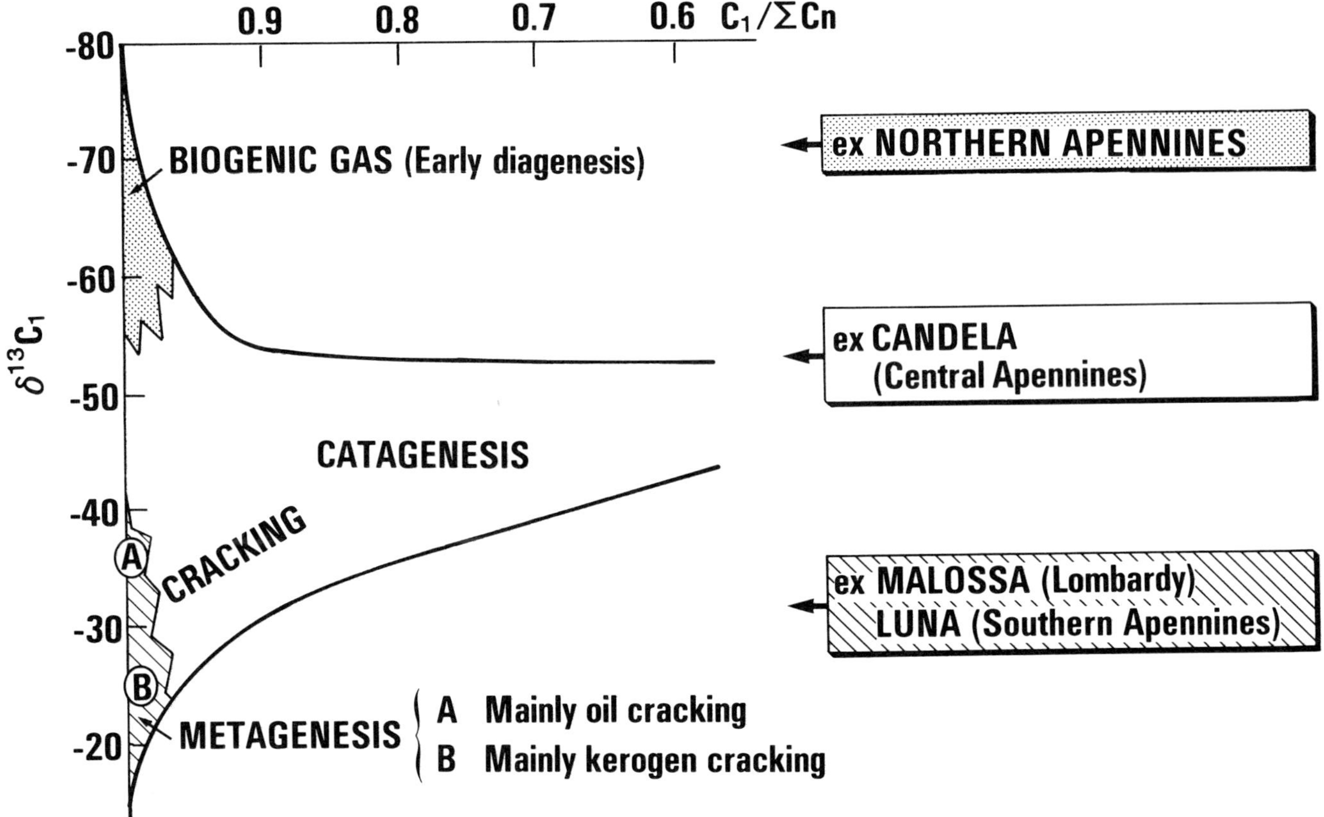

Figure 3. Abundance and isotopic composition of methane formed during the successive stages of kerogen evolution. Examples are given from Italy, in the zones of diagenesis, catagenesis, and metagenesis.

oils. Nevertheless, some significant molecular distinctions enable them to be correlated with Meride limestones, which are of Ladinian age (Riva et al., 1986; Novelli et al., 1987).

In order to determine the time of hydrocarbon formation in Malossa and in Gaggiano, the burial histories of both the Riva di Solto (Malossa source rock) and Meride (Gaggiano source rock) formations were reconstructed, using the MATOIL software. In the "backstripping" phase of the program, the compaction of the sedimentary pile is taken into account through the lithostratigraphic description of the series and porosity versus depth curves for basic standard lithologies. It is assumed that the volume variation with burial of a given layer is due only to the expulsion of water from porous sediments and that the volume of the solid matrix in a given layer remains constant through time. A representative borehole was chosen in each area (Figure 8). The Riva di Solto shales were deeply buried soon after their deposition, within the context of the Tethyan rifting. By contrast, the Meride limestones were not affected by strongly subsident conditions before the Alpine deformation at the end of the Cenozoic.

The cracking of kerogen into hydrocarbons is a chemical reaction in which the main controlling factor is the temperature. The quantities generated may be calculated according to chemical reaction kinetics, which can be calibrated by pyrolysis experiments for each type of kerogen. Several parallel reactions are modeled, which account for the diversity of chemical bonds in the source organic matter. The results of the calibration used are expressed as activation energy distribution histograms representing the potential yields of each of these reactions, classified by increasing activation energy. The narrowness of the distribution is inversely related to the variety of chemical bonds to be broken in the kerogen molecule. Histograms are narrow for type I, and wide for type III (Tissot et al., 1987). Such histograms were obtained for representative kerogens of Riva di Solto shales and Meride limestones (Figure 9) (Novelli et al., 1987). They are directly usable in integrated basin models that provide a reconstruction of the thermal history undergone by the source rocks.

In the second part of MATOIL reconstruction, the temperature evolution and the timing of hydrocarbon formation are computed. The heat flow existing at the bottom of the sedimentary column is estimated from present measurements and geodynamic considerations. In the model, heat is transmitted through the sediments by conduction and forced convection (due to compaction). For the Riva di Solto shales (Figure 10A), the thermal history is markedly characterized by a temperature peak linked to

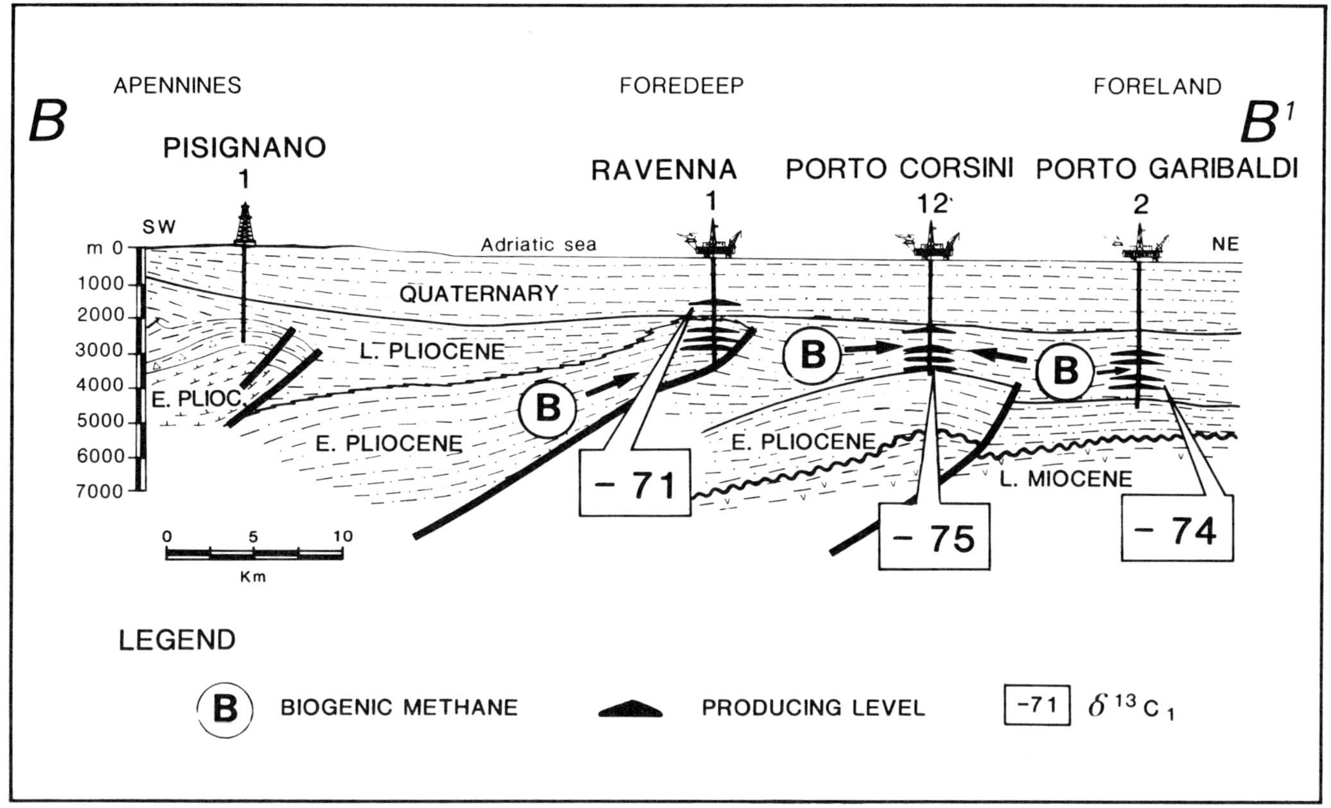

Figure 4. Cross section through the Ravenna area illustrating biogenic gas production. Pools of biogenic methane are widespread in middle late Pliocene sediments. High rates of sedimentation, synsedimentary tectonics, the turbiditic nature of the sediments, and cold thermal regimes are the favorable conditions called upon to explain the large accumulations encountered. See Figure 2 for the location of cross section B-B′.

Tethyan rifting. This temperature caused rapid oil formation. The catagenetic stage was reached as early as the end of the Lias. Then the situation remained stable as far as hydrocarbon formation is concerned until the Neogene Alpine movements, which carried maturation further toward metagenesis, with secondary cracking of liquid hydrocarbons remaining in the source rock and massive formation of gas.

By contrast, Gaggiano (Figure 10B) (and Villafortuna) oils are clearly related to recent high-subsidence Alpine tectonics (very late formation). In this particular case, a very deep occurrence of liquid oil, resulting from the late and short burial of the source rock, must be emphasized (42°API crude at 6200 m depth).

SICILY (IMMATURE OILS)

The last specific example from Italy described here is the Ragusa Basin, located in the southeastern part of Sicily (Figure 11). Within this area, which was found to be one of the most rewarding parts of Italy for petroleum exploration at the end of the 1950s, several major oil fields have been discovered. Among these are the onshore Ragusa and Gela fields and, more recently, the offshore Vega field. The area belongs to the foreland of the Neogene Apenninic/Maghrebidic orogenic belt (Figures 11, 12). During the opening of the Tethys, it was part of the African passive margin (Catalano and D'Argenio, 1982). The foreland is now represented by the Iblean Plateau. Distensive tectonics in the Rhaetian created a strongly subsident basin, with favorable conditions for the accumulation of organic matter.

The area under consideration has two specific features.

1. The main reservoirs are located in a dolomitic facies stratigraphically overlain by a Rhaetian anoxic facies (Figure 12). The source rocks, which do not crop out anywhere in the country, have been extensively cored and analyzed, notably from a sedimentological viewpoint (Patacca et al., 1979; Brosse et al., 1988 and in press).
2. The oils are heavy, with API gravities lower than 20° and a high sulfur content (Pieri and Mattavelli, 1986).

Two successive formations, Noto (mainly Rhaetian) and Streppenosa (mainly Hettangian), present interesting generating potentials (Figure 13). Noto

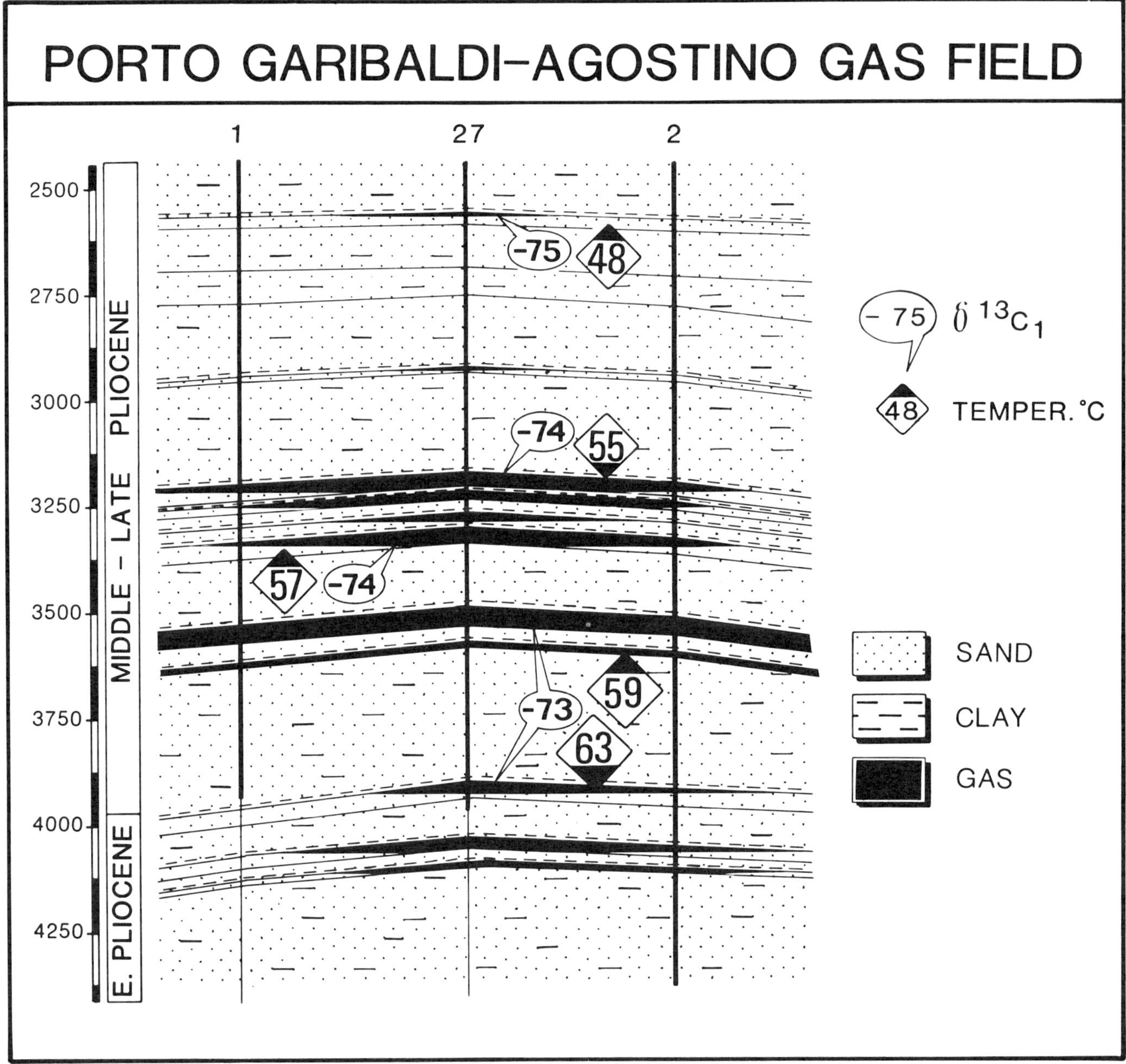

Figure 5. Multipay pools of biogenic gas in turbiditic sequences in the Porto Garibaldi-Agostino gas field. Recorded temperatures (about 60°C at 4000 m) indicate a cold thermal regime.

is considered as the best source rock. It contains several types of sedimentary sequences, where clayey and carbonate sediments have combined in different ways: marls, shales, and carbonates alternating both as thick beds and also as laminites. Interpretation in terms of paleoenvironments is difficult because such sequences lack informative structures. At least part of the laminites indicate tidal influences. Organic matter is associated with the argillaceous fraction of the sediments (Brosse et al., 1988 and in press). Highly sensitive geochemical tools, such as isotopic or GC-MS analyses, have established that all the oils in the Ragusa Basin are similar and may be correlated with Noto and Streppenosa black shales (Pieri and Mattavelli, 1986; Mattavelli and Novelli, in press). The presence of specific molecular markers, such as gammacerane, has been noted.

The kinetic parameters of cracking have been calibrated for two immature kerogens from the Noto formation. Their different chemical compositions are depicted in the classic Van Krevelen diagram, and they reveal significantly distinct distributions of activation energies (Figure 14) (Brosse et al., 1988; in press).

The variety of geological situations with respect to the allochthonous nappes leads to different Neogene burial histories from one point to another. Basin modeling shows that, in some cases, the Noto

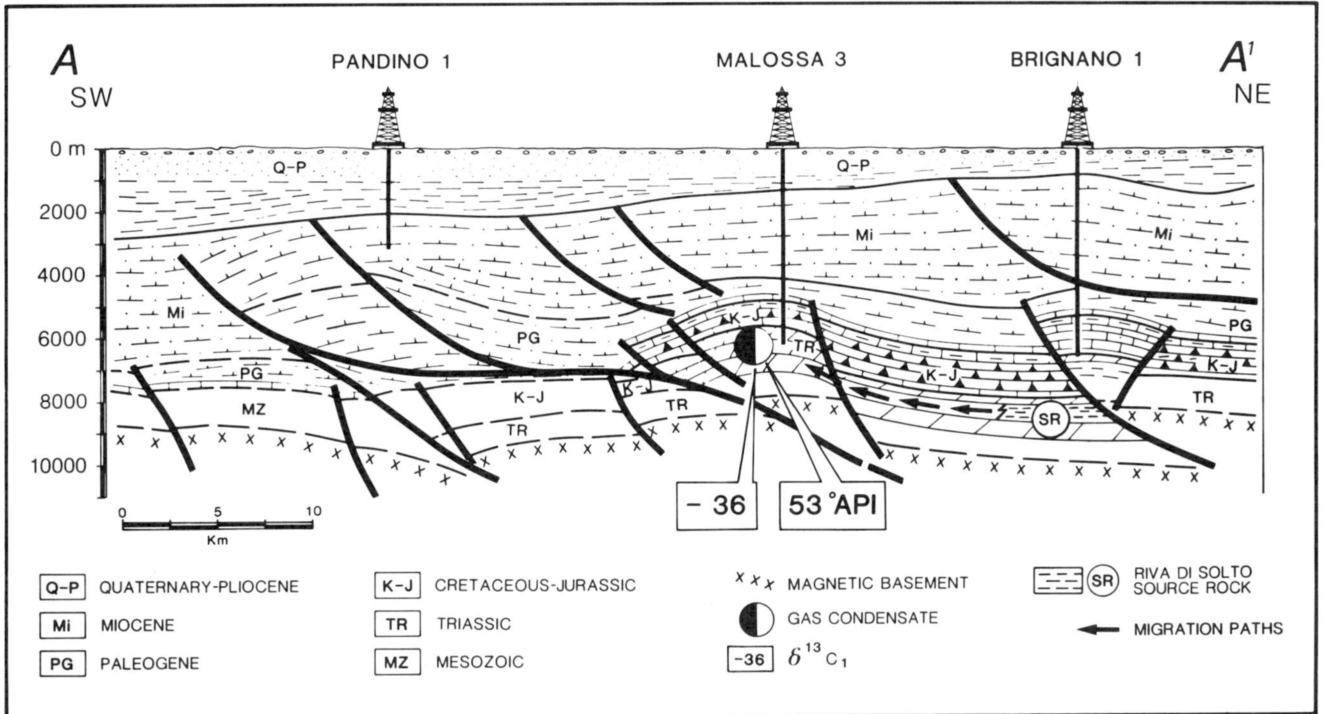

Figure 6. Cross section through the Malossa gas-condensate field in the Lombardy area. Malossa gas condensates accumulated in the Triassic/Liassic Hauptdolomit, and have their source in the Riva di Solto shales of Rhaetian age. See Figure 2 for location of cross section A-A′.

formation may have reached the beginning of oil formation in Jurassic time due to its rapid subsidence along the Apulian margin of the Tethys (Brosse et al., in press). However the greatest amount of immature heavy oil was generated as a consequence of the high subsidence rate of the Plio-Pleistocene series (Figure 15).

CONCLUSIONS

Important lessons in petroleum exploration may be learned from the study of frontier areas such as the various parts of Italy.

1. Thermal regimes are very important for understanding oil and gas generation.
2. The intrinsic nature of the original organic matter is also a basic parameter, as shown by the transformation ratio reached by different kerogens (Figure 16) and also by the absolute quantities of hydrocarbons they generate in response to a given thermal history. Forecasting possibilities in this domain depends largely on research on both source-rock sedimentology and the exact composition of cracking products, two targets of primary importance for organic geochemistry in the years to come.
3. Likewise, concepts and analytical methods of organic geochemistry appear to take on their full meaning when integrated within a modeling approach so as to consider all the geological constraints that determine hydrocarbon formation and accumulation.

In this respect, Italy is deeply marked by the Apennine and Alpine tectonic patterns. The distribution of gas fields by gas types provides particularly demonstrative illustrations of this dependence (Figures 17, 18). Thrust belts, rich in thermogenic gas formed in the metagenesis zone, and foreland areas, rich in bacterial gas, present occasionally prolific gas fields. The foredeep accounts for 77% of the gas reserves because in this area both biogenic and thermogenic products can mix together.

ACKNOWLEDGMENTS

We would like to thank P. Y. Chenet for his help with the modeling, F. Roure for his review of the manuscript, and AGIP and IFP draftsmen.

REFERENCES CITED

Behar, F., and M. Vanderbroucke, 1987, Chemical modelling of kerogens: Organic Geochemistry, v. 11, p. 15-24.

Bernard, B. B., J. M. Brooks, and W. M. Sackett, 1977, A geochemical model for characterization of hydrocarbon gas sources in marine sediments, *in* Offshore Technology Conference, OTC 2934, p. 435-438.

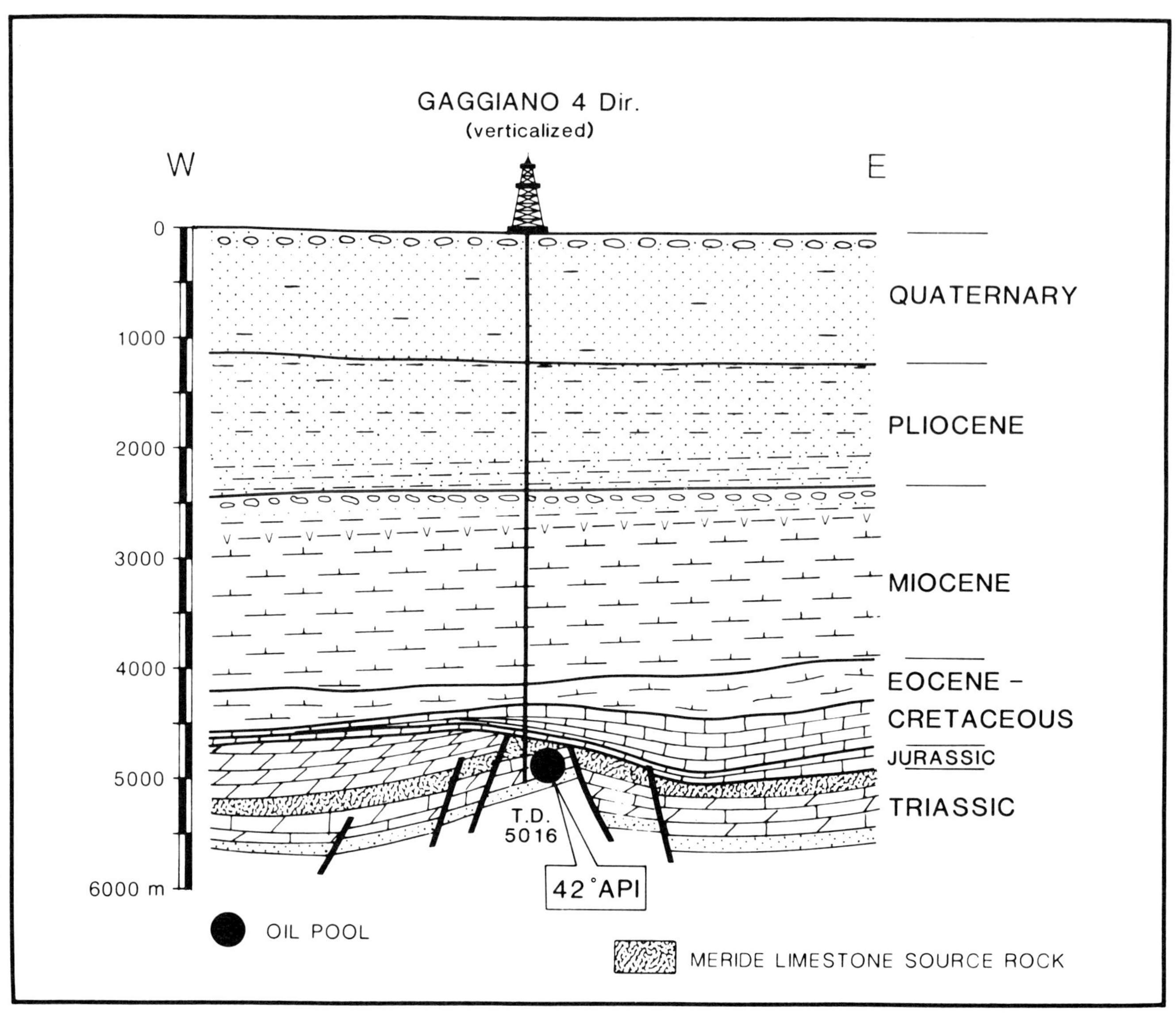

Figure 7. Cross section through the Gaggiano oil field in the Lombardy area. Gaggiano oil reservoirs are Middle Triassic dolomites, and the oils have their source in the Meride limestone of Ladinian age.

Berner, R. A., 1974, Kinetic models for the early diagenesis of nitrogen, sulfur, phosphorous, and silicon in anoxic marine sediments, *in* E. D. Goldberg, ed., The sea: New York, Wiley, p. 427-450.

Brosse, E., J. P. Loreau, A. Y. Huc, A. Frixa, L. Martellini, and A. Riva, 1988, The organic matter of interlayered carbonates and clays sediments, Trias/Lias, Sicily, *in* Advances in organic geochemistry 1987: Organic Geochemistry, v. 13, p. 433-443.

Brosse, E., A. Riva, J. P. Loreau, S. Santucci, M. Bernon, F. Laggoun-Defarge, and A. Frixa, in press, Some sedimentological and geochemical characters of the late Triassic Noto formation, source rock in the Ragusa basin (Sicily), *in* Advances in organic geochemistry 1989: Organic Geochemistry.

Castellarin, A., and G. B. Vai, 1982, Introduzione alla geologia strutturale del Sudalpino, in Guida alla geologia del Sudalpino centro-orientale: Guide geologia regionale, Societa Geologia Italiana, p. 1-22.

Catalano, R., and B. D'Argenio, 1982, Infraliassic strike-slip tectonics in Sicily and Southern Apennines: Rendiconte Societa Geologia Italiana, v. 5, p. 5-10.

Chiaramonte, M. A., and L. Novelli, 1986, Organic matter maturity in Northern Italy: Some determining agents, *in* Advances in organic geochemistry 1985: Organic Geochemistry, v. 10, p. 281-290.

Curtis, C. D., 1977, Sedimentary geochemistry: environments and processes dominated by involvement of an aqueous phase: Philosophical Transactions of the Royal Society of London, v. A286, p. 353-372.

D'Argenio, B., and W. Alvarez, 1980, Stratigraphic evidence for crustal thickness changes on the southern Tethyan margin during the Alpine cycle: GSA Bulletin, v. 1, p. 681-689.

D'Argenio, B., F. Horvath, and J. E. T. Channell, 1980, Paleotectonic evolution of Adria, the African promontory, *in* J. Aubouin et al., eds., Géologie des chaines alpines issues de la Tèthys: Paris Geological Congress Proceedings, Memoir BRGM, p. 331-351.

Gaetani, M., 1975, Jurassic stratigraphy of the Southern Alps: a review: Earth Science Society of the Libyan Arab Republic, Tripoli, v. 1, p. 377-402.

Laubscher, H. P., and D. Bernoulli, 1977, Mediterranean and Tethys, *in* A. E. M. Nairn and F. G. Stehli, eds., The ocean basins and margins, v. 4, The Eastern Mediterranean: New York, Plenum Press, p. 1-28.

Mattavelli, L., 1987, Origin of natural gases in Northern Italy deduced from stable isotope studies, *in* R. Rodriguez-Clemente

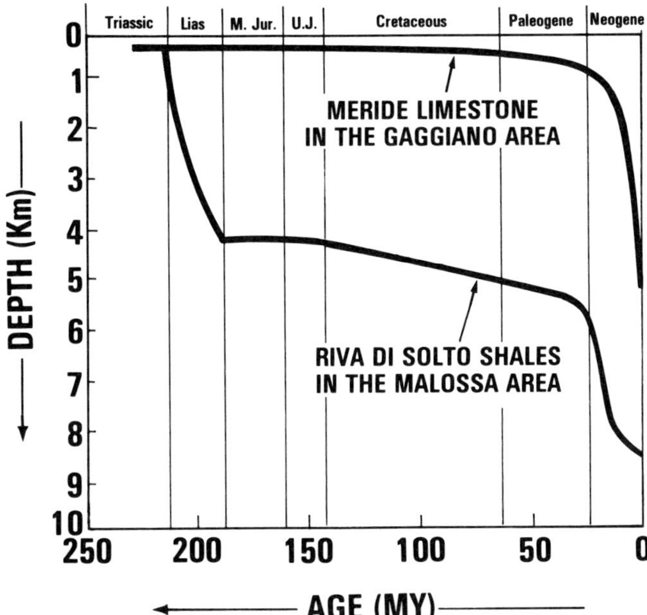

Figure 8. Modeling of the burial histories of the Meride and Riva di Solto source rocks. The burial histories of Meride and Riva di Solto source rocks were reconstructed using MATOIL software, with geological data obtained from two wells in the Gaggiano and Malossa areas.

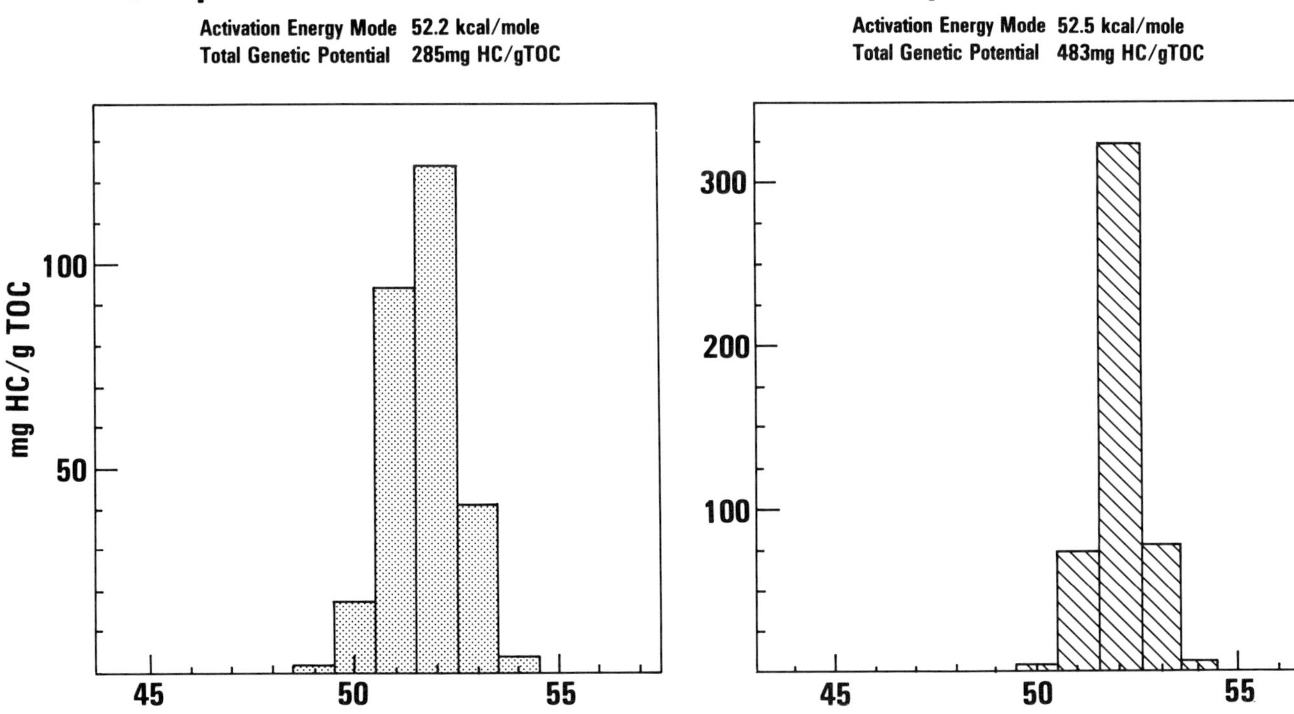

Figure 9. Distribution of kinetic parameters used to represent the primary cracking of Riva di Solto and Meride kerogens.

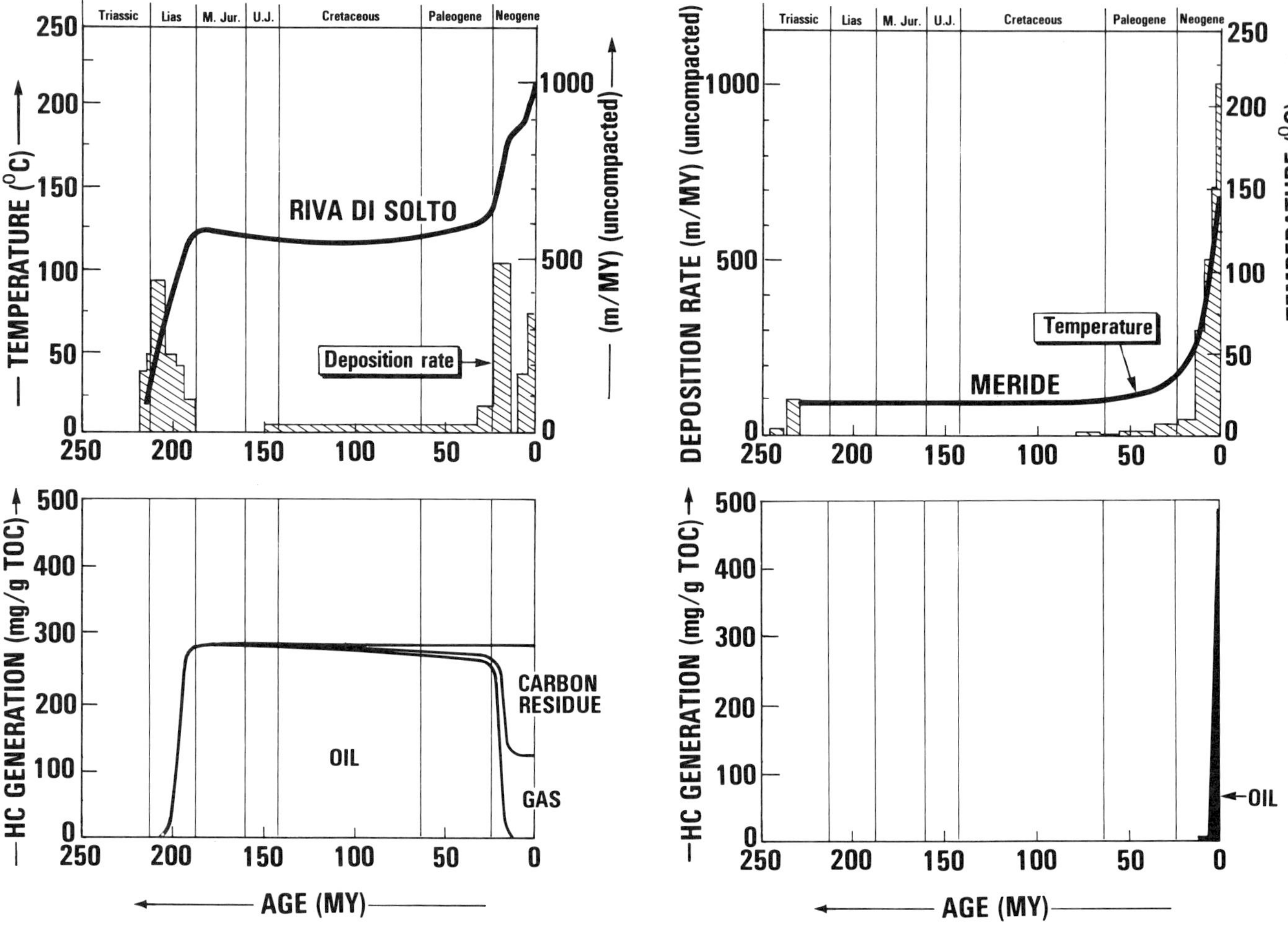

Figure 10. Reconstruction of the sedimentation rates, temperature evolution, and hydrocarbon-generation history by modeling. A. Malossa area (Riva di Solto source rock). B. Gaggiano area (Meride source rock). The MATOIL software was used to reconstruct hydrocarbon generation based on hypotheses concerning thermal regimes and organic-matter types. The results were checked by present-day measurements of temperature and kerogen maturity, and the hypotheses were adjusted if necessary. In both areas, a heat flow of 50 mW/m² was assumed to obtain the bottom-hole temperatures. For Malossa, the heat flow was disturbed by the Lias rifting phenomenon. The primary cracking of kerogen into oil was computed using the kinetic parameters shown in Figure 9. The secondary cracking of oil into gas was modeled by a single reaction (unique activation energy, 57 kcal/mole).

and Y. Tardy, eds., Geochemistry and mineral formation in the earth surface: Consejo Superior de Investigaciones Cientificas, Madrid, p. 399-420.

Mattavelli, L., and L. Novelli, 1987, Origin of the Po Basin hydrocarbons: Mémoire de la Société Géologique de France, N. Sp., n. 151, p. 97-106.

Mattavelli, L., T. Ricchiuto, D. Grignani, and M. Schoell, 1983, Geochemistry and habitat of natural gases in the Po Basin/Northern Italy: AAPG Bulletin, v. 67, p. 2239-2254.

Novelli, L., M. A. Chiaramonte, L. Mattavelli, G. Pizzi, L. Sartori, and P. Scotti, 1987, Oil habitat in the Northwestern Po Basin, *in* B. Doligez, ed., Migration of Hydrocarbons in Sedimentary Basins: 2d IFP Exploration Research Conference, TECHNIP, Paris, p. 27-58.

Orr, W. L., 1985, Kerogen/asphaltene/sulfur relationships in sulfur-rich Monterey oils: Organic Geochemistry, v. 10, p. 499-516.

Patacca, E., P. Scandone, G. Giunta, and V. Liguori, 1979, Mesozoic paleotectonic evolution of the Ragusa zone (Southeastern Sicily): Geologica Romana, v. 18, p. 331-369.

Pelet, R., 1984, A model for the biological degradation of recent sedimentary organic matter: Organic Geochemistry, v. 6, p. 317-325.

Pieri, M., and G. Groppi, 1981, Subsurface geological structure of the Po Plain, Italy: Consiglio Nazionale Richerche Progetto Finalizzato Geodinamica, v. 414, p. 1-13.

Pieri, M., and L. Mattavelli, 1986, Geologic framework of Italian petroleum resources: AAPG Bulletin, v. 70, p. 103-130.

Riva, A., T. Salvatori, R. Cavaliere, T. Ricchiuto, and L. Novelli, 1986, Origin of oils in Po Basin, Northern Italy, *in* Advances in organic geochemistry 1985: Organic Geochemistry, v. 10, p. 391-400.

Rocco, T., 1959, Gela in Sicily, an unusual oil field: Proceedings of the 5th Petroleum Congress, Geology and Geophysics, v. 1, p. 207-232.

Roure, F., P. Heitzman, and R. Polino, 1989, Deep structure of the Alps: Mémoire de la Societe Geologique de France, n. 156, Mémoire de la Société Géologique de Suisse, n. 1, Societa Geologia Italiana v.sp. 1.

Schoell, M., 1983, Genetic characterization of natural gases: AAPG

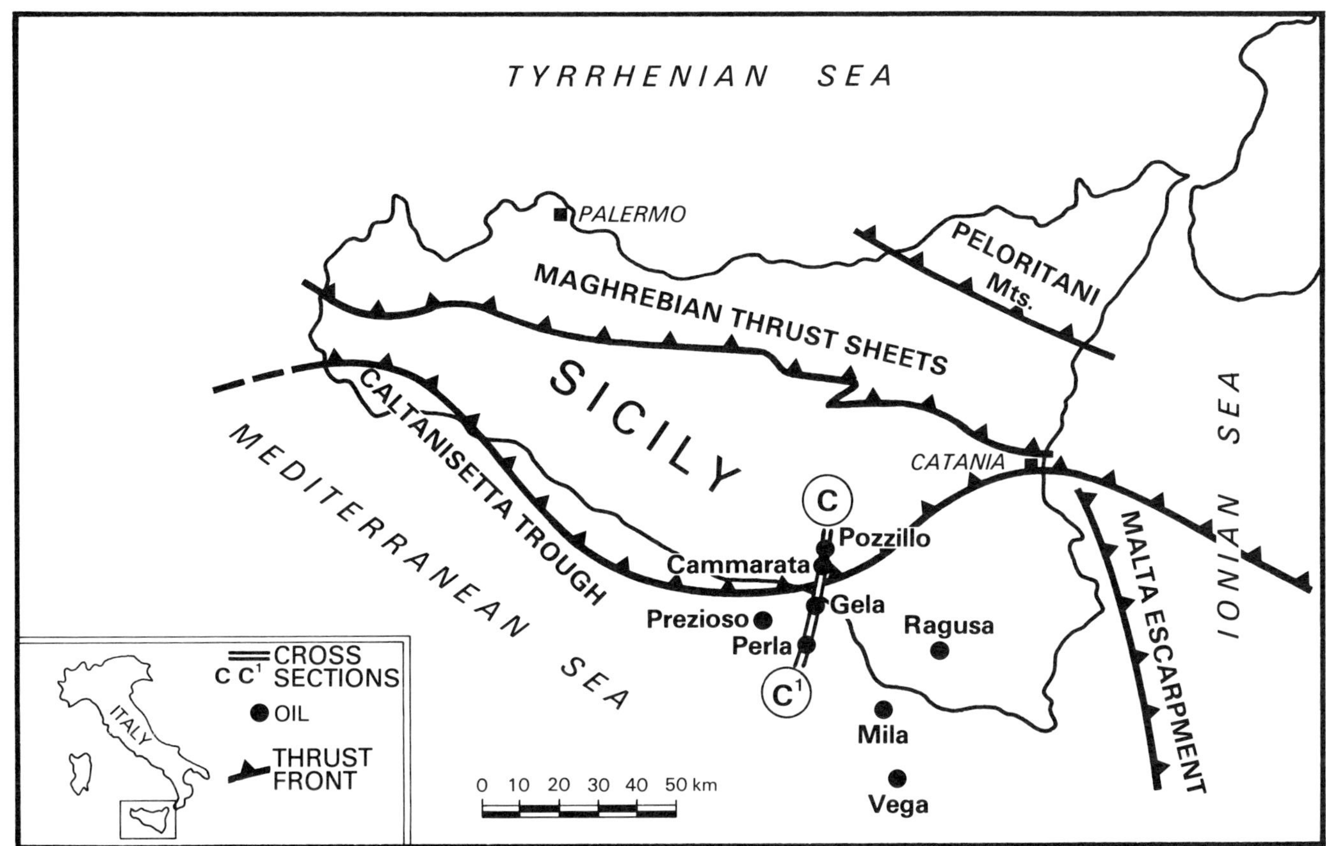

Figure 11. Tectonic map of Sicily showing the location of cross section C-C′ in the Ragusa Basin.

Bulletin, v. 67, p. 2225–2238.

Schoell, M., 1984, Stable isotopes in petroleum research, *in* J. Brooks and D. Welte, eds., Advances in petroleum geochemistry, London, Academic Press, v. 1, p. 215–245.

Sluijk, D., and J. R. Parker, 1986, Comparison of predrilling predictions with postdrilling outcomes, using Shell's Prospect Appraisal System, *in* D. D. Rice, ed., Oil and Gas Assessment: AAPG Studies in Geology 21, p. 55–58.

Tissot, B., and G. Bessereau, 1982, Géochimie des gaz naturels et origine des gisements de gaz en Europe Occidentale: Revue de l'Institut Franais du Pétrole, v. 37, p. 63–77.

Tissot, B. P., and D. H. Welte, 1984, Petroleum formation and occurrence, 2d ed: New York, Springer Verlag, 699 p.

Tissot, B. P., R. Pelet, and P. Ungerer, 1987, Thermal history of sedimentary basins, maturation indices, and kinetics of oil and gas generation: AAPG Bulletin, v. 71, p. 1445–1466.

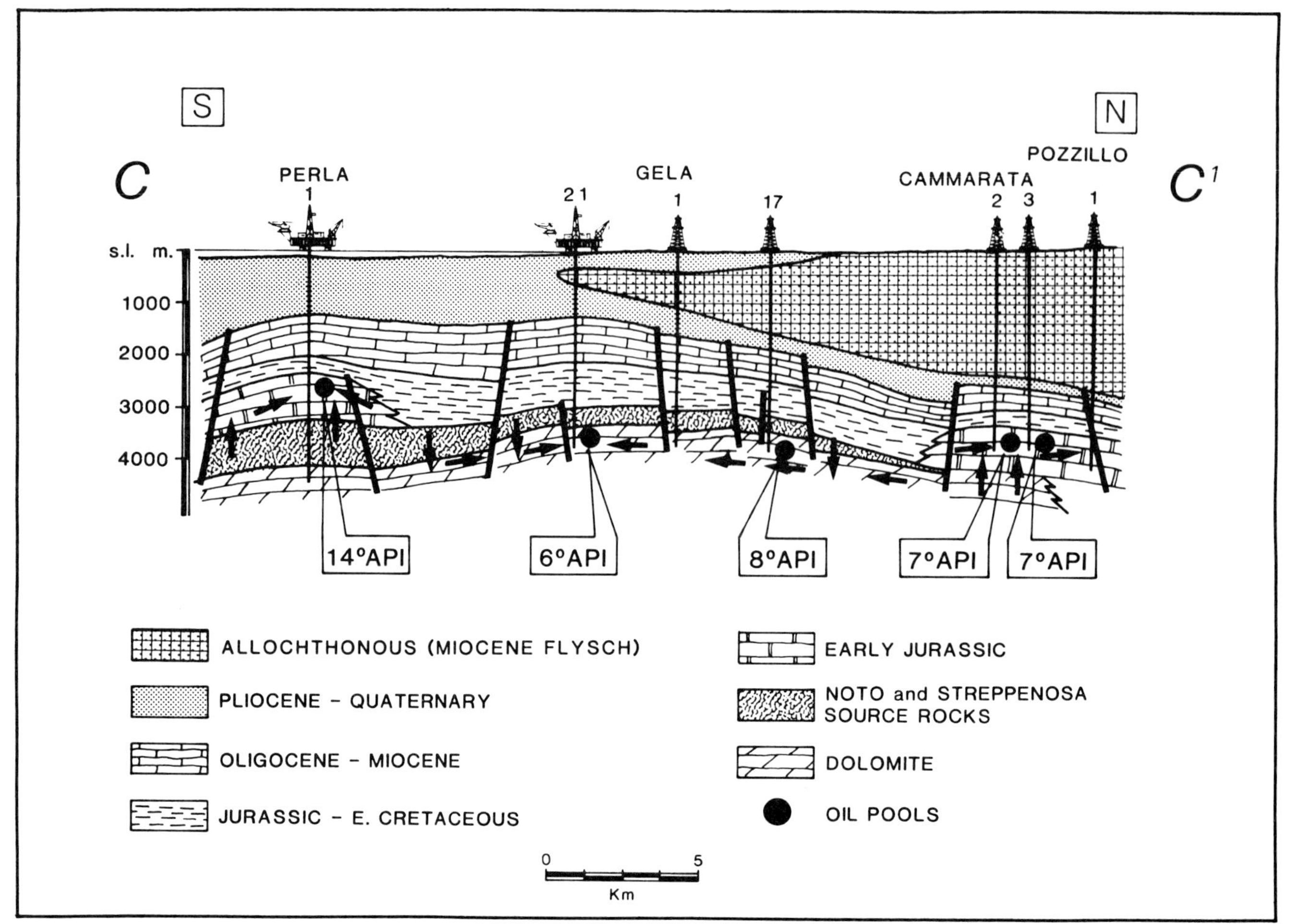

Figure 12. Cross section through the Gela field in the Ragusa Basin, Sicily, producing immature oils.

TOC AND ROCK-EVAL VALUES

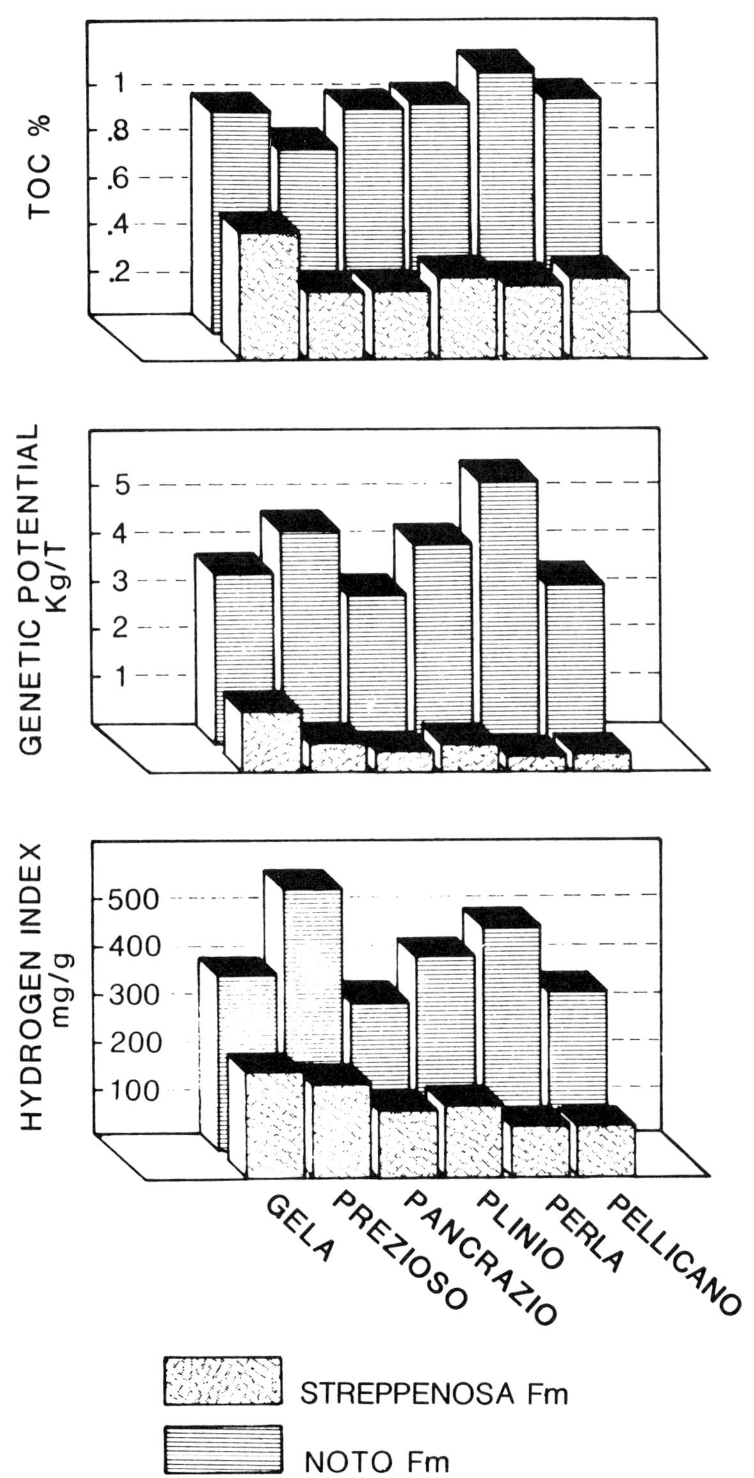

Figure 13. Results of evaluation of total organic carbon and of petroleum potential by Rock-Eval pyrolysis, of the Noto and Streppenosa formations from various locations in the areas of the Ragusa Basin, Sicily.

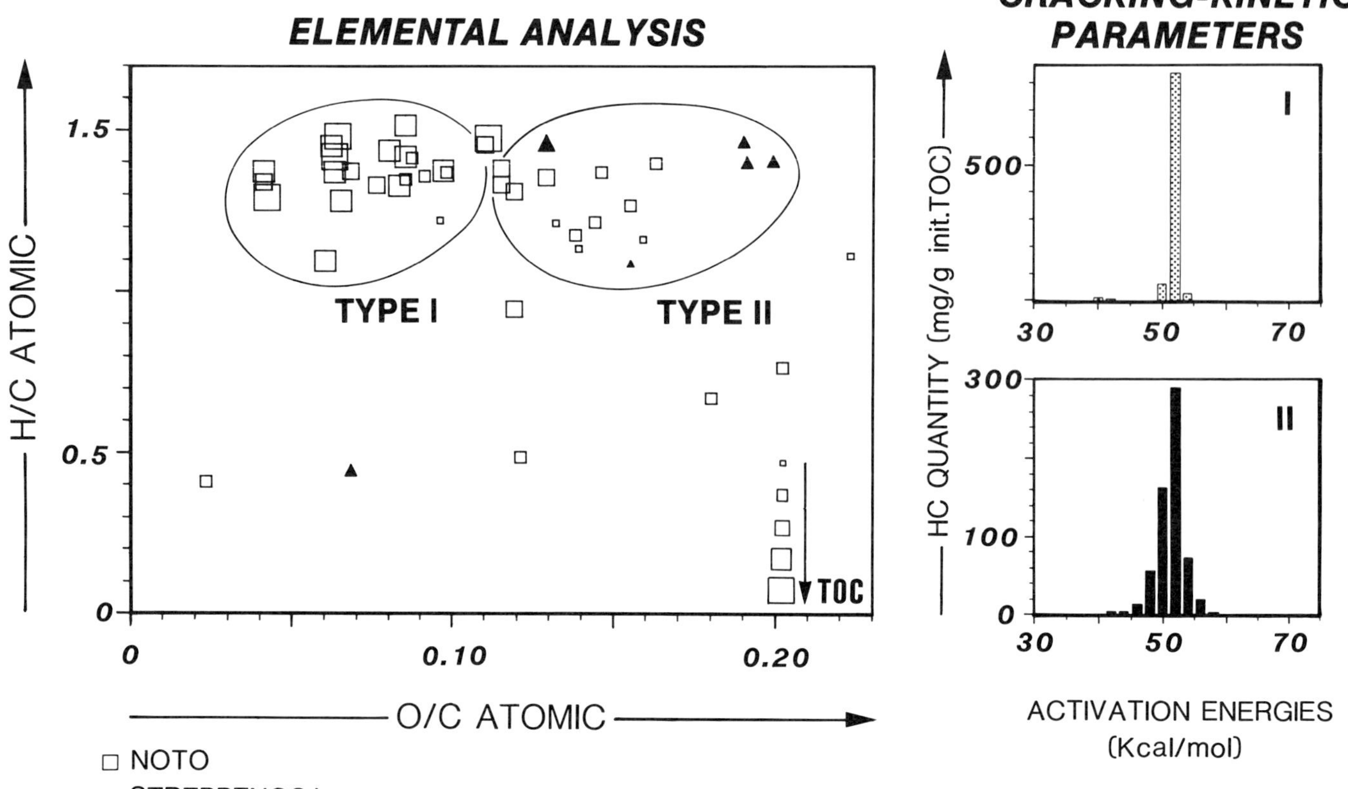

Figure 14. Distribution of kinetic parameters of cracking for two kerogens from the Noto source rock in the Ragusa Basin, Sicily. Kerogens isolated from immature Noto and Streppenosa source rocks present an elemental composition that varies between type I and type II areas. The parameters of thermal cracking into hydrocarbons have been calibrated for a kerogen representative of each of these two areas.

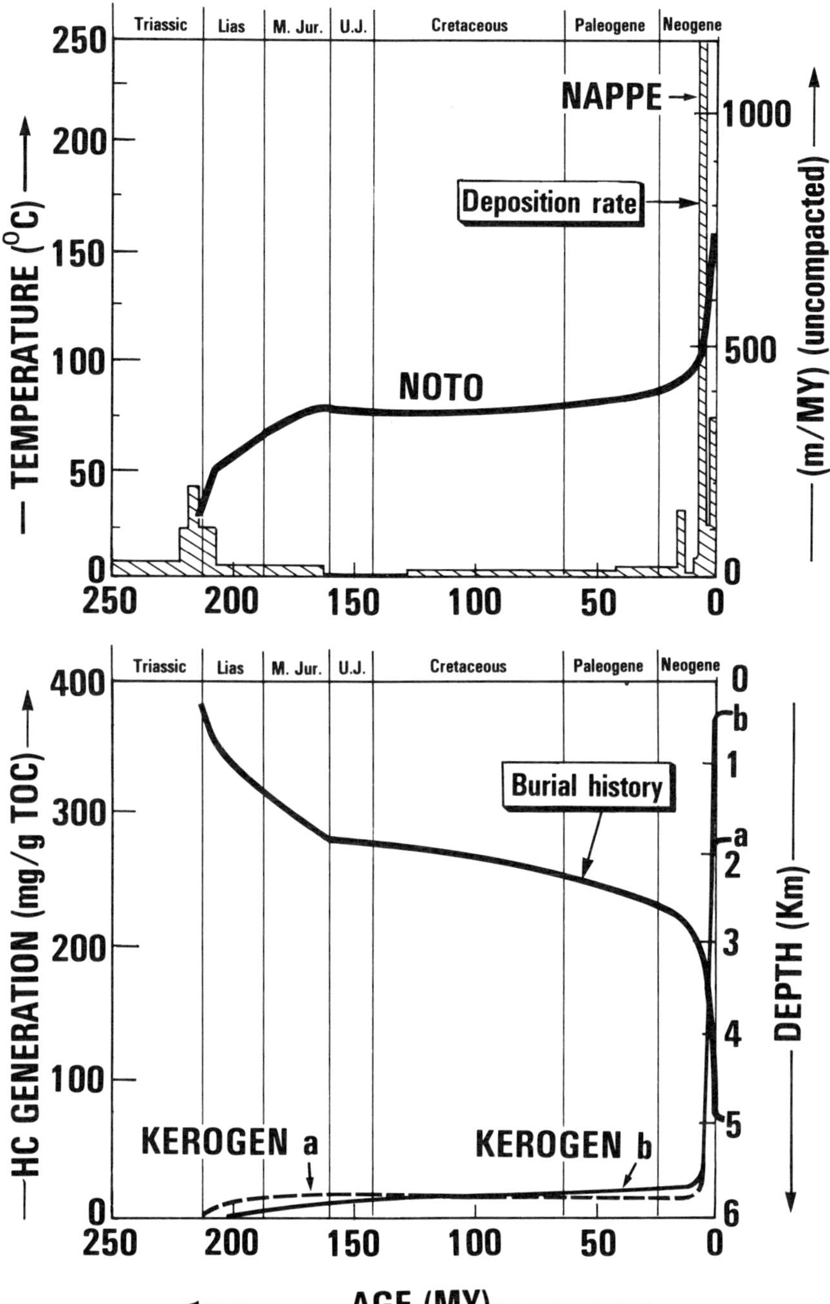

Figure 15. Reconstruction by modeling of the sedimentation rates, temperature evolution, and hydrocarbon-generation history in the Gela area of the Ragusa Basin, Sicily; Noto and Streppenosa are the source rocks. A heat flow of 50 mW/m^2 was assumed to obtain the bottom-hole temperatures. This heat flow was moderately disturbed by the Lias and Middle Jurassic rifting phenomenon. The primary cracking of kerogen into oil was computed using the kinetic parameters shown in Figure 14.

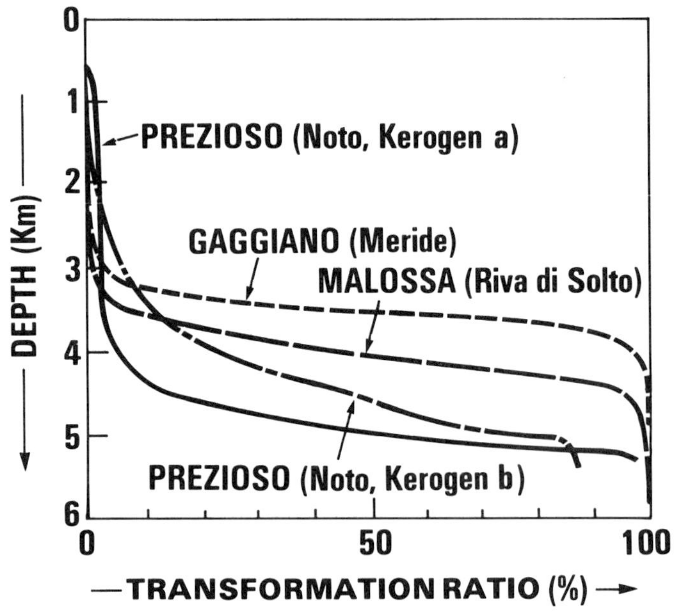

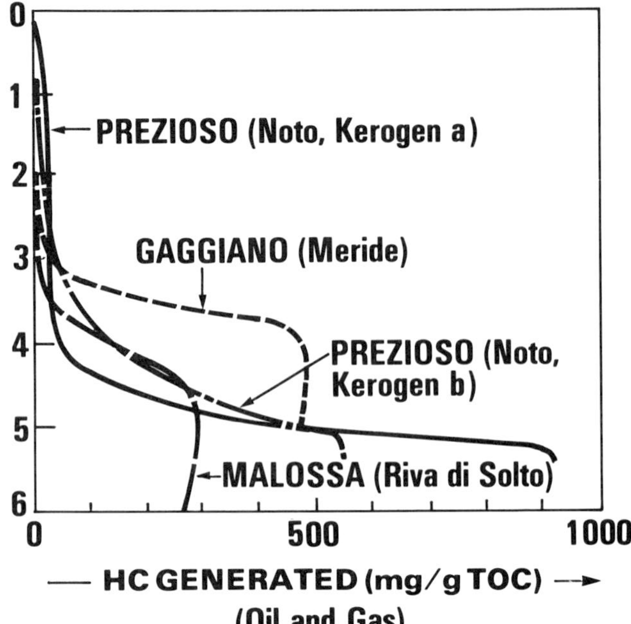

Figure 16. The role of kerogen type in the production of hydrocarbons. The maturation of the four kerogens described in Figures 9 and 14 was computed with the same burial and thermal histories in order to show the influence of the kinetic calibration used to characterize the kerogens intrinsically. Both the transformation ratio in the present state (and consequently the timing of oil formation) and the amounts of hydrocarbons generated are influenced.

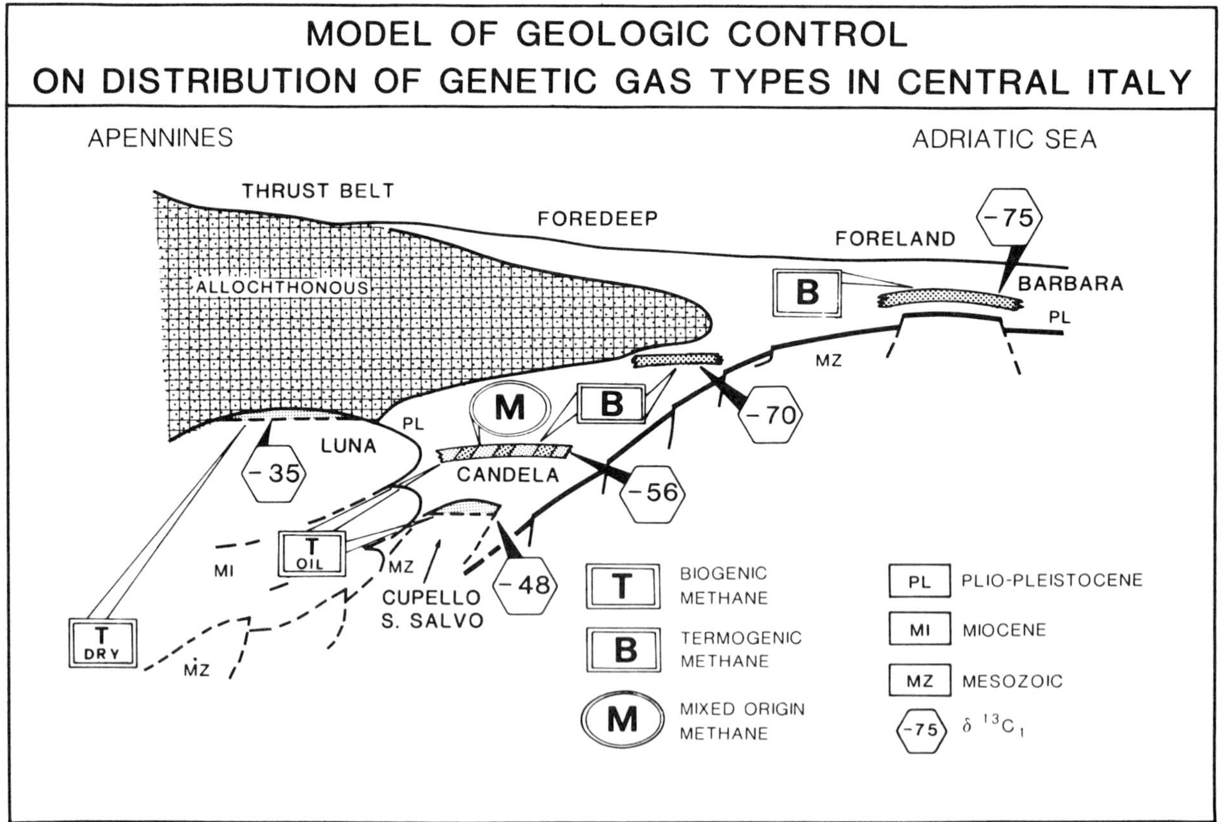

Figure 17. Geologic sketch showing the distribution of gas types (biogenic, thermogenic, mixed) in central Italy.

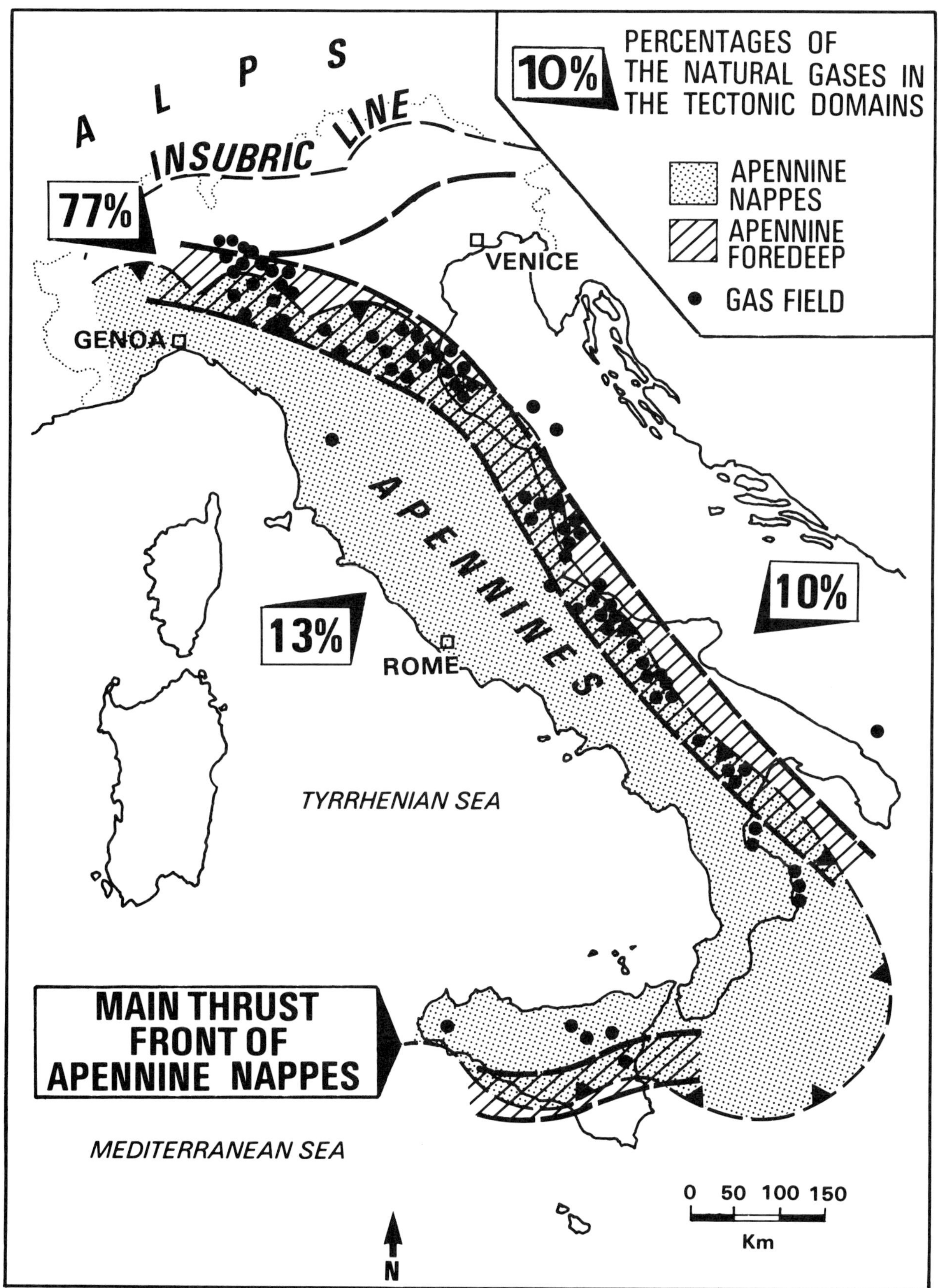

Figure 18. Main tectonic domains and distribution of natural-gas fields in Italy.

Geochemical Alteration of Organic Matter in Eutrophic Lake Greifen: Implications for the Determination of Organic Facies and the Origin of Lacustrine Source Rocks

D. Hollander
Geologisches Institut
ETH-Zentrum
Zürich, Switzerland
Current Address: Institut Francais du Pétrole
Rueil-Malmaison, France

F. Behar
Institut Francais du Pétrole
Rueil-Malmaison, France

M. Vandenbroucke
Institut Francais du Pétrole
Rueil-Malmaison, France

P. Bertrand
Unité de Recherche en Pétrologie Organique
Université d'Orléans, Orléans, France

J. A. McKenzie
Geologisches Institut
ETH-Zentrum
Zürich, Switzerland

The most recent sedimentary sequence (ca. <200 yr) of Lake Greifen in northeastern Switzerland shows evidence for an oxic/anoxic depositional-environment transition dated at 1930. Organic matter is composed predominantly of autochthonous algal and amorphous material with minor contribution of allochthonous terrestrial material. A change in the chemical composition and concentration of organic matter across the laminated-marl/seasonally varved organic-rich sediment transition is characterized by decreasing O/C and increasing H/C ratios, Hydrogen Indices, TOC, organic carbon accumulation rates, and petroleum potential (kg Hydrocarbons/ton bulk rock). Variations in the quality and quantity of organic matter deposited are synchronous with measured changes in water-column-nutrient concentrations, surface-water productivity, and, most importantly, the thickness and extent of the oxygen-depleted water mass. The lacustrine algal source material deposited under predominantly anoxic waters (seasonally varved sediments) is hydrogen rich and resembles chemically type I kerogens and organic facies "A" to "AB" according to Jones' classification (Jones, 1987), while organic matter deposited under primarily oxygenated waters (laminated-marl sediments) shows H/C and O/C ratios and HI values resembling those of type II or III kerogens and organic facies "B" to "BC". The correlation between changing composition and concentration of deposited organic matter and the historical water-column geochemical record documents how quality and quantity of organic matter can be significantly altered, depending upon the relative thickness of the oxygen-depleted water mass at the time of deposition. Variability in the efficiency of organic-matter decomposition during settling and at the sediment-water interface, in association with a varying redox potential of the oxygen-depleted water mass and in the sediments, is perhaps the ultimate control on the composition and concentration of sedimentary organic matter and the determination of organic facies (Jones, 1987).

The deposition of high-quality lacustrine organic facies "A" was constrained stratigraphically to a time interval when the water column was severely depleted in oxygen and sulfate and the sediments were permanently anoxic. Moreover, the occurrence of methane in the water column at this time suggests that fermentation processes in the permanently anoxic sediments and possibly in the stratified water-column were probably dominating organic-matter decomposition. Carbon isotopic analyses of individual branched and cyclic hydrocarbons further suggest that methanotrophic and chemoautotrophic bacteria were present in the uppermost sediment layers and the severely anoxic waters. We postulate that, during times of widespread depletion of oxygen and sulfate in the water column and sediments, the predominance of bacterial fermentation processes controlling organic-matter degradation and the addition of a lipid-rich bacterial biomass to the sediments resulted in the occurrence of a highly preserved organic facies "A" and/or sediments of an excellent petroleum potential. The predominance of an anoxic and sulfate-depleted water column and the existence of permanently anoxic sediments are perhaps the environmental conditions necessary to explain the common occurrence of highly preserved organic facies "A" in modern and ancient lacustrine settings and their rare occurrence in marine environments.

INTRODUCTION

Recently, the relative degree of organic-matter preservation in sediments has been set into a classification scheme of seven organic facies from "A" to "D" (Jones, 1987). This concept of organic facies is primarily based on hydrogen richness (Hydrogen Indices and H/C atomic elemental ratios) and secondarily on optical characteristics of organic matter in sediments. Organic facies "A" shows the highest degree of preservation while organic facies "D" has the lowest. Paramount to the preservation of organic matter is the oxygen content of the waters overlying the sediments (Demaison and Moore, 1980; Emerson, 1985). Environmental conditions affecting the oxygen content of the water column and the preservation of organic matter include productivity rates, sedimentation rates, and the liability of the organic matter to oxidative processes (Calvert, 1987; Tyson, 1987; Morris, personal communication). The occurrence of organic facies "A", which is chemically defined by HI>850 mg HC/g Org.-C, H/C>1.45 and optically characterized by an overwhelming dominance of algal, amorphous organic matter and minor amounts of terrestrial material, is common in lacustrine environments but rare in marine environments (Jones, 1987). Jones suggested that the persistence of sediment anoxia is necessary, but is not a sufficient condition for the deposition of an organic facies "A". A host of other hypotheses has been set forth in order to explain their occurrence and includes bacterial reworking, Eh and pH controls, and/or sulfate contents of the waters (Tissot et al., 1978; Tissot and Welte, 1984; Jones, 1987; Kelts, 1988). However, until now, these are hypotheses, and the environmental conditions which favor the deposition of an organic facies "A" source rock remain uncertain.

It has been shown in water-column studies that bacterial decomposition of organic matter in oxygenated waters can decrease the flux of organic matter to the sediments and can alter the original organic geochemical signatures of the biological material (Müller and Suess, 1979; Suess, 1980; Tissot and Pelet, 1981; Pelet, 1981; Pelet and Deroo, 1983; Pelet, 1984; Vandenbroucke et al., 1985; Calvert, 1987; Pelet, 1987; Tyson, 1987; Cole et al., 1987). Sedimentological studies have also shown that the degree of bioturbation in the sediments and the hydrological characteristics of a basin can cause measurable changes in the Hydrogen and Oxygen Indices and atomic elemental ratios of the deposited organic matter (Pratt, 1984; Huc 1988a, b). In determining organic facies and making source-bed evaluations, it is important to understand processes occurring in the water column and sediments that alter the Hydrogen Indices (mg hydrocarbons/g Org.-C) and the atomic elemental ratios (H/C and O/C) of an organic source material. Specifically, what role does water-column oxygenation play on the alteration of bulk organic geochemical parameters of an algal source material so that: (1) degradation of primary organic matter by bacterial communities becomes important, (2) incorporation of bacterial biomass to the sediments becomes significant, (3) there is a change in organic-facies classification and/or preservation of organic matter and, (4) deposited organic matter resembles, by its bulk chemical parameters, material of varying origin according to the type I, II, and III classification scheme proposed by Espitalié et al. (1976) and Durand and Monin (1980).

Over the past 100 years, Lake Greifen, an eutrophic lake in northeastern Switzerland, has experienced a major increase in surface-water productivity and eutrophication due to increased nutrient influx, a result of increased agriculture and industrialization in the drainage area. A 40-year historical record of water-column geochemistry indicates that lake productivity and eutrophication reached a maximum in 1974. Since then, stricter regulation on the input of nutrients has been imposed and, subsequently, lake productivity and eutrophication have progressively decreased. In this paper, a study of the organic geochemistry of the lacustrine sediments deposited during this historical period of productivity change and fluctuating thickness of the oxygen-depleted water mass is presented. This case study provides an opportunity to determine the importance that an expanding oxygen-depleted water mass and various bacterial communities have on the chemical composition of an original organic source material, the determination of organic facies, and the petroleum potential of sediments. In addition, the controls on the environmental conditions which existed in the water column and sediments of Lake Greifen allow us to determine the conditions which favor the deposition of a hydrogen-rich, organic facies "A" source rock.

METHODS

Filtered samples of biological material from Lake Greifen surface waters and sediment core samples were demineralized using 1N HCl and 40% HF for the preparation of proto-kerogens. Bulk samples were analyzed for TOC content using a Leco apparatus. Analyses to determine the elemental composition of proto-kerogens were made by ATX Inc., France, and infrared absorption spectra were obtained using methods of Rouxhet et al. (1980). Proto-kerogens from sediments were pyrolyzed using a minifurnace, and pyrolysates were further fractionated into saturated, branched-cyclic, unsaturated, and aromatic hydrocarbons, and NSO compounds following the methods of Behar and Pelet (1985). Saturated and unsaturated hydrocarbons were injected into a Varian 3500 gas chromatograph equipped with an autosampler and on-column injector on a 1 CB silica capillary column (30 m × 0.32 mm ID). Branched-cyclic hydrocarbons were analyzed by GC-mass spectroscopy using a Varian GC coupled to a Nermag R50 10C quadrapole spectrometer, and carbon-isotope ratios of individual compounds were determined using an improved method of isotope ratio monitoring-gas chromatography-mass spectroscopy (irm-GCMS) (Freeman et al., 1990). Methods used for the acquisition of all geochemical parameters measured in the water column (PO_4 and O_2 concentration, productivity, temperature) are obtainable through Dr. H. Ambühl, Limnology Research Group, EAWAG, Dübendorf, Switzerland.

RESULTS AND DISCUSSION

Water Column and Sediment Record of Increasing Eutrophication and Productivity

Presently, the annual cycle in Lake Greifen is as follows. Water-column overturn in winter, as a result of decrease in the surface-water temperature, replenishes nutrients to the surface waters and brings oxygen to the bottom waters (McKenzie, 1982; Hollander, 1989). During spring and summer, productivity blooms in the epilimnion consume these nutrients and produce a flux of settling organic matter. Decomposition of the sinking organic matter by bacteria in the hypolimnion results in the development of oxygen-depleted bottom waters in as little as 3 months after winter overturn. These 3 months when the bottom waters are oxygenated are insuffícent for the development of an aerobic bottom-dwelling community at the sediment-water interface. Thermal and chemical stratification of the water column prevails for approximately 75–80% of the annual cycle.

The annual cycles are recorded in the sediments as seasonal varves, in which fine clay layers represent the winter turnover, and organic-carbon-rich horizons and inorganically precipitated carbonates characterize spring and summer productivity blooms (Hollander, 1989). During the last 100 years, progressive eutrophication with the gradual increase in nutrient influx is clearly seen in sediment cores which show a transition from laminated marls to seasonally varved, organic-carbon-rich sediments (Figure 1). During the earlier period of marl deposition, the majority of the water column was oxic, as indicated by the historical water-column geochemical record (Figure 1; Table 1). However, during this episode of deposition, the sediment-water interface was probably dysaerobic to anoxic as shown by the preservation of undisturbed laminations. With progressive eutrophication since 1930, the sediment-water interface became permanently anoxic, and seasonally varved, organic-carbon-rich sediments were deposited. During this time interval of increased eutrophication, the water column experienced major fluctuations in the thickness of the oxygen-depleted water mass, and in 1974, up to 45% of the total water column was anoxic. Since 1974 and the reduction in nutrient input, the extent of the oxygen-depleted water mass has decreased to cover only 25% of the water column while the sediments have remained

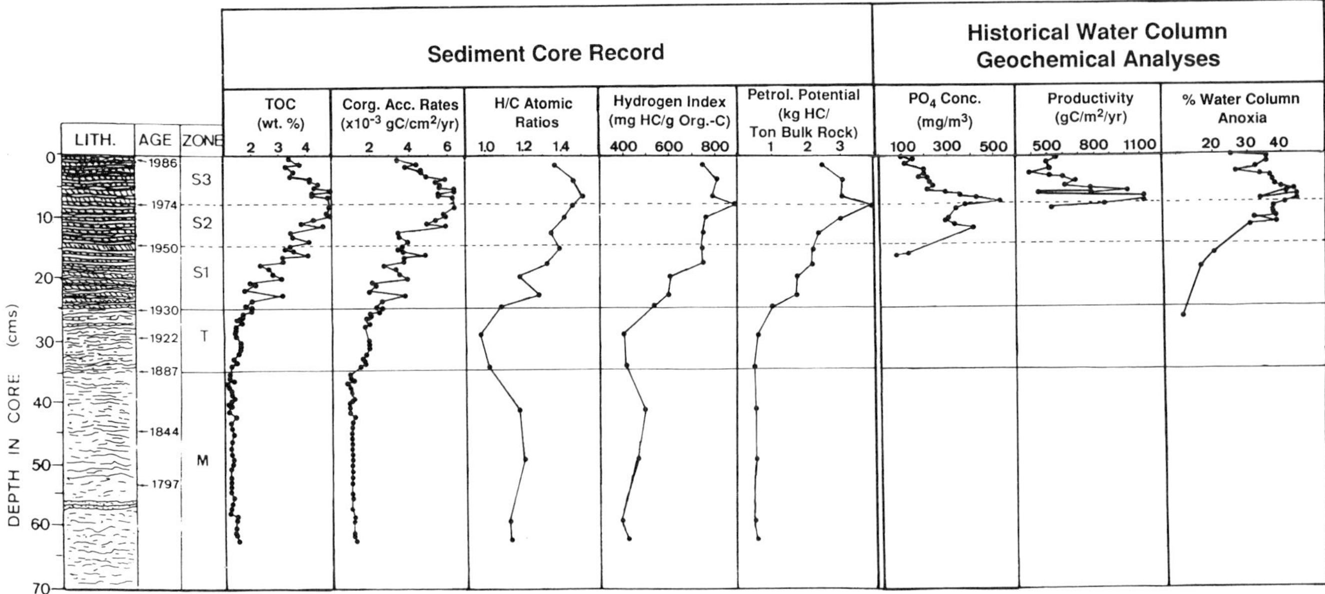

Figure 1. Schematic stratigraphic column, sediment zones (M, T and S), and age of sediments from the Lake Greifen sediment core. Historical water-column geochemical analyses (concentration of PO_4 and O_2 and the rate of productivity), dating back to 1917, are plotted adjacent to their year of acquisition. The sapropel zone (S) can be divided into three subzones relative to the severity of eutrophic conditions during the lake's history (see Hollander, 1989). Note that the year 1974 marks the maximum in lake eutrophication. Since 1974, the eutrophic condition of the lake has improved as indicated by decreasing PO_4 concentration, rate of productivity, and the percent of water column anoxia. PO_4 concentrations were measured at the time of winter turnover; the rates of primary productivity were calculated by the summation of monthly rates of productivity as determined by the C-14 method; O_2 concentrations were transformed into percent anoxic water column by integration of the O_2 concentration isopach of O mg/l over the annual cycle.

permanently anoxic (Figure 1; Table 1) (Ambühl, 1988; Hollander, 1989).

Source of Deposited Organic Matter

Optical characterization of the organic matter both above and below the laminated-marl/seasonally varved sediment transition indicates a predominantly algal and amorphous source material (85–95%) with minor contributions of allochthonous terrestrial material (5-15%). Throughout the entire sediment core, the occurrence of algal structures and minimal distribution of pollen and spores supports these estimates. The largest difference in the petrologic characteristics of organic matter through the sediment core is the relative proportion of amorphous organic matter, which is most abundant in the seasonally varved sediments. Gas chromatograms of alkanes and alkenes isolated from pyrolysates of proto-kerogens confirm the overall uniformity of the organic source material across the sediment transition (Figure 2). The lack of odd to even predominance for saturated hydrocarbons and minimal occurrence of the long-chained hydrocarbons (>C-25) also suggest a relatively small contribution of terrestrial organic matter to the sediments. Similar gas chromatograms of pyrolysate-extracted alkanes in recent anoxic sediments from thermally stratified Lake Tanganyika also show a predominance of carbon-chain lengths in the C-15 to C-21 range (Vandenbroucke et al., 1985). Analyses of branched-cyclic hydrocarbons from the Lake Greifen sediments indicate that higher molecular weight hydrocarbons are rich in hopanes with minor amounts of steranes (>C-27) (Figure 2). The consistency in the distribution of individual branched-cyclic hydrocarbons further emphasizes the uniformity of organic source materials throughout the Lake Greifen sediment core (Hollander, 1989).

Although the organic source material is predominantly of planktonic origin, variations in the relative populations of phytoplankton, such as blue-green algae, diatoms, and other green algae, must be quantified in order to assure that variations in chemical composition of the deposited organic matter are not the result of historical changes in phytoplankton population dynamics. A 60-year record of pigment (carotenoids) extracts and skeletal remains from the Lake Greifen sediments shows increasing amounts of pigments and skeletal remains, together with increasing lake eutrophication and productivity, until 1974 when these parameters reached a maximum (Züllig, 1982). Since 1974 and the subsequent reduction in productivity, the concentrations of total pigments and skeletal remains have decreased

Table 1. Historical water column geochemical measurements.

Age	Depth (cm)	PO_4 Concentration (mg/m^3)	Anoxic Water/Total Water Column (%)	Primary Productivity ($gC/cm^2/yr$)
1987	0.0	96	25.26	501.67
1986	0.5	150	35.48	556.28
1985	1.1	115	35.58	496.68
1984	2.1	198	32.20	513.61
1983	2.8	202	26.71	391.86
1982	3.3	176	33.46	513.75
1981	3.5	212	36.58	600.18
1980	4.2	226	37.32	680.99
1979	4.8	238	37.83	614.03
1978	5.3	213	39.87	774.31
1977	5.7	291	44.08	1016.74
1976	6.1	350	41.42	785.08
1975	6.3	351	35.55	430.18
1974	6.7	428	44.58	1126.43
1973	7.3	529	44.75	1118.96
1972	7.9	380	41.22	869.46
1971	8.5	335	37.35	525.00
1970	9.1	—	37.25	—
1969	9.7	—	37.28	—
1968	10.2	300	38.38	—
1967	10.24	290	31.90	—
1966	11.1	33	38.74	—
1965	11.5	408	30.62	—
1964	11.7	410	—	—
1953	15.7	131	—	—
1951	16	79	20.31	—
1941	18.3	—	16.41	—
1930	25	—	—	—
1917	26.4	—	11.22	—

(Züllig, 1982). In contrast to the systematic variations in total pigments and skeletal remains, there were no changes in the relative proportion of pigments (carotenoids) and skeletal remains which characterize the input of specific phytoplankton: e.g., blue-green algae (myxoxanthophyll), other green algae (zeaxanthin, diadinoxanthin, and leutein), and diatoms (centricae and pennatae) (Züllig, 1982). This consistency in the relative distribution of pigments and skeletal remains suggests that, although the trophic status of the lake had dramatically changed, there were no significant changes in the phytoplankton population dynamics within the lake.

Organic-Matter Quality and Quantity

The chemical compositions of proto-kerogens and concentrations of total organic matter from Lake Greifen sediments are presented in Table 2. "Proto-kerogens" extracted from filtered samples of biological materials from lake surface waters allow measurement of the initial atomic elemental ratios of the autochthonous "planktonic" organic matter. Plotted on a Van Krevelen diagram (Figure 3), elemental ratios (H/C and O/C) of proto-kerogens across the laminated-marl/seasonally varved sediment transition vary dramatically with H/C values ranging from 0.97 to 1.53 and O/C values from 0.28 to 0.15 (Table 2). Relative to the elemental ratios of the biological material from the surface water, the organic matter deposited under primarily anoxic waters (seasonally varved sediments) shows decreases of up to 50% in the O/C and 15% in the H/C ratios, whereas organic matter deposited under predominantly oxic waters (laminated-marl sediments) is characterized by only a 7% decrease in the O/C and a maximum decrease of 46% in the H/C ratios (Figure 3). According to the classification scheme proposed by Durand and Monin (1980), our results could be interpreted as a change in the source of organic matter from one dominated by terrestrial material to one dominated by algal material. However, optical petrologic data, gas-chromatographic results, and concentrations of pigment and skeletal remains all strongly suggest that the material has remained consistent through the sediment core and is dominated by algal and amorphous organic matter. Figure 3 portrays suggested degradation and/or alteration pathways for primary algal organic matter in permanently anoxic sediments where settling occurred through variably oxygenated waters, as shown by elemental-ratio variations in Lake Greifen proto-kerogens. These trends in elemental composition of the deposited organic matter are also supported by results of infrared spectroscopy, which show that proto-kerogens deposited under predominantly

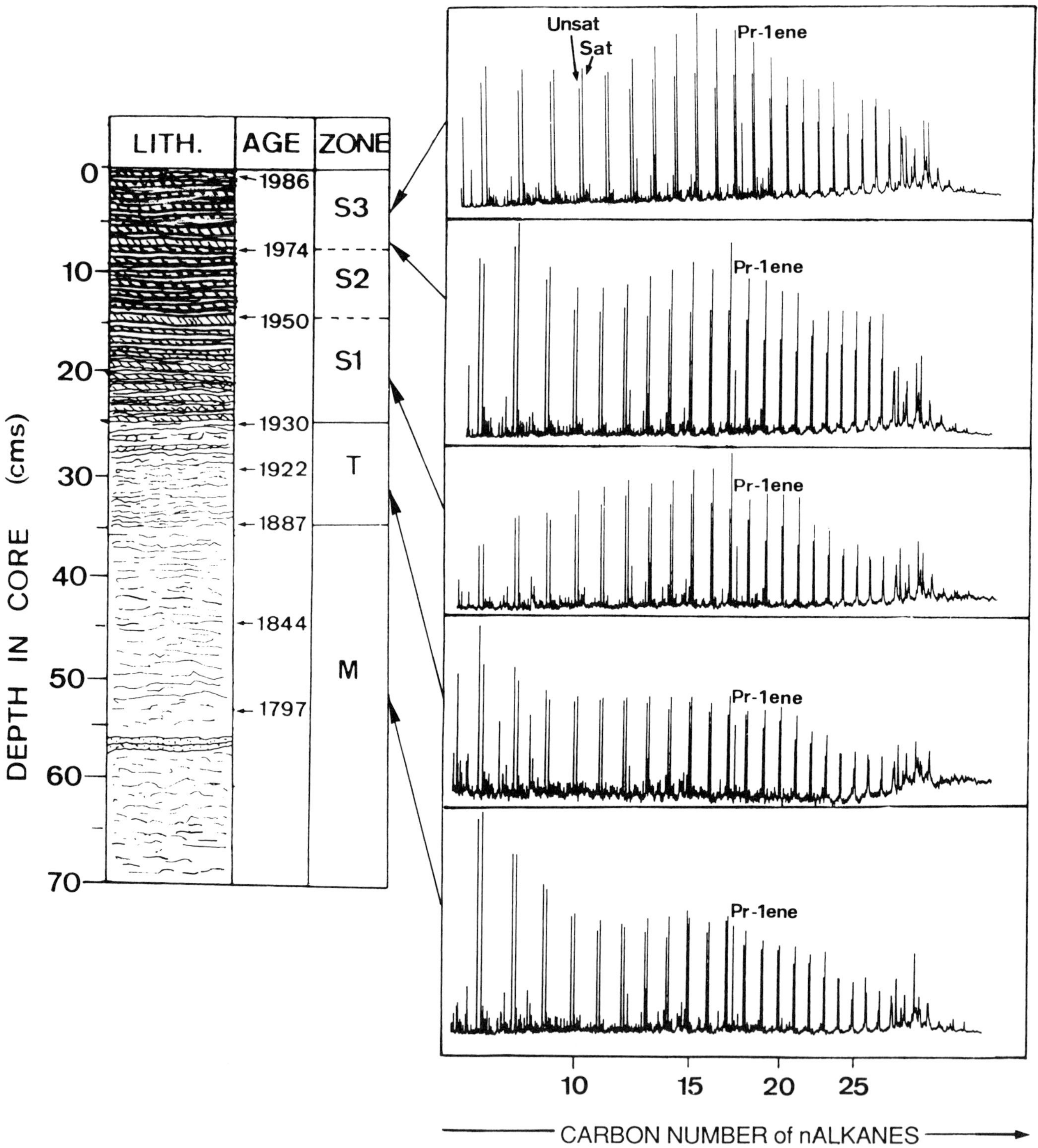

Figure 2. Chromatograms of saturated and unsaturated hydrocarbons isolated from C7+ pyrolysates of protokerogens from laminated marls (M), transition zone (T), and seasonally varved sediments (S) when the water column of Lake Greifen was predominantly oxic, dysaerobic, and anoxic, respectively. Similarity between the chromatograms from varying depositional environments suggests that the source material in Lake Greifen has not changed in accordance with the changing degree of water-column oxygenation. Throughout the entire sediment core, there is a predominance of short-chain hydrocarbons (<C-25) with no odd/even dominance. The higher molecular weight hydrocarbons (>C-27) are dominated by hopanes with minor amounts of steranes. The distribution of saturated and unsaturated hydrocarbons suggests that Lake Greifen sediments contain only a minor contribution of terrestrial material.

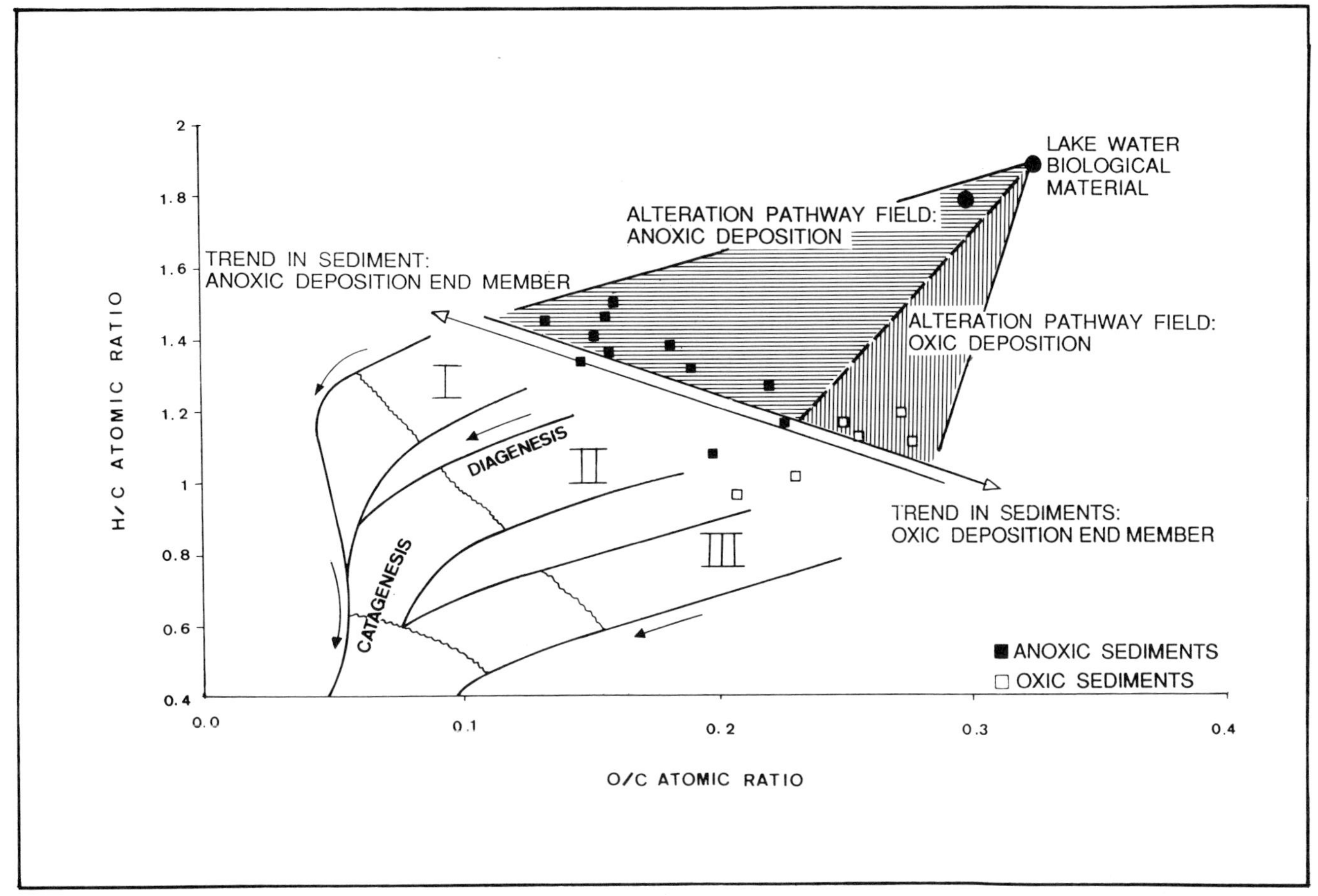

Figure 3. Atomic H/C versus O/C ratios of proto-kerogens from filtered surface-water biological material (filled circles) and deposited organic matter from predominantly oxic (open squares) and anoxic (filled squares) depositional environments. Horizontal and vertical hatched areas represent alteration pathways in anoxic and oxic water columns, respectively. Trends observed in sedimented proto-kerogens produce an apparent pathway relative to the thickness of the oxygen-depleted water mass, characterized by anoxic and oxic end member pathways. Evolution of kerogen composition during burial diagenesis and catagenesis processes for different reference kerogens (Durand and Monin, 1980) are marked by ranges and arrows for each particular path.

anoxic waters are relatively depleted in oxygen-rich carbonyl, alcohol, and ether functional groups and relatively enriched in aliphatic material when compared to proto-kerogens deposited under predominantly oxygenated waters (Figure 4).

The decomposition of organic matter in the water column and sediments proceeds through a series of reactions which are controlled by the concentration of dissolved oxidants (i.e., biological electron acceptors: O_2, NO_3^- and $SO_4^=$). This series of reactions is characterized by decreasing efficiency following the sequence: oxic degradation, nitrate reduction, sulfate reduction, and, finally, bacterial fermentation (Irwin et al., 1977; Froelich et al., 1979). In the case of a water column and sediments severely depleted in oxygen and other dissolved oxidants, fermentation processes dominate organic-matter decomposition (Froelich et al., 1979). In the water column and sediments of Lake Greifen, oxygen and sulfate concentrations are so low that degradation by aerobic and sulfate-reducing bacteria is not controlling decomposition of organic matter (Ambühl, personal communication). The presence of dissolved methane in intermediate and bottom waters of Lake Greifen indicates that bacterial fermentation is now and, at least since the onset of methane analyses in 1971, has been the dominant anaerobic process degrading organic matter both in the surface sediments and possibly in the water column (Ambühl, 1988; Hollander, 1989). This in turn suggests that, as the flux of organic matter settling through the hypolimnion increased in correlation with historical changes in productivity, organic-matter decomposition proceeded through the less efficient bacterial fermentation processes. This differential degradation of organic matter by various bacterial communities in the water column and sediments, in relation to the concentration of dissolved oxygen sources may, in part, explain the observed changes in the composition of deposited organic matter.

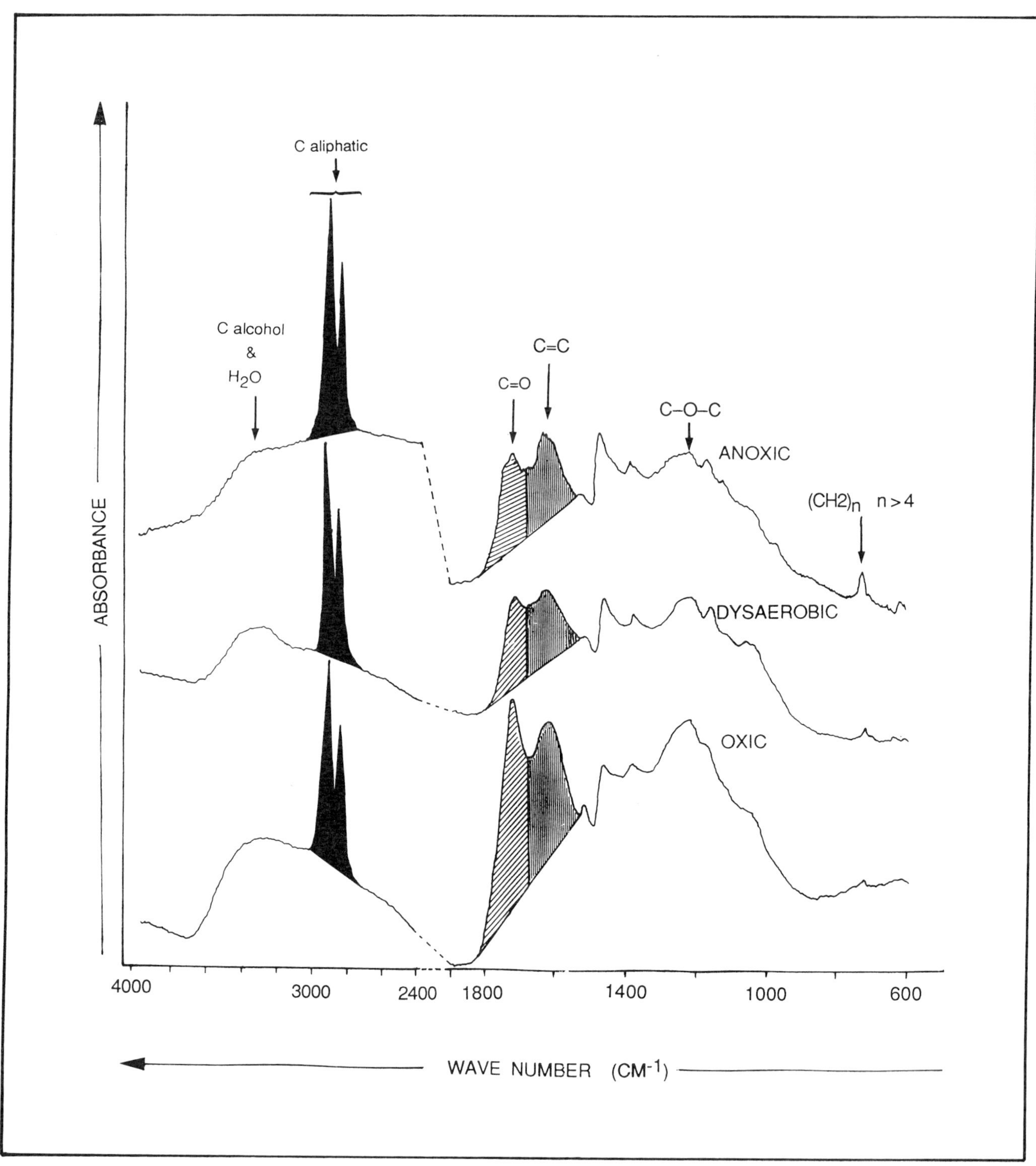

Figure 4. Infrared spectra of proto-kerogens from the laminated marls, transition-zone sediments, and seasonally varved organic-rich sediments when the water column was predominantly oxic, dysaerobic, and anoxic, respectively. The peak height for each particular functional group represents its relative abundance. Deposition of organic matter in a progressively more anoxic water column is characterized by decreasing the relative concentration of oxygen-rich carbonyl (1710 cm-1), alcohol (3400 cm-1), and ether (1250 cm-1) functional groups and increasing the relative concentration of aliphatic (2900 cm-1) functional groups and $(CH_2)_n$ $n>4$ aliphatic chains.

Methane formed during fermentation processes mediated by methanogenic bacteria carries a carbon-isotope signature which is extremely negative relative to coexisting dissolved CO_2 (Whiticar et al., 1986). Biogenic methane produced in fresh-water sediments has an average $\delta^{13}C$ value of -60 °/oo, with values ranging from -40 to -80 °/oo (Whiticar et al., 1986). Most recently, a similar range of carbon isotopic ratios has been observed in some hopanes from the Messel Shale (Eocene), from Germany, which was deposited in an ancient anoxic lacustrine environment (Freeman, 1989; Freeman et al., 1990). The origin of these hopanes was linked to bacterial communities which assimilated both methane and CO_2 depleted in C-13 during a time when the water column and sediments were severely anoxic (Freeman, 1989; Freeman et al., 1990). The incorporation of these bacterially derived hopanes provided evidence that the sedimentary organic matter in the Messel Shale was of multiple origin, including contributions from primary producers and bacterial biomass (Freeman, 1989; Freeman et al., 1990).

Preliminary results from isotope ratio monitoring-gas chromatography-mass spectroscopy (irm-GCMS) of individual branched-cyclic hydrocarbons isolated from pyrolysates of proto-kerogens from the Lake Greifen sediments indicate that hopanes have $\delta^{13}C$ values within the range -45 to -75 °/oo. As in the Messel Shale, carbon isotopic composition of these hopanes suggests that the sedimentary organic matter has multiple origins, including a contribution from bacterial methanotrophs and possibly chemoautotrophs using CO_2 depleted in C-13. This makes it clear that methanogens were present in the sediments and possibly in the water column. Bacterial lipids from such archaebacteria have been shown to be comprised predominantly of aliphatic-rich cell walls and membranes (Chappe et al., 1979). Although the exact contribution of hopanes in the Lake Greifen sediments is unknown, these lipids are found in a greater proportion in the seasonally varved sediments which were deposited under predominantly anoxic waters (Hollander, 1989). This suggests that the bacterial lipids could potentially contribute an additional hydrogen-rich source material to the sediments (Figures 3, 4; Table 2).

The Hydrogen Indices of proto-kerogens range from 395 mg HC/g Org.-C in the laminated-marl sediments (deposited under oxygenated waters) to a maximum value of 888 mg HC/g Org.-C in the seasonally varved sediments (deposited under oxygen-deficient waters) (Figure 5; Table 2). This more than two-fold increase in the Hydrogen Indices is stratigraphically correlated with a four-fold increase in organic carbon concentration (from 1.2% to 4.9%) and a six-fold increase in the organic-carbon accumulation rate (from 0.001 to 0.0065 g $C/cm^2/yr$) (Figure 5; Table 2). These trends of increasing quality and quantity of organic matter, in association with the laminated-marl/seasonally varved sediment transition, are synchronous with measured increases in nutrient supply (PO_4) and higher rates of surface-water productivity (Figure 5). This increased flux of primary organic matter obviously caused the increased oxygen consumption by bacteria below the photic zone during respiration and gave rise to the progressive expansion of an oxygen-depleted water mass (Figure 5). We propose that increases in the thickness of the oxygen-depleted water mass and decreases in the redox potential of the water column promoted a decrease in organic-matter decomposition because anaerobic bacteria are less efficient than aerobic bacteria, and that this is the probable cause for the majority of observed changes in the chemical composition, relative concentrations of functional groups, and, in part, total organic carbon content of deposited organic matter throughout the sediment core.

The sediment-water interface and the uppermost 10 cm of sediments are normally the most active sites of benthic decomposition and early-burial chemical alteration, respectively (Emerson et al., 1985; Emerson, 1985; Cole et al., 1987) and should, therefore, show systematic decreases in both quality and quantity of organic matter. In contrast, the uppermost 10 cm in the Lake Greifen core shows systematic increases in both the quality and quantity of deposited organic matter (Figure 5). Concerning the process of benthic decomposition, organic geochemical changes observed in Lake Greifen sediments resemble the bulk geochemical parameters observed in the Cenomanian/Turonian Boundary Event of the Western Interior Basin, North America (Pratt, 1984; Pratt et al., 1986) and the Miocene sediments of the Monterey Formation, California (Mertz, 1984). These studies were able to associate compositional changes of organic matter with the relative degree of bioturbation in the sediments. In the Lake Greifen sediment core, the preservation of seasonal varves since 1930 and laminations in the marl sediments before 1930 indicates that burrowing fauna were not active. Therefore, we can assume that benthic decomposition had little influence on the observed variation in the composition and concentration of deposited organic matter throughout the Lake Greifen sediment core.

The variations in the composition and concentration of deposited organic matter are also probably not the result of early-burial chemical diagenesis. The sediment deposited between the years 1974 and 1986 (corresponding to 0–7 cm sediment depth) shows systematic increases in both the quality and quantity of organic matter with increasing sediment depth, suggesting that early-burial chemical alteration did not significantly overprint the original geochemical composition and concentration of the deposited organic matter (Figure 5; Table 2). Moreover, the most dramatic increases in Hydrogen Indices, H/C ratios, TOC concentrations, and organic-carbon accumulation rates all occurred within the permanently anoxic burial environment of the seasonally varved sediments (Figure 5). The uniformity in

Table 2. Sediment geochemical record.

Depth (cm)	TOC (%)	Org.-C Accumulation Rate (g/cm²/yr)	H/C (elemental ratio)	Hydrogen Index (mg HC/g Org. C)	Petroleum Potential (kg HC/ton bulk rock)
1.5	3.81	0.0044	1.366	746	2.45
4.0	4.18	0.0054	1.464	809	3.06
6.5	4.23	0.0055	1.507	790	3.05
8.0	4.91	0.0064	1.456	888	3.97
10.0	3.84	0.0049	1.412	761	2.99
12.5	3.49	0.0034	1.339	746	2.36
15.0	3.28	0.0034	1.391	744	2.16
17.5	3.18	0.0037	1.325	747	2.15
19.5	2.81	0.0035	1.182	609	1.73
22.5	2.06	0.0026	1.279	597	1.72
24.5	1.86	0.0023	1.082	538	1.01
29.0	1.46	0.0019	0.975	404	0.60
34.0	1.31	0.0017	1.021	414	0.50
41.0	1.24	0.0011	1.178	498	0.54
49.0	1.37	0.0010	1.204	467	0.58
59.0	1.50	0.0011	1.126	395	0.52
62.0	1.56	0.0012	1.135	423	0.60

anoxic burial chemical environments within the seasonally varved sediments further suggests that early-burial chemical diagenesis could not independently generate the dramatic changes in the quality and quantity of organic matter observed in the Lake Greifen sediments.

Results of this study strongly imply that changing rates of surface-water productivity and subsequent expansion of the oxygen-depleted water mass have controlled quality, as well as quantity, of organic matter. Figure 5 allows comparison of the record of water-column conditions with the organic geochemical record. The observed correlation between increasing quality and quantity of deposited organic matter and increasing productivity and expansion of the oxygen-depleted water mass may be explained by: (1) increasing preservation of organic matter in association with decreasing redox potential of the water column and sediments and the predominance of bacterial fermentation processes for the degradation of organic matter and (2) perhaps the addition of a lipid-rich bacterial biomass to the surface sediments. The alteration of the original composition of an algal source material by various bacterial communities can be so significant that it has the potential to change the organic-facies determination and the petroleum potential of the sediments.

IMPLICATIONS

Controls on the Determination of Organic Facies

According to the classification scheme proposed by Jones (1987), the bulk composition of Lake Greifen sediments places them in organic facies "B" to "BC" from the laminated-marl sediments (HI=395 to 498 mg HC/g Org.-C; H/C= 0.98 to 1.2) and in organic facies "A" to "B" from the seasonally varved sediments (HI=598 to 888 mg HC/g Org.-C; H/C=1.18 to 1.51). In contrast, organic petrological analyses indicate that the Lake Greifen sediments are composed only of organic facies "A" to "AB" which are characterized by the predominance of algal and amorphous organic matter with only minor occurrence of terrestrial material (Jones, 1987). The changes in composition of deposited organic matter across laminated-marl/seasonally varved sediment transition, characterizing an increase in organic-matter preservation, have been previously postulated to be the result of a change in the decomposition of organic matter from predominantly oxic degradation to bacterial fermentation, and perhaps the contribution of a lipid-rich bacterial biomass to the sediments.

The implication of the Lake Greifen case study for the determination of organic facies, according to the scheme proposed by Jones (1987), is that the thickness and the extent of the oxygen-depleted water mass and the concentration of other dissolved oxidants will have a significant effect on the preservation and/or alteration of primary source material. The bulk organic geochemical parameters can be an excellent indicator of the relative preservation of this organic matter which, in turn, is a reflection on oxygen-depletion in the water column and sediments. In contrast, morphological characteristics of the organic matter seems best for characterization of the primary source. This case study indicates that a multidisciplinary approach to the studies of organic facies including gas chromatography (GC), GC-mass spectroscopy (GCMS), irm-GCMS, infrared spectroscopy, and detailed optical petrography can provide significantly more information than routine characterization of bulk chemical composition and basic optical properties of the deposited organic matter.

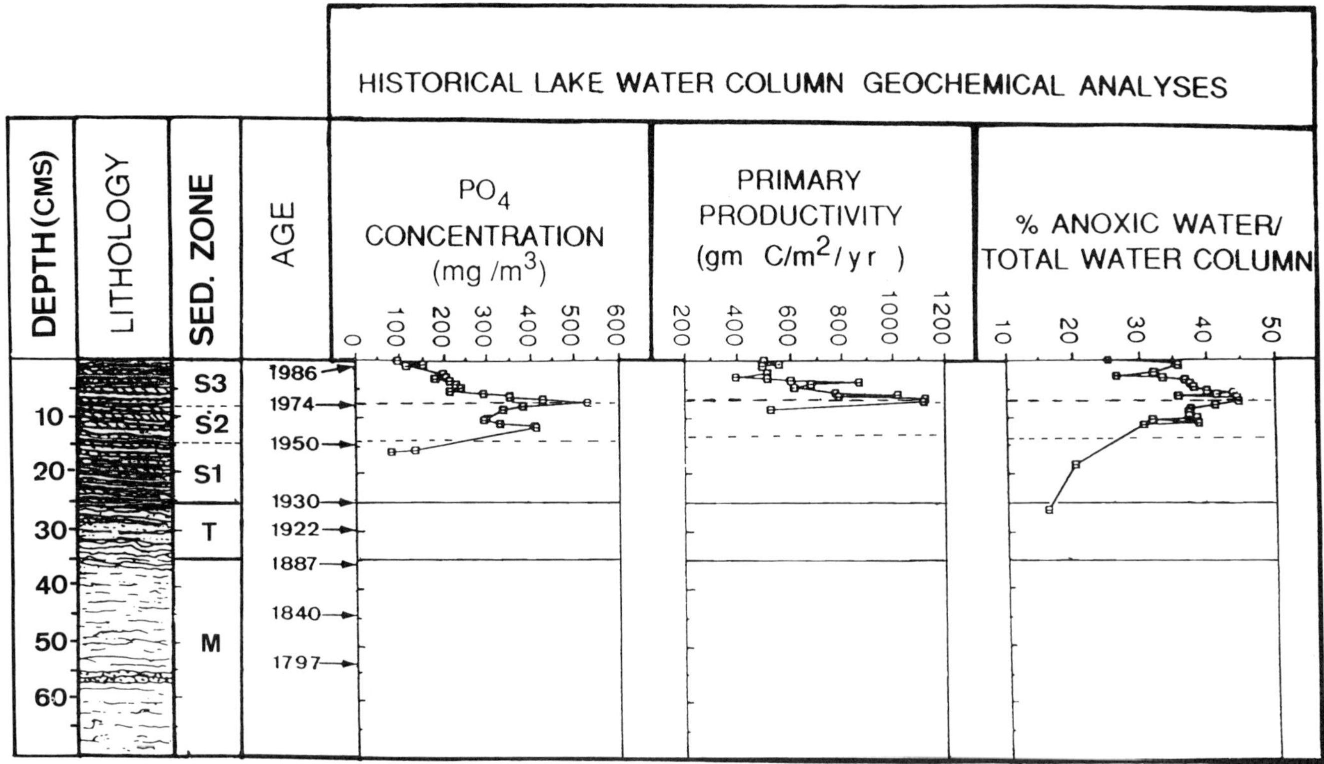

Figure 5. Stratigraphic column, total organic-carbon content, organic/carbon accumulation rates, H/C atomic ratios, and Hydrogen Indices of proto-kerogens and sediment petroleum potential for the Lake Greifen sediment core and the historical water column geochemical measurements from Lake Greifen. Note that the transition from laminated marls (M) (below 37 cm) to transition/zone sediments (T) (25-37 cm) to seasonally varved sediments (S1 and S2)(25-7.8 cm) is correlated with a two-fold increase in Hydrogen Indices, four-fold increase in total organic carbon, six-fold increase in organic-carbon accumulation rate, and eight-fold increase in petroleum potential. Since 1974 (S3) (above 7.8 cm), the PO_4 concentrations, rate of primary productivity, and percent of anoxic water column have all decreased and are correlated with decreases in the quality and quantity of deposited organic matter. The TOC contents and organic-carbon accumulation rates presented in Table 2 are an annotated data set and not the complete data set presented in this figure (see Hollander, 1989).

Environmental Controls on the Deposition of Organic Facies "A"

The controls of the water column on the geochemistry of depositional and burial environments from Lake Greifen provide a causal determination for the conditions which favor the deposition and accumulation of an organic facies "A" source rock. In the Lake Greifen sediments, the most significant changes in the hydrogen richness of the organic matter occur within the seasonally varved sediments where early-burial chemical environments were permanently anoxic, and primary sources of organic matter have remained uniform. During this time interval, the most dramatic environmental change occurred in the water column, as reflected by an increase in the thickness and extent of the oxygen-depleted water mass. We infer that chemical characteristics of the water column are the most important factors controlling the hydrogen richness of sedimentary organic matter. The deposition of an organic facies "A" (lacustrine source rock) in Lake Greifen is limited to a time when the majority of the water column was oxygen-deficient and the sediments were permanently anoxic (represented by approximately 7 cm of sediment core). This, in turn, suggests that the redox potential of the water column, which, in part, controls the efficiency of organic-matter decomposition during settling and the contribution of a bacterial biomass to the sediments, is the constraining environmental condition that favors the deposition of organic matter with the highest preservation.

In an environment where the sedimentation rate has remained constant, the degree of organic-matter decomposition during settling in the water column and at the sediment surface is not only related to the oxygen concentration but also, in part, to the concentration of other dissolved oxidants, such as NO_3^- and $SO_4^=$. It follows that in a water column which is predominantly anoxic but where these other oxidants are abundant, efficient decomposition of organic matter by nitrate- and sulfate-reducing bacteria can continue. The occurrence of dissolved methane both in intermediate and bottom waters of Lake Greifen testifies to the degradation of organic matter through fermentation processes in the sediments and, possibly, in the water column. This in turn suggests that once oxygen was consumed, subsequent degradation of organic matter, in the absence of other oxidants, proceeded through the less efficient bacterial fermentation processes (Froelich et al., 1979). During these times of low redox potential in the water column, a larger proportion of lipid-rich bacterial biomass may have been incorporated with the primary organic matter in the sediments. This scenario of environmental conditions may provide an explanation for why the deposition of an organic facies "A" source rock is constrained to such a small interval in the sediments of Lake Greifen. This explanation is perhaps also valid for understanding the occurrence of organic facies "A" found in the modern and ancient lacustrine environments from Africa (Lakes Tanganyika and Bogoria), USA (Green River Fm.), China (Songliao Basin), Germany (Messel Shale), West Africa (Rift Basins), and Scotland (Orcadia Basin) and their limited distribution in marine shales (e.g., Jurassic, North Sea; Galena Shale, USA) (Jones, 1987; Huc, 1988).

CONCLUSIONS

Decreasing oxygen content of the water column, recorded across the laminated-marl/seasonally varved organic-carbon-rich sediment transition, correlates with a trend in the sedimented algal organic matter which is characterized by decreasing O/C ratios and increasing H/C ratios, Hydrogen Indices, TOC concentrations, and organic-carbon accumulation rates. Changes in the chemical composition and concentration of sedimented organic matter from Lake Greifen appear to be largely dependent on the surface-water productivity and oxygen content of the water column at the time of deposition, and are not a consequence of changing primary source material, benthic decomposition, or overprinting by early-burial diagenetic processes. In Lake Greifen, the thickness and extent of the oxygen-depleted water mass, which controls the rate of decomposition of organic matter and the incorporation of bacterial lipids into the sediments, appears to be the most important environmental condition to influence the chemical composition of deposited organic matter and the determination of organic facies. The Lake Greifen case study suggests that given an initial organic source material, its original geochemical composition can be significantly altered in correlation with the oxygen content of the water column.

In Lake Greifen, the deposition of an organic facies "A" and the accumulation of a lacustrine source rock is limited to a time interval when the majority of the water column was anoxic and organic-matter degradation was dominated by bacterial fermentation processes. This environmental condition only occurs when there is an insufficient amount of other dissolved oxidants to allow for decomposition of organic matter at a faster rate. In these low redox-potential waters and surface sediments, bacterial populations develop (e.g., methanogenic and methanotrophic) and may contribute an additional lipid-rich biomass to the primary organic matter. This is perhaps a valid explanation for the common occurrence of organic facies "A" in modern and ancient lacustrine deposits and for their limited distribution in marine shales.

ACKNOWLEDGMENTS

We would like to thank Dr. John Hayes, Indiana University, for allowing one of us (DH) to use the irm-GCMS apparatus and for helpful discussions. We are grateful to Drs. M. Schoell, K. Kelts, R. Morris, R. Pelet, A. Y. Huc, and B. Durand for their fruitful discussion and review of the manuscript. A special thanks goes to M. Fabre, C. Leblond for their laboratory assistance and to D. Merritt for her assistance in running the irm-GCMS apparatus. We are indebted to H. Ambühl, EAWAG, for permission to use aqueous geochemical data from Lake Greifen. This work was supported by the ETH Research Fund and the Institut Francais du Pétrole, and is a contribution to the Laboratory of Experimental Geology at the ETH-Zentrum.

REFERENCES CITED

Ambühl, H., 1988, Unpublished data from monthly water-column sampling campaign of Lake Greifen, Limnology Research Group, EAWAG, ETH, Dübendorf, Switzerland.

Behar, F., and R. Pelet, 1985, Pyrolysis-gas chromatography applied to organic geochemistry; structural similarities between kerogens and asphaltenes from related rock extracts and oils: Journal of Analytical and Applied Pyrolysis, v. 8, p. 73–187.

Calvert, S. E., 1987, Oceanographic controls on the accumulation of organic matter in marine sediments, *in* J. Brooks and J. Fleet, eds., Marine petroleum source rocks: Geological Society of London Spec. Pub. 24, p. 137–152.

Chappe, B., W. Michaelis, and P. Albrecht, 1979, Molecular fossils in archaebacteria as selective degradation product of kerogens, *in* A. Douglas and J. Maxwell, eds., Advanced Organic Geochemistry: John Wiley, Newcastle, p. 265–274.

Cole, J. J., S. Honjo, and J. Erez, 1987, Benthic decomposition of organic matter at a deep-water site in the Panama Basin: Nature, v. 327, p. 703–704.

Demaison, G. J., and G. T. Moore, 1980, Anoxic environments and oil source bed genesis: AAPG Bulletin, v. 64, p. 179–1209.

Durand, B., and J. C. Monin, 1980, Elemental analyses of kerogens (C, H, O, N, S, Fe), *in* B. Durand, ed., Kerogen, insoluble organic matter in sediments: Paris, Technips, p. 13–33.

Emerson, S., 1985, Organic matter preservation in marine sediments, *in* E. T. Sunquist and W. S. Broecker, eds., The carbon cycle and atmospheric CO_2: Natural variation archean to present: American Geophysical Union Geophysical Monograph 32, p. 78–88.

Emerson, S., K. Fischer, C. Reimers, and D. Heggie, 1985, Organic matter dynamics and preservation in deep-sea sediments: Deep-Sea Research, v. 32, p. 1-21.

Espitalié, J., J. L. Laporte, M. Madec, F. Marquis, P. Leplat, J. Palt, and A. Boutefeu, 1976, Methode rapide de caracterisation des roches meres de leur potentiel petrolier et de leur degre d'evolution: Revue Institut Francaise Pétroleo , v. 32, p. 24-32.

Freeman, K. H, 1989, The carbon-isotopic composition of individual compounds in the Messel Shale (Eocene): M.S. Thesis, Indiana University, 58 p.

Freeman, K. H., J. M. Hayes, J. M. Trendel, and P. Albrecht, 1990, Evidence from GC-MS carbon-isotopic measurements for diverse origins of sedimentary hydrocarbons: Nature, v. 353, p. 254-256.

Froelich, P. N., G. P. Klinkhammer, M. L. Bender, N. A. Luedtke, G. R. Heath, D. Cullen, P. Dauphin, D. Hammond, B. Hartman, and V. Maynard, 1979, Early oxidation of organic matter in pelagic sediments of the eastern equitorial Atlantic: Sub-oxic diagenesis: Geochimica et Cosmochimica Acta, v. 43, p. 1075-1090.

Hollander, D. J., 1989, Carbon and nitrogen isotope cycling and organic geochemistry of eutrophic Lake Greifen: Implications for preservation and accumulation of ancient organic carbon-rich sediments: Ph.D. Thesis, ETH/Zentrum, Zürich, Switzerland, 318 p.

Huc, A. Y., 1988a, Aspects of depositional processes of organic matter in sedimentary basins, *in* Advances in organic geochemistry, v. 13, p. 433-443.

Huc, A. Y., 1988b, Sedimentology of organic matter, *in* F. H. Frimmel and R. F. Christman, eds., Humic substances and their role in the environment: New York, John Wiley, p. 215-243.

Irwin, H., C. D. Curtis, and M. Coleman, 1977, Isotopic evidence for sources of diagenetic carbonates formed during burial of organic-rich sediments, Nature: v. 269, p. 209-213.

Jones, R. W., 1987, Organic facies, *in* J. Brooks and D. Welte, eds., Advances in petroleum geochemistry: Berlin, Springer-Velag, v. 2, p. 1-90.

Kelts, K., 1988, Environments of deposition of lacustrine petroleum source rocks: An introduction, *in* A. J. Fleet, K. Kelts, and M. R. Talbot, eds., Lacustrine petroleum source rocks, Blackwell Scientific Publications, p. 3-26.

McKenzie, J. A., 1982, Carbon-13 cycle in Lake Greifen: A model for restricted ocean basins, *in* S. Schlanger and M. B. Cita, eds., Nature and origin of Cretaceous carbon-rich facies: London, Academic Press, p. 197-207.

Mertz, K. A., 1984, Origin and depositional history of the Sandholt Member, Miocene Monterey Formation, Santa Lucia Range, California: Ph.D. Thesis, University of California, Santa Cruz, 295 p.

Müller, P. J., and E. Suess, 1979, Productivity, sedimentation rate and sedimentary organic matter in oceans; 1. Organic carbon preservation: Deep-Sea Research, v. 27A, p. 347-1362.

Pelet, R., 1981, Preservation and alteration of present-day sedimentary organic matter, *in* Advances in Organic Geochemistry, John Wiley, p. 241-250.

Pelet, R., 1984, A model for the biological degradation of recent sedimentary organic matter: Organic Geochemistry, v. 6, p. 317-325.

Pelet, R., 1987, A model of organic sedimentation on present-day continental margins, *in* J. Brooks and J. Fleet, eds., Marine petroleum source rocks: Geological Society of London Special Publication 24, p. 167-180.

Pelet, R., and G. Deroo, 1983, Vers une sédimentologie de la matière organique: Bulletin de la Societe Geologique de France, v. 25, p. 483-493.

Pratt, L., 1984, Influence of paleoenvironmental factors on preservation of organic matter in middle Cretaceous Greenhorn Formation, Pueblo, Colorado: AAPG Bulletin, v. 68, p. 146-1159.

Pratt, L. M., G. E. Claypool, and J. D. King, 1986, Geochemical imprint of depositional conditions on organic matter in laminated-bioturbated interbeds from fine-grained marine sequences: Marine Geology, v. 70, p. 67-84.

Rouxhet, P. G., P. L. Robin, and G. Nicaise, 1980, Characterization of kerogens and their evolution by infrared spectroscopy, *in* B. Durand, ed., Kerogens, insoluble organic matter in sediments: Paris, Technips, p. 165-190.

Suess, E., 1980, Particulate organic carbon flux in the oceans: Surface productivity and oxygen utilization: Nature, v. 288, p. 260-263.

Tissot, B., and D. H. Welte, 1984, Petroleum generation and occurrence: Berlin, Springer-Verlag, 699 p.

Tissot, B., G. Deroo, and A. Hood, 1978, Geochemical study of the Uinta Basin: Formation of petroleum from the Green River formation: Geochimica et Cosmochimica Acta, v. 42, p. 1469-1485.

Tissot, B. P., and R. Pelet, 1981, Sources and fate of organic matter in ocean sediments, Geology of Oceans: Oceanologica Acta Special Publication, p. 97-103 .

Tyson, R. V., 1987, The genesis and palynofacies characteristics of marine plankton source rocks, *in* J. Brooks and J. Fleet, eds., Marine petroleum source rocks: Geological Society of London Special Publication 24, p. 47-68.

Vandenbroucke, M., R. Pelet, and Y. Debyser, 1985, Geochemistry of humic substances in marine sediments, *in* D. McKnight, ed., Humic substances in soil, sediments, and water: Geochemistry, isolation and characterization, p. 249-273.

Whiticar, M. J., E. Faber, and M. Schoell, 1986, Biogenic methane formation in marine and freshwater environments: CO_2 reductions vs. acetate fermentation—isotopic evidence: Geochimica et Cosmochimica Acta, v. 50, p. 693-709.

Züllig, H., 1982, Untersuchungen uber die Stratigraphie von Carotinoiden im geschichteten Sedimenten, vom 10 Schweizer Seen zur Erkundung fruherer Phytoplankton-Entfallungen, Schweizerische Zeitschrift fuer Hydrologie, v. 44, p. 1-98.

Climate Model Prediction of Paleoproductivity and Potential Source-Rock Distribution

Eddy Kruijs and Eric Barron
Earth System Science Center
Pennsylvania State University
University Park, Pennsylvania, U.S.A.

The utility of climate models to provide predictions of the distribution of upwelling regions and areas of high productivity is assessed by: (1) determination of the association between present-day upwelling and primary productivity, (2) determination of the capability of a General Circulation Model (GCM) to simulate present-day upwelling, (3) determination of the sensitivity of the GCM to changes in geography and climate, and (4) determination of the model capability to predict past regions of upwelling and productivity by comparison with a suite of productivity indicators using a middle Cretaceous example. The greatest level of success occurs for coastal upwelling when there is also knowledge of the seasonality of upwelling. To a high degree, this study demonstrates the value of physically based models in providing valuable data on paleoproductivity and potential source-rock distribution.

INTRODUCTION

Sediments associated with upwelling tend to be rich in biologically produced opaline silica, marine organic carbon, and phosphorus (Suess and Thiede, 1983; Thiede and Suess, 1983). The tendency for higher accumulation rates of organic carbon has generated substantial interest in upwelling regions as environments of marine petroleum-source-rock formation. The economic importance of "upwelling" deposits has led to several attempts to predict the distribution of wind-driven upwelling and productivity throughout Earth history (Parrish, 1982; Barron, 1985; Scotese and Summerhayes, 1986).

Parrish (1982) pioneered the prediction of paleoupwelling locations using a qualitative model based on analogies with the present-day climate and consideration of the role of land-sea distribution in modifying circulation patterns. Scotese and Summerhayes (1986) developed a parametric approach based on the methodology described by Parrish (1982). Barron (1985) predicted the pattern of middle Cretaceous upwelling, utilizing wind stresses calculated with a global numerical climate model.

Each of the previous efforts is characterized by a number of limitations. For example, each study assumed that areas of upwelling are related to high primary productivity, although this relationship has not been quantified. In addition, the model capability to predict the present distribution of wind stress or upwelling was not rigorously addressed. The availability of global data sets for primary productivity (Koblentz-Mishke et al., 1970) and wind stress (Hellerman and Rosentein, 1983), as well as compilations of the locations of upwelling deposits, allows a comprehensive assessment of current capabilities to predict paleoproductivity and potential upwelling-related source rocks (Kruijs, 1989).

Model predictive capabilities are assessed here through four steps: (1) determination of the actual association between present-day wind-driven upwelling and primary productivity in order to establish appropriate rules or relationships which can be applied to Earth history, (2) determination of the capability of current state-of-the-art General Circulation Models (GCM) to predict the present-day wind stress in order to establish the capability of the model to simulate the key variable required to compute upwelling, (3) determination of the sensitivity of the model to changes in geography and climate in order to establish the necessity of calculating wind stress using physically based models, and (4) determination of the capability of the model to predict past regions of upwelling and productivity by comparison with a suite of upwelling indicators in order to verify the sensitivity of the model to changes in geography and climate. Through this stepwise analysis, an ability to predict areas of high paleoproductivity is demonstrated.

ASSOCIATION BETWEEN UPWELLING AND PRIMARY PRODUCTIVITY

Primary productivity is limited by the availability of the essential nutrients—nitrogen, phosphorus, and silica. Within the photic zone these nutrients are rapidly utilized by organisms, and the availability of the limiting nutrient regulates the level of productivity. Nutrients are returned to the water column at depth by the oxidation of biologic material settling through the water column. The association between upwelling and primary productivity stems from the fact that the upward movement of water from depth to the surface re-supplies nutrients to the photic zone. This association is an underlying assumption in previous efforts to predict paleoproductivity (Parrish, 1982; Parrish and Curtis, 1982; Barron, 1985; Scotese and Summerhayes, 1986). However, the relationship between upwelling and primary productivity has not been fully evaluated.

Upwelling

Upwelling rates are not usually directly measured because vertical velocities are relatively small in the world ocean ($\sim10^{-2}$ cm/sec). Consequently, upwelling is typically calculated based on surface wind stress. Parrish (1982) and Barron (1985) give a full description of the wind-stress conditions which promote upwelling.

Hellerman and Rosenstein (1983) compiled a global wind-stress data set with a 2-degree by 2-degree grid resolution spanning 106 years (1870–1976) from data at the National Climatic Center of Asheville, North Carolina, and from the Polar Seas Atlas of the U.S. Navy Hydrographic Office.

The wind stress is utilized to compute wind-driven upwelling, or more appropriately the divergence of the Ekman transport, using the following formula for a spherical coordinate system

$$\nabla \cdot V_e = (af)^{-1} (1/\cos\phi \times \delta\tau^{\phi} / \delta\lambda - \delta\tau^{\lambda}/\delta\phi + \tau^{\lambda}/\sin\phi \times \cos\phi)$$

where a is the radius of the Earth (6371×10^6 m), τ is wind stress, ϕ is latitude, λ is longitude, f is the Coriolis force ($f = 2\Omega\sin\phi$, where omega is the Earth's rotation).

The wind-stress components are calculated from the u (eastward) and v (northward) components of the wind velocity using standard quadratic bulk aerodynamic formulations (e.g., Barron, 1985; Hellerman and Rosenstein, 1985).

Figure 1 illustrates the wind-driven upwelling calculated from the Hellerman and Rosenstein (1983) wind-stress data set using the above equations.

Productivity

Phytoplankton make up 95% of all the primary producers in the world ocean (Steeman-Nielsen, 1978). Koblentz-Mishke et al. (1970) compiled data on phytoplankton productivity (expressed in mg C/m^2/day) from more than 7000 stations. The values (Figure 2) are divided into five classes (<100; 100–150; 150–250; 250–500; and >500). The data set is somewhat problematic. First, the spatial and temporal distributions of data are uneven, and the original data are not available. Second, the data set includes four different methods (radiocarbon method, concentration of chlorophyll a, oxygen bottle, and algological methods), and these methods probably missed primary productivity resulting from photoautotrophic picoplankton (Li et al., 1983).

Unfortunately, modern satellite measurements, although promising as a globally consistent data source, lack sufficient temporal coverage to provide a long-term average. Consequently, the values of productivity given in Figure 2 are utilized as the only available global data set suitable for comparison with the distribution of upwelling.

Comparison of Upwelling and Primary Productivity

Since the primary productivity and Ekman divergence data sets are different in resolution, and point data are unavailable from Koblentz-Mishke et al. (1970), the two data sets were digitized over an evenly spaced grid (5° × 5°). The Ekman divergence was assigned to be high upwelling (>20 cm/day), low upwelling (0–20 cm/day), or downwelling (<0 cm/day). The primary productivity was assigned to four ranges (mg C/m^2/day): <150, 150–250, 250–500, and >500. The Ekman divergence maps were digitized for January and July and combined as an "annual" data set. Primary productivity data were not separable by season.

Results

The relationship between upwelling and primary productivity is not obvious. Only 18% of the "annual" upwelling regions are associated with primary productivity in excess of 250 mg C/m^2/day. Figure 3a illustrates that a large number of gridpoints in the open ocean or associated with seasonal and annual sea-ice cover are characterized by wind stresses conducive to upwelling, but lack high primary productivity. The locations of upwelling

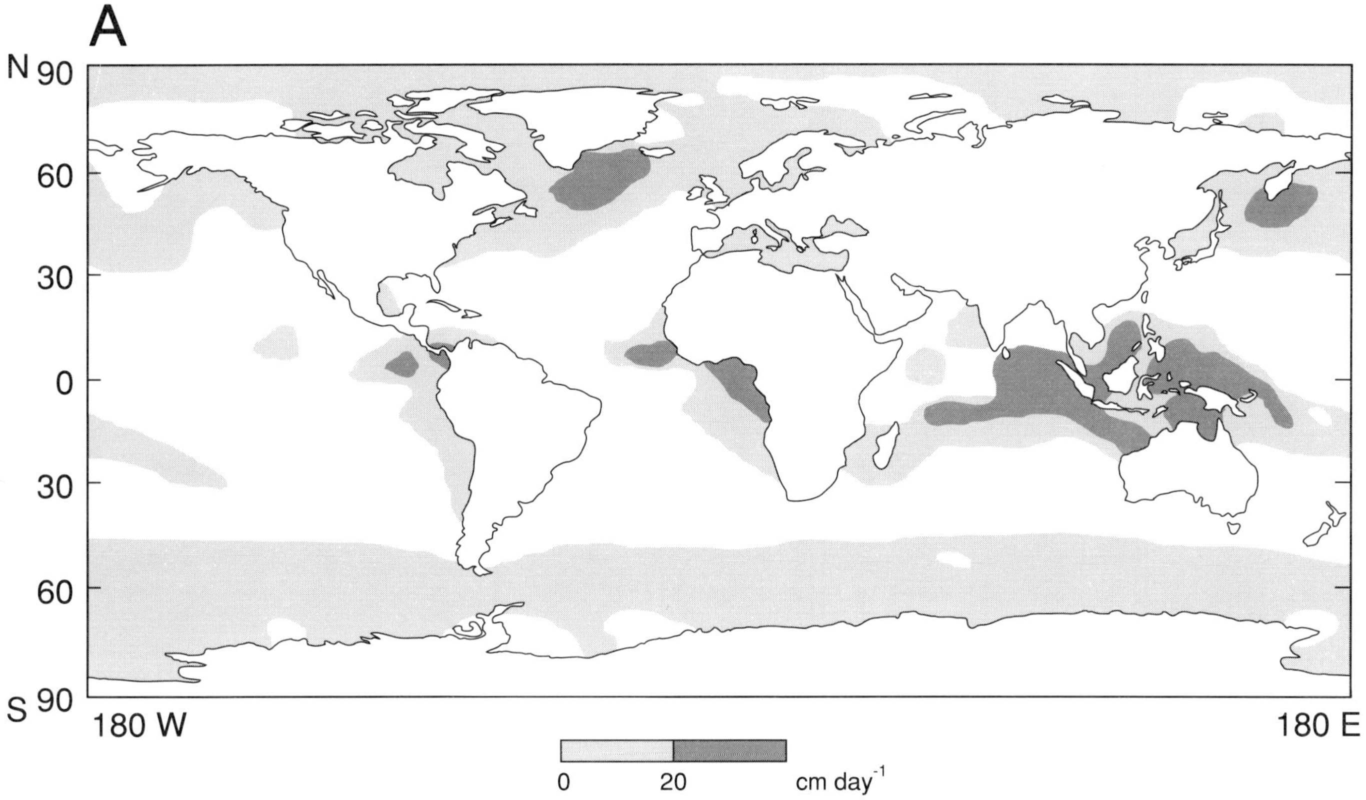

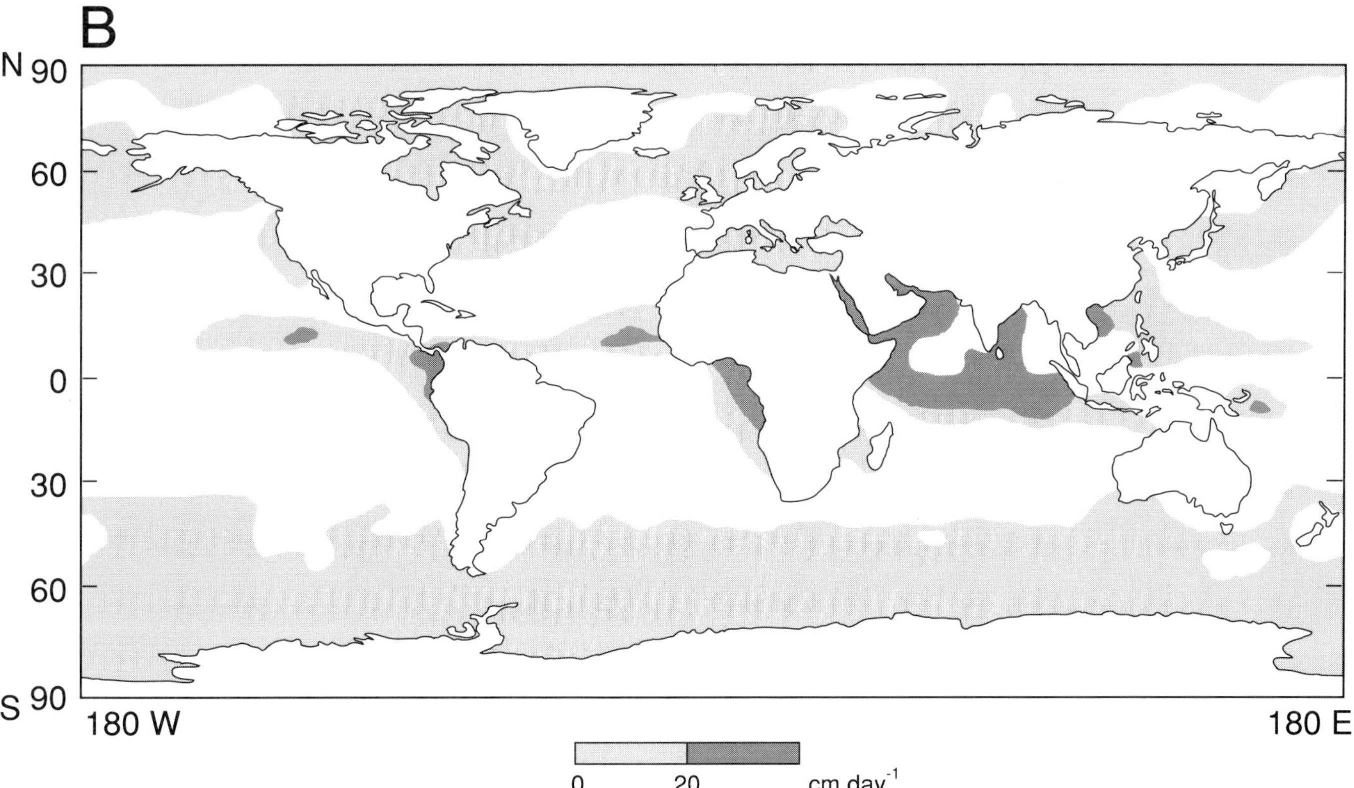

Figure 1. January (a) and July (b) wind-driven upwelling calculated from "observed" wind stress from Hellerman and Rosenstein (1983).

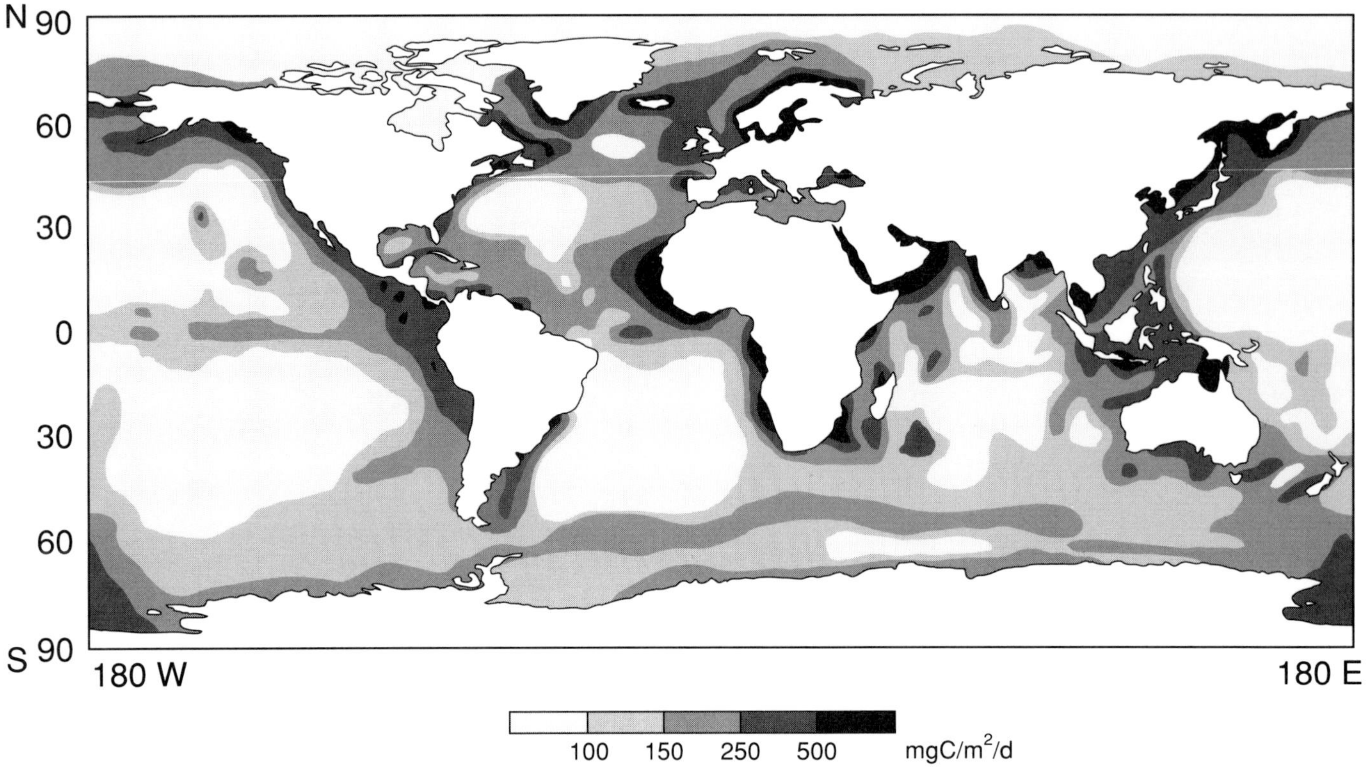

Figure 2. Marine primary productivity (mgC/m²/day) after Koblentz-Mishke et al. (1983).

sites which are associated with high primary productivity increase to 63% if regions of sea-ice cover and open ocean (more than one gridpoint from land) are excluded. The distribution of coastal sites with upwelling and high productivity is given in Figure 3b. Interestingly, there are very few sites of high primary productivity which are not associated with upwelling (Figure 3c). This fact suggests that high primary productivity is an indication of upwelling; however, the converse (and typically the assumption used) is not true without qualification.

Seasonal information on upwelling greatly improves the association between productivity and upwelling. In the best case, points with one season of strong upwelling and one season of weaker upwelling were associated with high primary productivity at the 91% level (after open-ocean and sea-ice sites were excluded).

The 63% association for upwelling and productivity after excluding open ocean and sea ice, and the additional distinctions between those gridpoints of strong upwelling one season and weak upwelling the other season (resulting in a 91% association), provide the strongest basis for modeling paleoproductivity. The next step is to determine if current climate models successfully simulate surface wind stress over the ocean.

COMPARISON BETWEEN MODEL UPWELLING AND OBSERVED UPWELLING

The Community Climate Model (CCM), utilized by Barron (1985) to predict upwelling for the middle Cretaceous, is a comprehensive three-dimensional numerical model of the atmosphere capable of investigating climate change as a function of geography and atmospheric circulation. Descriptions of the model and variety of simulations may be found in Ramanathan et al. (1983), Barron (1985), Barron and Washington (1984), and Washington and Meehl (1984).

Two types of model simulations were compared with the observed data set of Hellerman and Rosenstein (1983).

1. A mean annual model simulation which employs an annual average solar insolation and an energy-balance ocean formulation. This energy-balance formulation is also referred to as a swamp ocean because it does not include heat storage, transport, or diffusion. Sea ice forms if the temperature falls below 271.2 K. This type of model simulation has been successfully compared against the observed

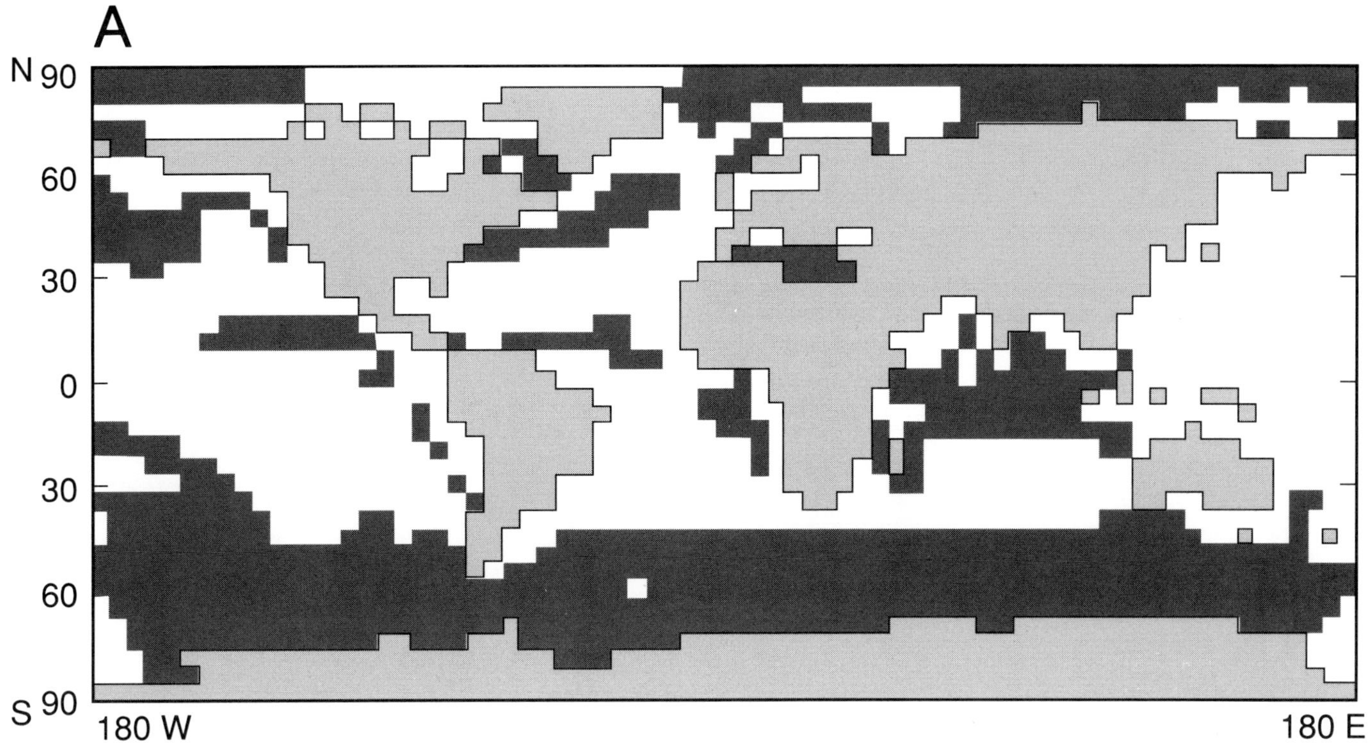

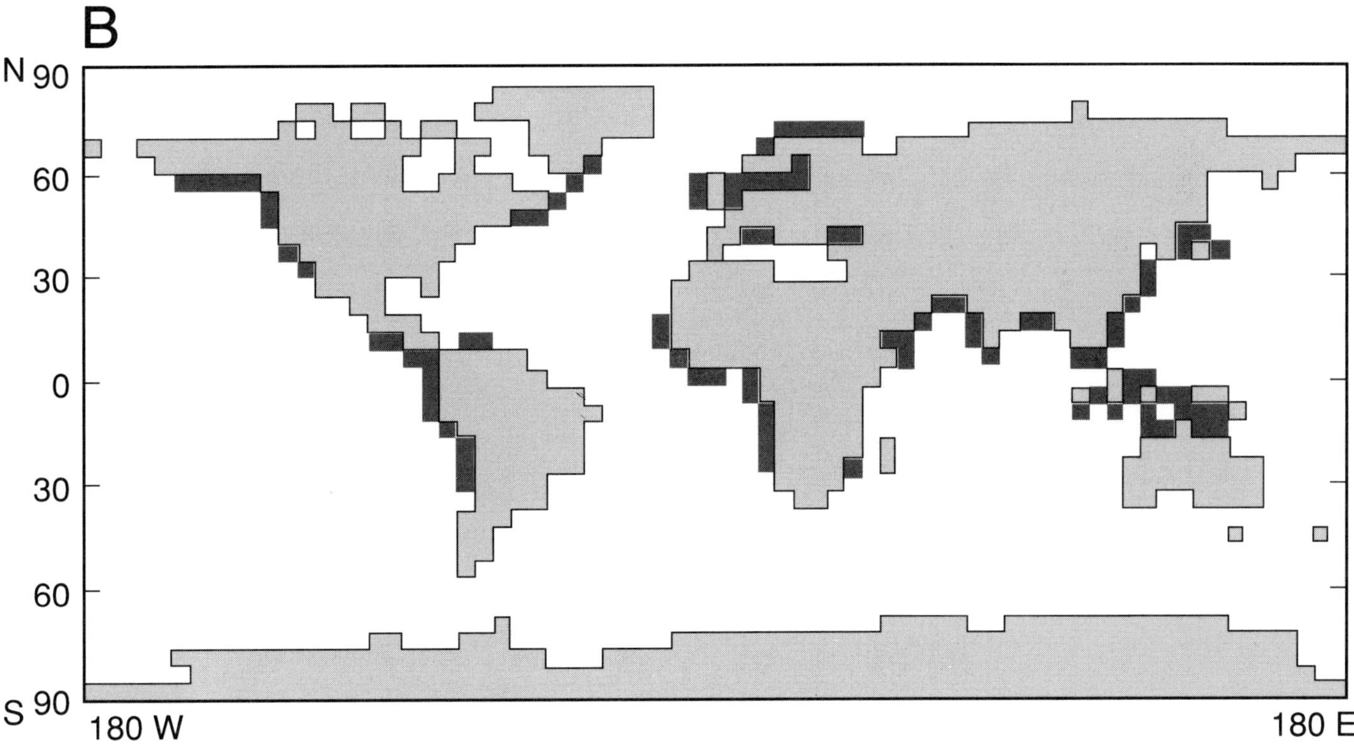

Figure 3. Associations between upwelling (Figure 1) and primary productivity (Figure 2): (a) upwelling gridpoints which are not associated with high (>250 mgC/m²/day) productivity, (b) upwelling gridpoints which are associated with high primary productivity, and (c) high primary productivity gridpoints which are not associated with wind-driven upwelling.

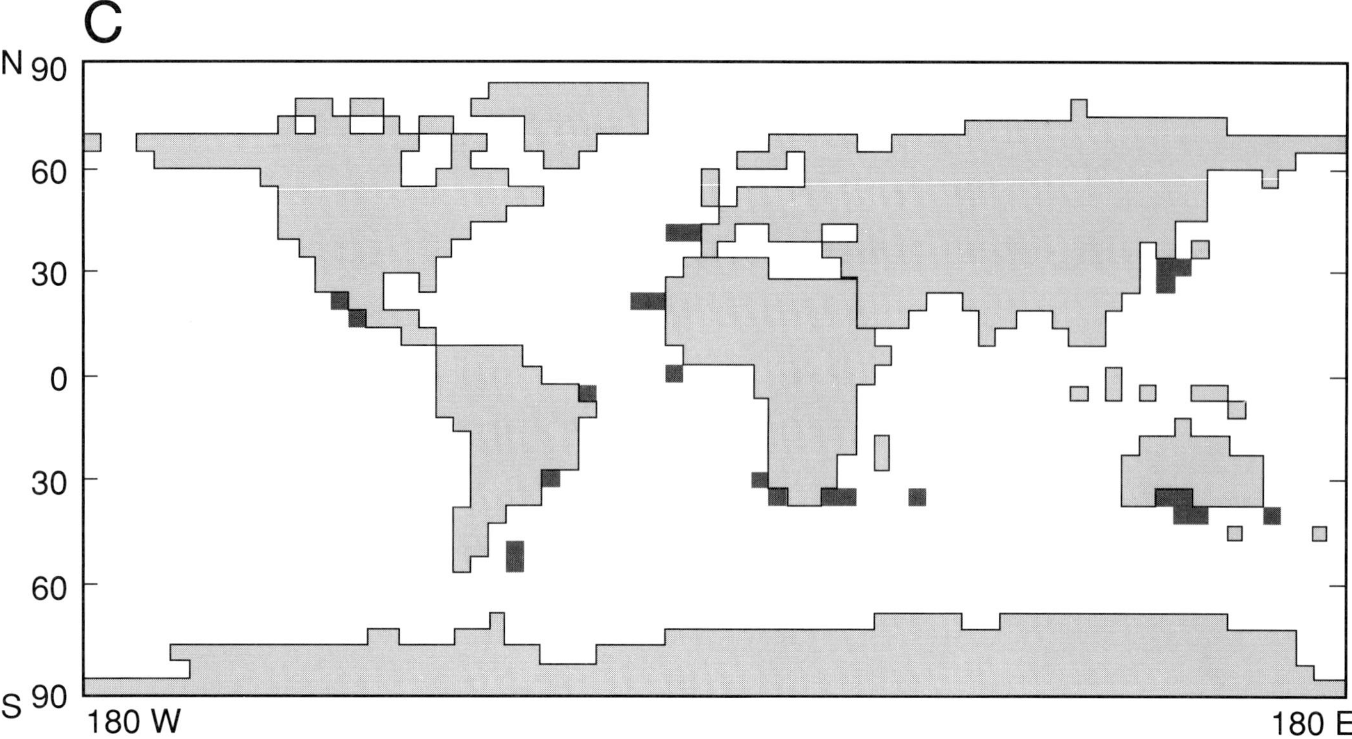

Figure 3. (Continued)

mean annual temperature and zonal wind distribution (Washington and Meehl, 1983). The advantage of a mean annual model simulation is the relative short time before the model reaches equilibrium (Barron and Washington, 1984).

2. A seasonal model simulation which employs a mixed-layer ocean. This seasonal mixed-layer model simulation differs from the above-described mean annual model simulation in that the solar insolation varies with the time of the year, and the oceans include heat storage. A major advantage of this type of model simulation is the ability to model the seasons. A disadvantage is the longer time before the model reaches equilibrium because of the varying solar input and heat storage. A comparison of a present-day seasonal model simulation with observations is given by Washington and Meehl (1984).

The CCM computes wind stress in the same format as Hellerman and Rosenstein (1983). Consequently, equation (1) can be used to calculate the divergence of the Ekman transport for both model and observations. An average of the mean annual simulation and an average of three January and three July months were used for comparison with the observed data set.

Results

Comparison of the wind-stress fields (Figure 4), the mean annual divergence of the Ekman transport (Figure 5), and the January and July wind stresses (Figures 6, 7) for the model with observations indicate a remarkable ability of the CCM to simulate present-day conditions. A gridpoint-by-gridpoint comparison was made based on the rules developed in the previous section (gridpoints adjacent to land and excluding sea ice). In the mean annual model, 73% of all upwelling and downwelling were predicted correctly, while 82% of all the observed upwelling sites were correctly simulated. For January, 83% of all observed upwelling sites were predicted, and for July, 86% were correctly simulated. X^2 tests demonstrated that these results are statistically significant (0.1% level).

Problems associated with incompleteness in the observations and the fact that the CCM is a limited representation of the atmosphere may explain the differences between model simulation and observations. The relatively high success ratios provide a reasonable basis to attempt simulations for different time periods. The results also should be considered as an indication of the limits of the application of state-of-the-art climate models to the prediction of wind-driven upwelling.

PREDICTED MIDDLE CRETACEOUS UPWELLING AND PRODUCTIVITY

If the major atmospheric-circulation features were static in time or were altered in a simple manner as a function of changing geography, models of paleoproductivity and upwelling would be relatively

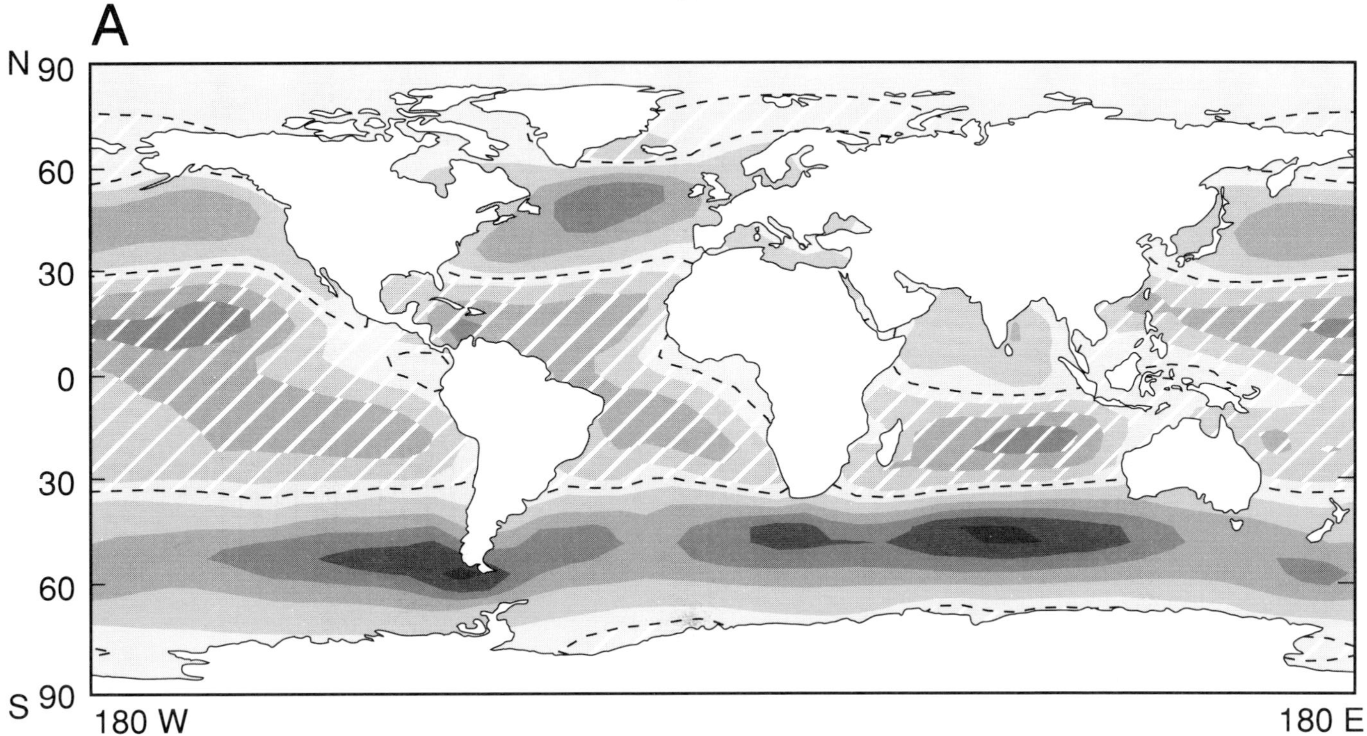

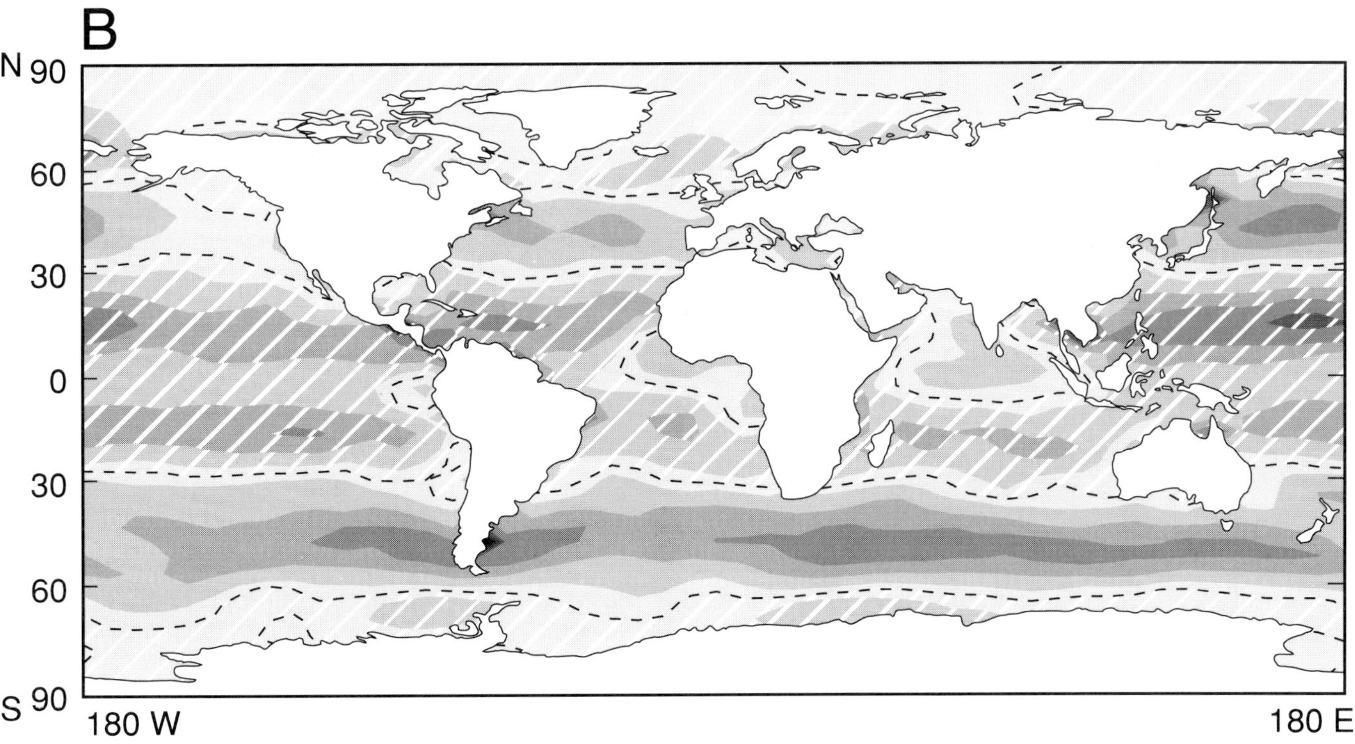

Figure 4. Comparison of mean annual model-simulated and observed components of the wind stress. (a) observed τ^x, (b) model calculated τ^x, (c) observed τ^y, (d) model calculated τ^y. Dashed line is the zero contour. Fields with diagonal white lines are less than zero. Contour interval is 0.04 N/m².

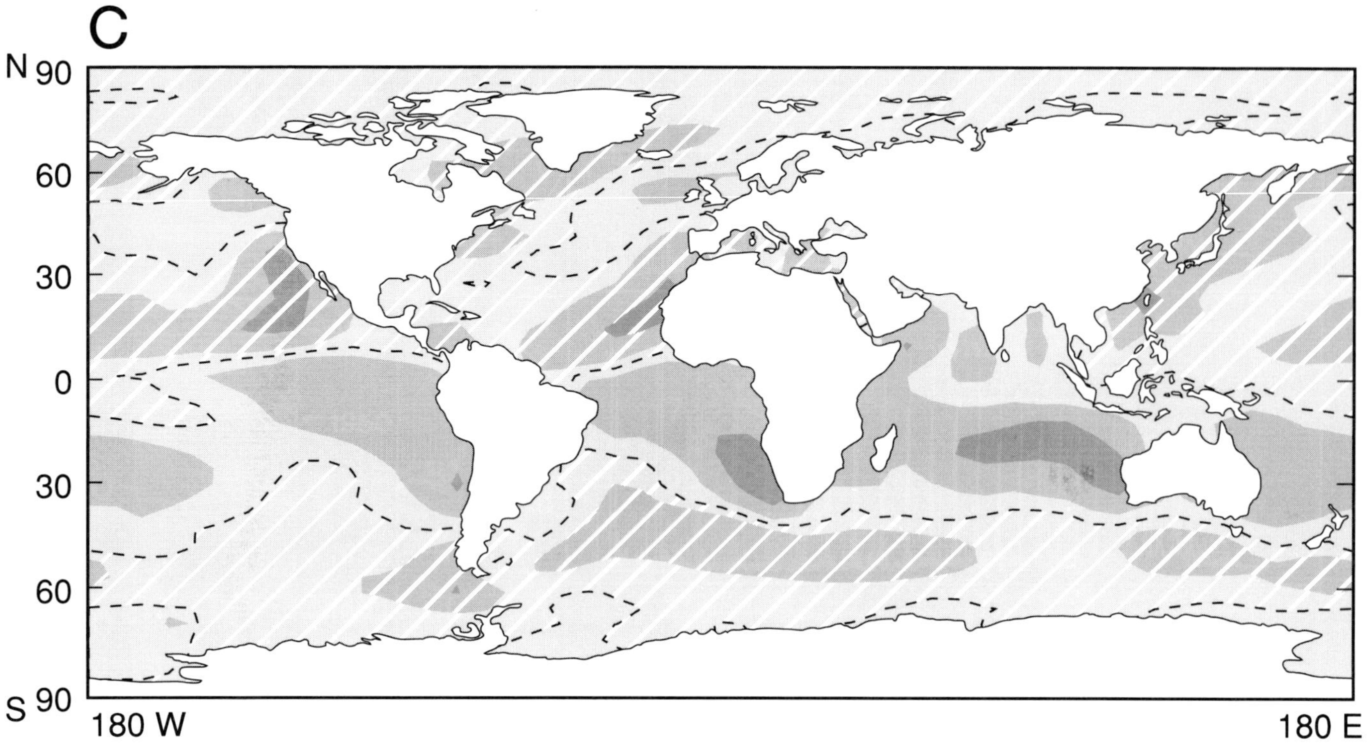

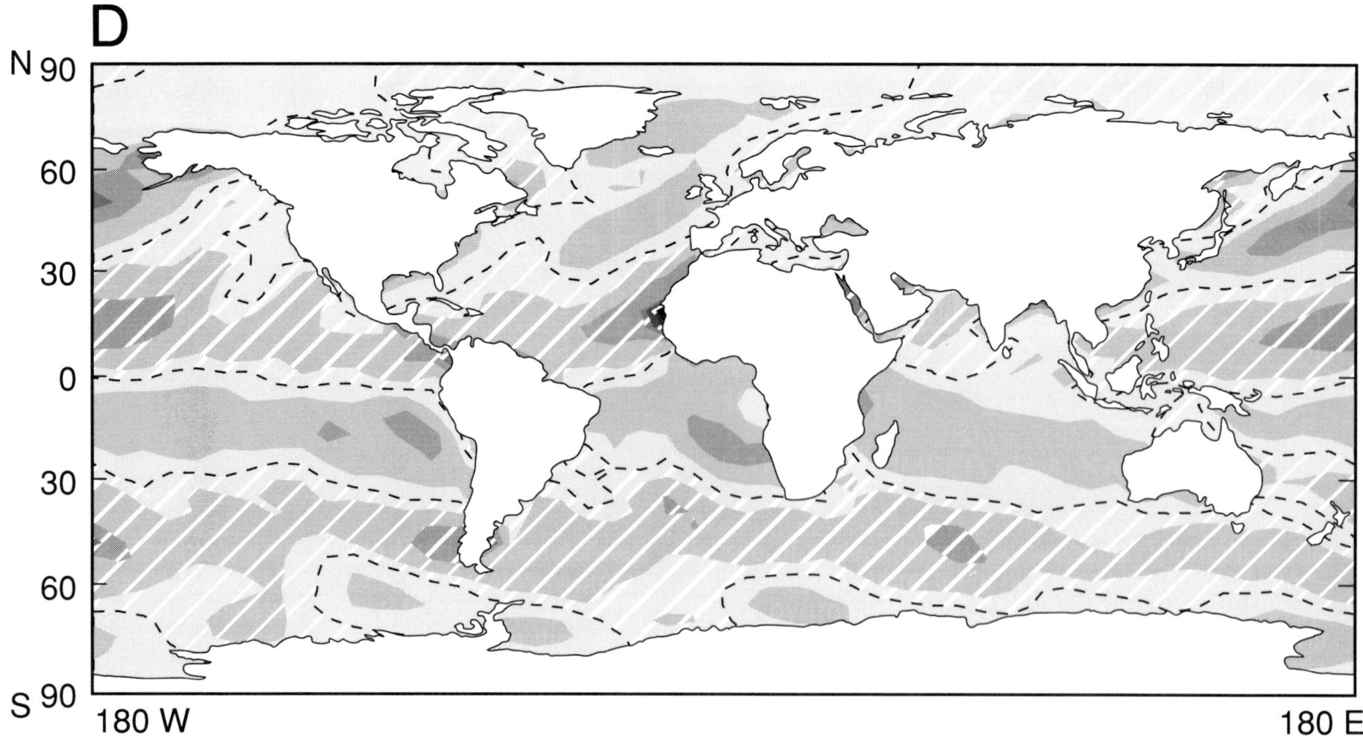

Figure 4. (Continued)

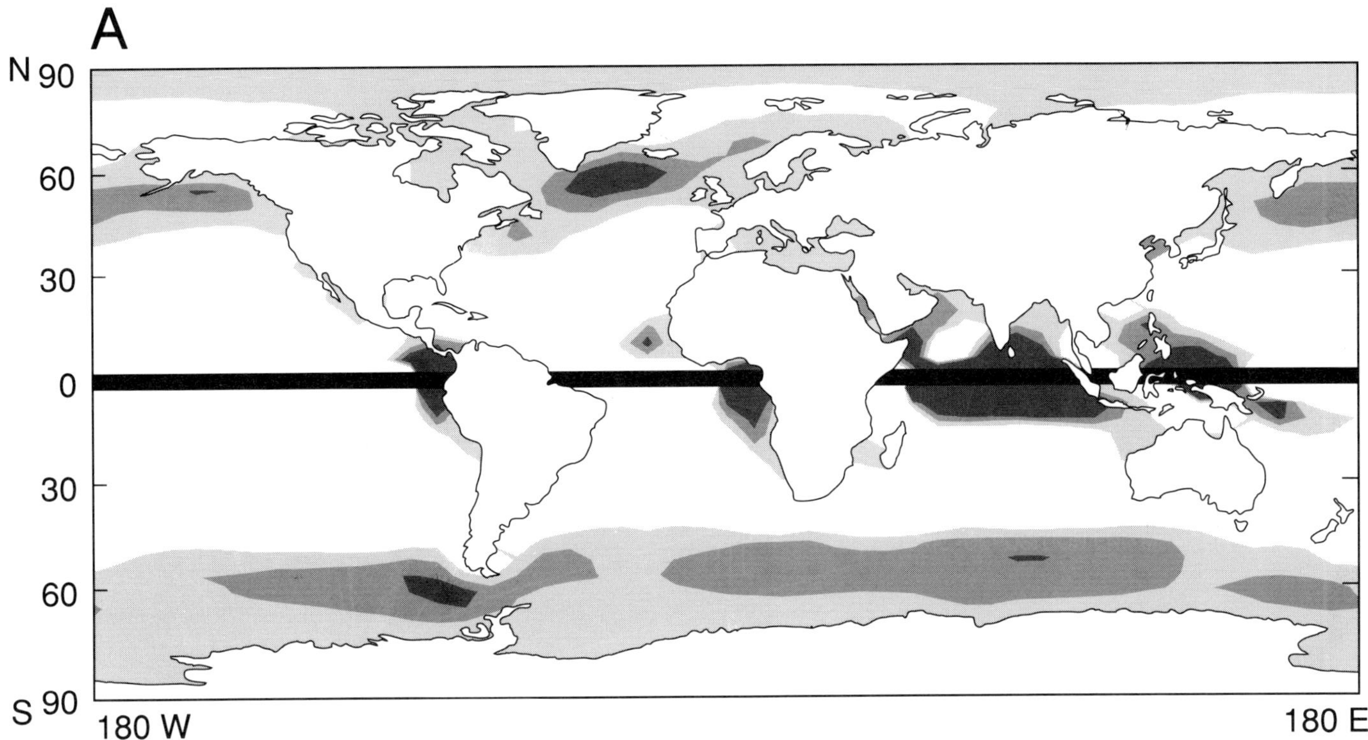

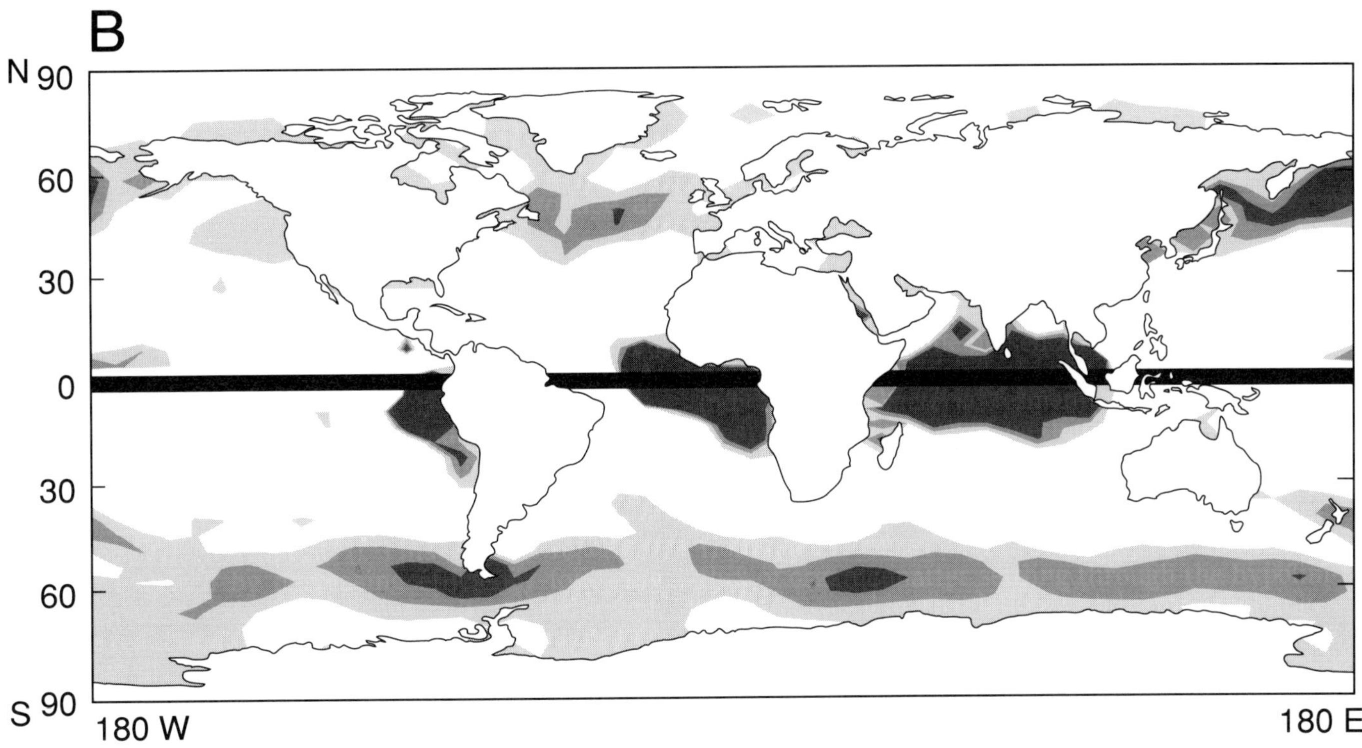

Figure 5. Comparison of mean annual model-simulated and observed divergence of the Ekman transport (wind-driven upwelling): (a) calculated from observed wind stress, (b) calculated from model-simulated wind stress. Contour interval is 5.0 cm/day. Darkest shade is everything greater than 10 cm/day.

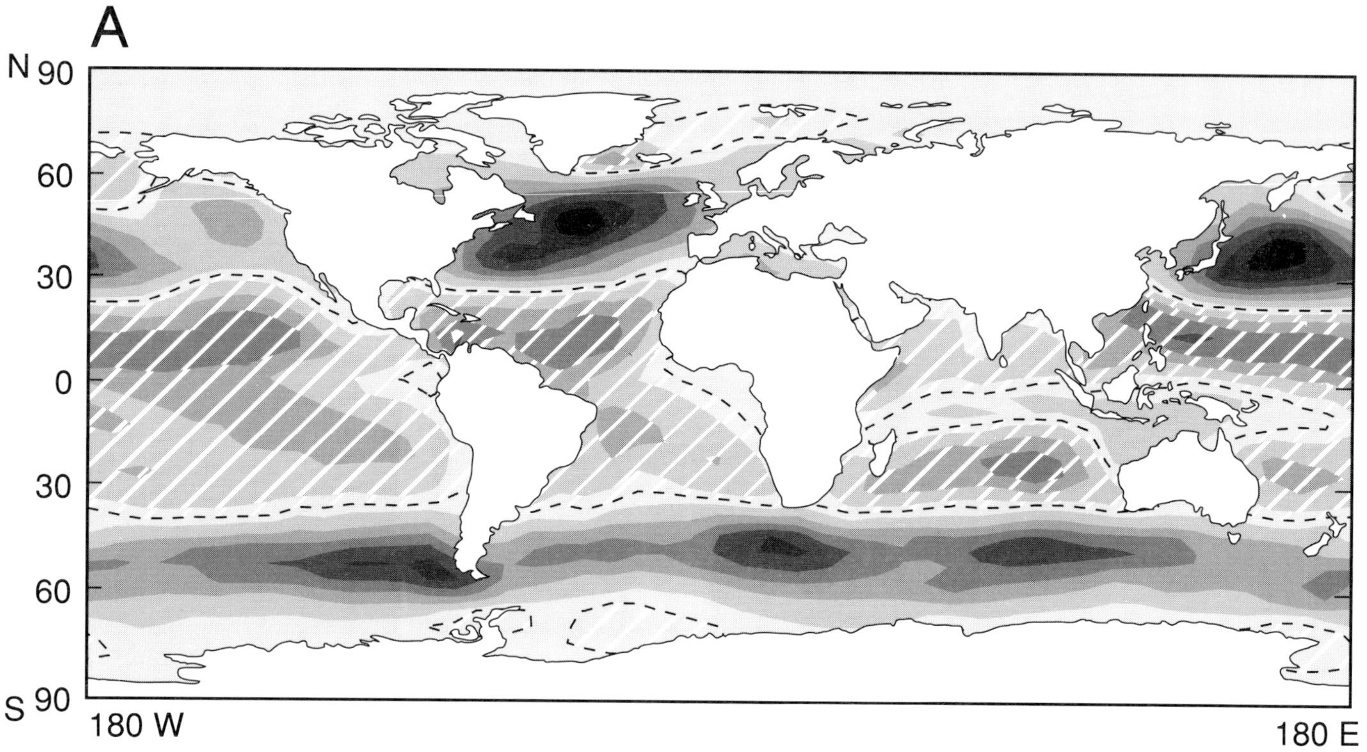

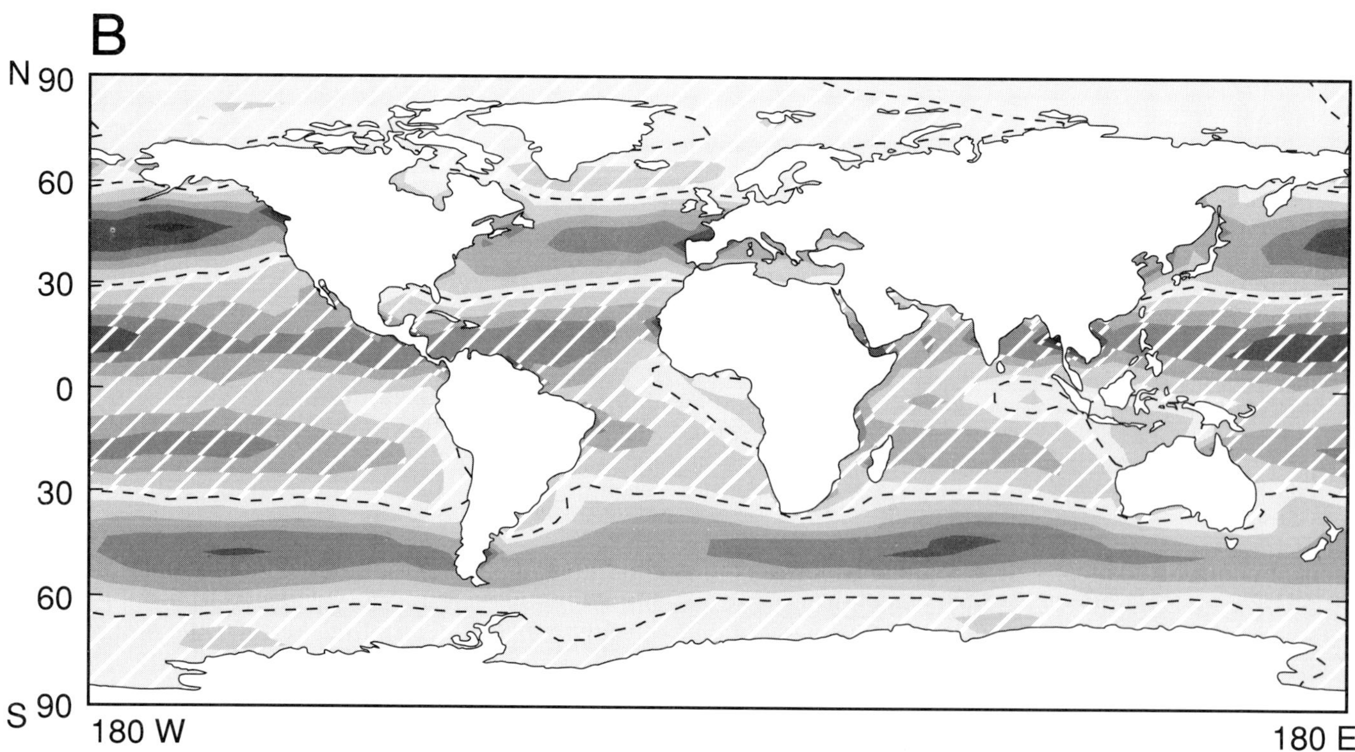

Figure 6. Comparison of January model-simulated and observed components of the wind stress: (a) observed τ^x, (b) model-calculated τ^x, (c) observed τ^y, (d) model-calculated τ^y. Fields with diagonal white lines are less than zero. Contour interval is 0.04 N/m².

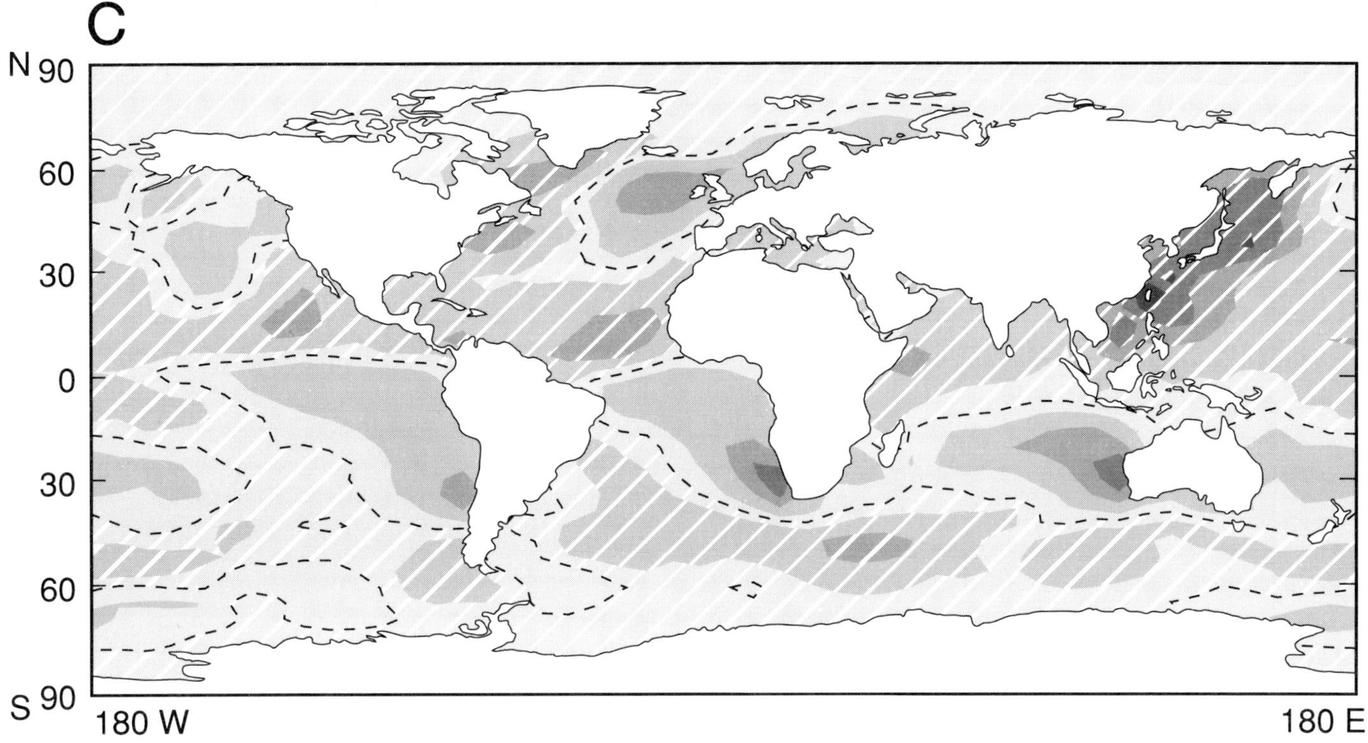

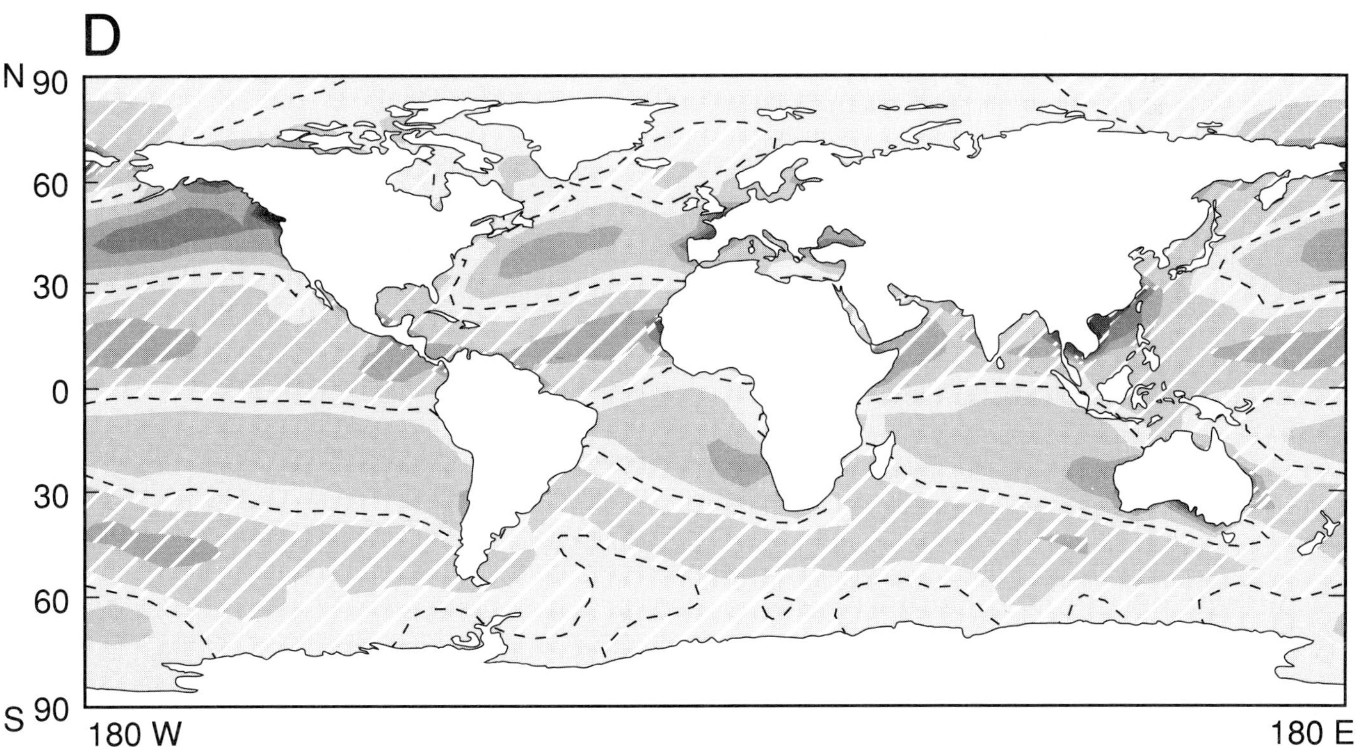

Figure 6. (Continued)

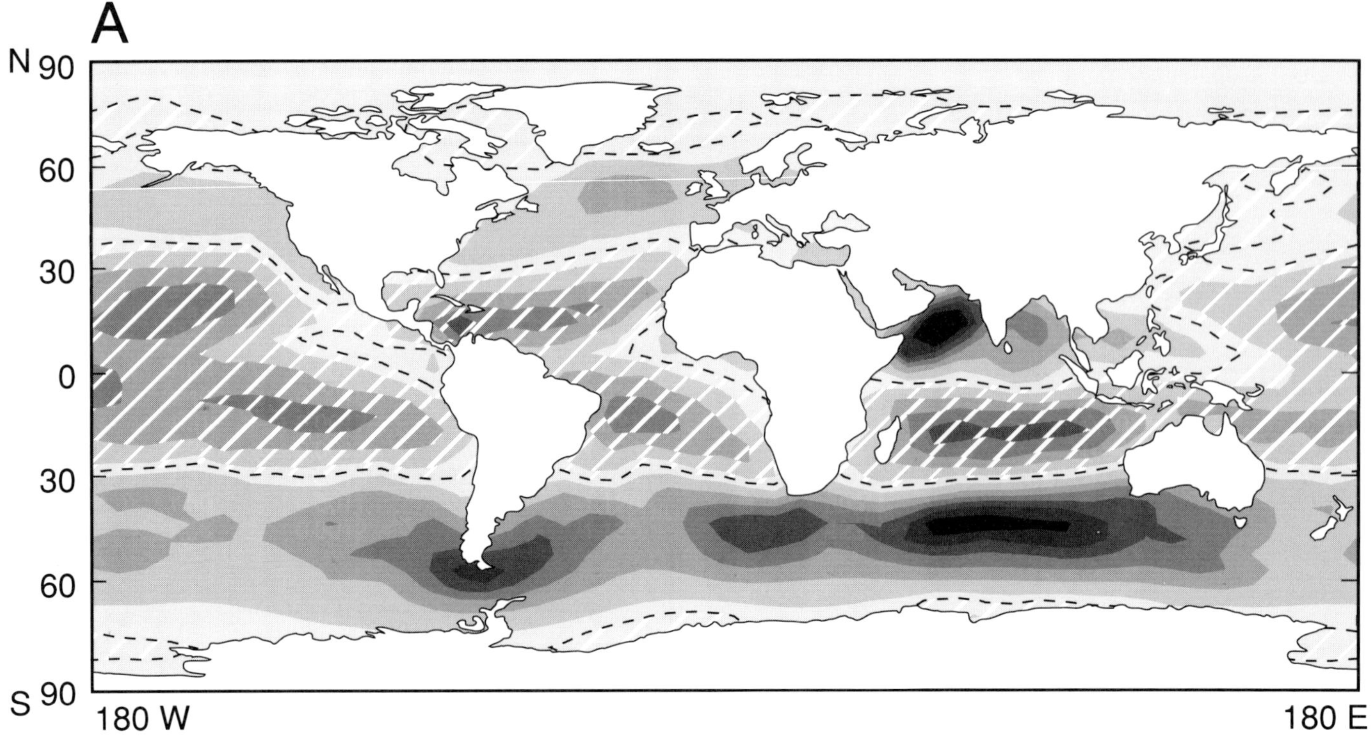

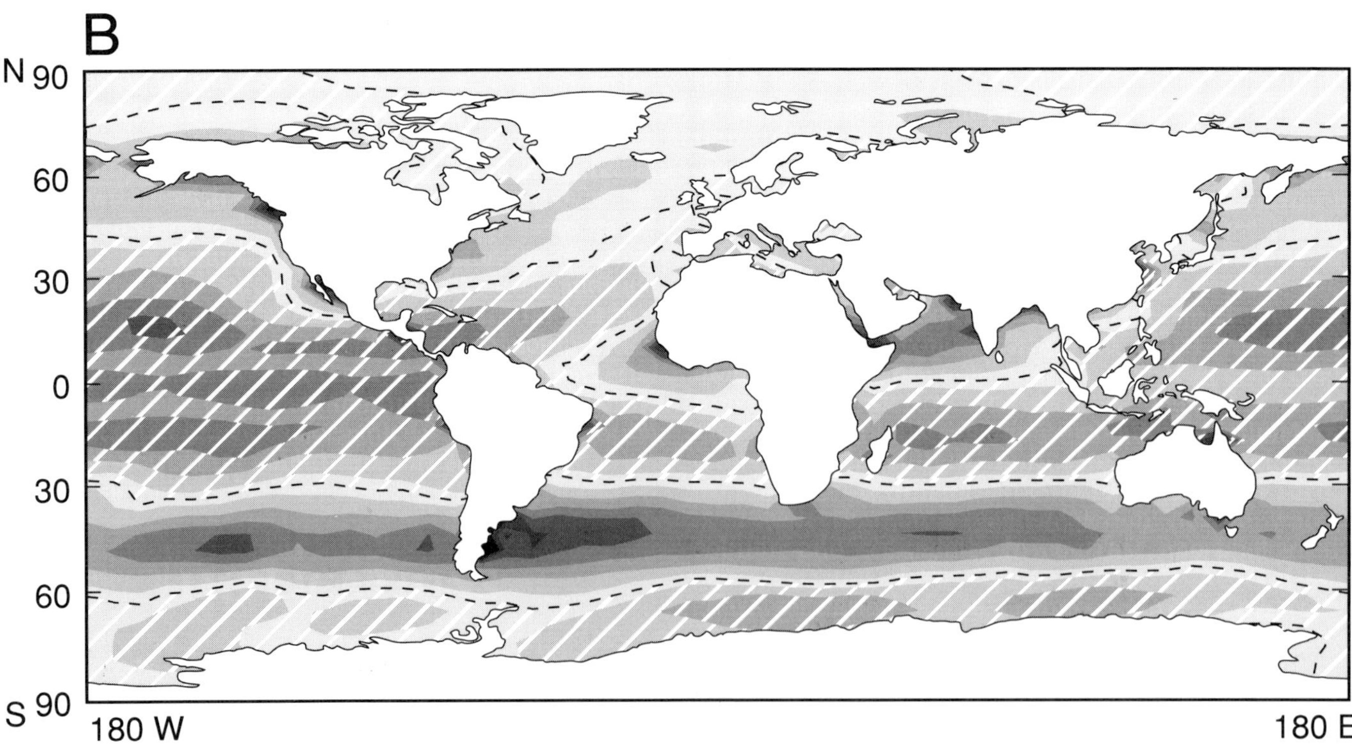

Figure 7. Comparison of July model-simulated and observed components of the wind stress: (a) observed τ^x, (b) model-calculated τ^x, (c) observed τ^y, (d) model-calculated τ^y. Fields with diagonal white lines are less than zero. Contour interval is 0.04 N/m².

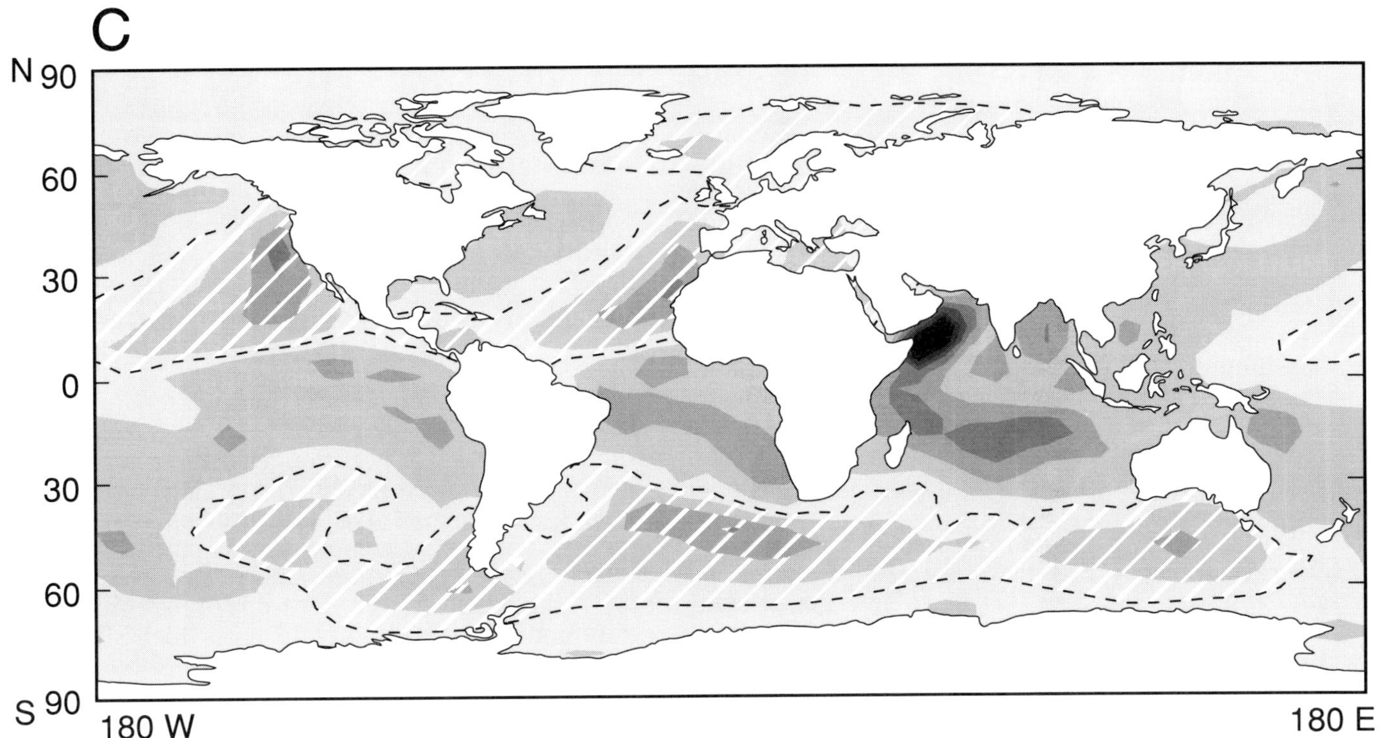

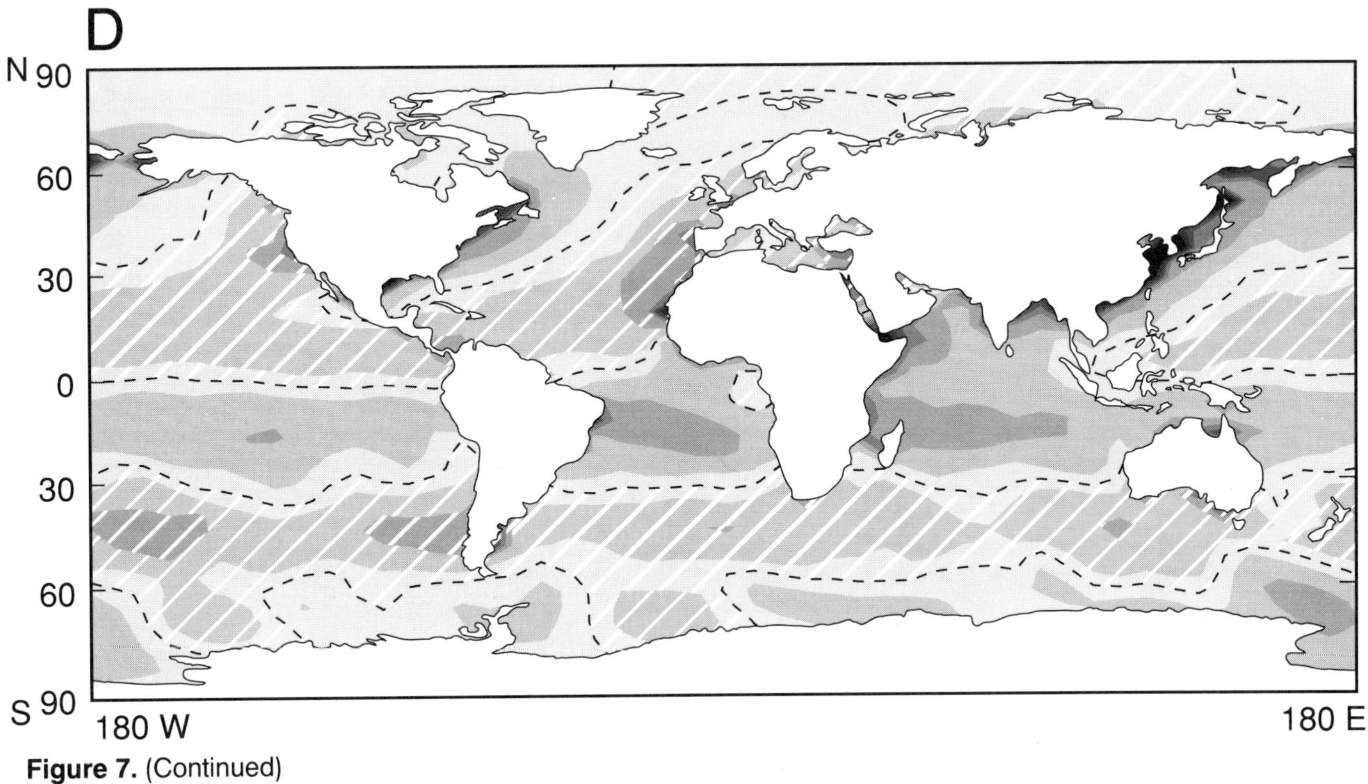

Figure 7. (Continued)

straightforward. The necessity of calculating wind stress using physically based models depends on demonstrating the sensitivity of the climate to changes in geography and climate.

Wind-stress data were obtained from three middle Cretaceous CCM experiments: (1) January and July averages from a full annual model simulation of the seasonal cycle, (2) a mean annual model simulation described by Barron and Washington (1984), and (3) a mean annual model simulation in which the atmospheric CO_2 was increased four-fold with respect to the present day as described by Barron and Washington (1985). Higher CO_2 has been postulated as an explanation of Cretaceous warmth (Berner et al., 1983). Each of the experiments used the geographic-boundary conditions in model resolution given in Figure 8 (see Barron, 1985, for additional description).

Again, equation (1) was utilized to compute the divergence of the Ekman transport. Following the guidelines developed in the above sections, coastal upwelling for each simulation is given in Figures 9, 10, and 11. These figures illustrate a number of important differences. First, the middle Cretaceous simulations are characterized by a substantially larger number of gridpoints adjacent to land associated with upwelling (larger by close to a factor of two). The suggestion here is that the nature of Cretaceous geography and the characteristics of the atmospheric circulation conspire to promote extensive coastal upwelling in comparison with the present. Taken a step further, if validated, models may indicate time periods in Earth history which may be particularly conducive to coastal upwelling and with a greater potential for high paleoproductivity. Second, the distribution of Cretaceous upwelling has a number of distinctive features, including a strong association of upwelling-conducive wind stresses on the margins of the Tethys Ocean and the margins of Australia, and major differences in high-latitude upwelling as a function of season and planetary warmth (CO_2 level). A few regions are characterized by upwelling regardless of specified CO_2 level or type of model (e.g., northwest Africa).

The final step is to verify the predictions of middle Cretaceous upwelling by comparison with sedimentary indicators.

EVALUATION OF MODEL CAPABILITY TO PREDICT PAST REGIONS OF UPWELLING

Areas of present-day upwelling and high productivity, such as the coasts of Namibia and Peru, are underlain by sediments enriched in organic matter, biogenic silica, phosphorus, and a suite of trace elements (Baturin, 1983; Diester-Haass, 1978). However, present-day upwelling sites also are characterized by large differences in sedimentary characteristics as a function of shelf morphology, energy levels, bottom current directions, level of terrigenous sedimentation, and bioturbation level. Consequently, clear-cut indicators of upwelling may or may not be apparent in the geologic record. Further, analysis of the middle Cretaceous has other limitations. As one example, diatoms were not important as siliceous sediments until the Cenozoic (Hein, 1987). Despite these limitations, considerable data are available to assess model capability.

Three data sets from the middle Cretaceous provide the opportunity to evaluate the middle Cretaceous simulations on a global basis. First, Miskell (1983) calculated the silica accumulation rates for middle Cretaceous Deep Sea Drilling Program (DSDP) sites. Miskell (1983) used two approaches to indicate productivity levels—comparison of percent organic carbon and percent biogenic silica and comparison of Cretaceous and present-day biogenic silica accumulation rates. Figure 12 gives 14 DSDP sites which fulfill the requirements described by Miskell (1983) as representative of high productivity.

Second, Roth and Krumbach (1986) identified a number of specific nannofossils indicative of high surface fertility (*Biscutum constans* and *Zygodiscus erectus*), and species which are resistant to dissolution (*Watznauria* spp). Figure 13 gives areas of high productivity based on the relative abundance of dissolution-resistant species (*Watznauria*) versus the two species indicative of high fertility.

Third, the middle Cretaceous was a time of widespread deposition of organic carbon-rich rocks (Schlanger and Jenkyns, 1976; Jenkyns, 1980). The occurrence of these "black shales" has been explained as a product of organic-matter supply, high rates of sedimentation, and variation in bottom-water oxygen content (Arthur et al., 1984). Anoxic stratified basins, high primary productivity resulting from upwelling, and high influx of terrestrial organic matter are potential explanations for high organic-carbon deposition. Figure 14 illustrates the location of middle Cretaceous (Albian-Cenomanian) "black shales" characterized by marine organic matter, within one gridpoint from a coastline (following the guidelines described earlier).

The paleoproductivity indicators described above were derived from 79 DSDP sites and 4 Ocean Drilling Project sites which recovered middle Cretaceous strata. These 83 sites yield 32 independent data points adjacent to continents based on the three global data sets of productivity indicators described previously. Assigned to the model resolution grid, 24 points are available for comparison with the locations of upwelling predicted by the CCM. For the seasonal model, 18 gridpoints had model-predicted upwelling confirmed by upwelling indicators, 2 gridpoints of model-predicted upwelling were not confirmed by observations, 3 gridpoints were predicted to be characteristic of downwelling and no evidence of

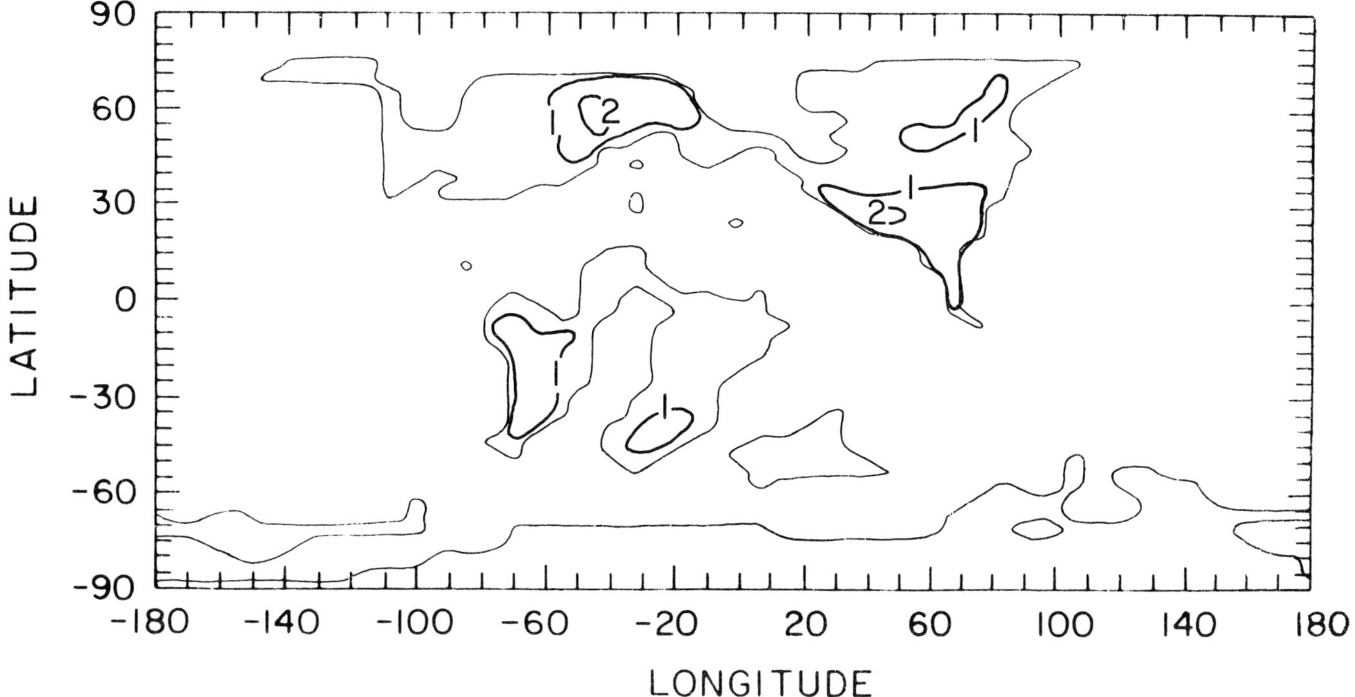

Figure 8. Middle Cretaceous climate model boundary conditions for geography and topography (1 km contours) after Barron (1985).

upwelling was found, and 1 gridpoint contained evidence for upwelling but was not predicted. In summary, 90% of the model predictions of upwelling were confirmed. The success ratio of the model in predicting both downwelling and upwelling is 88%. The comparison of model predictions with observations for the middle Cretaceous is given in Figure 15 for the seasonal simulation.

In the mean annual model (Figure 16) and the mean annual model with four times present-day CO_2 (Figure 17) the success rates in predicting upwelling points are 74% and 68%, respectively. The seasonal model has a substantially larger success ratio. Although all of the upwelling or productivity observations are not predicted, interestingly, the sites of model-predicted upwelling are consistently confirmed. Of model-predicted sites of upwelling, the seasonal model has a 95% success ratio, the mean annual model has a 93% success ratio and the high CO_2 simulation has a 93% success ratio, in confirmation of the model prediction. This point suggests that although model-predicted upwelling does not explain every site with indicators of organic-carbon deposition or high fertility, the cases where the model predicts upwelling are consistently confirmed by data.

Figures 15, 16, and 17 also illustrate those gridpoints for which model predictions are available, but for which observations are not readily available to confirm the predictions. Obviously, the majority of sites fall within this category. Still, the results presented here are a strong indication that state-of-the-art climate models have considerable potential as tools to predict upwelling and paleoproductivity.

DISCUSSION

The ability to predict areas of high paleoproductivity within certain limits has been demonstrated through a stepwise analysis. First, a comparison of Ekman divergence computed from observed wind stresses with the distribution of high productivity demonstrated a poor relationship between upwelling and productivity. However, if open-ocean sites are excluded and regions of seasonal sea ice are excluded, a strong association is found between coastal upwelling and high productivity. Second, a comparison of model-predicted wind stresses for the present day with observed wind stresses indicated a substantial capability of the Community Climate Model to simulate the key variable (wind stress) required to predict upwelling. Third, a middle Cretaceous example demonstrated that the distribution and area of upwelling sites is a function of geography and climate. Fourth, the model capability to predict middle Cretaceous upwelling was strongly confirmed by comparison with observations.

There remains considerable room for improvement in this analysis. The data set for primary productivity (Koblentz-Mishke et al., 1970) lacks consistency and lacks spatial and temporal resolution. The degree of error introduced, or the degree to which an improved

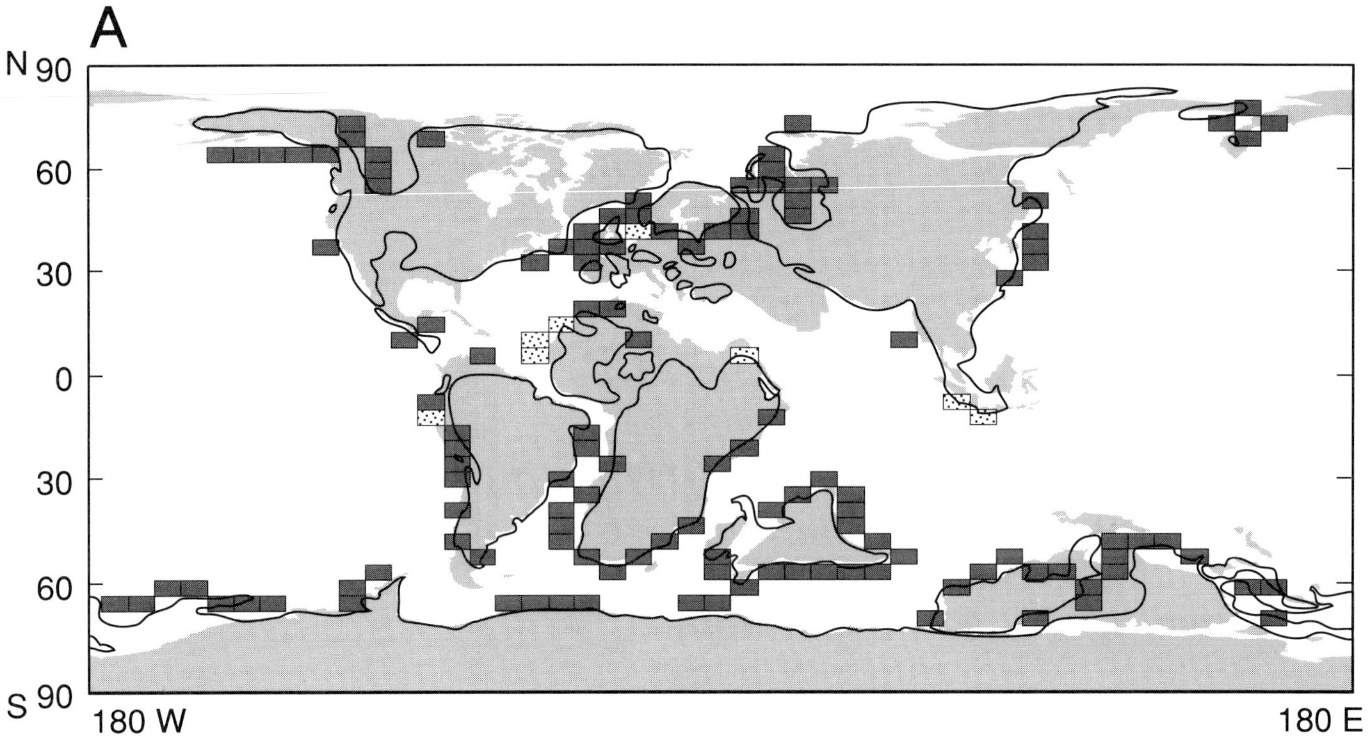

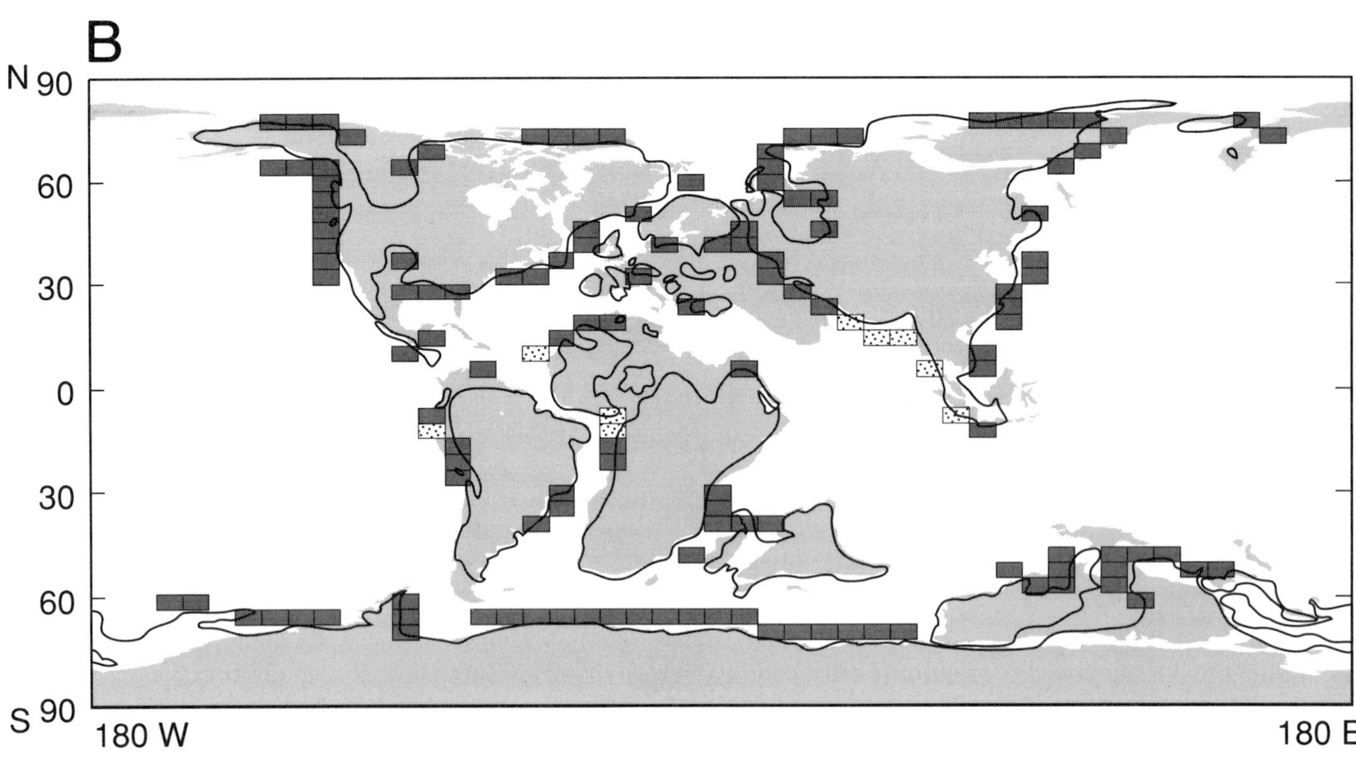

Figure 9. Predicted middle Cretaceous January (a) and July (b) coastal upwelling gridpoints (Dark boxes—low upwelling of 0-20 cm/day; stippled boxes—high upwelling of greater than 20 cm/day).

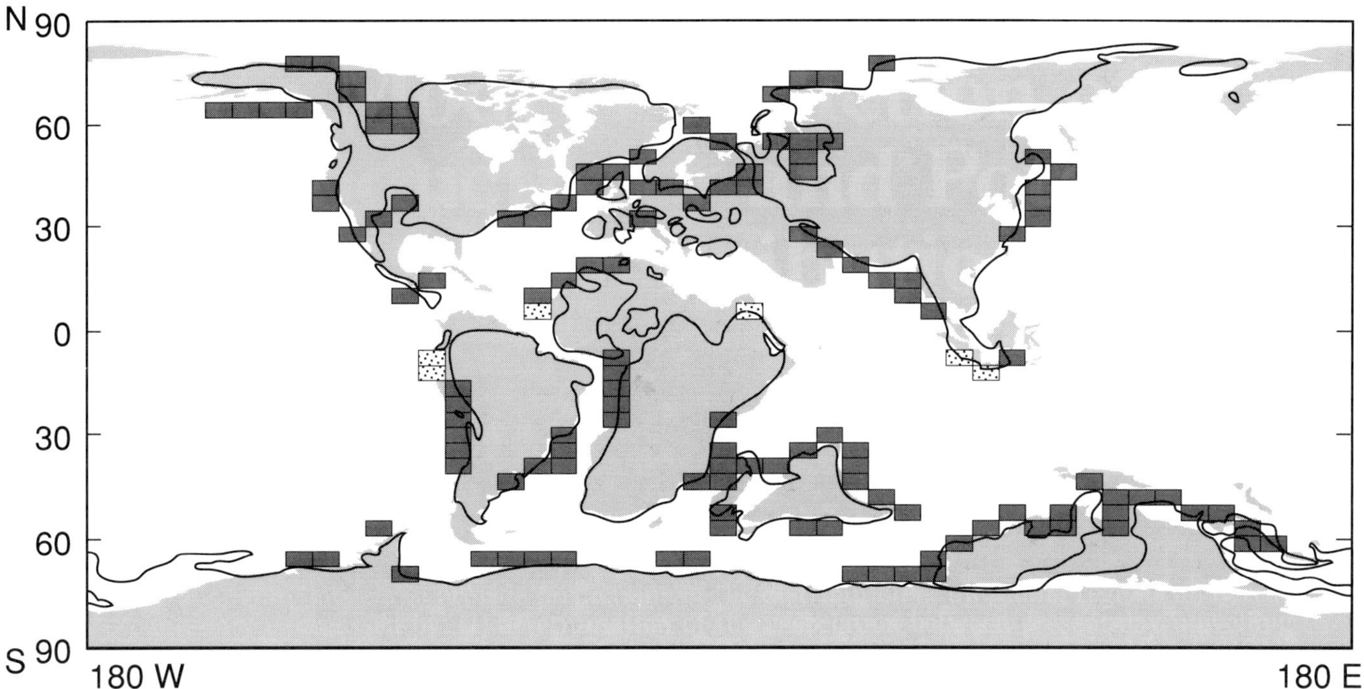

Figure 10. Predicted middle Cretaceous mean annual coastal upwelling gridpoints (Dark boxes—low upwelling of 0-20 cm/day; stippled boxes—high upwelling of greater than 20 cm/day).

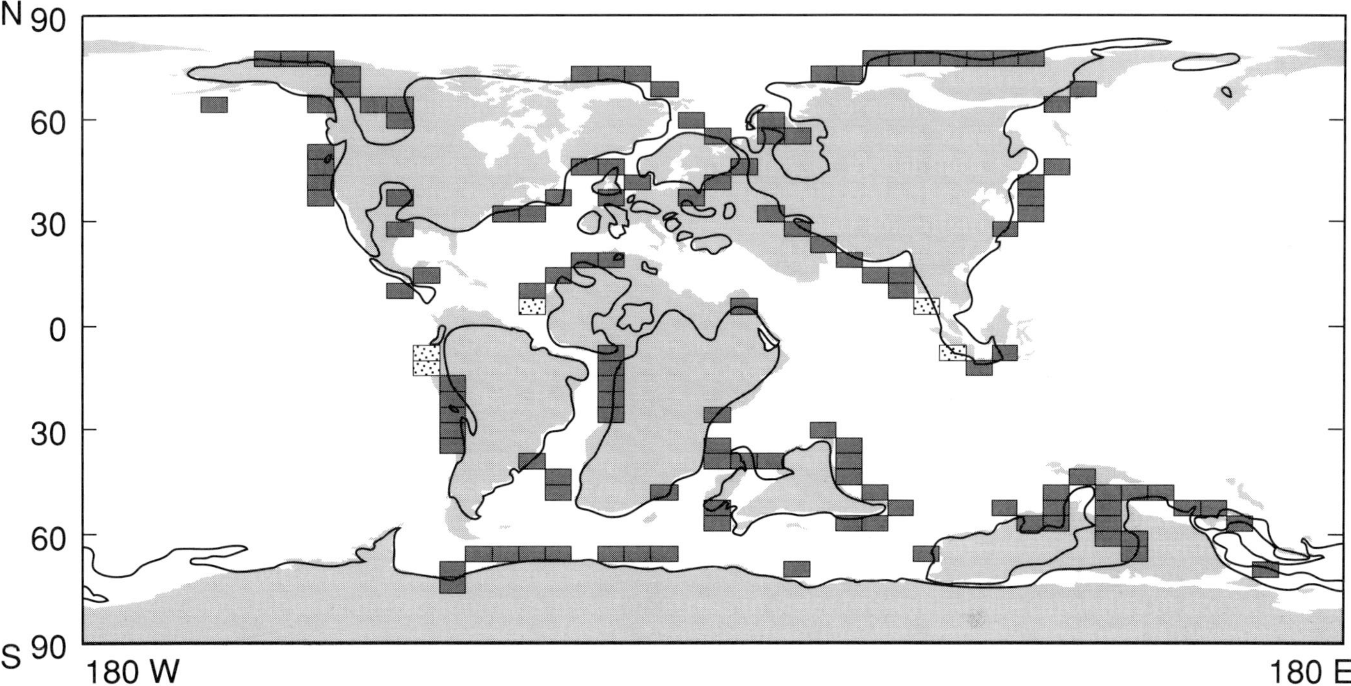

Figure 11. Predicted middle Cretaceous mean annual coastal upwelling gridpoints for the case with 4x present CO_2 levels (Dark boxes—low upwelling of 0-20 cm/day; stippled boxes—high upwelling of greater than 20 cm/day).

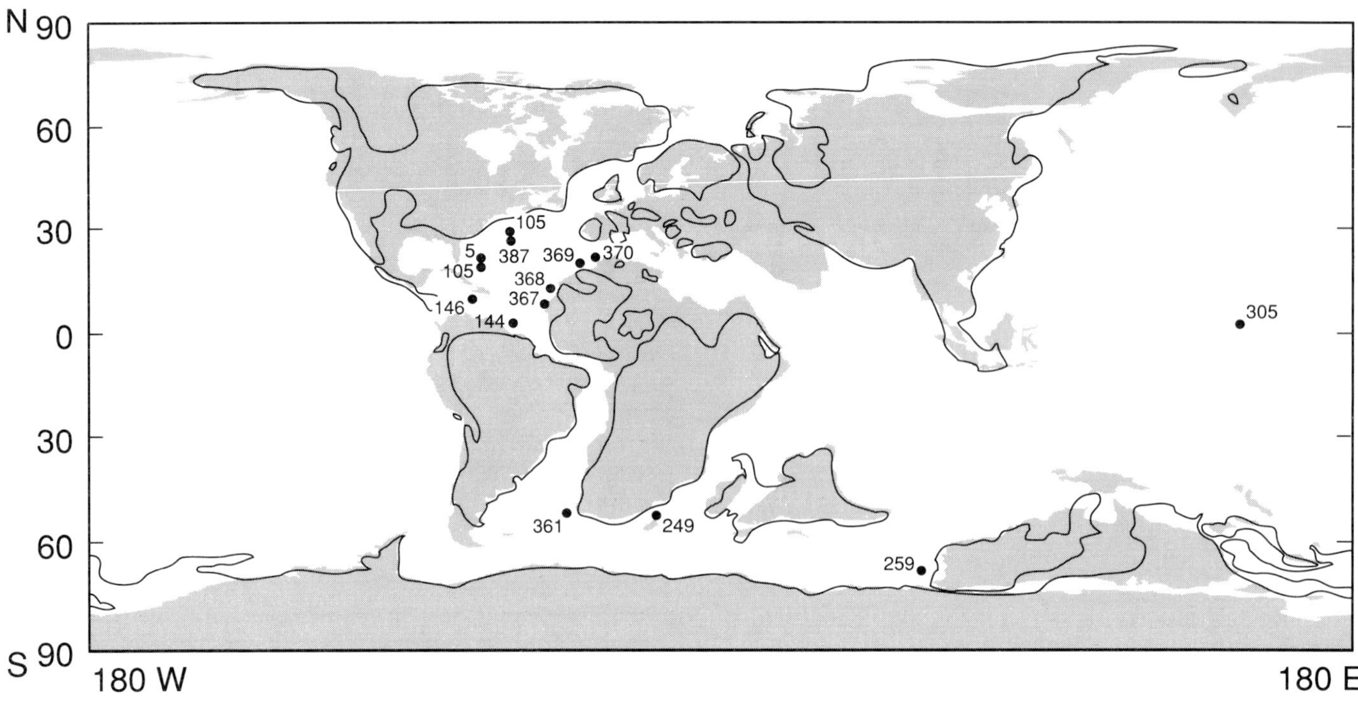

Figure 12. DSDP locations of middle Cretaceous samples with high productivity based on biogenic-silica content and organic-carbon deposition as indicated by Miskell (1983).

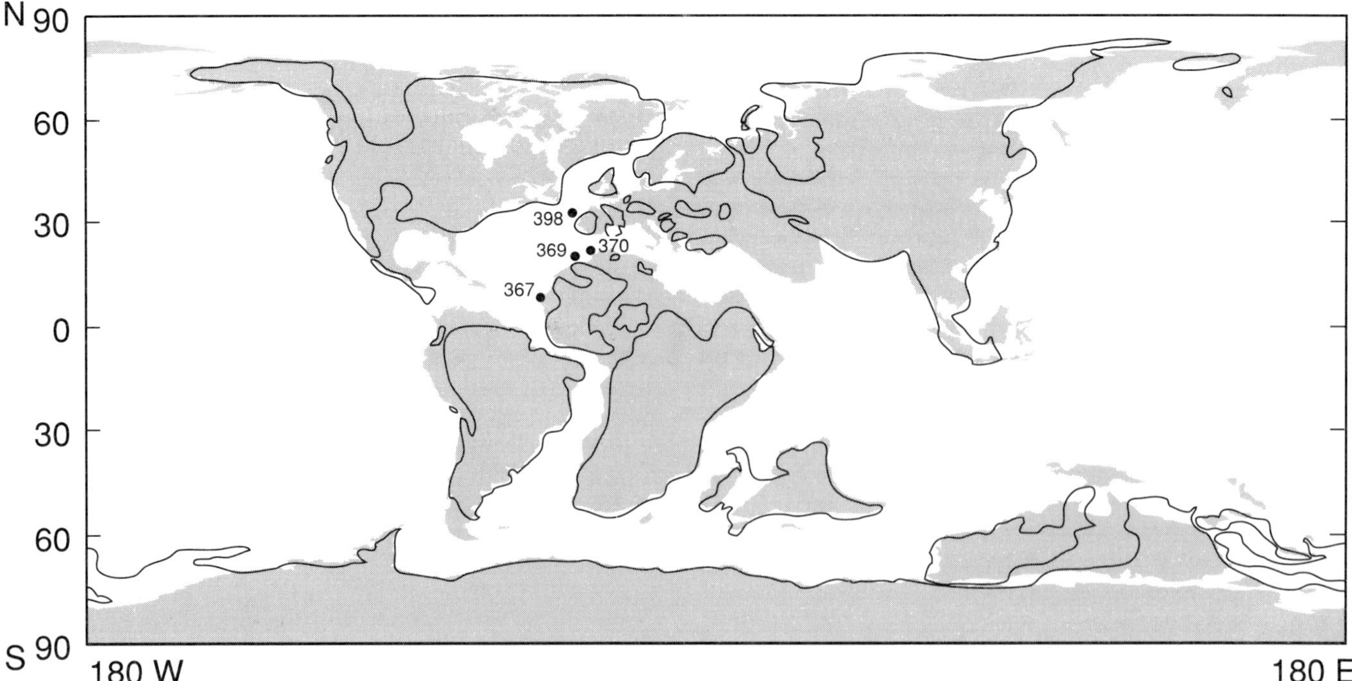

Figure 13. DSDP locations of middle Cretaceous samples with high productivity based on an analysis of nannofossils by Roth and Krumbach (1986).

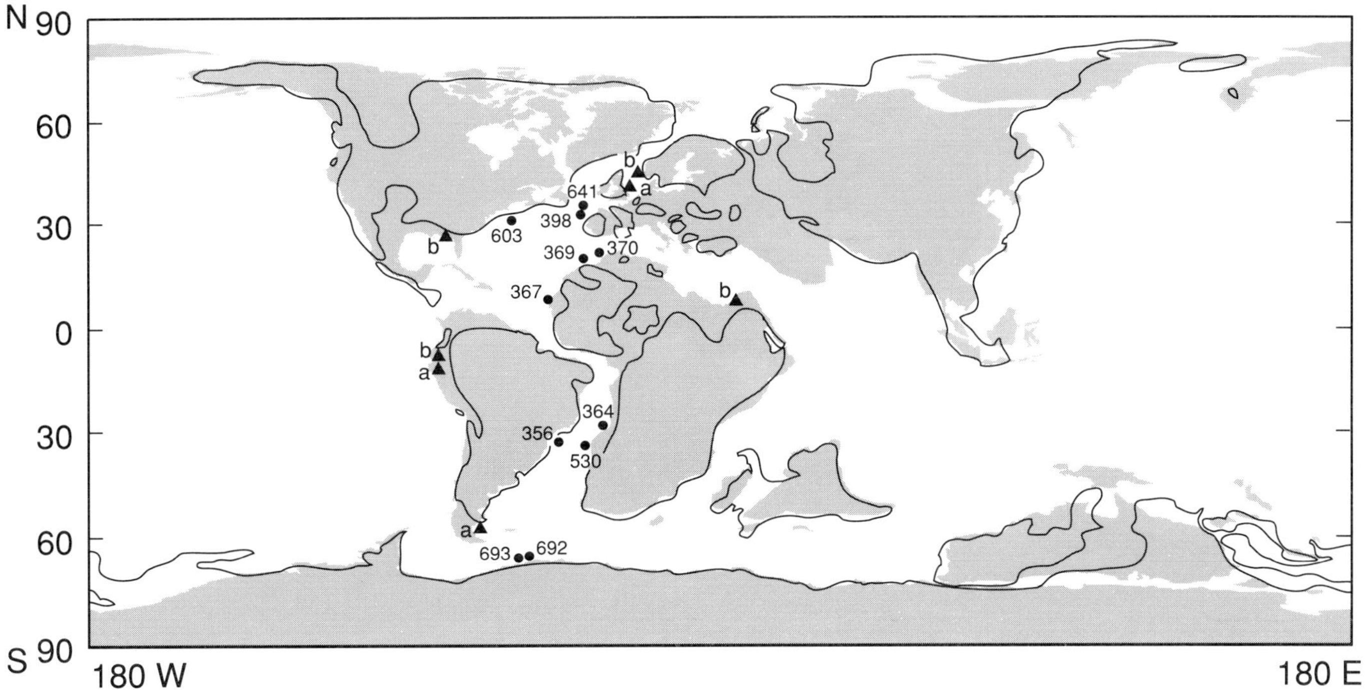

Figure 14. DSDP, ODP, and other data locations of middle Cretaceous samples adjacent to land that contain concentrations of marine organic matter. (a) high productivity (b) oil source beds. From Parrish and Curtis (1982).

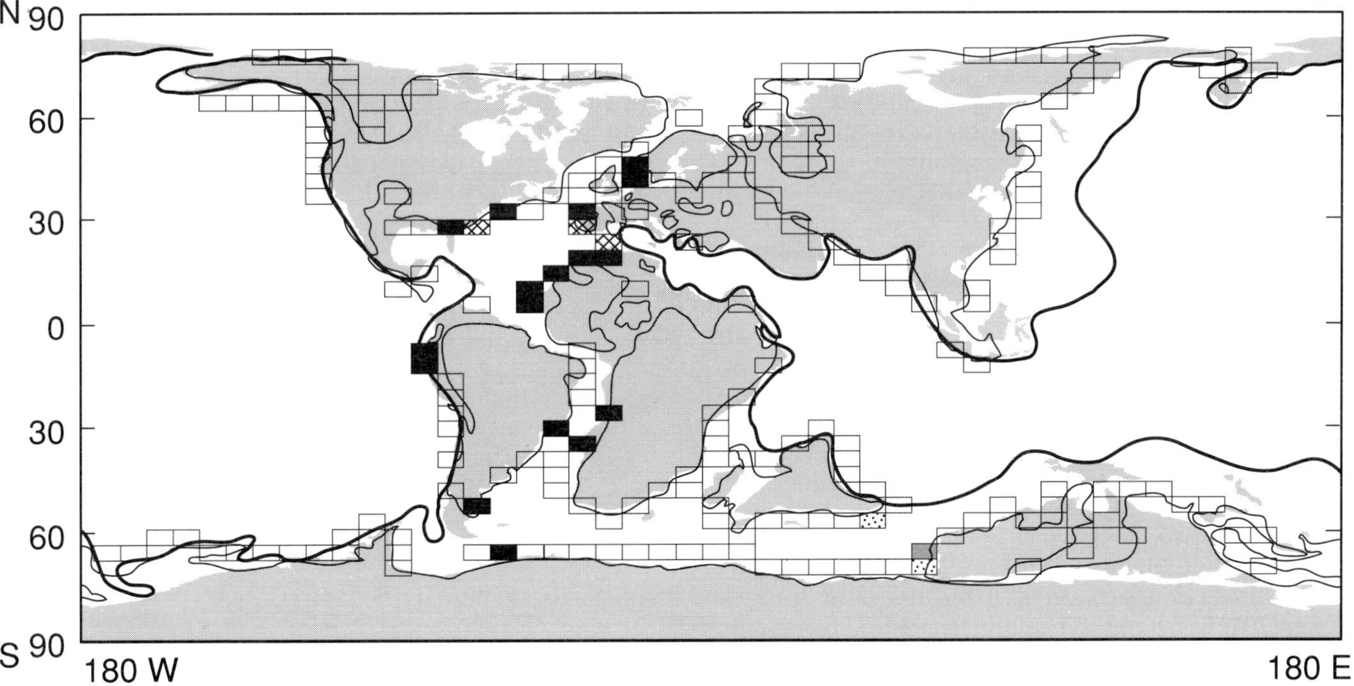

Figure 15. Comparison between seasonal model-predicted locations of coastal upwelling gridpoints and gridpoints for which data are available. Heavy solid line indicates areas of ocean floor that have been subducted or obducted. Black boxes, gridpoints in which predicted upwelling is confirmed by data; cross-hatched boxes, predicted downwelling is confirmed by data; heavy-stippled boxes, gridpoints of predicted upwelling but data do not confirm upwelling; light-shaded boxes, gridpoints for which model predicts downwelling but data indicate upwelling; empty boxes, upwelling predicted but no data available.

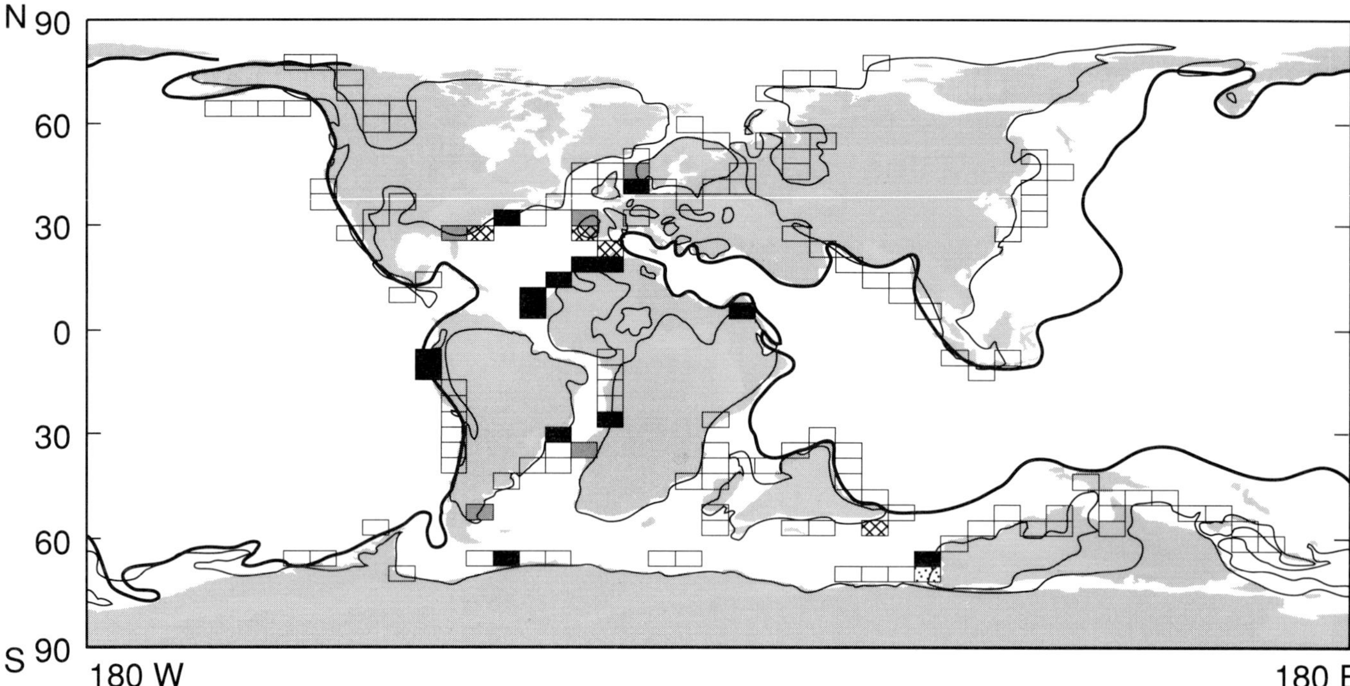

Figure 16. Comparison between mean annual model-predicted locations of coastal upwelling gridpoints and gridpoints for which data are available. Heavy solid line indicates areas of ocean floor that have been subducted or obducted. Black boxes, gridpoints in which predicted upwelling is confirmed by data; cross-hatched boxes, predicted downwelling is confirmed by data; heavy stippled boxes, gridpoints of predicted upwelling but data do not confirm upwelling; light-shaded boxes, gridpoints for which model predicts downwelling but data indicate upwelling; empty boxes, upwelling predicted but no data available.

data set would further support or discredit this analysis, is unknown. Clearly, comprehensive data sets on global productivity are essential if we are to establish the rules for predicting paleoproductivity.

Seasonal model simulations are obviously preferred, although computationally more expensive, over mean annual simulations because the solar forcing is more realistic, the control of winds by land-sea thermal contrasts are better simulated, and because many areas of upwelling are distinctly seasonal in nature.

Model predictions for past time periods introduce added uncertainty. First, model sensitivity to large changes in climate forcing are uncertain. Even the most complex model introduces a number of simplified physical processes (see Parrish and Barron, 1986, for discussion), and consequently the capability of a model to simulate a climate very different from the modern is uncertain. Each geologic data set which can be used for comparison reduces this uncertainty. Second, a multitude of factors, in addition to geography and atmospheric composition, may have influenced the climates of the past. Prediction of past climate characteristics are therefore dependent on knowledge of the appropriate forcing factors or a lack of sensitivity of the climate system to the factors which are unknown.

The capability of the model to predict sites of high productivity during the middle Cretaceous builds substantial confidence. Here again, many limitations are evident. Not all upwelling sites will have preserved indicators of upwelling or high productivity. Not all upwelling situations are a direct product of the wind stress (e.g., topographic upwelling). The climate of a coastal region (e.g., fresh-water input), water depth, basin morphology, and ocean circulation influence upwelling and sediment characteristics. In short, a one-to-one correspondence between wind-induced upwelling and productivity indicators is not expected. The conditions are sufficiently complicated that a consistently high success ratio is probably not an obtainable goal. The success for the middle Cretaceous may even be anomalous. Furthermore, the vast majority of predicted points are untested, in large part because of a lack of data.

Despite these limitations, the model simulations are highly successful. The results indicate that the CCM, combined with maps of paleogeography, provide a valuable basis to predict areas of ancient upwelling, productivity, and associated sedimentary deposits.

CONCLUSIONS

1. A comparison between present wind-driven upwelling and primary productivity indicates that only 18% of all upwelling gridpoints have high productivity (> 250 mgC/m^2/day). Evidently

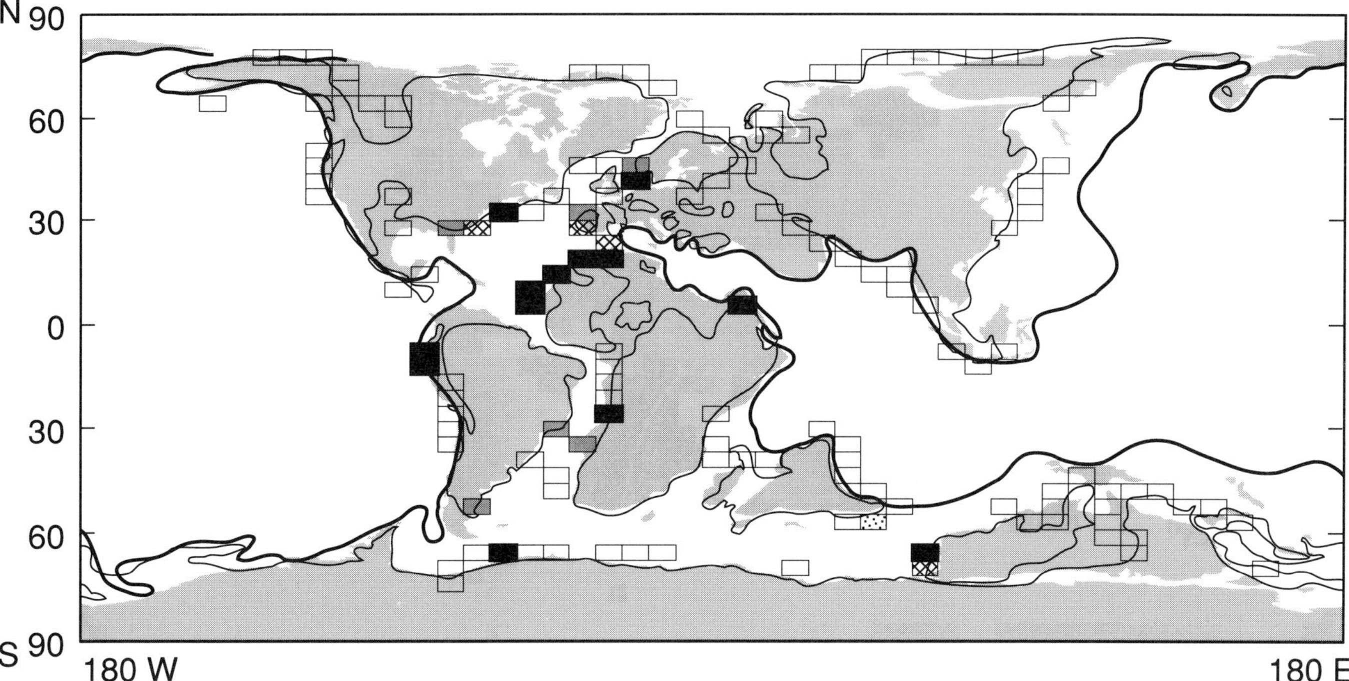

Figure 17. Comparison between mean annual with four times the present-day concentration of CO_2 model-predicted locations of coastal upwelling and gridpoints for which data are available. Heavy solid line indicates area of ocean floor that has been subducted or obducted. Black boxes, gridpoints in which predicted upwelling is confirmed by data; cross-hatched boxes, predicted downwelling is confirmed by data; heavy-stippled boxes, gridpoints of predicted upwelling but data do not confirm upwelling; light-shaded boxes, gridpoints for which model predicts downwelling but data indicate upwelling; empty boxes, upwelling predicted but no data available.

upwelling is not a good indicator of productivity, although productivity is a good indicator of upwelling.

2. The elimination of open-ocean regions and areas covered by seasonal sea ice substantially improved the association between upwelling and productivity. In coastal regions, 63% of the wind-driven upwelling areas are associated with high productivity.
3. Knowledge of the season of upwelling further improves the association between upwelling and productivity. Strong upwelling in one season and weak upwelling in the other season had the best association (91%) with high productivity.
4. Modern state-of-the-art climate models (evaluated using the NCAR Community Climate Model) accurately predict areas of present-day upwelling (mean annual and annual model versions have success ratios of 73%, 76% for July, and 80% for January, respectively).
5. Specific time periods, such as the middle Cretaceous, may be conducive to coastal upwelling. Importantly, upwelling is a complex function of geography and climate. Consequently, physically based climate models are required to evaluate fully the distribution of upwelling during Earth history.
6. Based on the distributions of siliceous sediments, marine organic-rich sediments, and specific nannofossils, model predictions of upwelling during the middle Cretaceous are largely confirmed. Although wind-induced upwelling apparently does not explain all the data sites indicative of high productivity, the model predictions of upwelling were substantially confirmed (> 90%) for both mean annual and annual model simulations.
7. To a high degree, the capability of physically based climate models to provide valuable data on the distribution of upwelling and high productivity for past time periods has been demonstrated.
8. Modern data sets of the global distribution of productivity and the lack of geologic data to test model simulations fully are considered to be the major limitations of this study.

ACKNOWLEDGMENTS

The authors acknowledge Al Guber and Rudy Slingerland for their critical comments, and William Peterson for critical comments and programming assistance. This research was supported by the National Science Foundation (Grant No. ATM-87-15499). Computer time was provided by the National Center for Atmospheric Research (NCAR). NCAR is supported by the National Science Foundation.

REFERENCES CITED

Arthur, M. A., W. E. Dean, and D. A. V. Stow, 1984, Models for the deposition of Mesozoic-Cenozoic fine-grained organic-carbon-rich sediments in the deep sea, *in* D. A. V. Stow and D. J. W. Piper, eds., Fine grained sediments: Deep water processes and facies: Geological Society of London Special Publication 15, p. 527-560.

Barron, E. J., 1985, Numerical climate modeling, a frontier in petroleum source rock predictions: Results based on Cretaceous simulations: AAPG Bulletin, v. 69, p. 448-459.

Barron, E. J., and W. M. Washington, 1984, The role of geographic variables in explaining paleoclimates: Results from Cretaceous climate model sensitivity studies: Journal of Geophysical Research, v. 89, no. D1, p. 1267-1279.

Barron, E. J., and W. M. Washington, 1985, Warm Cretaceous climates: High CO_2 as a plausible mechanism, *in* E. T. Sundquist and W.S. Broecker, eds., The carbon cycle and atmospheric CO_2: Natural variations Archean to present: American Geophysical Union Geophysical Monograph Series, v. 32, p. 546-553.

Baturin, G. N., 1983, Some unique sedimentological and geochemical features of deposits in coastal upwelling regions, *in* J. Thiede and E. Suess, eds., Coastal upwelling, its sediment record, Part B: Sedimentary records of ancient coastal upwelling: New York, Plenum Press, p. 11-28.

Berner, R. A., A. C. Lasaga, and R. M. Garrels, 1983, Carbonate-silicate geochemical cycle and its effect on atmospheric carbon dioxide over the last 100 million years: American Journal of Science, v. 283, p. 641-683.

Diester-Haass, L., 1978, Sediments as indicators of upwelling, *in* R. Boje and M. Tomczak, eds., Upwelling ecosystems: Berlin, Springer-Verlag, p. 261-281.

Hein, J. R., 1987, Fine grained siliceous deposits: Host for multifarious mineral deposits, *in* J. R. Hein, ed., Siliceous sedimentary rock-hosted ores and petroleum: New York, Van Nostrand Reinhold, p. 3-9.

Hellerman, S., and M. Rosenstein, 1983, Normal monthly windstress over the World with error estimates: Journal of Physical Oceanography, v. 13, p. 1093-1104.

Jenkyns, H. C., 1980, Cretaceous anoxic events: from continents to oceans: Journal of Geological Society of London, v. 137, p. 171-188.

Koblentz-Mishke, O. J., V. V. Volkovinsky, and J. G. Kabanova, 1970, Plankton primary production of the World ocean, *in* W. S. Wooster, ed., Scientific exploration of the South Pacific: Washington, DC, National Academy of Sciences, p. 183-193.

Kruijs, E., 1989, Predicting the locations of mid-Cretaceous wind-driven upwelling and productivity: A critical evaluation: Pennsylvania State University, University Park, Pennsylvania, Master's thesis, 166 p.

Li, W. K. W., S. V. Subba Rad, W. G. Harrison, J. C. Smith, J. J. Cullen, B. Irwin, and T. Platt, 1983, Autotrophic picoplankton in the tropical ocean: Science, v. 219, p. 292-295.

Miskell, K. J., 1983, Accumulation of opal in deep sea sediments from mid-Cretaceous to the Miocene: A paleocirculation indicator: University of Miami, Coral Gables, Florida, Master's thesis, 168 p.

Parrish, J. T., 1982, Upwelling and petroleum source beds with reference to the Paleozoic: AAPG Bulletin, v. 66, p. 750-774.

Parrish, J. T., and E. J. Barron, 1986, Paleoclimates and economic geology: SEPM Short Course, v. 18, 162 p.

Parrish, J. T., and R. Curtis, 1982, Atmospheric circulation, upwelling and organic-rich rocks in the Mesozoic and Cenozoic eras: Palaeogeography, Palaeoclimatology, Palaeoecology, v. 40, p. 31-66.

Ramanathan, V., E. J. Pitcher, R. C. Malone, and M. L. Blackmon, 1983, The response of a spectral general circulation model to refinements in radiative processes: Journal of Atmospheric Sciences, v. 40, p. 605-630.

Roth, P. H., and K. R. Krumbach, 1986, Middle Cretaceous calcareous nannofossil biogeography and preservation in the Atlantic and Indian Oceans: Implications for paleoceanography: Marine Micropaleontology, v. 10, p. 235-266.

Schlanger, S. O., and H. C. Jenkyns, 1976, Cretaceous oceanic anoxic events: Causes and consequences: Geology Mijnbouw, v. 55, p. 179-184.

Scotese, C. R., and C. P. Summerhayes, 1986, Computer model of paleoclimate predicts coastal upwelling in the Mesozoic and Cenozoic: Geobyte, Summer, p. 28-42.

Steeman-Nielsen, E., 1978, Growth of plankton algae as a function of N-concentration measured by means of a batch technique: Marine Biology, v. 46, p. 185-189.

Suess, E., and J. Thiede, 1983, eds., *in* Coastal upwelling, its sedimentary record, Part A: Responses of the sedimentary regime to present coastal upwelling: New York, Plenum Press, 604 p.

Thiede, J., and E. Suess, 1983, eds., *in* Coastal upwelling, its sedimentary record, Part B: Responses of the sedimentary regime to present coastal upwelling: New York, Plenum Press, 610 p.

Washington, W. M., and G. A. Meehl, 1983, General circulation model experiments on the climate effects due to a doubling and quadrupling of carbon dioxide concentrations: Journal of Geophysical Research, v. 88, p. 6600-6610.

Washington, W. M., and G. A. Meehl, 1984, Seasonal cycle experiments on the climate sensitivity due to a doubling of CO_2 with an atmospheric circulation model coupled to a simple mixed-layer ocean model: Journal of Geophysical Research, v. 89, no. D6, p. 9475-9503.

Facies Evolution of Late Cretaceous Black Shales from Southeast Egypt

H. Herwig Ganz
Peter Luger
Eckart Schrank
Technical University of Berlin
Berlin, Federal Republic of Germany

Paul W. Brooks
Martin G. Fowler
Institute of Sedimentary and Petroleum Geology
Calgary, Alberta, Canada

Newly developed and conventional organic/inorganic geochemical and sedimentological techniques, as well as micropaleontological studies (palynomorphs, foraminifera), resulted in the reconstruction of the depositional environment and an estimation of the resource potential of the Upper Cretaceous black shale/phosphorite sequences in southeastern Egypt. The depositional environment is characterized by a broad, shallow shelf which was separated from the continental hinterland by restricted coastal swamps. Depending on structures on the sea floor and depth of the water column, reducing conditions prevailed. The Campanian/Maastrichtian transgression was interrupted by multiple regressive phases which caused intensive reworking of the marine sediments and enrichment of phosphatic layers. Minor terrestrial influx occurred during the formation of oyster bioherms. Open-marine conditions in the upper part of the section (Dakla Formation) indicate prograding marine transgression.

INTRODUCTION AND REGIONAL STRATIGRAPHY

The Cretaceous dark pelitic sediments which accompany the phosphate beds (Duwi Formation) in the Quseir/Safaga area in the eastern part of Egypt have been known for many years (Said, 1962). To date, however, relatively few analytical studies have been published on these so-called black shales (e.g., Malak et al., 1977; Gindy and Tamish, 1985). The present study is based on organic and inorganic geochemistry, sedimentology, and micropaleontology of palynomorphs, as well as foraminifera. By the combined use of a number of newly developed geochemical techniques and paleontological facies indicators, it was possible to recognize ecologic-facies/geochemical changes within the sections and to reconstruct the depositional environments.

In the area around Quseir and Safaga (Figure 1) the Tertiary sequence overlies Maastrichtian to early Tertiary bituminous marls—the Dakhla Formation—and the late Campanian-early Maastrichtian phosphoritic Duwi Formation (Youssef, 1957; Said, 1962). The late Campanian to early Maastrichtian age of the main phosphorites of Egypt has recently been confirmed in the Wadi Qena about 100 km west of Quseir by Hendriks and Luger (1987) and Luger and Gröschke (1989). Ammonite assemblages from the lower phosphorite horizon (Hamadat bed) are of early late Campanian age, whereas the middle horizon (Duwi bed) was determined as late late Campanian. In the present study *Rosita* cf. *patelliformis* (Gandolfi), *Archaeoglobigerina blowi* Pessagno, *Rugoglobigerina rugosa* (Plummer), and heterohelicids occurred in a sample from the uppermost part of the phosphoritic Duwi Formation (Atshan bed). Thus the age of this horizon is late Campanian to early Maastrichtian. The base of the Dakhla Formation is characterized by the planktonic foraminifera *Globotruncanita angulata* (Tilev), *Globotruncana aegyptica* (Nakkady), and *Globotruncana gagnebini*

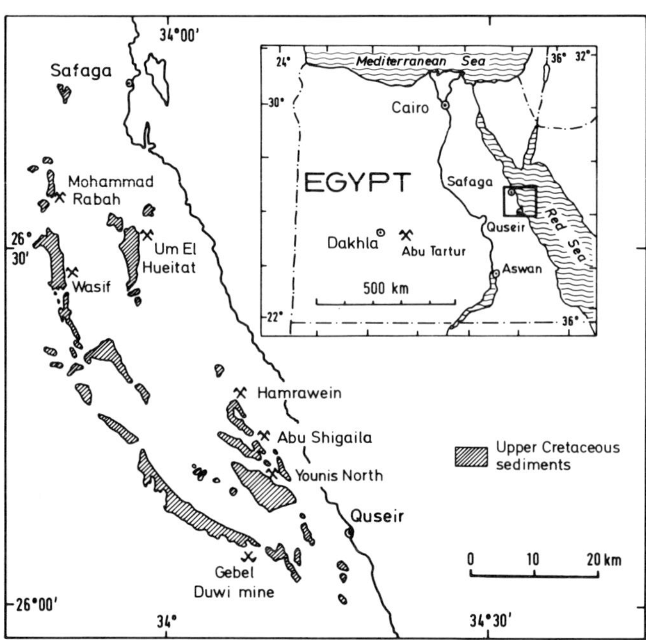

Figure 1. Location map of the study area showing sample locations.

(Tilev). Therefore, the sample can be attributed to the upper part of the *Globotruncana falsostuarti* zone (*Globotruncana gagnebini* subzone, sensu Bellier, 1983), the late early Maastrichtian.

In the area of Quseir and Safaga, from top to bottom, the following stratigraphic units can be distinguished: Thebes Formation (early Eocene, up to 120 m); Esna Formation (late Paleocene to early Eocene, up to 50 m); Tarawan Formation (late Paleocene, 6–7 m); Dakhla Formation (early Maastrichtian to middle Paleocene, 40–70 m); Duwi Formation (late Campanian to early Maastrichtian, 30–60 m); Quseir Formation (early Campanian, up to 70 m). The Quseir Formation overlies a 100–300-m thick series of pre-Campanian sediments (formerly "Nubian Group") which are subdivided into units of Late Jurassic to Late Cretaceous age (Klitzsch, 1986).This study deals with samples of the Quseir Formation ("Variegated Shale"), the Duwi (phosphate) Formation, and the early-middle Maastrichtian part of the Dakhla Formation (Figure 2).

SAMPLED MATERIAL

In the area around Quseir and Safaga, phosphorite has been mined since the beginning of this century (Hume, 1927), and a number of mines are still active today. This offered a good opportunity to obtain fresh and unweathered material, which is essential for studies involving organic geochemistry and palynology. Seven localities (Figure 1) were sampled and investigated in the course of our work. Two of these sections (Abu Shigaila and Younis North, Figure 3) are discussed in detail. The facies characteristics of the two sections are comparable to the other localities studied and give a representative picture of the facies development in the Quseir and Safaga area.

GEOCHEMICAL METHODS

The samples were ground in a vibratory cup mill to a grain size smaller than 10μ (Ganz and Kalkreuth, 1989). Organic-carbon contents were determined by use of a LECO C/S 244 analyzer. All samples were extracted by soxhlet apparatus using CH_2Cl_2 as a solvent, and aliphatic, aromatic, NSO, and asphaltene fractions were isolated by standard chromatographic procedures. The aliphatic hydrocarbons were analyzed by gas chromatography (GC) and computerized gas chromatography-mass spectrometry (GC-MS), according to previously published methods (Fowler and Brooks, 1987). In this study only C27/29 sterane distributions (C27/C27+C29) are shown in Figure 2. A detailed discussion of the biomarker composition of the Egyptian black shales will be the subject of another paper.

After bitumen extraction, kerogen is isolated by a hot-acid and centrifuging technique, followed by a new infrared method for the determination of kerogen type and thermal maturation (Ganz et al., 1987; Ganz and Kalkreuth, 1989). The new method is more accurate than Rock-Eval pyrolysis and much faster than the common elemental analysis. Infrared spectra of the isolated organic material were produced using a Perkin Elmer 598 IR system in tandem with the PE datastation 3600, and the relative ratios of the intensities of aliphatic and aromatic bands (A-factor), as well as carboxyl/carbonyl- and aromatic bands (C-factor), were determined. Plotting these ratios in a diagram results in an excellent differentiation of the organic matter which is readily comparable to the van Krevelen diagram (Figure 4). Thermal maturation of kerogen can be predicted by a vitrinite-reflectance-equivalent grid, the IR-aromaticity, and the shift of the aromatic C=C bands toward the minimum wavenumber (Wmin). Since the latter is not influenced by weathering, Wmin offers a valuable tool for the study of such effects. Oil and gas potential of organic-rich rocks can be predicted by the A-factor derived from infrared spectroscopy.

Mineralogical and sedimentological characteristics are also determined by a new infrared method (Ganz et al., 1987; Ganz and Kalkreuth, 1989). The amounts of quartz, calcite, dolomite, kaolinite, chlorite, illite, smectite, and pyrite in the original homogenized sample are caculated. Accuracy is comparable to that obtained from the combination of x-ray diffraction and x-ray fluorescence analyses.

Whole-rock and kerogen-decomposition analyses were performed on the samples, and elemental contents were determined with the aid of a Perkin Elmer ICP/AES 5000 in conjunction with a PE 7500 laboratory computer. By the use of factor analysis for

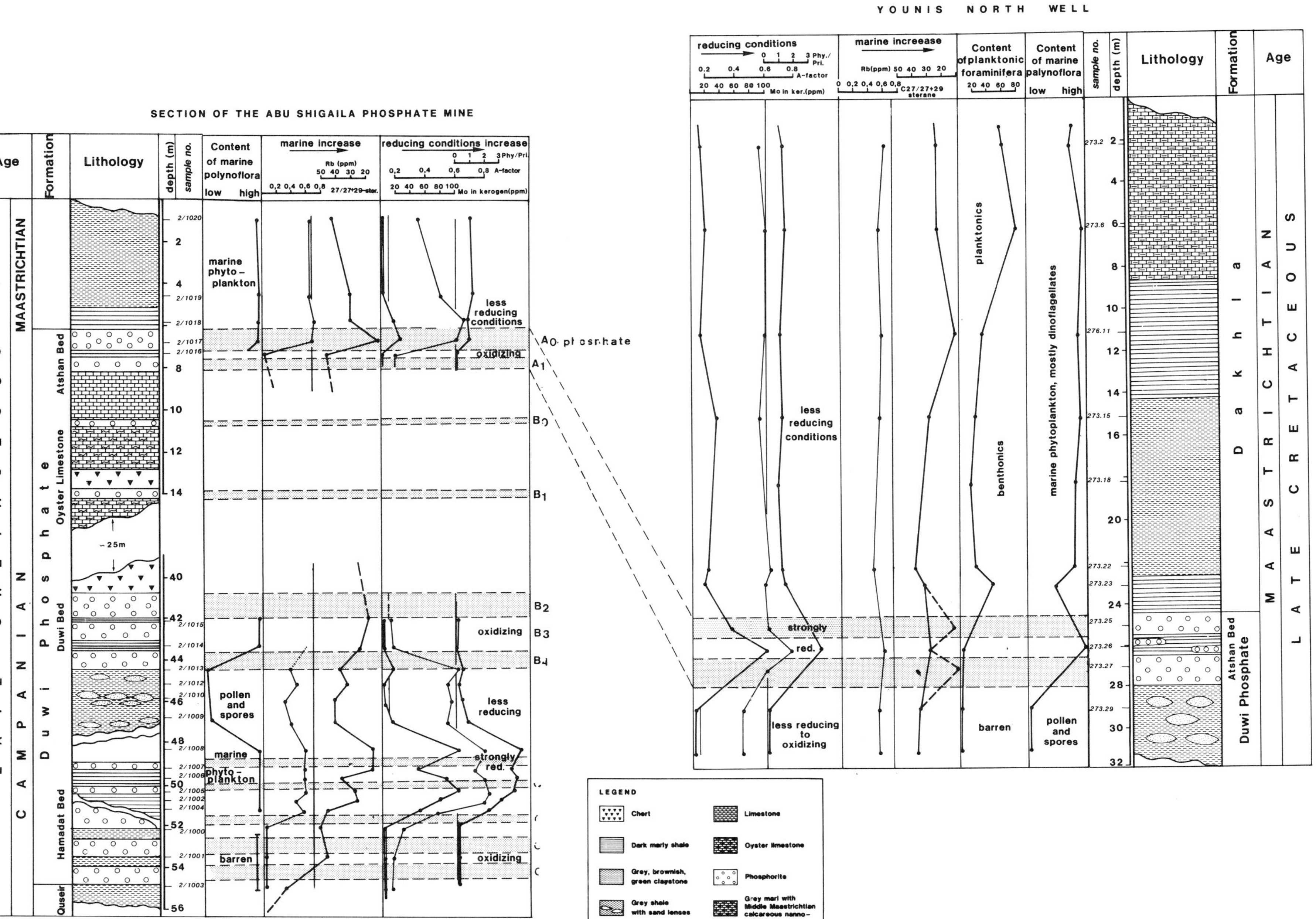

Figure 2. Geochemical and paleontological characterization of the Abu Shigaila and Younis North profiles.

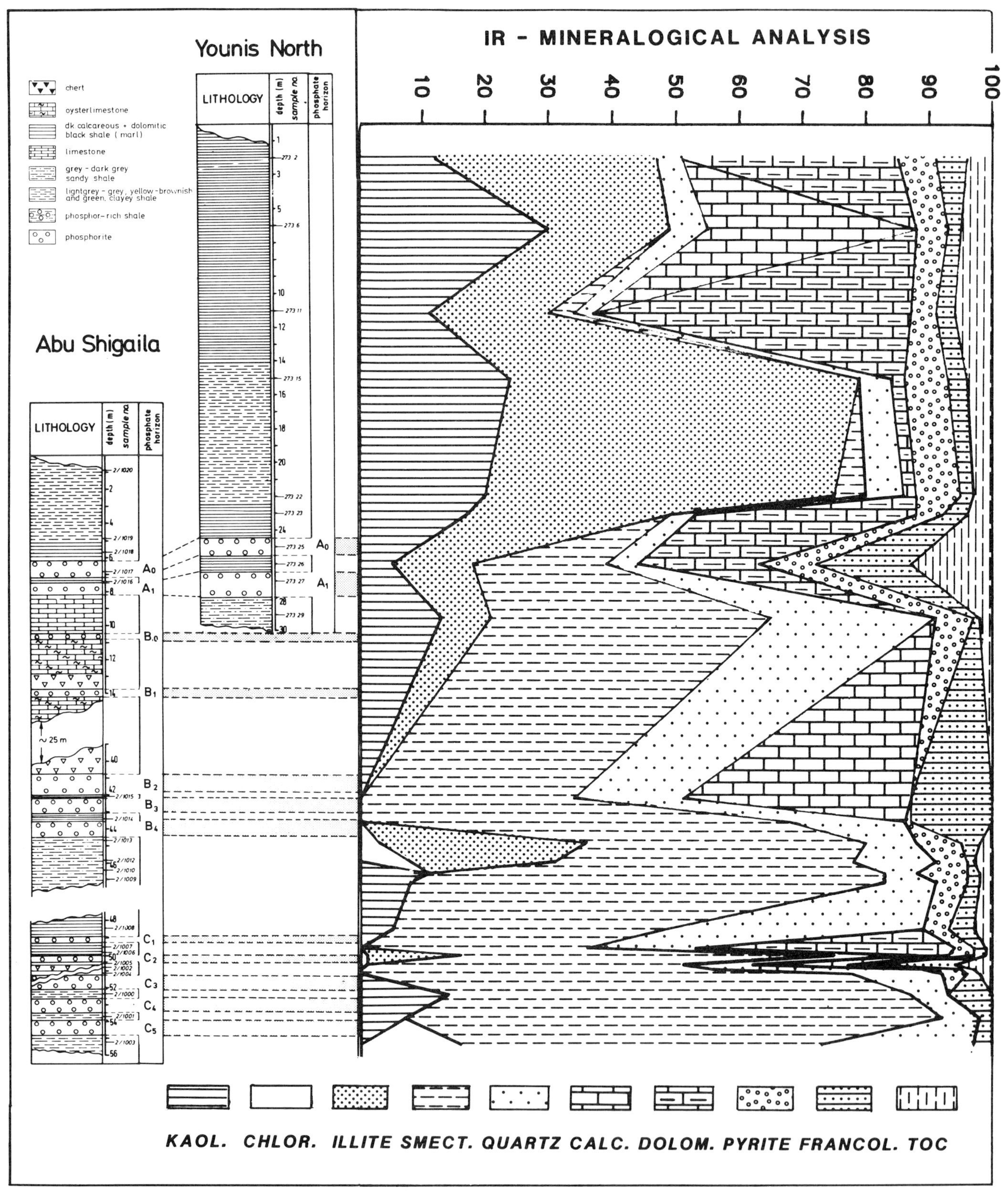

Figure 3. Sample position and mineralogical composition of the Abu Shigaila and Younis North samples. Kaol.= kaolinite; chlor.= chlorite; smect.= smectite; calc.= calcite; dolom.= dolomite; francol.= francolite.

45 variables and nearly 200 samples, a number of significant factors were extracted. The Rb, Ti, and Zr contents determined from the whole-rock analyses have proved to be good indicators of the relative degree of terrestrial input into the sediment; Mo contents of kerogen are excellent parameters for the degree of

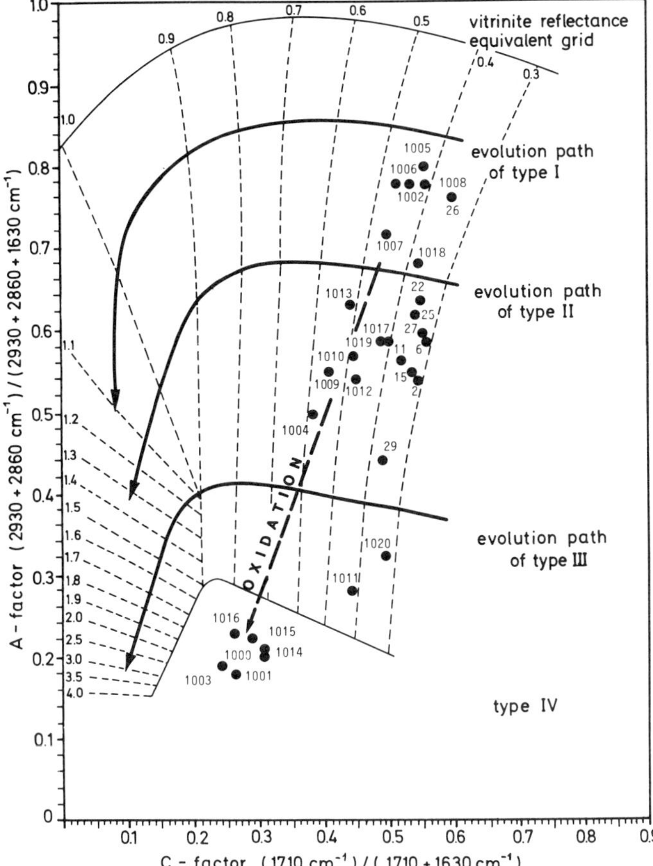

Figure 4. IR diagram for the classification of kerogen type and thermal maturation. Hydrocarbon potential (mg HC/g of rock) is also calculated with the help of the A-factor (A-factor × TOC × 10).

marine reducing environment (Ganz, 1986). Metal ratios for geochemical facies analysis (e.g., V/Cr, V/Mo, Ni/Co, and /Ni) can be misleading, due to the clay and heavy-mineral content of the sediment. Also, Mo contents from the whole-rock analysis are influenced by heavy minerals (Ganz, 1986).

RESULTS

Mineralogy

The results of IR-mineralogical analysis are shown in Figure 3. The most characteristic trend in this profile is marked by the clay minerals. Chlorite is nearly absent and smectite is the most prominent phase in the lower part of the sequence. Quartz is present in moderate amounts, while kaolinite and illite, as well as carbonates (dolomite and calcite), are only minor constituents. In the upper part of the section (Younis north profile) the mineral composition is totally different. Smectite and quartz contents are drastically lower, whereas kaolinite and illite are dominant, along with significant amounts of carbonates.

Organic-Carbon Content, Kerogen Classification, and Hydrocarbon Potential

The organic-carbon content of the black shales is below 6%, except for sample 273.26 (Table 1), which is significantly enriched in organic matter (13%). Some samples are also sufficiently enriched in P_2O_5 (273.25,26, 27, 2/1017, 2/1005;Table 1). Kerogen type and stage of thermal maturation are shown in the IR diagram (Figure 4). The hydrocarbon potential (mg HC/g of rock) is determined by multiplying the A-factor × TOC × 10. All samples are immature, with a fair to good hydrocarbon potential (kerogen types II-III); however, some of them are heavily degraded due to reworking in an early stage of sedimentation (kerogen type IV). The composition of the black shales is very heterogeneous. Generally the beds which yield a high energy potential (up to 6000 kJ/kg) are only isolated thin layers. However, the average values of a 40-m Dakhla shale profile indicate that this sediment is an immature, potentially good source rock for oil and might be suited as a raw material for energy and cement production (Ganz et al., 1987a). Since the average trace-element contents of the Dakhla shale are rather low, final leaching of the combustion residue is not worthwhile.

The total vertical thickness of the Abu Shigaila and Younis North profiles is about 80 m. Thus all samples should have about the same maturation. Vitrinite-reflectance measurements on selected samples revealed 0.38–0.42% Ro, and all samples should follow more or less the 0.4% vitrinite reflectance equivalent line in the IR diagram. It has been discovered, however, that the position in the IR diagram and the chemical composition are highly affected by weathering processes (Ganz and Kalkreuth, 1989) and cannot be used in this case for the proper determination of maturity. With increasing maturation, the peak maximum intensity of the aromatic C=C band near 1630 cm-1 is shifted toward a minimum wavenumber (Wmin). The minimum wavenumber, Wmin, is unchanged by weathering effects and reflects the true maturation of all types of kerogen. Wmins of the Egyptian black shales are between 1625–1630 cm-1, indicating maturation of < 0.4 %Ro (Figure 5). Thus the effects of weathering and oxidation on those samples may be recognized by comparison of the position in both the IR and the Wmin diagrams. Samples 2/1004, 1009, 1010, 1012, 1013, and particularly 2/1000, 1001, 1003, 1014, 1015, and 1016 show the effects of oxidation processes.

Geochemical Facies Parameters

The results of geochemical analyses are shown together with paleontological data in Figure 2. The

Table 1. Geochemical data for the studied samples.

Sample No.	TOC (%)	P_2O_5 (%)	Mo (ppm)	Rb (ppm)	A-Fact.	C-Fact.	27/27+29 ster.	Phy/Pr	Ro (%)
273.2	4.50	1.92	12	25	0.55	0.54	0.58	1.08	—
273.6	5.23	0.75	19	27	0.59	0.55	0.51	1.34	—
273.11	4.84	1.67	12	14	0.56	0.51	0.55	0.94	—
273.15	4.21	1.19	34	31	0.55	0.53	0.52	1.08	—
273.22	2.76	0.68	107	41	0.63	0.54	0.54	1.09	0.38
273.23	4.14	1.86	17	34	0.59	0.55	0.55	1.32	—
273.25	1.63	9.61	52	14	0.61	0.53	—	—	—
273.26	13.00	6.46	719	30	0.76	0.60	0.59	3.71	—
273.27	0.26	25.68	—	2	0.60	0.55	—	—	—
273.29	2.16	0.28	—	37	0.44	0.48	0.61	0.33	0.35
2/1020	0.84	1.46	0	44	0.32	0.49	0.63	0.90	—
2/1019	1.77	1.28	0	31	0.59	0.49	0.61	1.06	—
2/1018	3.55	1.37	13	31	0.68	0.54	0.66	0.76	—
2/1017	3.01	14.08	22	12	0.59	0.48	0.66	0.82	—
2/1016	0.31	9.08	—	47	0.23	0.26	—	—	—
2/1015	0.20	5.36	—	30	0.21	0.31	—	—	—
2/1014	0.27	5.55	—	25	0.20	0.31	—	—	—
2/1013	1.69	1.39	12	39	0.63	0.43	0.39	0.39	—
2/1012	2.75	0.56	0	34	0.54	0.44	0.47	0.20	—
2/1011	15.3	4.51	12	7	0.28	0.45	—	—	0.42
2/1010	2.18	0.78	6	43	0.57	0.43	0.30	0.26	0.38
2/1009	2.05	0.79	13	42	0.55	0.40	0.49	0.67	—
2/1008	2.70	1.59	260	17	0.78	0.55	0.59	4.22	—
2/1007	1.03	0.76	46	17	0.72	0.48	0.54	3.56	—
2/1006	1.30	0.87	84	38	0.78	0.50	0.55	3.91	—
2/1005	5.74	5.34	163	30	0.80	0.55	0.56	3.66	—
2/1002	2.56	0.93	73	29	0.78	0.52	0.48	2.85	—
2/1004	1.84	0.73	48	47	0.50	0.38	—	2.02	—
2/1000	0.14	3.05	—	52	0.22	0.29	—	—	—
2/1001	0.18	0.69	—	47	0.18	0.26	—	—	—
2/1003	0.10	1.31	—	75	0.19	0.24	—	—	—

Rb contents of the whole- rock analysis and the C27/29 sterane distribution are reliable indicators for the relative proportion of terrestrial detritus in the sediment. First Huang and Meinschein (1976, 1979) proposed the use of sterane distribution as a source indicator of organic matter. C27 steranes were regarded as mainly marine, whereas C29 steranes were thought to be terrestrially derived. Later it was found that particularly Paleozoic oils had high C29 sterane contents, which could not be derived from land-plant material (Arefev et al., 1980; Rullkötter et al., 1986; Moldowan et al., 1985). However, Grantham and Wakefield (1988) showed that there are increases in the relative content of C28 steranes and decreases in the relative C29 steranes through geological time, whereas no consistent variations in the relative content of C27 steranes have been observed. The variations reflect the chemical evolution of the sterols according to the increased diversification of phytoplanktonic assemblages through time. Since Late Cretaceous time, only minor changes occurred, and therefore C27/29 sterane ratios can be used in this study to provide valuable information as ecological indicators.

The A-(aliphatic) factor of the IR kerogen analysis serves here, above all, to differentiate between terrestrial and marine organic material. Moreover, from the height of the A-factor certain conclusions may be drawn regarding the prevailing Eh of the environment. In order to differentiate between reducing and oxidizing conditions, however, the following parameters are necessary in addition: the Mo contents of kerogen, and the C27/29 sterane distribution. Low A-factor values may imply either terrestrial organic matter or may be the result of oxidation. Whereas the sterane distribution in many cases still allows the differentiation between marine and terrestrial material, the Mo content in the kerogen unmistakably reflects the degree of reducing environment. In the following pages the most

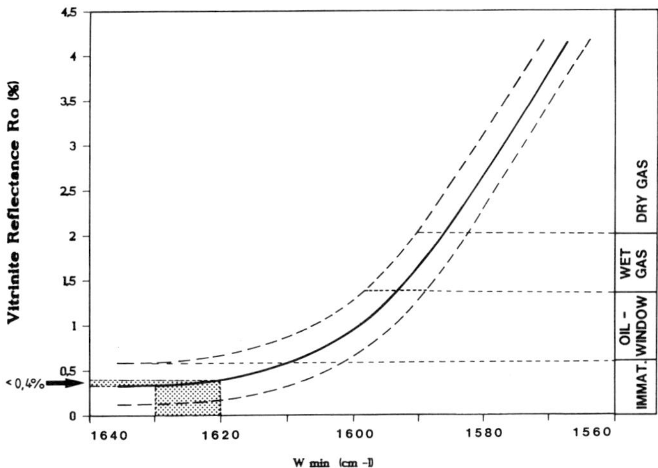

Figure 5. Average value of Ro % as determined by means of the minimum wavenumber (Wmin) of the same samples shown in Figure 4 which are characterized through the IR diagram.

important geochemical, palynological, and micropaleontological results are presented in ascending stratigraphic order.

Quseir Formation (Variegated Shale)

The sedimentary sequence of the Abu Shigaila section, which includes units ranging from the Campanian Quseir Formation (variegated shales) to the Maastrichtian lower part of the Dakhla Formation, reflects the pulsating transition from nearshore terrestrial conditions to an open-marine shelf environment. In the lower part of the section, foraminifera are too poorly preserved to permit identification, and only palynological and geochemical data could be used. Samples 2/1003, 1001, and 1000 (Figure 2) did not yield any palynomorphs due to intensive reworking and oxidation (kerogen type IV in the IR diagram). Thus, only inorganic data such as Rb values, which are highest in this part of the section, were applied.

Duwi Formation

Hamadat Bed (Lower Phosphatic Unit)

The organic-rich marls between the lowermost phosphorite horizons (Hamadat C bed) are characterized by marine-derived organisms (dinoflagellates) with marked contributions of terrestrially derived organic matter. High C27/29 sterane ratios and low Rb values coincide well with the palynological data. The upper part of the Hamadat bed and the immediately overlying shale contain palynomorph assemblages with the dinocyst species *Odontochitina operculata* which characterizes the Campanian dinoflagellate zone of the same name (Williams and Bujak, 1985). Peridinoid dinoflagellates such as *Andalusiella* and *Senegalinium* are present throughout the lower part of the Duwi Formation (samples 2/1004 to 2/1014). A peridinioid maximum in the basal Hamadat shale (sample 2/1004) coincides with the occurrence of the agglutinated foraminifer *Ammoastuta*. Both may be used to assume the presence of a brackish, marginally marine environment. Dinoflagellate assemblages rich in peridinioids have been regarded as characteristic of low-salinity conditions (Harland, 1977; Hultberg, 1986).

The organic-rich layer just above the uppermost Hamadat phosphate (sample 2/1008) has an exceptionally high phytane/pristane ratio and an unusual predominance of the saturated n-alkanes. These features point to a strongly reducing, hypersaline, evaporitic environment (Welte and Waples, 1973), which is also indicated by high Mo values for the kerogen. The associated dinoflagellates show dominance of spineless cysts of the *Chytroeisphaeridia* type, together with unusually short-spined variants of the peridinioid *Andalusiella*.

The lenticular bedded, silty shales overlying the Hamadat phosphate (samples 2/1009–2/1013) indicate a progressively increasing terrestrial influence in their geochemical and palynological parameters. High Rb contents, low C27/29 sterane ratios, and increasing percentage of terrestrial pollen and spores and land debris can be observed, although the marine dinoflagellates remain as a subordinate element. In the terrestrially influenced facies below the Duwi bed, the green alga *Pediastrum*, typically an inhabitant of fresh-water lakes, is also present. This might also explain the position of this sample in the IR diagram (kerogen type II).

Duwi Bed (Middle Phosphatic Unit)

The claystones intercalated in the Duwi bed yielded only a few types of palynomorphs. In sample 2/1014 some land-plant debris and a few dinoflagellates were observed. Kerogen is of type IV, and no other organic geochemical parameter could be obtained. Rb values, however, indicate prograding marine transgression.

Oyster Limestone

Between the middle and upper phosphate beds, thick accumulations of chert, bioclastic debris, and, particularly, monospecific oyster limestones (*Ambigostrea villei*, COQUAND) occur (N. Malchus, personal communication). The limestone beds thin toward the east and northwest and are absent north of the Quseir/Safaga area (El Tarabili, 1969). In situ oyster assemblages are largely restricted to southwestern exposures. Low-angle crossbedding and the inclusion of bioclastic debris suggest that these reefal communities lived in a relatively high-energy, oxidizing environment (Glenn and Mansour, 1979). Organic-matter content of these sediments is very low, and shales comparable to the other samples are totally missing. Phosphate beds B0 and B1 are only thin intercalations with minor local P_2O_5 enrichment.

Atshan Bed (Upper Phosphatic Unit)

Growing marine influence is palynologically reflected at the base of the Dakhla shale in the composition of the dinoflagellate assemblages by a growing percentage of spiny cysts which reach their maximum 1.5 m above the Atshan bed (A phosphate) in sample 2/1019 where *Spiniferites* is extremely abundant, associated with *Cyclonephelium*, *Cordosphaeridium*, *Cannosphaeropsis*, and others. Although phosphate beds A1 and A0 are only minor enrichments, sample 2/1017 yielded more than 14% P_2O_5, and this sample is very important for the understanding of the principles of phosphorite genesis. Sufficient organic material remains in that sample to permit detailed organic geochemical classification. The gas chromatograms indicate oxidation or degradation, and the sterane distribution points to very strong marine influence. In the IR diagram this sample is in the position of a type II kerogen. Very low Rb content reflects minimum terrestrial influence. Thus, this sample is interpreted to be the result of certain oxidation or reworking processes. The original reducing conditions are indicated by moderate Mo contents in the kerogen fraction. The highly polar structure of the kerogen seems to prevent metals from being removed during the early stages of oxidation and reworking, when most other chemical parameters, such as Ph/Pr, have already been affected.

It is important for the understanding of these sediments to mention that sample 273.29 from the neighboring Younis North profile yielded very high Rb contents, and also very high C27/29 sterane ratio and hopane contents. This is even more interesting because at the base of the Younis North section predominantly terrigeneous palynomorphs occur. Samples 273.25 and 273.27 are also examples of the upper phosphorite horizons from the neighboring Younis North well. These samples have very high C27/29 sterane ratios, together with low Rb contents and moderate reducing conditions. At 273.26 geochemical data are in striking contrast to those of the phosphorites, indicating here highly reducing marine conditions. At 26 m the palynofacies indicates a marine environment with a low-diversity phytoplankton community composed of small acritarchs, the prasinophycean alga *Pterospermella*, and the calcareous nannofossil *Micula decussata* (Schrank and Perch-Nielsen, 1985). A single scolecodont was also found, but no terrigeneous palynomorphs. In general prasinophytes are regarded as "disaster species," being most abundant in the absence of the other phytoplankton (Tappan, 1980, p. 814). This might also be true for the small acritarchs which are present as planktonic background nearly everywhere in the phosphatic and shale sequence of Egypt. Therefore, some kind of environmental stress may be assumed for the microplankton community at 26 m. Interestingly, 95% of the foraminifera assemblages are small benthonic forms, although species diversity is rather low. This points to shallow-marine conditions during the deposition of these sediments. The benthic association is mainly characterized by small rotaliids (buliminids, gavelinellids).

Dakhla Formation

As was the case above the lower and middle phosphorite horizons, terrestrial influences also increase above the uppermost A phosphate bed (see Rb contents of samples 2/1020, 273.23, and 273.22). Only a single phytoplankton genus, the prasinophycean *Pterospermella*, was found in the Abu Shigaila section (sample 2/1020), reflecting some kind of environmental stress for the marine organic-walled plankton. It is noteworthy that Rb, Ti, and Zr contents of equivalent stratigraphic units decrease from south to north in the area studied. This is consistent with the southerly shallowing of the Egyptian shelf sea as already described by other authors (e.g., Ward et al., 1979; Luger, 1985). In the Younis North well the Dakhla shale is characterized by dominance of dinoflagellates with the exception of the 23-m sample, where 54% of the counted grains were mostly small land-derived palynomorphs, indicating the possibility of wind transport. In the sample at 22 m, among the larger land-derived palynomorphs, the fresh-water fern *Azolla* was observed, which indicates influx of fresh-water. The kerogen of the Dakhla shale is of the amorphous type associated with varying quantities of land-plant debris. A minimum of terrestrial influx is observed between 15 and 6 m. This is reflected in the dinoflagellate assemblages by relatively abundant spinose cysts (*Florentinia*, *Spiniferites*, *Cordosphaeridium*). The same is indicated by very low Rb contents of sample 273.11.

The foraminiferal assemblages of the Dakhla Formation in the Younis North well probably reflect different sedimentary environments in an open euhaline shelf sea. At the base (23 m) abundant planktonic foraminifera (44%; see Figure 2) occur together with a relatively highly diverse benthonic foraminiferal assemblage, which points to a middle- to outer-shelf environment of deposition. Reduced numbers of planktonic foraminifera are observed in the samples between 22 and 15 m. Here the planktonic association is strongly dominated by heterohelicids; globotruncanids are very rare. The diversity of benthonic species is generally relatively high. The reduced plankton content points to a shallowing of the sea level, and these assemblages may be interpreted to be of middle- to inner-shelf origin (Sliter, 1972; Luger, 1985). The high amount of simple agglutinated foraminifera at a depth of 22 m may be due to partial decalcification. IR lithologic analysis clearly indicates that dolomite is the main carbonate phase in the Younis North section.

Calcareous nannofossils, members of the middle Maastrichtian *Lithraphiditres quadratus* zone, occur in the upper part of the section between 8 and 1 m (Schrank and Perch-Nielsen, 1985). They reflect a marine environment which favors the development of microplankton, although the terrestrial influx increases to more than 25% at the top of the sequence (Figure 2).

Highly diverse planktonic associations are observed between 14 m and 1 m. Heterohelicids are common in the planktonic associations. The first appearance of *Gansserina gansseri* (Bolli) is in the sample at 14 m. Therefore this and the overlying assemblages may be attributed to the *Gansserina gansseri* zone of the middle Maastrichtian. Between 12 and 4 m, the benthonic foraminiferal associations generally have a reduced species diversity. These associations mainly consist of small buliminids. Nodosariids, generally common in euhaline shelf sediments, are rare to absent. The diverse planktonic associations and their high percentage of the total foraminiferal assemblage point to a rise in sea level combined with an opening of the basin. Although the low diversity of benthonic species might indicate a reduced oxygen level at the sea floor, this is not supported by the geochemical data. Mo content, A-factor, and Ph/Pr ratio all show rather constant values throughout the upper part of the Dakhla shale section of Younis North. The maximum transgression is probably marked by the high percentage of planktonic foraminifera in the interval from 9 to 4 m. However, the great abundance of heterohelicids in the planktonic association indicates a shelf environment, with the highest planktonic/benthonic ratios probably reflecting outer-shelf environments.

DISCUSSION

Paleobiogeographic and Paleoclimatic Implications of Palynological and Mineralogical Studies

During Late Cretaceous, the study area was in a nearly equatorial position (Smith et al., 1982), which is in accordance with its assignment to the Senonian Palmae province of Herngreen and Chlonova (1981). Typical members of this province have been found in the Abu Shigaila and Younis North sections (Schrank, 1987). They include pollen such as *Spinizonocolpites* and *Proxapertites* which can be referred to the extant mangrove palm *Nypa*. *Nypa* mangrove is presently distributed in humid, tropical areas of Southeast Asia from Sri Lanka and the Ganges delta via Indonesia to northern Australia. *Nypa* localities comprise tidal shores and estuaries where *Nypa* mangrove swamps may extend over hundreds of square miles (Tralau, 1964). Similar vegetation may have existed in Egypt during the Late Cretaceous and Early Tertiary.

It is interesting to note that clay-mineralogy data may also be explained by the presence of mangrove swamps. As a general rule, nonmarine rocks are richer in kaolinite and montmorillonite than marine rocks, which tend to have more illite and chlorite. Generally, warm, wet, tropical and subtropical climates produce kaolin-rich soils, whereas warm, dry climates produce smectite-rich soils. However, detailed studies in localized areas have shown that clay minerals in modern oceanic sediments are not a simple reflection of source and climate on the adjacent landmass. Because of their small size and ease of transport, clays are part of large, integrated, sediment-dispersal systems. These systems can only be understood by studying the entire association (Eslinger and Peavear, 1988). Irion and Petr (1979) studied recent mangrove swamps in the Gulf of Papua. The sediments of that area are clearly differentiated according to their relative geographical position. Sediments deposited by large-channel river systems are characterized by high contents of kaolinite and quartz and subordinate amounts of smectite, chlorite, feldspars, illite, and heavy minerals. Sediments of the broad coastal area, however, yield tremendous amounts of smectite, which was formed in aquitate soils by increased salinity during high tides.

A similar situation may have existed during the Late Cretaceous of Egypt. According to the palynological results, a mangrove environment is suggested for a wide areal extent along the shoreline. Also, the unusual composition of clay minerals in the Abu Shigaila and Younis North profiles can easily be explained by such an environment. In the lower part of the section, terrestrial influence is very high, and clays are dominated by smectite. Thus, these sediments may have accumulated in environments similar to the recent nearshore mangrove-swamp deposits of Papua. Increasing transgression, which is particularly recognized in the upper part of the Younis North profile, flooded the coastal areas, and kaolinite- and illite-rich mineral detritus from the continental hinterland was deposited in the former lagoonal area, together with reworked nearshore sediments. Geochemical and palynological data also indicate that repeated transgressive events occurred in the lower part of the Abu Shigaila section (Duwi Formation). Clay mineralogy, however, clearly shows the difference in the relative degree of marine transgression, indicating different depositional environments.

The Campanian/Maastrichtian transgression came from a northwesterly direction. The depositional area was probably separated from the open sea by a local high, located about 200 km north of the Quseir/Safaga area (Bartov and Steinitz, 1977). This is also indicated by sedimentological investigations (Schröter, 1986; Bock, 1987). According to the present paleontological evidence, a broad shallow shelf sea may have existed between a northern high and the southern coastal swamp area during the transgressive peaks in the late Campanian to early Maastrichtian, the time of formation of the black shale/phosphorite sequences. During the transgressive peaks, the buildup of oyster bioherms in this shallow shelf sea probably caused a differentiation into local highs and lows, favoring the formation of local reducing conditions. Therefore, not only mangrove swamps, but also organic buildups increasingly sheltered the depositional environment against strong terrestrial input. The regressive periods were probably characterized by a northerly migration of the mangrove-swamp facies.

Phosphorite/Black Shale Genesis

The existence of phosphorites, black shales, and chert is often regarded as the product of "upwelling mechanisms" (Sheldon, 1980). Recently a computer program has been developed which models paleoclimate to predict areas of coastal upwelling and the formation of organic-rich, phosphoritic muds (Scotese and Summerhayes, 1986). In contrast to earlier predictions of Parrish and Curtis (1982), northeast Africa is not indicated as an area of coastal upwelling during the Late Cretaceous. Although upwelling may not totally be excluded, this mechanism is probably not necessary to explain the composition of the late Campanian/early Maastrichtian sediments. A coastal area with mangrove swamps and tropical climate, together with intensive humid weathering, would have produced sufficient quantities of terrestrially derived organic material and phosphorus to initialize blooming of microorganisms in a shallow shelf embayment (Valeton, 1983). The high rainfall which accompanied the tropical climate resulted in strong offshore currents of fresh water and partly brackish conditions. The low density of fresh water may have stabilized the formation of a halocline. Similar conclusions were drawn by Kolodny (1980), explaining the Upper Cretaceous phosphorites of Israel. As described in recent sedimentary environments, dip and depth of haloclines are highly sensitive to climatic and geographic conditions (Reineck, 1984). Regarding the very shallow shelf of the Late Cretaceous Egyptian coastline, even minor changes of the halocline may have resulted in considerable shifting of oxygen-minimum zones. This might explain the presence of benthonic foraminifera in sample 273.26, which is surprising because geochemical parameters indicate a strongly reducing environment. However, the species diversity is low, and benthonic foraminifera are extremely small. A similar situation occurs in dark laminae of sediments in the Santa Barbara Basin (Harman, 1960). Hülsemann and Emery (1961) concluded that small changes in oxygen content of the water in the Santa Barbara Basin cause relatively large changes in benthic populations, which in turn permit seasonal changes in type of sediment that is either recorded and preserved as laminae or mixed and destroyed as homogeneous sediment. The latter is probably true for sample 273.26 of the Younis North well, where the study of slides did not reveal laminated textures.

According to Loughman (1984) phosphate authigenesis happens preferentially at the base of the oxygen-minimum zone, rather than in highly anoxic sediments. Thus, the formation of phosphorite-rich beds may be governed by meandering of the oxygen-minimum zone. However, to get thick, economic phosphorite accumulations such as those in the Quseir/Safaga area, reworking and winnowing of the primary enriched phosphatic muds is necessary. This results from high wave energy due to regressive phases and subsequent lowering of the fair-weather wave base. Organic geochemical data clearly indicate reworking processes on the phosphorite-rich samples. Some of them (particularly samples of the uppermost phosphate bed) yield enough TOC to permit organic geochemical studies. They show high C27/29 sterane ratios and degradation of the n-alkanes in the saturate fraction, together with extremely low Rb contents reflecting the marine character (e.g., sample 2/1017). A reducing environment is indicated by moderate Mo contents of kerogen, thus reflecting the depositional environment of the primary enriched phosphatic muds. Schröter (1986) proved reworking and oxidation of phosphorite rocks with petrographic data. Garrison et al. (1979) interpreted the sedimentary structures of the Egyptian phosphorites as offshore sandbars with characteristic features such as erosional unconformities and cross-bedding. Other indications of an oxygenated, shallow-water depositional environment is the abundance of bioturbations, which in most cases occur at the base of the phosphorite layers. Phosphatic particles commonly fill the *Thalassinoides*-type burrows that penetrate the underlying finely laminated marls and shales (Germann et al., 1987).

Short-term regressive tendencies such as regional or local tectonic uplifts (Dominik and Schaal, 1984) within the major Late Cretaceous transgression are sometimes cited as the major reason for the reworking and enrichment of phosphatic layers. Hendriks and Luger (1987) described features of repeated synsedimentary tectonism which controlled the phosphorite enrichment in an area about 80 km east of Safaga. However, distinct lithologic differences as compared to the coeval Duwi Formation on the Red Sea coast are also noted. Comparison of worldwide eustatic curves of Haq et al. (1987) and the fluctuations of sea level according to the present study (Figure 6) shows close similarity for early late Campanian to the Maastrichtian. In many parts of the world the Late Cretaceous is characterized by various transgressive events (cycles UZA-4, 4.2–4.5 in Haq et al., 1987) which can also be identified in the Quseir/Safaga area (cycles I–IV, Figure 6). According to the chronology of Haq et al. (1987) and the stratigraphic correlations of the present study, the Abu Shigaila and Younis North sequences were deposited between 69 and 79 million years ago.

Facies Development in the Red Sea Area

Based on the results of geochemical and paleontological analysis, a facies model is proposed for the

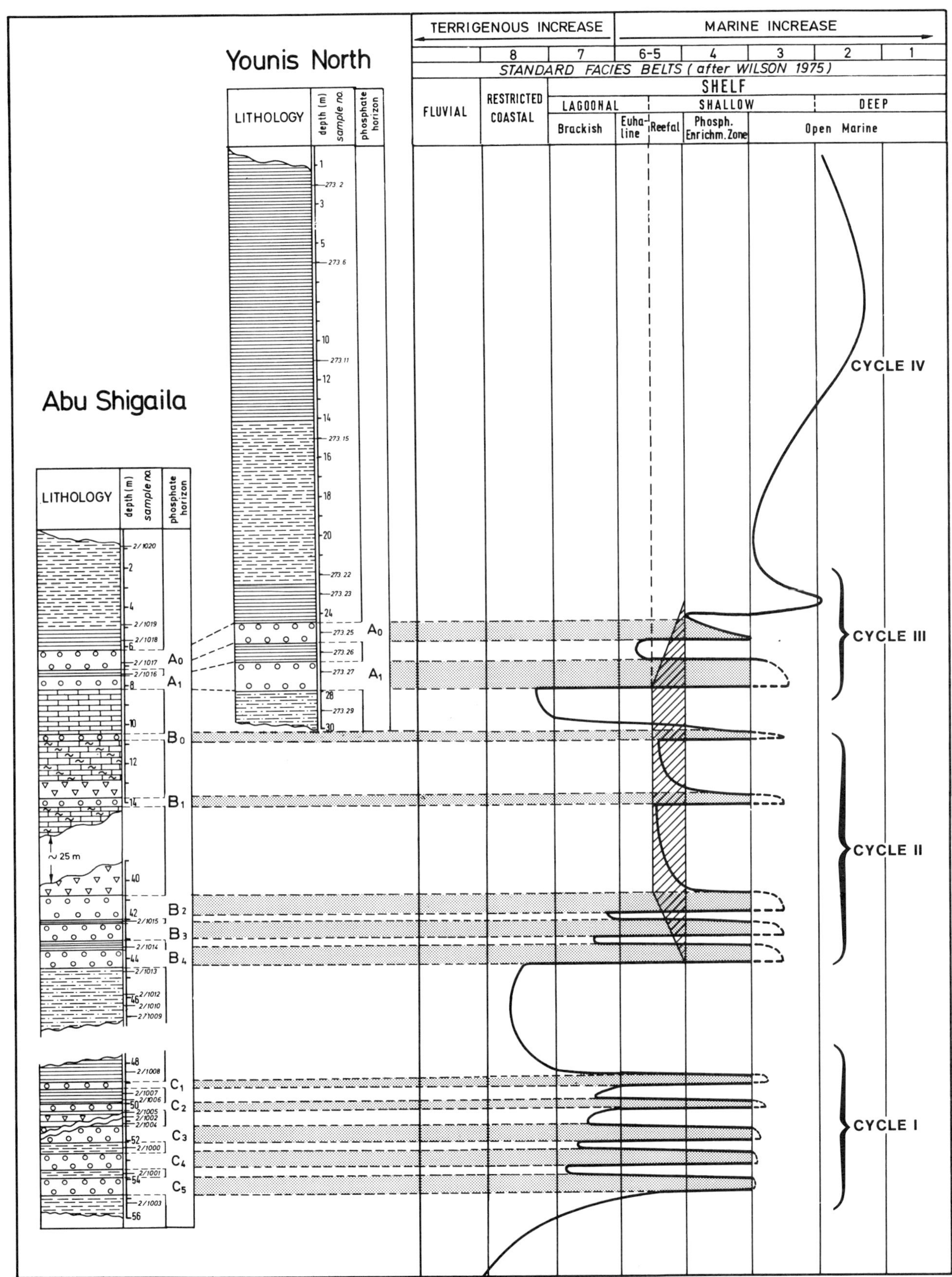

Figure 6. Interpretation of facies evolution based on data in Figure 2. Note that phosphatic beds (dashed horizons) are the product of reworked marine sediments (shaded units; facies belt 3). Cycles I-IV may correspond to the 3rd-order cycles 4.2-4.5 (supercycle UZA-4) of Haq et al. (1987).

Red Sea area black shale/phosphorite sequences (Figure 6). The model is based on the scheme of standard facies belts which is adopted with minor changes from Wilson (1975).

The transgressive sequence starts from the restricted coastal facies of the Quseir Formation (variegated shale) which were deposited under dominantly oxidizing, but locally reducing, conditions. The terrigeneous influx was relatively high at that time, and extensive mangrove swamps may have acted as sediment traps in the southern coastal areas. Phosphate enrichment was initiated in this nutrient-rich estuarine environment by extensive biomass accumulation. Strong currents with high wave energy contributed to the selective concentration of phosphate particles during times of low sea-level stand. The lower sequences, including the Hamadat C bed, retained more or less their typical terrestrial geochemical fingerprints. During deposition of the middle (B) and upper (A) phosphate beds, the shallow shelf morphology was interrupted, mainly by the buildup of oyster bioherms which protected the depositional environment from terrestrial input. Consequently, the geochemistry of these stratigraphically higher phosphorites and black shales is marked by a more or less normal marine character with very subordinate (but still recognizable) terrigeneous features. In the organic-rich marl and black shale host rocks of the phosphorites, in which the detrital terrestrial components were preserved, the geochemical markers of terrestrial influence on marine conditions were maintained. In areas where circulation was more restricted, a strongly reducing environment may have developed, in which organic-rich sediments of oil-shale character were produced.

In the higher levels of the sequence, the phosphate content diminished due to a rise in sea level, and the deeper marine black shales of the Dakhla Formation were deposited. In the Younis North section in a vertical distance of about 15 m from the uppermost phosphate horizon, the maximum development of marine characteristics was recorded. Together with the palynological and micropaleontological evidence, this indicates that the maximum marine transgression occurred within the Maastrichtian part of the Dakhla Formation.

CONCLUSIONS

From the study of geochemical, sedimentological, and paleontological characteristics of subsurface and core samples from Upper Cretaceous black shale/phosphorite sequences in southeast Egypt, we reached the following conclusions.

1. Rb content of the whole-rock analysis and C27/29 sterane distribution are reliable numerical indicators for the relative proportion of terrestrial detritus in the sediment. Both correspond largely to the quantity of terrestrially derived palynomorphs and other land-plant debris.
2. Mo content of kerogen, Ph/Pr ratio, and A-factor of IR-kerogen analysis allow differentiation between the various degrees of reducing conditions.
3. Palynological and clay-mineralogy investigations suggest the existence of coastal mangrove swamps during the time of black shale/phosphorite formation.
4. Through the application of all disciplines, facies analysis may be performed with very high resolution to study complex geological problems such as the formation of black shale/phosphorite sequences (or any other cyclicity in sedimentary systems). Good correlation between relative-sea-level fluctuations during the Late Cretaceous of southeast Egypt and worldwide eustatic curves of Haq et al. (1987) indicates that the process of phosphorite formation might be controlled mainly by eustasy and the local paleogeographic conditions, including a wide shallow shelf embayment. The nutrient-rich coastal swamp area is particularly regarded as the main factor for initialization of enrichment of organic matter and phosphate. Upwelling mechanisms, although not totally excluded, probably did not play a major role in the Quseir/Safaga area.

REFERENCES CITED

Arefev, O. A., M. N. Zabrodina, V. M. Makushina, and A. A. Petrov, 1980, Relic tetra- and pentacyclic hydrocarbons in the old oils of the Siberian Platform: Izvestiya Akademii Nauk SSSR, Seriya Geologicheskaya, v. 3, p. 135-140.

Bartov, Y., and G. Steinitz, 1977, The Judea and Mount Scopus Groups in the Negev and Sinai with trend surface analysis of the thickness data: Israel Journal of Earth-Sciences, v. 26, p. 119-148.

Bellier, J. P., 1983, Foraminifera, planktoniques du Crétacé de Tunisic Septentrionale: Systematique, biozonation, utilisation stratigraphique de l'Albien au Maastrichtian: Universite Curie Paris, Thése de Docteur Etat Mémoir Science de Terre, 250 p.

Bock, W. -D., 1987, Geochemie und Genese der oberkretazischen Phosphorite Ägyptens: Berliner geowissenschaftliche Abhandlungen (A), v. 82, 138 p.

Dominik, W., and S. Schaal, 1984, Notes on the stratigraphy of Cretaceous phosphorites (Campanian) of the Western Desert, Egypt: Berliner geowissenschaftliche Abhandlungen (A), v. 50, p. 153-175.

El Tarabili, E. E., 1969, Paleogeography, paleoecology, and genesis of the phosphatic sediments in Quseir-Safaga area, U.A.R.: Economic Geology, v. 64, p. 172-182.

Eslinger, E., and D. Pevear, 1988, Clay minerals for geologists and engineers: SEPM short-course notes no. 22.

Fowler, M. G., and P. W. Brooks, 1987, Organic geochemistry of Western Canada Basin tar sands and heavy oils. 2. Correlation of tar sands using hydrous pyrolysis of asphaltenes: Journal of Energy and Fuels, v. 1, p. 459-467.

Ganz, H., 1986, Organisch und anorganisch geochemische Untersuchungen an ägyptischen Schwarzschiefer/Phosphoritsequenzen—Methodenentwicklung und genetisches Modell: Berliner geowissenschaftliche Abhandlungen, (A), v. 70, 113 p.

Ganz, H., and W. Kalkreuth, in press (1989), Infrared spectroscopical classification of kerogen type, thermal maturation, hydrocarbon potential and lithological characteristics in potential source and reservoir rocks: Journal of Southeast Asian Earth Sciences, Pergamon Press.

Ganz, H., W. Kalkreuth, F. Öner, M. J. Pearson, and J. S. Small, 1987, Enhanced geochemical screening analysis—accurate determination of petroleum source rock characteristics and comparison with standard techniques, *in* J. Brooks and K. Glennie, eds., Petroleum geology of North West Europe: Graham and Trotman Ltd, p. 847-851.

Ganz, H., U. Tröger, and F. Öner, 1987a, Energy and resource potential of the Upper Cretaceous black shales of SE-Egypt: Berliner geowissenschaftliche Abhandlungen, (A), v. 75.3, p. 691-698.

Garrison, R. F., C. R. Glenn, P. D. Snavely, and S. E. A. Mansour, 1979, Sedimentology and origin of Upper Cretaceous phosphorite deposits at Abu Tartur, Western Desert, Egypt: Cairo, Annals of the Geological Survey of Egypt, v. 9, p. 261-281.

Germann, K., W. D. Bock, H. Ganz, T. Schröter, and U. Tröger, 1987, Depositional conditions of Late Cretaceous phosphorites and black shales in Egypt: Berliner geowissenschaftliche Abhandlungen, (A), v. 75.3, p. 629-668.

Gindy, A. R., and M. M. O. Tamish, 1985, Some major and trace constituents of Phanerozoic Egyptian mudrocks and marls: Journal of African Earth Sciences, v. 3, p. 303-320.

Glenn, C. R., and S. E. A. Mansour, 1979, Reconstruction of the depositional and diagenetic history of phosphorites and associated rocks of the Duwi Formation (Late Cretaceous), Eastern Desert, Egypt: Cairo, Annals of the Geological Survey of Egypt, v. 9, p. 388-407.

Grantham, P., and L. L. Wakefield, 1988, Variations in the sterane carbon number distribution of marine source rock derived crude oils through geological time: Organic Geochemistry, v. 12, p. 61-73.

Harland, R., 1977, Dinoflagellate cysts from the Bearpaw Formation (Upper Campanian to Maastrichtian) of Montana: Palaeontology, v. 20, p. 179-193.

Haq, B. U., J. Hardenbol, and P. R. Vail, 1987, Chronology of fluctuating sea levels since the Triassic: Science, v. 235, p. 1156-1167.

Harman, R. A., 1960, Foraminiferal variation in sediment layers of the Santa Barbara Basin, California: SEPM, Pacific Section, Annual Meeting, Los Angeles, p. 22-23.

Hendriks, F., and P. Luger, 1987, The Rakhiya Formation of the Gebel Qreiya area: evidence of Middle Campanian to Early Maastrichtian synsedimentary tectonism: Berliner geowissenschaftliche Abhandlungen, (A), v. 75.1, p. 83-96.

Herngreen, G. F. W., and A. F. Chlonova, 1981, Cretaceous microfloral provinces: Pollen et Spores, v. 23, p. 441-555.

Huang, W. -Y., and W. G. Meinschein, 1976, Sterols as source indicators of organic materials in sediments: Geochimica et Cosmochimica Acta, v. 40, p. 323-330.

Huang, W. -Y., and W. G. Meinschein, 1979, Sterols as ecological indicators: Geochimica et Cosmochimica Acta, v. 43, p. 739-745.

Hülsemann, J., and K. O. Emery, 1961, Stratification in recent sediments of Santa Barbara Basin as controlled by organisms and water character: Journal of Geology, v. 69, p. 279-290.

Hultberg, S. U., 1986, Danian dinoflagellate zonation, the C-T boundary and the stratigraphical position of the fish clay in southern Scandinavia: Micropaleontology, v. 5, p. 37-47.

Hume, W. F., 1927, The phosphate deposits in Egypt: Cairo, Geological Survey of Egypt Paper 41, 53 p.

Irion, G., and T. Petr, 1979, Mangrovensedimente am Golf von Papua: Senckenberg am Meer Nr.365: Natur und Museum, v. 109, p. 290-296.

Jarzen, D. M., 1980, The occurrence of *Gunnera* pollen in the fossil record: Biotropica, v. 123, p. 117-123.

Klitzsch, E., 1986, Plate tectonics and cratonal geology in NE Africa (Egypt/Sudan): Geologische Rundschau, v. 75, p. 753-768.

Kolodny, Y., 1980, Carbon isotopes and depositional environment of a high productivity sedimentary sequence—the case of the Mishash-Ghareb Formations, Israel: Israel Journal of Earth-Sciences, v. 29, p. 147-156.

Loughman, D. L., 1984, Phosphate authigenesis in the Aramachay Formation (Lower Jurassic) of Peru: Journal of Sedimentary Petrology, v. 54, p. 1147-1156.

Luger, P., 1985, Stratigraphie der marinen Oberkreide und des Alttertiärs im südwestlichen Obernil-Becken (SW-Ägypten) unter besonderer Berücksichtigung der Mikropaläontologie, Palökologie und Paläogeographie: Berliner geowissenschaftliche Abhandlungen, (A), v. 63, 151 p.

Luger, P., and M. Gröschke, 1989, Late Cretaceous ammonites from the Wadi Qena area in the Egyptian Eastern Desert: Palaeontology, v. 32, p. 355-407.

Malak, E. K., A. Philobbos, and M. M. Ashry, 1977, Some petrographical, mineralogical and organic geochemical characteristics of black shales from Quseir and Safaga, Red Sea area, Egypt: Cairo, Desert Institute Bulletin ARE, v. 27, p. 1-15.

Moldowan, J. M., W. K. Seifert, and E. J. Gallegos, 1985, Relationship between petroleum composition and depositional environment of petroleum source rocks: AAPG Bulletin, v. 69, p. 1255-1268.

Parrish, J. T., and R. Curtis, 1982, Atmospheric circulations, upwelling and organic rich rocks in the Mesozoic and Cenozoic: Palaeogeography, Palaeoclimatology, Palaeoecology, v. 40, p. 67-101.

Reineck, H. E., 1984, Aktuogeologie klastischer Sedimente: Frankfurt, Senckenberg-Buch 61, 348 p.

Rullkötter, J., P. A. Meyers, R. G. Schaefer, and K. W. Dunham, 1986, Oil generation in the Michigan Basin: A biological marker and carbon isotope approach: Organic Geochemistry, v. 10, p. 359-375.

Said, R., 1962, The geology of Egypt: New York, Elsevier, 377 p.

Schrank, E., 1987, Paleozoic and Mesozoic palynomorphs from Northeast Africa (Egypt and Sudan) with special reference to Late Cretaceous pollen and dinoflagellates: Berliner geowissenschaftliche Abhandlungen (A), 75.1, p. 249-310.

Schrank, E., and K. Perch-Nielsen, 1985, Late Cretaceous palynostratigraphy in Egypt with comments on Maastrichtian and Early Tertiary calcareous nanofossils: Newsletters on Stratigraphy, v. 15, p. 81-99.

Schröter, T., 1986, Die lithofazielle Entwicklung der oberkretazischen Phosphatgesteine Ägyptens—ein Beitrag zur Genese der Tethys-Phosphorite der Ostsahara: Berliner geowissenschaftliche Abhandlungen, (A), v. 67, 105 p.

Scotese, C. R., and C. P. Summerhayes, 1986, Computer model of paleoclimate predicts coastal upwelling in the Mesozoic and Cenozoic: Geobyte, summer 1986, p. 28-42.

Sheldon, R. P., 1980, Episodicity of phosphate deposition and deep ocean circulation—an hypothesis: SEPM Special Publication 29, p. 239-247.

Sliter, W. V., 1972, Cretaceous foraminifers—Depth habitats and their origin: Nature, v. 239, p. 514-515.

Smith, A. G., A. M. Hurley, and J. C. Briden, 1982, Paläokontinentale Weltkarten des Phanerozoikums: Stuttgart, F.Enke Verlag, 102 p.

Tappan, H., 1980, The paleobiology of plant protists: San Francisco, Freeman & Co, 1028 p.

Tralau, H., 1964, The genus *Nypa* Van Wurmb: Kungliga Svenska Vetenskapsakademiens Handlingar, 4. ser., v. 10, p. 5-29.

Valeton, I., 1983, Klimaperioden lateritischer Verwitterung und ihr Abbild in den synchronen Sedimentationsräumen: Zeitschrift der Deutschen Geologischen Gesellschaft, v. 134, p. 413-452.

Ward, W. C., K. C. McDonald, and S. E. I. Mansour, 1979, The Nubia formation of the Qusseir-Safaga area, Egypt: Cairo, Annals of the Geological Survey of Egypt, v. 9, p. 420-431.

Welte, D. H., and D. Waples, 1973, Über die Bevorzugung geradzahliger n-Alkane in Sedimentgesteinen: Naturwissenschaften, v. 60, p. 516-517.

Williams, G. L., and J. P. Bujak, 1985, Mesozoic and Cenozoic dinoflagellates, *in* H. M. Bolli, J. B. Saunders, and K. Perch-Nielsen, eds., Plankton stratigraphy: Cambridge, Mass., Cambridge University Press, p. 847-964,

Wilson, J. L., 1975, Carbonate facies in geologic history: Berlin, Springer Verlag, 471 p.

Youssef, M. I., 1957, Upper Cretaceous rocks in Qusseir area: Cairo, Desert Institute Bulletin ARE , v. 2, p. 25-53.

Index